T0206125

Floral Diagrams

Second Edition

Floral morphology is key for understanding floral evolution and plant identification. Floral diagrams are two-dimensional representations of flowers that replace extensive descriptions or elaborate drawings to convey information in a clear and unbiased way. Following the same outline as the first edition, this comprehensive guide includes updated and relevant literature, represents the latest phylogeny, and features twenty-eight new diagrams. Diagrams are presented in the context of the most recent classifications, covering a variety of families and illustrating the floral diversity of major groups of plants. A strong didactic tool for observing and understanding floral structures, these diagrams are the obvious counterpart to any genetic study in flowering plants and a contribution to the discussion of major adaptations and evolutionary trends of flowers. This book is invaluable for researchers and students working on plant structure, development and systematics, as well as an important resource for plant ecologists, evolutionary botanists and horticulturists.

LOUIS P. RONSE DE CRAENE is a botanist at the Royal Botanic Garden Edinburgh, and the director of the MSc course in the biodiversity and taxonomy of plants jointly organised with the University of Edinburgh. He has published more than 130 peer-reviewed papers and edited four books, and his main research interests are centred on floral morphology, the evolution of flowers and the use of floral characters in plant phylogeny. He has developed an internationally acclaimed expertise in floral structural morphology encompassing a broad range of angiosperm families.

Floral Diagrams

An Aid to Understanding Flower Morphology and Evolution

Second Edition

Louis P. Ronse De Craene
Royal Botanic Garden Edinburgh

CAMBRIDGE
UNIVERSITY PRESS

University Printing House, Cambridge CB2 8BS, United Kingdom

One Liberty Plaza, 20th Floor, New York, NY 10006, USA

477 Williamstown Road, Port Melbourne, VIC 3207, Australia

314–321, 3rd Floor, Plot 3, Splendor Forum, Jasola District Centre, New Delhi –110025, India

103 Penang Road, #05–06/07, Visioncrest Commercial, Singapore 238467

Cambridge University Press is part of the University of Cambridge.

It furthers the University's mission by disseminating knowledge in the pursuit of education, learning, and research at the highest international levels of excellence.

www.cambridge.org
Information on this title: www.cambridge.org/9781108825733
DOI: 10.1017/9781108919074

First published 2010
Second edition 2022

A catalogue record for this publication is available from the British Library.

Library of Congress Cataloging-in-Publication Data
Names: Ronse Decraene, L. P. (Louis Philippe), author.
Title: Floral diagrams : an aid to understanding flower morphology and evolution / Louis P. Ronse De Craene.
Description: Second edition. | New York : Cambridge University Press, 2022. | Includes index.
Identifiers: LCCN 2021050296 (print) | LCCN 2021050297 (ebook) | ISBN 9781108825733 (paperback) | ISBN 9781108919074 (ebook)
Subjects: LCSH: Angiosperms – Morphology – Charts, diagrams, etc. | Flowers – Evolution. | Angiosperms – Morphology. | Flowers – Evolution – Charts, diagrams, etc. | BISAC: SCIENCE / Life Sciences / Botany
Classification: LCC QK495.A1 R66 2022 (print) | LCC QK495.A1 (ebook) | DDC 583–dc23/eng/20211108
LC record available at https://lccn.loc.gov/2021050296
LC ebook record available at https://lccn.loc.gov/2021050297

ISBN 978-1-108-82573-3 Paperback

To Catherine, Camille and Alexandre, with love

Contents

Preface

Flowers are extremely attractive to us as a source of inspiration and happiness. It is no wonder that various technical textbooks in plant science tend to enhance their front pages with glamorous illustrations of flowers. Despite this wide interest, our knowledge about the diversity of floral structures is still limited and often relies on research carried out more than 100 years ago.

Technically a floral diagram is a schematic cross-sectional drawing of a flower. However, a floral diagram is more than just a two-dimensional representation. There are more than 250,000 species of angiosperms, and their flowers vary in many ways. The arrangement of flowers in inflorescences, the number, position, identity and shape of floral organs, and the symmetry of the flower as a whole are rarely identical between different families, genera or even species, and even when they look superficially similar, important details differentiate them. Floral diagrams create a rich source of data for identification purposes and for understanding structures, but also to express a hypothesis of evolution. The information contained in floral diagrams is potentially immense and replaces complex descriptions.

Students often struggle with the identification of flowers, mainly because they fail to look at the structures hidden in the bud. However, comparable to the plan of a house created by an architect, the *Bauplan* of a flower – that is, the spatial arrangements of organs in the flower – is essential to understanding systematic relationships. This information is particularly important for identifying plants in the field and tells us much about the key characters of a specific group of plants. The educational merits of floral diagrams in the classroom are obvious for all ages. Used together with floral formulae, they convey information in a clear and rigorous way. They are important for systematists or evolutionary botanists in providing information for databases dealing with morphological data (e.g. Morphbank: www .morphbank.net) or for clarifying phylogenetic questions. In paleobotanical research floral diagrams are a useful resource in reconstructing the shape of fossilized flowers. Researchers in evolutionary developmental genetics will find appropriate questions about the nature of floral organs to investigate. Horticulturists or amateur botanists will find this book valuable to understand general patterns of flower construction and floral diversity. Finally, artists will find inspiration in the floral diagrams in creating various aesthetic interpretations of flowers (Keto Logua, 2020. Exhibition Studio Berghain Berlin).

It has been more than 140 years since August Wilhelm Eichler (1839–87), then a professor of botany in Kiel, produced a book on floral diagrams in two parts, the first published in 1875 and the second published in 1878. This book, entitled *Blüthendiagramme construirt und erlaütert*, is a major reference work concentrating the information about flowers known at that time. As such, it represents a treasure trove, detailed and often accurate, and even today extremely valuable as a source of data. Eichler's work was an inspiration for later generations of morphologists, such as Arthur H. Church (1865–1937) and Agnes Arber (1879–1960). A particularly fine example of a book using floral diagrams is *Types of Floral Mechanism* published by Church (1908) and intended as a series, but limited to a single volume by lack of interest and funds (Mabberley, 2000). Since Eichler's book was published, much progress has been made in documenting flower morphology, especially from the last decades of the twentieth century, when there was a renewed interest in floral morphology coupled with the use of the scanning electron microscope. This approach has increased tremendously in recent years.

However, information about flowers is often scattered in scientific papers that are not readily accessible, providing little scope for a broad overview of the flowering plants. Alternatively, it dates from important work carried out in the nineteenth century that is in danger of being forgotten. Floral diagrams were used sparingly in different textbooks as illustrative material (e.g. Baillon, 1867–95; Engler and Prantl, 1887–1909), but never to the extent of Eichler's book. More recent examples include Melchior (1964), Sattler (1973), Graf (1975), Stützel (2006), Leins and Erbar (2010) and Simpson (2016). The more recent major textbooks on angiosperm phylogeny (e.g. Soltis et al., 2005, 2018; Simpson, 2006) lack any floral diagrams. Spichiger et al. (2002) did include diagrams for major families, but the diagrams are oversimplified and riddled with mistakes.

The system of classification Eichler used is outdated, as it is based on the Englerian concept that simple, unisexual, catkin-like flowers are ancestral and that more elaborate bisexual flowers are derived. Recent changes in the phylogeny of flowering plants based mainly on molecular evidence have created the need for a better understanding of morphology and its relation to any molecular phylogeny. This new edition of *Floral Diagrams* based on the most recent synopsis of Soltis et al. (2018) should fulfil this purpose.

Acknowledgements

It is now more than ten years since the first edition of *Floral Diagrams* was published. I have taken much pride in this achievement, which has extended my understanding of the world of plants and has allowed me to meet many new colleagues and friends. Studying and understanding flowers is my great passion, and I think that floral diagrams are the best tool to express the wonderful intricacies of flowers. It is a real pleasure to see my book cited in many publications and to hear from colleagues that they keep it close to their desks as reference material for their teaching. I also use *Floral Diagrams* as the backbone for my teaching of angiosperm biodiversity at the Royal Botanic Garden Edinburgh.

However, I think that now is the time for revisiting the book. As with any first edition of a textbook, one finds tiny mistakes and imperfections that disappear with maturity. During the past ten years, botanical science has progressed tremendously, with much more emphasis on flower morphology, and important new information has become available that could be incorporated in this book. Lastly, I realized that adding a few diagrams would make the book more complete.

Several colleagues were inspirational and encouraging, such as Julien Bachelier, Richard Bateman, Peter Endress, Gabriele Galasso, Greg Kenicer, Alex Kocyan, Peter Linder, Michael Müller, Darin Penneys, Gerhard Prenner, Margarita Remizowa, Paula Rudall, Rolf Rutishauser, Rolf Sattler, Dmitry Sokoloff, Dennis Stevenson, Wolfgang Stuppy and Livia Wanntorp, among others. I had the pleasure of meeting new friends and colleagues and collaborating with them, including Arne Anderberg, Catherine Damerval, Thierry Deroin, Xu Fengxia, Karina Gagliardi, Wang Hengyang, Florian Jabbour, Bruce Kirchoff, Wei Lai, Zhao Liang, Cao Limin, Mariana Monteiro, Sophie Nadot, Ioan Negrutiu, Fernanda Pérez, Dietmar Quandt and Andrey Sinjushin, among others, and also to develop research projects with my students, Celina Barroca, Mauricio Cano, Jia Dong, Wang Junru,

Britta Kümpers, Carmen Puglisi, Pakkapol Thaowetsuwan and Zhang Yutong. The set-up of the FLO-RE-S network (see https://flores-network.com) was a great opportunity to share botanical knowledge and stimulate research in flowers. Many thanks to my friends and colleagues Kester Bull-Hereñu, Javiera Chinga, Regine Claßen-Bockhoff, Patricia Dos Santos, Juliana El Ottra, Akitoshi Iwamoto, Julius Jeiter and Joao Toni for their enthusiasm and inspirational discussions.

I also thank Howard Wills for analysing some diagrams from the first edition and pointing to errors that have now been corrected, and Peter Endress and Peter Stevens for their generous reviews of the first edition. Without the vast living collections at the Royal Botanic Garden Edinburgh and other botanical gardens, the extent of my book would be meagre indeed. I thank members of the horticultural staff, especially Sadie Barber, Pete Brownless, Fiona Inches, Gunnar Ovstebo and David Tricker, for their help. Most importantly, the writing of this book would not have been possible without the constant support and patience of my dear wife, Catherine.

PART I INTRODUCTION TO FLORAL DIAGRAMS

1

Introduction to Flower Morphology

1.1 Definition of Flowers

There is no general agreement nor any rule about how a flower should be defined. Since the end of the nineteenth century two main contrasting hypotheses have been provided and the discussion is still ongoing (reviewed in Bateman, Hilton and Rudall, 2006). The pseudanthial hypothesis accepts that flowers evolved from a branched, multiaxial structure – that is, a condensed compound inflorescence (e.g. Eichler, 1875; Eames, 1961). This means that a flower is an assemblage of separately functioning entities that became grouped together. The euanthium hypothesis states that the flower evolved from a simple uniaxial (euanthial) structure – that is, a condensed sporophyll-bearing axis with proximal microsporophylls and distal megasporophylls (e.g. Arber and Parkin, 1907). However, reconstructions of the early angiosperm flower (e.g. Sauquet et al., 2017) suffer from the absence of clear transitional forms between ancestral prototypes and angiosperms. Floral organs all have attributes of leaves, and leaf-like elements, such as stipules, leaf bases, petioles and blades, occasionally appear in flowers (e.g. Arber, 1925; Guédès, 1979). The stamen is recognized as equivalent to a microsporangiophore (an axis) or microsporophyll (a leaf) bearing microsporangia (the pollen sacs), while the ovary is described as a grouping of folded megasporophylls (the carpels) enclosing the megasporangia (ovules) (Endress, 2006). However, the origin of stamens is unclear, with a greater diversity of stamen structures in other seed plants (Endress, 2006). Developmental and genetic evidence supports the fertile organs of flowers to be a combination of axes and subtending leaves (at least for the ovary: e.g. Endress, 2019), reflecting a modular vegetative system of a main axis with lateral branches arising in the axil of leaves. However, not all researchers unanimously accept the concept of flowers evolving from a vegetative shoot. Claßen-Bockhoff (2016; see also Claßen-Bockhoff and Arndt, 2018; Claßen-Bockhoff and Frankenhaüser, 2020) pointed to the fundamental difference between flowers and shoots, including the lack of apical growth and expansion in flowers, leading to a strong differentiation from vegetative shoots.

Recent phylogenetic studies have supported the theory that flowers evolved once and that all flowers are thus homologous (the 'anthophyte hypothesis' reviewed in Bateman, Hilton and Rudall, 2006). The theory is supported by evolutionary developmental evidence that the same genes are acting on the flower and vegetative shoot, and that the flower is best interpreted as a short shoot with specialized leaves (Glover, 2007).

More specifically, a flower can be defined as a determinate structure with a generally defined number of organs (modules); it bears both staminate and pistillate parts and organs are set in four series: sepals, petals, stamens and carpels. However, several angiosperm flowers lack these defining features, and differentiation of a perianth or limits of flowers and inflorescences can be unclear. The definition of flowers implicitly refers to angiosperms, but should also include gymnosperms. The gymnosperm cone could also be described as a flower, though the organization of the cone is generally unisexual. A defining character for angiosperms is enclosure of ovules by carpels (angiospermy), separating the flowering plants from their closest relatives (gymnosperms). Despite any occurring variations, most flowers are conservative, with a well-defined ground plan (also called Bauplan) that is genetically fixed (see Smyth, 2018). Flowers are usually grouped into inflorescences that may be simple to highly complex, and are – at least in bud – often subtended by a specialized leaf, the bract, together with one or more (generally two) smaller leaves (bracteoles) placed in a lateral position. Bracts and bracteoles are absent in some species, and this can influence the position of organs in the flower. In some plant groups the limits between bracts and perianth parts are unclear. The inflated axis, called a receptacle, bears floral organs in a spiral or in whorls (or a mixture of both). The outer floral organs are sterile leaves called a perianth. When undifferentiated they are described as tepals. More often, there is differentiation into an outer whorl (calyx or sepals) and an inner whorl (corolla or petals). The androecium, or the totality of stamens, can be organized in a single series or into several whorls, with a specific position relative to the petals. The gynoecium consists of carpels bearing ovules. Carpels can be free or, more often, fused into an entity enclosing the ovules or seeds. Besides carpels and stamens, some flowers have sterile structures (staminodes or carpellodes) or other emergences of the receptacle. These are often developed as nectaries and can be conspicuous.

Flowers share another characteristic besides leaves with the vegetative parts of the plant, which is the phyllotaxis or order of initiation of floral organs (see p. 00). The transition in phyllotaxis from the vegetative shoot to the flower can be gradual or abrupt and is mediated by bracts and sepals. The calyx usually continues the same spiral sequence as vegetative leaves. There is generally a disruption in the initiation sequence between sepals and petals, leading to an alternation of whorls. The stabilization of numbers and the position of floral parts relative to each other is fundamentally important in understanding and interpreting the structure of the flower.

1.1.1 *Complex versus Reduced Flowers*

The evolution of flowers is correlated with the mode of pollination. There is a marked difference between flowers with a biotic (animal) pollination syndrome and those with an abiotic (wind or water) syndrome. Animal pollination is accompanied by a series of adaptations to attract and offer rewards to specific pollinators and to protect the floral parts from damage. Differentiation of protective sepals and carpels, nectaries, showy petals and stamens are part of the arsenal leading to effective fertilization. Depending on the pollinating animal, different strategies were developed to increase the success of pollination, occasionally leading to complex flowers or inflorescences, such as an increase in the number of floral parts (especially stamens), the development of an attractive perianth and reward system, and the evolution of highly specialized spatial interactions with pollinating organisms (Proctor, Yeo and Lack, 1996). Several examples illustrate the close connection between the pollinator and the evolution of floral traits, such as the development of spurs in *Aquilegia* (Whittal and Hodges, 2007), petaloid staminodes and stamen fertility in Zingiberales (Specht et al., 2012), or the corolla length in Polemoniaceae (Rose and Sytsma, 2021). Wind or water pollination is accompanied by a syndrome of derived characters, such as smaller, unisexual flowers, loss of petals or reduction of the perianth, long styles and filaments, production of a large amount of pollen, lack of viscin in pollen and reduction in the number of ovules (e.g. Linder, 1998; Friedman and Barrett, 2008; Friedman, 2011).

Reversals in the pollination syndrome occur frequently in the angiosperms. Secondary wind-pollinated flowers evolved in all major clades with a predominance of insect pollination (e.g. *Thalictrum* in Ranunculaceae, *Poterium* in Rosaceae, *Macleaya* in Papaveraceae, some *Acer* in Sapindaceae, some *Erica* in Ericaceae, *Fraxinus* in Oleaceae, *Xanthium* in Asteraceae, *Theligonum* in Rubiaceae, *Leucadendron* in Proteaceae). Larger, predominantly wind-pollinated clades including Fagales and Poales show the occasional reversal to insect pollination (e.g. *Castanea* in Fagaceae, most Buxaceae, *Euphorbia* in Euphorbiaceae).

Specific elaborations of petals and/or staminodes are clearly linked with pollination syndromes and are triggered by the kind of pollinators that evolved with the flowers. Secondary stamen and carpel increases are widespread, and are linked with the potential for higher pollen or ovule supply. Very often there is a close mechanical correlation between perianth and stamens, or stamens and style, in the release of pollen or the protection of the anthers. Different elaborations on the petals are also directly linked to the pollinator (e.g. the building of landing platforms, nectar containers and nectar guides: see Endress and Matthews, 2006b).

An important condition for increased complexity in flowers is the attainment of greater synorganization (Endress, 2006), allowing a concerted change and interaction of floral organs. Therefore, events such as merism change, shifts in symmetry, fusions, and stamen and carpel increases depend on a close interaction of different organs in flowers, but also on interactions within the inflorescence. Flowers are morphologically highly dynamic entities with a potential for evolution reaching far beyond our preconceived ideas of the limitations of floral evolution. This flexibility is closely linked with what is available at the organ level at a given time, as well as inherent mechanical processes controlling the development of flowers (Ronse De Craene, 2018). Subtle shifts in the timing of organ initiation or internal pressures during the floral development can cause dramatic morphological changes in the flower with consequences for pollinator interactions (see Ronse De Craene, 2016, 2018; Chinga et al., 2021; Wei and Ronse De Craene, 2020).

1.2 Floral Organs

1.2.1 Perianth

> *If treated in isolation, there is no character combination which could stringently prove an organ's nature as a petal or sepal.* (Endress, 1994: 26)

The perianth is the envelope of sterile leaves enclosing the fertile organs of the flower. The perianth is either differentiated into sepals and petals, or undifferentiated (perigone or tepals). In the latter case the perianth can be green (sepaloid) or pigmented (petaloid). A distinction between sepals and petals is applicable only when two different series of perianth parts are found. In cases with more than two whorls the transition between sepals and petals can be progressive with blurred limits.

In some cases there is unclear distinction between the perianth and enclosing bracts. In other cases bracts can be variably associated with the flower, often in the form of an epicalyx (e.g. Malvaceae), or sometimes as a fused cap or calyptra (e.g. Papaveraceae, Aextoxicaceae). Bracts can be distinguished from the perianth by presence of axillary buds (never in floral organs) and differences of plastochron (transition of a decussate to spiral phyllotaxis: Buzgo, Soltis and Soltis, 2004). Endress (2003a) suggested that bracts should be considered as phyllomes with a lower complexity than tepals. However, I believe that a distinction between bracts and tepals is sometimes impossible to make, given the existence of intermediate organs and the easy incorporation of bracts in the flower. Inclusion of bracts at the base of the flower makes an originally undifferentiated perianth biseriate, as in Magnoliaceae. Some taxa have transitional organs between bracts and tepals (in German called *Höchblätter*), as in Myrothamnaceae and some Ranunculaceae. There are known cases where bracts replace sepals that have been previously lost (*Quinchamalium* in Santalaceae) or act as a secondary calyx when petals are lost (*Mirabilis* in Nyctaginaceae).

The distinction between sepals and petals is not always straightforward (for a discussion and review, see Endress, 2006; Ronse De Craene, 2007, 2008; Ronse De Craene and Brockington, 2013). While core eudicots currently have a bipartite perianth of sepals and petals, the perianth is rarely differentiated in basal angiosperms, monocots and early diverging eudicots with variable homologies. If only a single whorl of perianth parts is present, it is sometimes difficult to categorize members of this whorl as sepals or petals, because one whorl may have been lost. A distinction can be made between primary apetaly (as in basal angiosperms with tepals and no distinction between perianth parts) and secondary apetaly (apopetaly: Weberling, 1989), in cases where evidence exists that petals have been present and have been lost during evolution. Sepals are petaloid in several families or can have a mixed nature (partly green and pigmented; e.g. *Impatiens*, *Polygala*). One of the reasons for this variability is that the perianth can change function at different stages of the development of the flower. In general, the calyx tends to protect inner organs and is photosynthesizing. Later, it can become attractive for the dispersal of fruits (e.g. *Physalis* in Solanaceae). Additionally, the pigmentation of the sepals can be regulated by other factors such as variable genetic shifts (e.g. petaloid calyx of *Tulipa*: Kanno et al., 2007, and *Rhodochiton*: Landis, Barnett and Hileman, 2012), or the influence of light on the developing bud (e.g. *Nymphaea*: Wagner, Rudall and Frohlich, 2009; *Kewa*: Brockington et al., 2013). The main purpose of petals is to attract pollinators, but they can become transformed into protective organs or dispersal units (e.g. in *Coriaria*). Petals are occasionally indistinguishable from sepals in Pentapetalae, but this probably represents a derived condition (e.g. Dipentodontaceae: Byng, 2014; *Prunus*: Wang et al., 2021).

The origin of the perianth was discussed in several textbooks and papers, with various interpretations for the origin of petals: either from stamens, from bracts, from both structures, or as something totally new with different mechanisms leading to the differentiation of the perianth (for recent reviews, see Irish, 2009; Ronse De Craene and Brockington, 2013; Glover et al., 2015; Monniaux and Vandenbussche, 2018). In general, recent evidence from morphological and evo-devo studies suggests that petals in the majority of angiosperms are derived from bract-like structures and are homologous to the sepals in the flower. Contrary to a general assumption, cases where petals are unequivocally derived from stamens are rare in the angiosperms. Petals are inserted between outer bract-like organs and inner stamens and undergo influences from both sides. In extreme cases petals can take over stamen characteristics so as to become confused with stamen-derived structures. This is caused by a delayed initiation of the petals, leading to their absorption in stamen tissue and a stronger influence of genes affecting stamens. Morphologically this results in petals resembling staminodes (small insertion base, single vascular bundle, bilobed lamina, etc.) and may be the reason for the persistent belief that petals represent transformed stamens (see Ronse De Craene and Brockington, 2013; Wei and Ronse De Craene, 2019). In several groups, petals have variously evolved by insertion of bracts in the confines of the flower and their differentiation into two functional whorls. An undifferentiated perianth was reconstructed as ancestral for the angiosperms in recent phylogenetic studies (Soltis et al., 2005; Sauquet et al., 2017). In basal angiosperms, attraction and protection are combined with pigmentation of the entire perianth. A differentiated perianth has evolved independently several times, at least once at the base of Pentapetalae (Ronse De Craene, 2008; Litt and Kramer, 2010).

The presence of a perianth (as well as surrounding organs, such as bracts) has a stabilizing effect on the flower in causing pressure and regulating the phyllotaxis, merism and symmetry (Ronse De Craene, 2018). Loss or reduction of the perianth is generally associated with a breakdown of a regular floral arrangement (e.g. *Achlys* in Berberidaceae: Endress, 1989; *Euptelea*: Ren et al., 2007; *Theligonum*: Rutishauser et al., 1998).

Undifferentiated Perianth (Tepals or Perigone)

An undifferentiated perianth tends to be concentrated in the basal angiosperms and monocots, and is usually associated with spiral or trimerous flowers. A distinction needs to be made between a primary undifferentiated and a secondary undifferentiated perianth. In the first case, tepals have evolved from bracts that became associated with reproductive organs and have acquired secondary functions of protection and attraction of pollinators. This kind of perianth is usually spiral with a gradual differentiation from outer bract-like to inner petaloid tepals (e.g. Austrobaileyaceae, Calycanthaceae). Alternatively, two trimerous perianth whorls of several monocots are undifferentiated and petaloid. However, the switch between sepals and petals can be easy in these cases (Ronse De Craene, 2007). A secondarily undifferentiated perianth arises by loss of either the calyx (e.g. Santalaceae, Apiaceae) or the corolla (e.g. *Geissos* in Cunoniaceae, *Rodgersia* in Saxifragaceae) and should be referred to as a reduction.

Calyx

Sepals have a spiral initiation sequence with rapid growth, a broad base, three vascular traces and an acuminate (pointed) tip. The homology of sepals with leaves is based on similar anatomy as well as on several characteristics such as the presence of stipules and stomata. Sepals are often compared with the petiole of a leaf due to their broad shape and the occasional presence of a small appendage or a dorsal crest (Arber, 1925; Guédès, 1979). On the contrary, petals often arise nearly simultaneously and have a delayed growth. They have a narrow base with only a single vascular trace and the tip is bifid or emarginated. Characteristics are more closely comparable to stamens than to leaves (Ronse De Craene, 2007).

Sepals can be fused together (gamosepaly). This fusion is often congenital (see p. 41) at their margins, leaving free calyx teeth. Sepals are more often persistent than caducous; very often, they increase in size after pollination and function in fruit dispersal. Reduction of sepals or their transformation into small scales or bristles is occasionally found in some families (e.g. in Asteraceae, Caprifoliaceae). The calyx may also vanish completely (e.g. in Santalaceae), leaving a single petal whorl combining attraction with protection of the flower.

Corolla

The corolla, or petals, represent the inner perianth whorl of the flower and are usually pigmented (petaloid). The number of petal whorls can be high, as in Annonaceae or Berberidaceae (up to four whorls). Petals can sometimes be highly distinctive, with a claw (a narrow base, compared with a broadened base in sepals), or they can be indistinct from sepals.

The corolla can be highly elaborate by development of ventral appendages, fimbriate margins or extended tips. Petals are often trilobed, with a protruding middle lobe (e.g. Saxifragaceae, Sapotaceae) or extensive lateral lobes (e.g. Elaeocarpaceae). Bilobed petals can be formed by reduction of the central lobe (Endress and Matthews, 2006b).

In some families, petals are highly elaborate in relation to specific pollination mechanisms. For example, Byttnerioideae (Malvaceae) have petals as inverted spoons, with a hood-like base and an extended apex. More examples are given by Endress and Matthews (2006b), who suggested a correlation between the smaller size or reduction of petals and elaborations on the petal surface. However, petal reduction or loss is linked with other factors, such as a shift of attraction to the sepals and the development of a hypanthium (Ronse De Craene, 2008). Fusion of petals (sympetaly) is a frequent phenomenon in angiosperms that occurs independently or is closely linked with the evolution of the androecium. Petal tubes appear to regulate access to flowers by a variable extent of development.

The nature and origin of petals remains a contentious subject, with uncertainty about the homology of petals (bracteopetals or andropetals), especially in Pentapetalae (Endress, 2006; Ronse De Craene and Brockington, 2013; Monniaux and Vandenbussche, 2018). However, floral developmental evidence tends to demonstrate that petals are influenced by neighbouring organs during their development. A delayed petal development can lead to them being overtaken by the stamens and their absorption in the meristematic tissue of the androecium (stamen-petal primordia: p. 54). As a result petals will resemble stamens (narrow base, single vasculature) and might even be confused with staminodial structures (Ronse De Craene, Clinckemaillie and Smets, 1993; Ronse De Craene and Bull-Hereñu, 2016; Wei and Ronse De Craene, 2019). Rapidly developing petals will often continue the spiral sequence of sepals and bear more resemblance to bracteopetals (e.g. Paeoniaceae, Clusiaceae, Pentaphyllaceae).

1.2.2 Androecium

The androecium consists of stamens, which make up the male part of the flower. Stamens are relatively uniform. They mostly consist of four pollen sacs arranged in two lateral thecae grouped in an anther that is linked by a connective to a filament. The orientation of anthers can be inward (introrse), lateral (latrorse) or outward (extrorse). Stamens can be basally connected into a tube (e.g. Meliaceae, Malvaceae), or connected with the petals in a common stamen-petal tube (e.g. Caricaceae, Rubiaceae). Anthers may become laterally connivent or fuse postgenitally (e.g. Asteraceae, Balsaminaceae). The number and position of stamens appear to be the most variable in the flower compared to other floral organs, but this variation is never randomized and is relatively conservative in angiosperms. Therefore, stamen position represents one of the most significant characters in flower phylogeny (see Ronse De Craene and Smets, 1987, 1996a, 1998a). Flowers with more than two stamen whorls are rare and largely restricted to basal angiosperms and some basal monocots. In other instances they are the result of a secondary increase (see p. 00: e.g. Rosaceae).

When there is only a single whorl, stamens are either inserted opposite the sepals (haplostemony: e.g. Gentianaceae, Violaceae), or less frequently opposite the petals (obhaplostemony: e.g. Vitaceae, Primulaceae). When two stamen whorls are present, they arise separately and are often spatially separated (Figure 1.1). The number of stamens is ten in pentamerous flowers or eight in tetramerous flowers, a common pattern among rosids. Diplostemony is the condition where the outer stamen whorl is situated opposite the sepals and the inner whorl is situated opposite the petals (e.g. Coriariaceae, Burseraceae). Obdiplostemony is the opposite condition, with outer stamens opposite the petals (e.g. Geraniaceae, Saxifragaceae). The distinction between diplostemony and obdiplostemony is often the result of developmental constraints and shifts

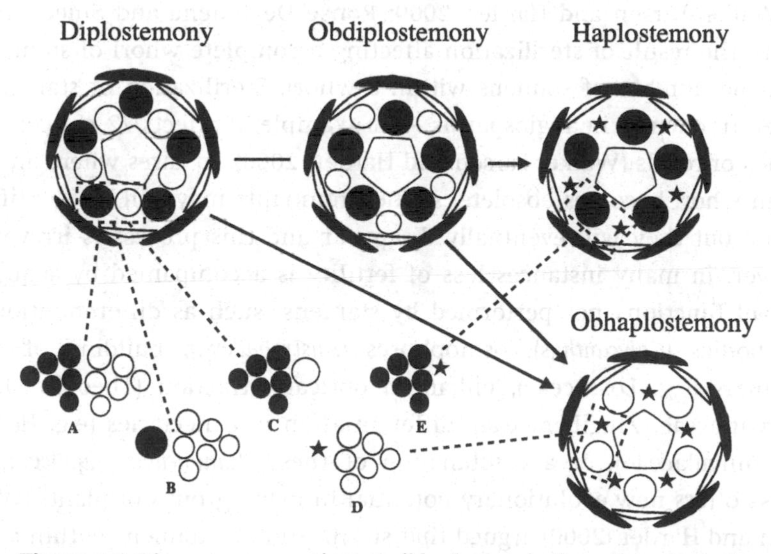

Figure 1.1 Diagram representing possible changes in the androecium of a pentamerous flower with two stamen whorls. Black dots, antesepalous stamens; white dots antepetalous stamens; asterisk, lost stamen position or staminodes. A. increase of antesepalous and antepetalous stamens; B. increase of antepetalous stamens only; C. increase of antesepalous stamens only; D. sterilization or loss of antesepalous stamens and increase of antepetalous stamens; E. sterilization or loss of antepetalous stamens and increase of antesepalous stamens

(for a review, see Ronse De Craene and Bull-Hereñu, 2016). Obdiplostemony may be caused by the pressure of developing carpels resulting in more outward orientation and shift of the antepetalous stamens (e.g. Ericaceae, Rutaceae, Geraniaceae: Leins, 1964a; Eckert, 1966; Ronse De Craene and Smets, 1995b; Endress, 2010a). In some groups of plants, the antepetalous stamens arise before the antesepalous stamens; this inverted development is often linked with weakening or sterilization of the antesepalous stamens leading to obhaplostemony (e.g. Malvaceae, Primulaceae). The importance of obdiplostemony should not be overestimated, although it represents a stable transitional condition in the evolution of a single whorl from two stamen whorls (Ronse De Craene and Bull-Hereñu, 2016). This trend may be related to a conversion of one whorl into staminodes.

Staminodes

Staminodial structures are defined as sterile stamens. The main criterion in recognizing staminodial structures is that of position, as any emergences in the flower can arbitrarily be called staminodial (Leins and Erbar, 2010). Rarely, the presence of sterile anthers is an indication of a former function as stamens. Staminodes are widely scattered in the angiosperms (e.g. Walker-Larsen and Harder, 2000; Ronse De Craene and Smets, 2001a), and are the result of sterilization affecting a complete whorl of stamens, or a variable number of stamens within a whorl. Sterilization of stamens has evolved frequently in angiosperms – for example, it affects 72 per cent of all families of rosids (Walker-Larsen and Harder, 2000). In cases where an entire stamen whorl becomes obsolete, stamen remnants may still be identified in flowers, but they will eventually disappear and this process is irreversible. However, in many instances loss of fertility is accompanied by acquisition of novel functions not performed by stamens, such as differentiation into food bodies (*Calycanthus*), osmophores (*Austrobaileya*), building of nectar containers (e.g. Loasaceae), enhanced optical attraction (Theophrastaceae, Eupomatiaceae, Zingiberaceae), differentiation into nectaries (e.g. *Helleborus* in Ranunculaceae), or a combination of these (*Ranunculus*, *Aquilegia*). This process offers new evolutionary potential in many groups of plants. Walker-Larsen and Harder (2000) argued that sterilization of stamens within a whorl is evolutionarily reversible, such as in Lamiales, where the adaxial staminode can be restored to a fully fertile stamen by a reversal to polysymmetry. However, there is a threshold beyond which such a reversal is impossible (Ronse De Craene, 2018).

The presence of staminodes is important as it indicates evolutionary transitions in stamen numbers or changes in flower configurations, and this can be clearly shown in floral diagrams.

Polyandry

In cases where the number of stamens is higher than double the number of petals or sepals the androecium is polyandrous. However, various non-homologous forms of polyandry exist in the angiosperms. Polyandry can be *primary*, by initiation of a high number of stamens in a spiral or whorls, or *secondary*, by division of primary (common) primordia into secondary stamens (Figures 1.1, 1.2; Ronse De Craene and Smets, 1987, 1992a, 1998a). In the past, a fundamental distinction was made between many and few stamens, the former being interpreted as ancestral. However, the mere distinction between many and few stamens is simplistic and is not reflected in the development of polyandry.

Primary polyandry (true polyandry) implies that a high number of stamen primordia arise in a specific sequence in the flower, whereas each primordium develops into a single stamen. Primary polyandry can be spiral or whorled, depending on the phyllotaxis of the flower (see p. 34). Trimerous or dimerous flowers with multistaminate androecia in several whorls often show an alternation of paired and unpaired stamens (e.g. Papaveraceae, Annonaceae). With primary polyandry, initiation of stamens is usually centripetal, although it can be reversed to a centrifugal development (e.g. Alismatales, Aristolochiaceae) due to spatial constraints.

In the case of secondary (complex) polyandry, the number of stamens is increased by developmental multiplication of stamen primordia. In fact, each stamen primordium functions as a fractionating unit, leading to a higher number of stamens by meristem expansion (Figure 1.1; cf. Claßen-Bockhoff, 2016). This development is fundamentally different from spiral or whorled polyandry, which is characterized by genesis of one stamen per primordium. A stamen primordium can divide into a pair or higher number of stamens, which may initiate in different directions (centrifugally, laterally, or centripetally), depending on the shape of the floral primordium and spatial constraints (Ronse De Craene and Smets, 1987, 1992a). The development of complex polyandry proceeds from one or two whorls of stamens. In some cases, the boundaries of the initial stamen primordium (primary primordium) can be identified and the secondary stamens appear fused into groups or fascicles. The stamens remain united at the base and they have a common vascular supply ('trunk bundles'). Secondary polyandry can be linked with diplostemony, when one whorl remains simple while the other is multiplied (e.g. *Reaumuria* in Tamaricaceae: Ronse De Craene, 1990), or when the other whorl becomes staminodial (e.g. several Malvaceae). Very often, only a single initial whorl is present in the flower (e.g. *Hypericum*). Polyandry becomes much more complicated when the identity of separate primordia is lost by development of a ring primordium (ring wall). Stamen groups may still be discernible on the ring (e.g. Malvaceae, *Hibbertia* in Dilleniaceae, *Stewartia* in Theaceae, Cistaceae), or a high number of stamens arise in girdles without clear boundaries between stamen groups (e.g. Bixaceae, *Camellia* in Theaceae, *Dillenia* in Dilleniaceae). The advantage of a ring primordium is that a very high number of small stamens can develop, considerably increasing the pollen load and creating opportunities for heteranthy (see Endress, 2006). Stamen initiation becomes decoupled from the rest of the flower and can be ongoing when the carpels are initiated. The growth of the receptacle or hypanthium can influence the extent of the secondary increase of stamens and the direction of development, which is often centrifugal (see p. 24; Ronse De Craene and Smets, 1991a, 1992a; Claßen-Bockhoff, 2016).

The general evolution of the androecium in the angiosperms probably evolved from a moderate number of stamens arranged in a spiral (e.g. *Amborella*). Different evolutionary lines led to an increase or reduction in stamen number, with several transitional steps. An outline of this evolutionary progression is shown in Figure 1.2 (Ronse De Craene and Smets, 1998a). As stamen number decreases and positions become fixed, it is possible to characterize specific 'types' of androecia. From primary polyandrous flowers, further increases or reductions can evolve, leading to much higher or lower numbers of stamens. In the case of polycycly (the case of several whorls of stamens), the number of whorls can be variable or decrease to two or one. These cases were described as dicycly or monocycly because they have a different origin from diplostemony and haplostemony (Ronse De Craene and Smets, 1987, 1998a). The two-whorled androecium of monocots is dicyclic and is not homologous with diplostemonous androecia. Even in groups with primary polyandry, the number of stamens can be secondarily increased, by extending the ontogenetic spiral on an extended cone, by centrifugal expansion of the floral apex, or by development of much smaller stamen primordia (e.g. Annonaceae, Papaveraceae). However, in most cases secondary polyandry develops from an androecium with an initially low number of stamens.

In the case of secondary polyandry, it is assumed that the complex stamen primordia were initially single stamens with a specific position in the flower (cf. Figure 1.1). Taxa with such secondary stamen increase are confined to Pentapetalae and monocots and are nested in clades with diplostemonous (dicyclic) or (ob-)haplostemonous (monocyclic) androecia. The number of

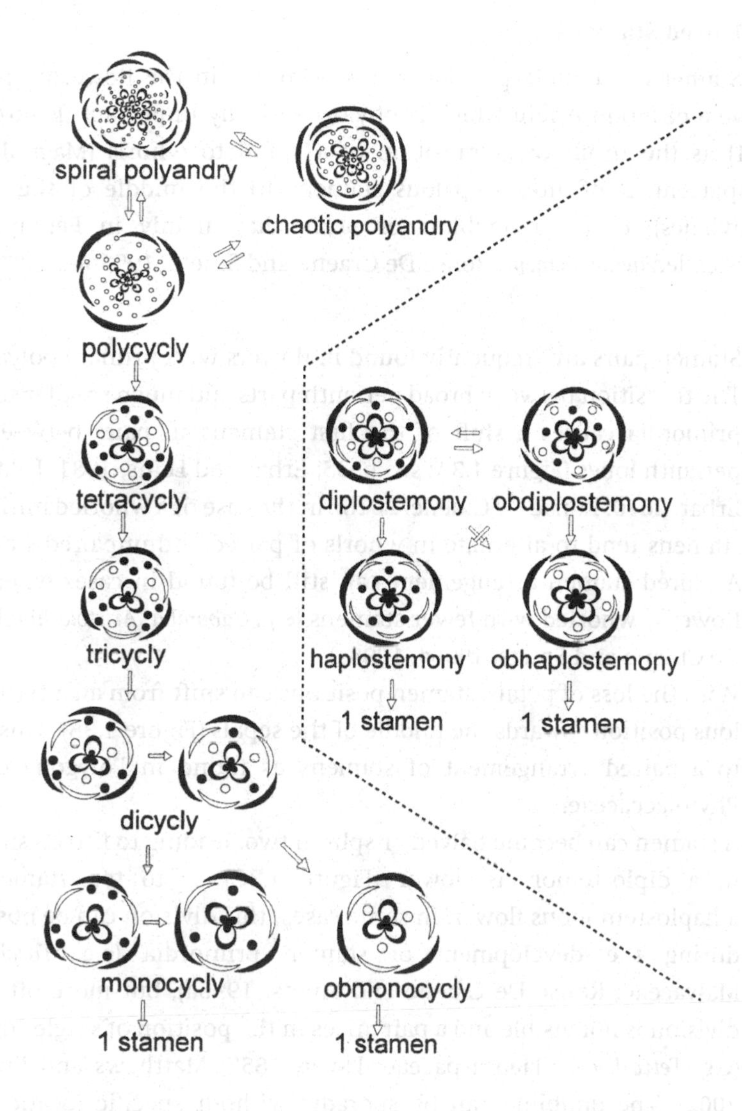

Figure 1.2 Diagrammatic presentation of androecium evolution in the angiosperms (from Ronse De Craene and Smets, 1998a, modified with permission of Plant Biology). The terminology used refers to different convergent origins of flowers with lower stamen numbers. Dotted line gives separation between polymery (left) and oligomery (right). For lower stamen numbers: black dots, antesepalous stamens; white dots, antepetalous stamens

stamens in the flower appears to be highly flexible with multiplications arising frequently in the angiosperms, especially Pentapetalae, as a means to increase the amount of pollen on offer to pollinators.

Paired Stamens

Stamens in double position are widespread in the flowering plants. They have a different origin which is phylogenetically important (Figure 1.3), either (1) as the result of a transition of a spiral to whorls (Magnoliales), (2) a displacement of alternisepalous stamens to the middle of the sepals (Caryophyllales), or (3) a doubling of primordia, mainly in Pentapetalae (a process called *dédoublement*: Ronse De Craene and Smets, 1993b).

(1) Stamen pairs are frequently found in flowers with primary polyandry. The transition between broad perianth parts and much smaller stamen primordia causes a shift of the first stamens in pairs between the perianth lobes (Figure 1.3A; see p. 35; Erbar and Leins, 1981; Leins and Erbar, 2010; Ronse De Craene, 2018). In the case of a whorled initiation, stamens tend to alternate in whorls of paired and unpaired stamens. A paired stamen arrangement can still be found in cases where the flower is whorled with fewer stamens (e.g. *Cabomba*, *Aristolochia*; Ronse De Craene and Smets, 1996a, 1998a).

(2) With the loss of petals, stamen positions can shift from an alternisepalous position towards the middle of the sepals (Figure 1.3B). This leads to a paired arrangement of stamens as found in Polygonaceae or Phytolaccaceae.

(3) A stamen can become halved or split in two, leading to fifteen stamens in a diplostemonous flower (Figure 1.3C) or to ten stamens in a haplostemonous flower. In some cases, this division can be observed during the development of stamen primordia (e.g. *Theobroma*, Malvaceae: Ronse De Craene and Smets, 1996a), but more often the division is not visible and a pair arises in the position of single stamens (e.g. *Tetratheca*, Elaeocarpaceae: Payer, 1857; Matthews and Endress, 2002). The doubling can be sporadic without specific location (e.g. some Lythraceae: Tobe, Graham and Raven, 1998). It can affect the antepetalous stamens only (e.g. *Sarcocaulon*, Geraniaceae), the antesepalous stamens (e.g. *Geissos*, Cunoniaceae), or more rarely both whorls (e.g. *Ginoria*, Lythraceae). A list of families with double stamens is given in Ronse De Craene and Smets (1996a). Doubling of the antepetalous stamens is often linked with obdiplostemony; the doubling of stamens probably results in the retardation and smaller size of the stamen pair. The occurrence of fifteen stamens in a pentamerous flower can be confusing, as it can be difficult to know whether paired stamens are the result of the splitting of one stamen whorl of a diplostemonous

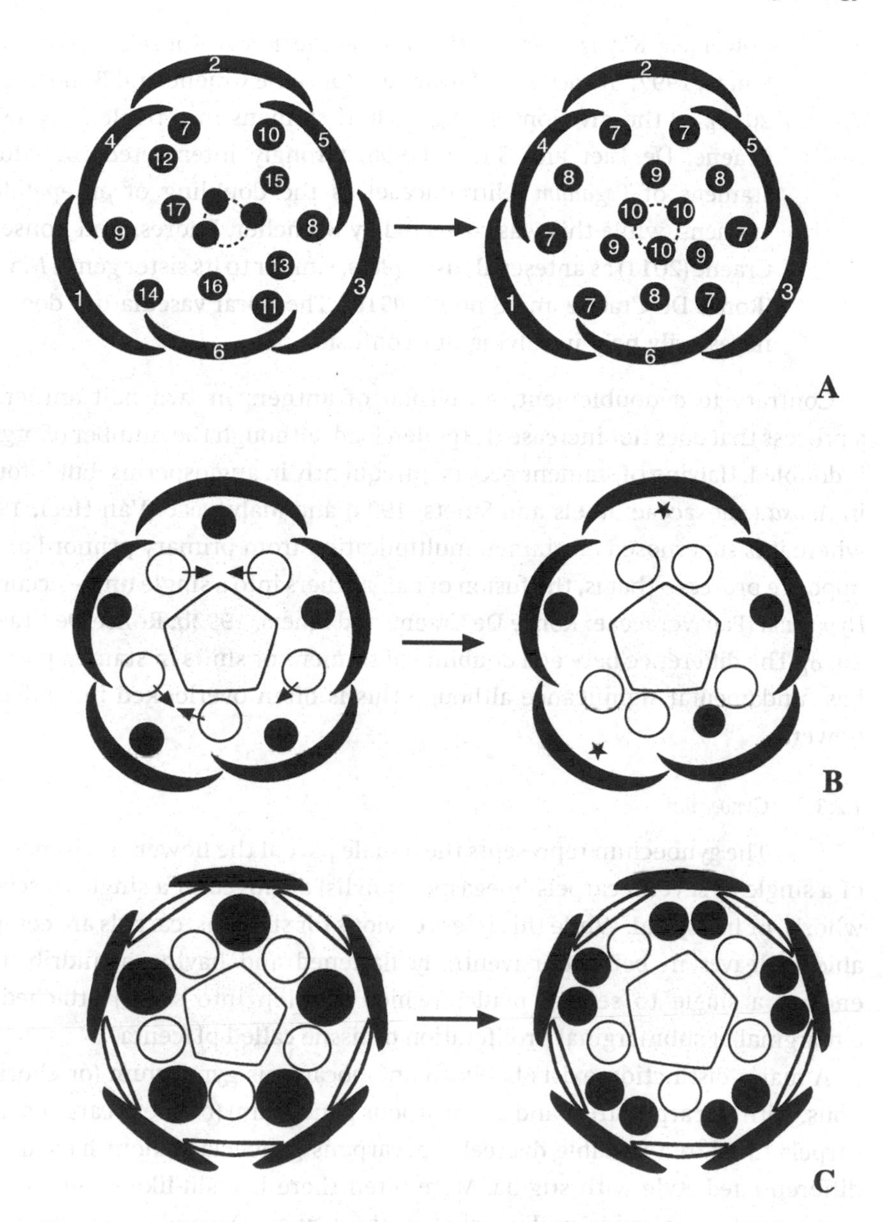

Figure 1.3 Different origins of stamen pairs. A. Through the cyclization of a spiral flower, leading to the paired arrangement of the outer stamens (based on Leins and Erbar, 2010); B. Through a shift of alternipetalous stamens in pairs linked with a shift of antesepalous stamens to the periphery and their eventual loss; C. through dédoublement of one of the stamen whorls. Dotted circle shows area of differentiating floral apex; black dots, antesepalous stamens; white dots, antepetalous stamens

flower (e.g. *Kirengeshoma* in Hydrangeaceae: Roels, Ronse De Craene and Smets, 1997; *Byttneria* in Malvaceae: Ronse De Craene and Bull-Hereñu, 2016), or the division of single-whorl stamens in a triplet. Ronse De Craene, De Laet and Smets (1996) wrongly interpreted the fifteen stamens of *Peganum* (Nitrariaceae) as the doubling of antepetalous stamens, while this was corrected by Bachelier, Endress and Ronse De Craene (2011) as antesepalous triplets, similar to its sister genus *Nitraria* (Ronse De Craene and Smets, 1991e). The floral vasculature does not necessarily help in solving this confusion.

Contrary to dédoublement, a division of anthers in two half anthers is a process that does not increase the pollen load, although the number of organs is doubled. Halving of stamens occurs infrequently in angiosperms, but is found in *Adoxa* (Adoxaceae: Roels and Smets, 1994) and Malvaceae (Van Heel, 1966) where it is superposed on stamen multiplication from primary primordia. The opposite process – that is, the fusion of half-anthers into a single unit – occurs in *Hypecoum* (Papaveraceae: Ronse De Craene and Smets, 1992b; Ronse De Craene, 2018). The difference between doubling of stamens or shifts in stamen position has fundamental significance although this is often overlooked in studies of flowers.

1.2.3 *Gynoecium*

The gynoecium represents the female part of the flower and is made up of a single or several carpels (megasporophylls) arranged in a single or several whorls, or in a spiral. While this is less obvious for stamens, carpels are comparable to leaves in being dorsiventrally flattened and having a midrib; they enclose a single to several ovules (which develop into seeds) attached on a marginal or submarginal proliferation of tissue called placenta.

A major distinction exists between an apocarpous gynoecium (or choricarpous, with all carpels free) and a syncarpous gynoecium (or coenocarpous, with carpels fused to a variable degree). Apocarpous gynoecia seldom have a well-differentiated style with stigma. More often there is a slit-like opening filled with secretion to guide pollen tubes to the ovules. Depending on the shape during early development, one distinguishes between ascidiate (urn- or bottle-shaped) and plicate (folded) carpels (Endress, 1994; Endress and Igersheim, 1997, 2000a; Leins and Erbar, 2010; Sokoloff, 2016), although there are intermediate shapes. Apocarpous gynoecia are found mainly in basal angiosperms. (Endress and Doyle, 2015).

In syncarpous gynoecia there is usually a distinction between an ovule-bearing zone (the ovary with a differentiation of a synascidiate and a symplicate zone), and styles with stigmas. The synascidiate zone is the area of complete septation of carpels, mainly bearing ovules, while the symplicate zone is the area where septa become detached from each other, eventually leading to the separation of different styles. The proportional size of the synascidiate and symplicate zones can fluctuate strongly among angiosperms and is the result of the extent of development of different parts of the ovary (see Leinfellner, 1950; Endress, 1994; Leins and Erbar, 2010; Ronse De Craene, 2021). Pollen is collected on the stigma and a pollen tube grows through the stylar tissue to reach the ovules. Styles may be separate and connect to individual carpels, or be basally fused with apical style branches (stylodes), or there is a common style with single apical stigma connecting all carpels. The limits between style and stigma can be blurred when stylodes present a receptive area over their whole adaxial surface, as is often found in Caryophyllales. When carpels are laterally fused there is a common zone shared by all carpels (compitum). This has the selective advantage for the flower to control fertilization much more efficiently by selecting for the fittest pollen and spreading pollen tubes equally over all carpels (Endress, Jenny and Fallen, 1983; Erbar, 1998; Endress, 2006). Stigmas can reflect the number of carpels in syncarpous gynoecia, or they are undivided. The position of styles and/or stigma lobes can be opposite the carpels (carinal), or alternating with the carpels (commissural). Styles are generally terminal, but become occasionally basal (gynobasic styles) by the more pronounced growth of the abaxial side of the carpels so that the style appears to emerge from the base of the ovary (e.g. Lamiaceae, Chrysobalanaceae).

The number of carpels in apocarpous gynoecia can be highly fluctuating with a strong tendency for a stabilization to a single carpel (monomery, Figure 1.4B–G). In syncarpous gynoecia the number of carpels is relatively stable and generally fluctuates between two and five. In some groups there is a tendency for one to four carpels to become sterilized with the retention of a single fertile carpel (pseudomonomery, Figure 1.4F; see p. 23). There are different degrees of fusion of carpels: (1) carpels may arise separately and only fuse postgenitally (frequently in monocots), (2) carpels may arise separately but are soon unified by a ring meristem (e.g. Rosaceae), and (3) carpels arise on a ring meristem (congenital fusion, see p. 41; e.g. Primulaceae).

There is a general evolutionary trend from apocarpy to syncarpy (Figure 1.4). More than 80 per cent of gynoecia are syncarpous, 10 per cent are apocarpous, and 10 per cent are monocarpellate (Endress, 1994). Apocarpous gynoecia are generally found in the basal groups of angiosperms and magnoliids with a high incidence of monocarpellate gynoecia (derived by reduction of carpels except one). Syncarpous gynoecia may rarely become secondarily apocarpous, and this is usually associated with postgenital fusion of the stylar parts. This evolved

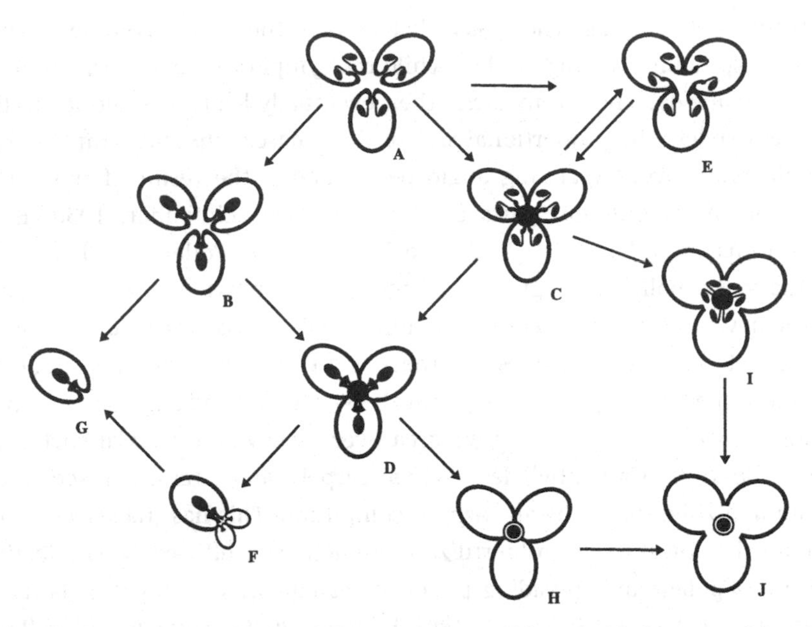

Figure 1.4 Carpel fusion and placentation types (shown with three carpels). A. apocarpous with marginal placentation; B, apocarpous with axile placentation; C. syncarpous, axile placentation with paired ovules; D. axile placentation with single ovules; E. parietal placentation; F. pseudomonomerous; G. single carpel; H. basal or apical placentation with single ovule; I. free-central placentation; J. basal or free-central placentation with single ovule

independently at least in three different orders of Pentapetalae but also in monocots (Endress, Jenny and Fallen, 1983; Sokoloff, 2016). Secondary apocarpy has the advantage of economy if only one carpel is fertile and there is no need to develop all carpels in fruit. In some basal angiosperms an external compitum is formed by other parts of the flower such as the floral receptacle (e.g. *Cananga* in Annonaceae, *Illicium* in Illiciaceae, *Tambourissa* in Monimiaceae or reduced perianth parts (Monimiaceae, *Siparuna*: Endress and Igersheim, 2000a; Staedler and Endress, 2009; Leins and Erbar, 2010). The central apex is usually used up in the development of the gynoecium but in some basal groups (e.g. *Illicium, Nymphaea*) or in eudicot flowers with increased carpel number (e.g. *Medusagyne, Sheffleria*), it remains prominent.

Syncarpy and associated placentation probably evolved multiple times in angiosperms. Endress and Doyle (2009) recognize two kinds of syncarpous ovaries, parasyncarpous (united carpels with unilocular ovary: parietal placentation) and eusyncarpous (with septa fused in the centre: axile placentation).

Placentation

In apocarpous gynoecia, ovules are generally lined along the margins of carpel lobes (marginal placentation, mostly with plicate carpels), or subapically (on the cross-zone in ascidate carpels) (Endress, 1994; Leins and Erbar, 2010). The placenta is occasionally laminar-diffuse, in cases where ovules are scattered over the inner margin of the carpel (e.g. Nymphaeaceae, Butomaceae).

In syncarpous gynoecia the lateral walls of individual carpels may develop in the ovary as septa or divisions, separating individual locules or cells (multilocular ovary), or septa may not develop and there is only a single cell or locule (unilocular). The main types of placentation in the angiosperms are axile or parietal. In cases where septa are fully formed, placentation often tends to be axile with ovules clustered in a central position (Figure 1.4C). During development of carpels the entire carpel margin may contribute to the formation of a septum, developing marginal placentae with ovules.

Changes between different placentation types are frequent and can go in different directions (Figure 1.4). Disappearance or breakdown of septa leads to a free-central placentation (Figire 1.4I). This can be progressive by a breakdown of septa during development (e.g. Caryophyllaceae: Ronse De Craene, 2021), or septa vanish altogether with occasional remnants present in some species (e.g. Primulaceae). In some groups the septa are restricted to the lower portion of the ovary (basal septum), the upper part of the ovary (apical septum), or both. Reduction of the septum may lead to a basal or apical placentation (Figure 1.4H, J). The placentation and carpels are physiologically and phylogenetically strongly connected and generally arise as a unit (Endress, 2019). However, in certain groups there was a developmental disconnection between placenta and carpel, with ovules arising away from the carpel wall on a meristematic floral apex (e.g. Cyperaceae, Amaranthaceae). Such gynoecia were described as acarpellate by Sattler (1974) among others, but this is misleading and an expression of different growth processes and a weakening of the carpel wall relative to the ovule bearing tissue (Ronse De Craene, 2021). One could say that ovule and carpel wall are phylogenetically connected, but developmentally disconnected, leading to much confusion.

A parietal placentation develops by contraction of the septa to the periphery (Figure 1.4D). Intermediate conditions exist with septa reaching halfway into the ovary (e.g. *Papaver*, Papaveraceae), or septa do not develop at all and the placenta develops on the ovary wall (e.g. Passifloraceae). While axile placentation is widespread in Pentapetalae, parietal placentation tends to be restricted to specific families. There are probably multiple origins of parietal placentation. The most common pathway is an increase in the symplicate zone and a retraction of the synascidate zone of the ovary. Several families of Pentapetalae show gynoecia that are apically incompletely septate and basally septate (Ickert-Bond, Gerrath and Wen, 2014). A second pathway is a direct derivation of ovaries with parietal placentation by merging of several carpels (cf. parasyncarpous ovaries). Such derivation is found in *Berberidopsis* (Ronse De Craene, 2017a).

Secondary partitions (false septa) may occasionally arise within the locule, either on areas of the carpel wall that are not linked to carpel margins (e.g. Linaceae), or from the placenta (e.g. Brassicaceae). The development of false septa has a function of separating developing ovules in partitions (e.g. Lamiaceae). Ronse De Craene and Smets (1998b) suggested that false septa can evolve into real partitions and can ultimately increase the number of carpels, although this is not supported by developmental evidence (Endress, 2014).

Ovules are often arranged in two series on the placenta (linear arrangement); they may also spread out in double series or without clear order (diffuse arrangement). By contraction of the locular space two series can merge in a single line (e.g. Oxalidaceae). There is much variation in the number of ovules, ranging from a very high number (e.g. Orchidaceae) to a pair (e.g. Vitaceae) or just a single ovule per carpel (e.g. Anacardiaceae). Ovules are attached to the placenta by a funiculus. The ovules are generally curved (anatropous) with the opening (micropyle) downwards by bending of the funiculus, more rarely erect (orthotropous or atropous), or curved around the funiculus (campylotropous). Endress (1994) recognized the importance of the orientation of the curvature of the ovule in relation to the ovary wall, distinguishing between ovules that follow the curvature of the carpels (syntropous) and those that are curved in opposite direction (antitropous).

The transfer of pollen from the style to the ovule via the placenta is made possible by secretion or bridging devices: this can be an obturator as an extension of the placenta (e.g. Euphorbiaceae, Sapindaceae), protrusions of the funiculus (e.g. Anacardiaceae), or apical extensions of the placental column in ovaries with free-central placentation (e.g. Primulaceae-Myrsinoideae). The gynoecium can be isomerous with the petals, which means that the number of carpels is the same as the petal whorl. In that case carpels are situated opposite the sepals or petals. Very often, the number of carpels is lower (three or two) and their position tends to be influenced by the position of the flower on the inflorescence (Ronse De Craene and Smets, 1998b).

Pseudomonomerous Gynoecia

In several groups of Pentapetalae and commelinids (monocots) the gynoecium is reduced to a single fertile carpel. A pseudomonomerous gynoecium can be defined as a gynoecium that consists of more than one carpel of which only one is fertile and fully developed while the other(s) are empty or absent at maturity. Reduction series can easily be constructed in some groups of plants, such as Restionaceae ranging from three fertile carpels to a single median carpel (Ronse De Craene, Linder and Smets, 2002; Figure 1.4F), or 'Urticales' with a series running between different families from two carpels to a single one (Bechtel, 1921).

Pseudomonomerous gynoecia were variably defined in the past. Guédès (1979) defined pseudomonomery as a pluricarpellate ovary with a single ovule. The ovule can be on the margins of the sterile carpel or is basal. Eckardt (1937) and Weberling (1989) interpreted it as a pluricarpellate ovary with rudimentary carpels. The fertile carpel can have more than one ovule. The latter interpretation is the correct one, as different degrees of sterilization of carpels and ovules are possible. Pseudomonomery is widespread among angiosperms and can develop by different degrees of reduction of carpels with different merisms (see Ronse De Craene and Smets, 1998b; González and Rudall, 2010). However, it can be difficult to distinguish monomerous and pseudomonomerous gynoecia without a phylogenetic context when evidence of lost carpels is missing (Sokoloff, 2016).

Pluricarpellate Gynoecia

As for the androecium the number of carpels can be considerably increased relative to the other whorls of the flower. However, the extent of increase is more limited than for the androecium, because the gynoecium occupies a distal position in the flower with little space for expansion. Carpel increase is linked with an extension of the diameter of the floral apex, which remains morphologically undifferentiated and can remain exposed at maturity (Endress, 2014; Ronse De Craene, 2016, 2017b). Because of lack of space carpels tend to be deformed and are closing in an irregular fashion. *Tupidanthus* (Araliaceae) represents an extreme in lateral polymerization resulting in a convolute flower (Sokoloff et al., 2007). The number of carpels is rarely increased by proliferation from five common primordia (e.g. *Kitaibelia*, *Malope* in Malvaceae: Van Heel, 1995; Endress, 2014). A higher number of carpels in a whorl is often correlated with a polymerization of the androecium or increased merism of the entire flower (p. 36; e.g. *Araliaceae*, *Actinidia*, *Citrus*). Carpels and stamens tend to be increased laterally in a girdle in most cases. In some rare cases a second whorl of carpels is initiated (e.g. *Citrus* in Rutaceae, *Pavonia* and *Urena* in Malvaceae, *Punica* in Lythraceae: Ronse De Craene and Smets, 1998b; Endress, 2014). In Rosaceae, the swelling of the receptacle permits the development of several tiers of uniovulate carpels (Kania, 1973). Also in basal groups extension of the cone-shaped receptacle can lead to an increase in numbers of carpels which are arranged in several series. This is clearly visible in some Ranunculaceae (e.g. *Myosurus*, *Laccopetalum*), or Alismataceae.

1.3 The Floral Axis and Receptacle

The floral receptacle (axis or torus) is the central part of the flower on which floral parts are inserted and is morphologically undifferentiated being neither homologous to the stem nor leaf (Endress, 2014, 2019; Claßen-Bockhoff, 2016). Floral organs develop by the differentiation of the floral apex, which can be used up completely or remain present as a small residue. When floral organs have been differentiated, the receptacle can be inconspicuous or strongly developed. A conical (strobilus) shape is characteristic for some basal angiosperms such as Magnoliaceae, while a cup-shaped receptacle (hypanthium) is more widespread. Flowers are hypogynous with superior ovary, in cases where the perianth and stamens are clearly inserted below the gynoecium (Figure 1.5A–F). In cases where the receptacle expands in a tube or a cup by intercalary meristematic growth, a hypanthium is formed which may place the perianth and stamens at a higher level than the gynoecium (Figure 1.5C, E). During development of the flower the gynoecium can sink within the receptacle by intercalary growth occurring at the level of the gynoecium. The flower becomes perigynous (with a half inferior ovary: Figure 1.5G) or epigynous (with an inferior ovary: Figure 1.5H, I) as a result. Differences between superior and inferior ovaries are often only a matter of degree. A large part of the twentieth century was dominated by the controversy between protagonists of the appendicular (American School) versus the axile origin of the receptacle (German School). This fruitless discussion centred on the question of whether the hypanthium represents the congenital fusion of the bases of different organs, or the result of an expansion of the receptacle. Transitions, however, can go in both directions with the possibility of reversals. Examples of changes in the position of the ovary are cited in Endress (2003a) and Costello and Motley (2004).

A hypanthium can have the same colour as the perianth, leading to it being mistaken for a calyx or corolla tube (e.g. Grossulariaceae, Tropaeolaceae, Onagraceae). The presence of a hypanthium is often accompanied by petaloid sepals and a tendency for reduction and loss of petals, especially in rosids (Ronse De Craene, 2008). Depending on the level of meristematic growth, calyx, corolla

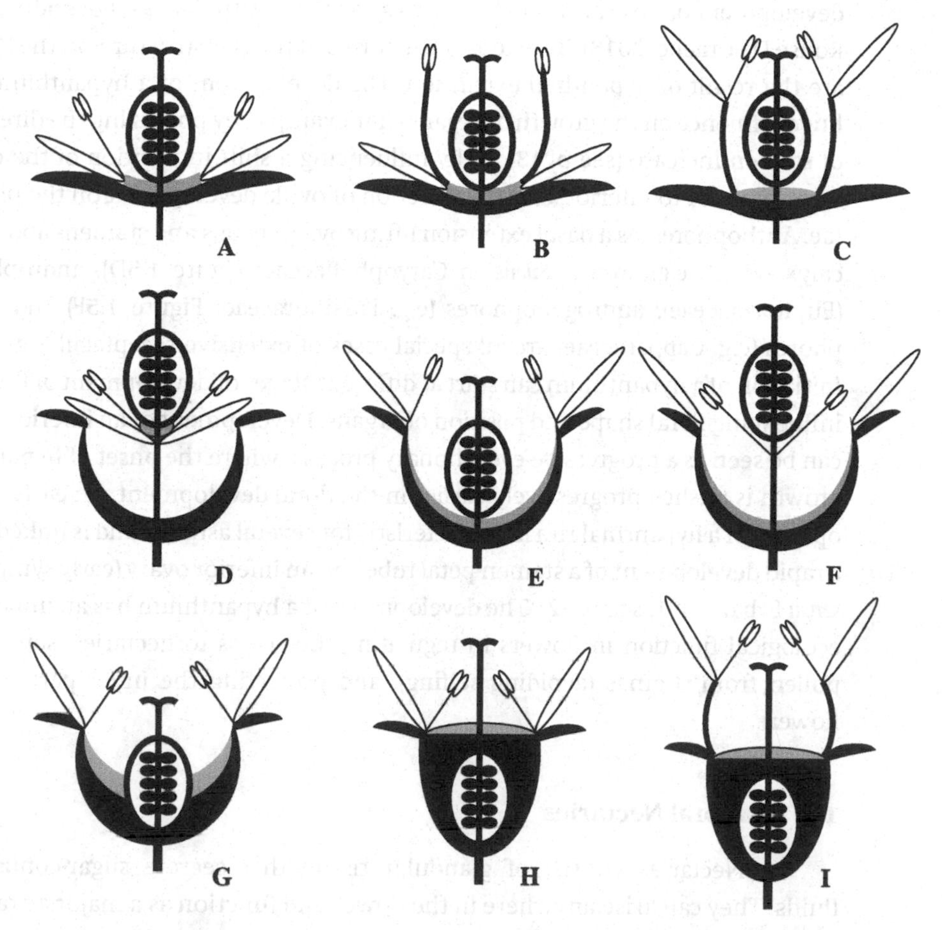

Figure 1.5 Hypanthium types and ovary position. Nectary shown by grey area. Sepal lobes and hypanthium shown in black; petals shown in white. A. flower with superior ovary and without hypanthium; B. flower with superior ovary and with stamen tube; C. flower with superior ovary and with stamen-petal tube; D. flower with sepal tube and anthophore; E. flower with cuplike hypanthium; F. flower with cuplike hypanthium and androgynophore; G. flower with half inferior ovary and hypanthium; H. flower with inferior ovary; I. flower with inferior ovary and stamen-petal tube

and stamens can become connected by a hypanthial tube (e.g. Myrtaceae: Figure 1.5E), or only calyx and corolla (e.g. Cucurbitaceae: Figure 1.5F). Stamen-petal tubes are a specific kind of hypanthium arising by formation of a ring meristem at the base of corolla and stamen lobes (Figure 1.5C, I; see p. 42).

Developmental evidence demonstrates that the establishment of intercalary expansion (zonal growth) within the receptacle is responsible for the development of a hypanthium (see Leins and Erbar, 2010; Claßen-Bockhoff, 2016; Ronse De Craene, 2018). By extension all forms of tubular structures in the flower are the result of hypanthial expansion. The development of a hypanthium has huge influence on the growth of organs – for example, by polarizing the direction of stamen increase (see p. 13), or by influencing a shift in position of the ovary from superior to inferior, and the direction of ovule development on the placentae. Anthophores (as a basal extension lifting ovary, petals and stamens above the calyx whorl: e.g. *Lychnis*, *Silene* in Caryophyllaceae: Figure 1.5D), androphores (Euphorbiaceae), androgynophores (e.g. Passifloraceae: Figure 1.5F) and gynophores (e.g. Capparaceae) are all special cases of extensive receptacular growth. Initiation of a hypanthium can start at different stages of development of flowers, influencing floral shape and position of organs. Development of an inferior ovary can be seen as a progressive evolutionary process, where the onset of hypanthial growth is pushed progressively earlier in the floral development. An early development of a hypanthial rim is characteristic for several asterids and is linked with a rapid development of a stamen-petal tube and an inferior ovary (early sympetaly *sensu* Erbar, 1991; see p. 42). The development of a hypanthium has an important ecological function in flowers in regulating the access to nectaries, separating pollen from stigmas (avoiding selfing), and protecting the inner parts of the flower.

1.4 Floral Nectaries

Nectaries consist of glandular tissue that secrete sugar-containing fluids. They can arise anywhere in the flower and function as a major attractant to pollinators. Nectaries do not make up an organ category of their own but represent a specific tissue not necessarily confined to particular floral organs. They rarely comprise whole organs (staminodial structures or carpellodes), more often appendages situated on the receptacle or on floral organs. There were several attempts to classify floral nectaries, either on shape, position, histology and function (reviewed in Pacini, Nepi and Vesprini, 2003, Bernardello, 2007 and Erbar, 2014). Vogel (1977, 1998c) distinguished three types of nectary based on their histological properties: mesenchymatous (usually a disc with underlying nectar tissue and secretion through transformed stomata), epithelial (secretion through the epidermis) and trichomatic nectaries (through transformed hairs). Smets (1986, 1988) distinguished between *nectaria caduca*, occurring on accessory structures in the flower (stamens and petals), and *nectaria persistentia*, occurring on parts of the flower that persist at fruiting (sepals, receptacle, ovary). The latter comprises two main types: disc nectaries common to Pentapetalae and septal nectaries of the monocots.

Nectaries are frequently associated with specialized compartments (nectar containers or spurs) that are independent of the nectar glands (e.g. Loasaceae, *Viola*, *Aconitum*, *Linaria*), or perform both functions (e.g. *Impatiens*, *Tropaeolum*, *Aquilegia*). In other instances the nectaries are inconspicuous and present as an epithelial layer on the hypanthium (e.g. Myrtaceae), thickenings at the base of the stamens (several Caryophyllales, including Polygonaceae, Geraniaceae), trichomes (e.g. Caprifoliaceae, Cucurbitaceae), or septal nectaries arising where septa are postgenitally formed in the monocots. Disc nectaries, although common in eudicots, are absent from monocots. Some authors discourage the use of the word 'disc' for nectaries, as it is indiscriminately employed for many non-homologous structures in flowers (see Bernardello, 2007). However, the use of 'disc' reflects a structure in the flower with a clearly identifiable shape. Disc nectaries are receptacular in origin, but can undergo important shifts in the flower, usually in a centripetal direction (e.g. Polygonaceae: Ronse De Craene and Smets, 1991b). The shift can run concomitantly with invagination of the ovary: from disc nectary to gynoecial and further to stylar nectary. Different forms of disc nectaries are illustrated in Figure 1.5.

The origin of nectaries is to be sought in the vegetative parts of the plant, in the form of transformed hydathodes or water pores (Vogel, 1998c; Smets et al., 2000). Outside the angiosperms nectaries are found in some ferns and gymnosperms (Pacini, Nepi and Vesprini, 2003). Association of nectaries with pollination evolved in *Ephedra* and *Welwitschia* (Gnetales, gymnosperms) and flowers of angiosperms. Some groups of plants have developed oil-secreting glands (elaiophores) instead of nectar glands (e.g. *Diascia* in Scrophulariaceae, calyx glands of Malpighiaceae). In some angiosperms with strongly reduced flowers that have switched from wind pollination to insect pollination, nectary tissue may develop outside the flower and become associated with the reproductive organs. The inflorescence of *Euphorbia* (cyathium) is a cup of bract origin with marginal glands. Pseudonectaries represent a special category of structures that imitate nectar drops to attract pollinators (Endress, 1994; Endress and Matthews, 2006b). They can occur on stamens (e.g. *Memecylon*, Melastomataceae), staminodes (e.g. *Parnassia*, Celastraceae), or petals (e.g. *Lopezia*, Onagraceae; *Gillbea*, Cunoniaceae: Endress, 1994).

1.5 Relationship of Flowers with Inflorescences

1.5.1 *Terminal and Lateral Flowers*

Flowers are seldom solitary. They are mostly grouped in inflorescences, which can be highly complex. Several attempts were made to morphologically interpret inflorescences, mostly with emphasis on describing inflorescences typologically (e.g. Weberling, 1989), and more rarely developmentally (see Tucker, 1999a; Claßen-Bockhoff and Bull- Hereñu, 2013). However, the complexity of the terminology can be overwhelming and without clear introduction one is easily lost or mistaken (Endress, 2010b). Moreover, the distinction between flowers and inflorescences can sometimes be difficult or almost impossible to make. Claßen-Bockhoff and Bull-Hereñu (2013) understood the importance of the transition of a vegetative system to a floral system in recognizing three levels of ontogenetic complexity in inflorescences, referring to the way meristems differentiate in inflorescences, flower unit meristems (giving rise to more flowers), or flower meristems, depending on the size and the extent of fractionation of an original meristem.

Inflorescences can be basically subdivided in two categories, polytelic (racemose or monopodial) versus monotelic (cymose or sympodial) inflorescences (Tucker, 1999a). A polytelic inflorescence has one principal growing axis producing flowers acropetally and is indeterminate. Lateral branches are a reiteration of the main branches. Indeterminate inflorescences are theoretically indefinite but they eventually decline in activity. A monotelic inflorescence is determinate with the principal axis developing in a flower and a secondary axis developing basipetally in its axil. The distinction between polytelic and monotelic inflorescences is not always clear as both forms can easily shift into one another. A racemose structure can evolve into a cymose inflorescence and vice versa, the former by a dominance of apical flowers (bothryoids), or terminalization (grouping of terminal pseudanthia into an apical flower), the latter by loss of apical dominance (loss of a terminal flower, called truncation) (Sokoloff, Rudall and Remizowa, 2006). This change, though quite frequent, is triggered by genetic shifts and can be the cause of evolutionary transitions from one type of inflorescence to another. A racemose inflorescence can resume vegetative growth after developing flowers (e.g. *Callistemon*, Myrtaceae). Lateral flowers can also shift in a terminal position in cases where an apical residuum is lost. Alternatively, a pseudanthium (see p. 00) is formed by the loose grouping of carpels and stamens in cases where the space for flower inception is too limited or flower identity breaks down (e.g. *Potamogeton*, *Piper*, *Triglochin*: Sokoloff, Rudall and Remizowa, 2006; Buzgo et al., 2006). In some cases it can be difficult to distinguish flowers from inflorescences because of the aggregation of highly reduced flowers (e.g. *Cercidiphyllum*: Endress, 1986; *Davidia*: Claßen-Bockhoff and Arndt, 2018; *Euptelea*: Ren, Zhao and Endress, 2007; *Ricinus*: Claßen-Bockhoff and Frankenhäuser, 2020).

More complex inflorescences can be derived from monotelic and polytelic inflorescences, with several lateral branches repeating the patterning of the main branch (e.g. panicles, umbels, compound capitula), or racemes with lateral cymose branching can be combined in a thyrse. However, it remains essential to link these inflorescences to their mode of initiation to recognize their initial form, which is either polytelic or monotelic, and this can best be achieved by concentrating on the subunits of inflorescences as done in this book.

1.5.2 *Pseudanthia*

The definition of a pseudanthium differs among authors. The traditional definition implies that pseudanthia are inflorescences that mimic flowers (e.g. Eames, 1961; Weberling, 1989; Endress, 1994). For Rudall and Bateman (2003) it can be something in between, neither a true flower nor a true inflorescence. Two kinds of pseudanthia can be identified: those that retain the identity of individual flowers and those in which flower identity is lost (Sokoloff, Rudall and Remizowa, 2006). The first case is often associated with small flowers grouped in compact inflorescences, either with a functional division between outer sterile attractive flowers and inner fertile flowers (e.g. Hydrangeaceae with sterile outer flowers), or with a contribution of bracts external to the inflorescence (e.g. Cornaceae with showy bracts or Marcgraviaceae with cuplike nectariferous bracts). More examples are presented by Weberling (1989) and Claßen-Bockhoff (1990).

Small, highly reduced flowers can also be aggregated in inflorescences. In this case it is far more difficult to distinguish inflorescence from flowers, as one can be fooled by flower-like cyathia of *Euphorbia* (Prenner and Rudall, 2008) or male 'flowers' of *Ricinus* (Claßen-Bockhoff and Frankenhäuser, 2020). As mentioned earlier, some flowers have probably evolved from pseudanthia, as shown by Sokoloff, Rudall and Remizowa (2006) for the alismatids Zannichelliaceae and Cymodoceaceae.

1.5.3 *Bracts and Bracteoles*

The transition between leaves and flowers is usually intermediated by bracts. These are appendages that can be leaf-like, or are often much smaller, sometimes intermediate between leaves and perianth parts. Bracts (pherophylls) subtend a lateral shoot, and bracteoles (prophylls) represent the first leaves of a lateral shoot; both phyllomes are not floral organs, although they may become closely associated with flowers and secondarily included as part of the perianth (Ronse De Craene, 2007). Most eudicot flowers have one subtending bract and a pair of lateral bracteoles. In the monocots there is usually one abaxial bract and one adaxial or latero-adaxial bracteole (Arber, 1925; Remizowa et al., 2012). Numbers of bracts can be much higher without differentiation of bracteoles, or bracts and bracteoles can be secondarily lost. It could be argued that a distinction between bracts and bracteoles is artificial. In compound monotelic inflorescences with bracts and bracteoles, the bracteoles of the main flowers act as bracts of the lateral flowers, and this is continued with the further initiation of more flowers. A clear distinction between bracts and bracteoles is therefore impossible. For descriptive purposes one can distinguish between first-order bract, second-order bract, etc.

Bracts and bracteoles are usually well delimited from flowers. However, especially in more basal groups the transition between bracts and perianth parts is progressive. Bracteoles often regulate the transition from a decussate arrangement of the vegetative leaves to a helical phyllotaxis in the flower (Eichler, 1875; Prenner, 2004a; Ronse De Craene, 2008; see p. 35). The first two sepals tend to be arranged pairwise in alternation with the bracteoles. When bracteoles become lost, the first sepals tend to occupy a lateral position of lost bracteoles. In some monocots the bracteole is transversal, not adaxial, and on the same level as the first sepal (Remizowa, Sokoloff and Kondo, 2008). Bracts and bracteoles have a strong influence on developing flowers by their position and the pressure they exercise on developing buds (see Ronse De Craene, 2018). Bracts are occasionally showy attractive organs simulating a perianth (e.g. *Cornus*), or contrasting with the flowers in colour (e.g. *Mussaenda* in Rubiaceae; *Melampyrum* in Orobanchaceae). Bracts and bracteoles are often strongly associated with the flower. They can occur high on the pedicel, close to the flower (e.g. Phytolaccaceae, Berberidopsidaceae) or may become part of the flower, enclosing sepals as a secondary calyx (epicalyx; e.g. *Aextoxicon* in Aextoxicaceae, *Afzelia* in Leguminosae, *Sarracenia* in Sarraceniaceae). This implies that in some cases bracts can have contributed phylogenetically to the differentiation of the perianth in calyx and corolla by a spatial shift (Albert, Gustafsson and Di Laurenzio, 1998; Ronse De Craene, 2008).

1.5.4 *Epicalyx*

Appendages may occur below the calyx as an extra whorl of sepals or as smaller structures. They are described as the epicalyx (calicle). Epicalyx members can be leaf-like, sometimes indistinguishable from sepals (e.g. *Potentilla*, *Malva*) or bristle-like (e.g. *Agrimonia*, *Neurada*, *Coris*). Origins of the epicalyx are varied and often not homologous, either derived from bracts or bracteoles that became closely associated with the flower (e.g. Dirachmaceae, Malvaceae, Caprifoliaceae, Passifloraceae, Convolvulaceae), as stipules of the sepals (Rosaceae), or commissural emergences formed by folds of the calyx (e.g. Lythraceae) (Pluys, 2002; Barroca, 2014; Bello et al., 2016). In some Rosaceae the two outer sepals bear stipules, and the third sepal has only a single stipule, which is missing in the inner sepals (Trimbacher, 1989).

1.6 Symmetry and Orientation of Flowers

Floral symmetry is one of the main structural factors affecting the Bauplan of the flower besides merism (see p. 36). The symmetry of flowers is the result of the initiation of floral organs and their subsequent growth and differentiation (Endress, 1994, 1999, 2012; Tucker, 1999b; Citerne et al., 2010; Bukhari et al., 2017; Naghiloo, 2020). A regular floral symmetry results from equal growth of organs within whorls. Such flowers have a radially symmetrical plan (polysymmetry or actinomorphy) and are accessible to insects from all directions. The flower can be divided from any angle in two equal halves relative to the axis. In cases where organs develop unequally within a whorl, one or two sides of the flower can become differentiated from the other sides. If two sides develop differently, a disymmetric flower with two lines of symmetry is formed. This is relatively rare and can be found in Fumarioideae of Papaveraceae or in Brassicales (Cleomaceae, Bataceae, Brassicaceae). The unequal development of one side of the flower leads to monosymmetry (zygomorphy or bilateral symmetry) with only one possibility to divide the flower in two equal halves. This pattern of development is widespread among angiosperms and is aimed at a specific access for pollinators (see Endress, 2012 for an overview). Most monosymmetric flowers have their symmetry line running along a median line from the main axis to the bract. Transversal monosymmetry is much rarer (e.g. *Corydalis* in Papaveraceae), while oblique monosymmetry (with the symmetry line neither transversal nor radial) characterizes several families (e.g. Moringaceae, Sapindaceae, Malpighiaceae).

Monosymmetry affects flowers by degrees and is an evolutionary progressive development. The timing of the onset of monosymmetry can be either manifested at early organ initiation, at organ growth and enlargement, or as a late differentiation of organs (Tucker, 1999b; Naghiloo, 2020). The expression of monosymmetry is strongly influenced by developmental constraints of inflorescence, bracts and neighbouring organs (Citerne et al., 2010; Naghiloo, 2020). Bract and the inflorescence axis act as two opposing gradients in shaping the floral bud during development (Endress, 1999). As a result, symmetry may shift during different developmental stages from inception to maturity. A secondary stamen and carpel increase in flowers (p. 23) tends to be less compatible with monosymmetry, as very few polyandric flowers are monosymmetric (e.g. some Cactaceae, Lecythidaceae), implying that a stamen increase may affect the dorsoventrality of the flower (Citerne et al., 2010). There is also a strong correlation between monosymmetry and a lower merism (being mostly restricted to trimerous or pentamerous flowers), as a higher merism appears incompatible with the development of bilateral symmetry. Additionally, gravity and light will affect the development of monosymmetry, as terminal flowers are generally polysymmetric and monosymmetric flowers are in a lateral position. Gravity is probably responsible for a different gradient in gene expression leading to monosymmetry (p. 59; see Koethe, Bloemer and Lunau, 2017).

There are different degrees in the extent of monosymmetry. Organs in flowers may be variously affected, and the extent can vary from a subtle displacement of organs or differences in size and shape, to an elaboration, reduction or suppression of organs (Endress, 1999, 2012; Rudall and Bateman, 2004; Citerne et al., 2010). The orientation of stamens and style and a size difference of the petals between adaxial and abaxial side of the flower may lead to weak monosymmetry (e.g. *Chamaenerion* in Onagraceae; *Gladiolus* in Iridaceae). More elaborate monosymmetry (e.g. Orchidaceae, Scrophulariaceae) is linked with a strong synorganization and various degrees of reduction (Endress, 2012). Initiation of organs can be unidirectional and associated with suppression or loss of some organs (one side of the flower develops more extensively at the expense of the other side: e.g. Leguminosae: Tucker, 1984, 1996, 2003a). This development can be accompanied by shifts in function between the anterior and posterior side of flowers (e.g. in the androecium of Commelinaceae). Monosymmetry may also be caused by extreme reduction. Some families have flowers reduced to a single carpel and one to few unilateral stamens (e.g. Chloranthaceae, Menispermaceae, Callitrichaceae). Such flowers are monosymmetric because most organs are lost. However, they may regroup in pseudanthial structures (see p. 29).

Tucker (1997, 1999b) demonstrated a strong correlation between the degree of monosymmetry and the developmental stage at which it is first apparent. This correlation also affects major taxonomic groups: in groups with a majority of polysymmetric taxa, monosymmetric flowers tend to differentiate late in floral development (e.g. *Tropaeolum* in Brassicales: Ronse De Craene and Smets, 2001b), while groups which are predominantly monosymmetric have unidirectional development and expression of zygomorphic characteristics much earlier in floral development (e.g. Lamiales: Endress, 1999). However, this is not universal and there are several exceptions (see Naghiloo, 2020). Monosymmetry can affect a single organ in groups, which are mainly polysymmetric. For example, a functional differentiation between different kinds of stamens (heteranth(er)y: differentiation of feeding stamens and pollen stamens) often leads to monosymmetry and a reduction of stamens, or also to highly monosymmetric flowers with elaborate androecia (e.g. Lecythidaceae, Commelinaceae). Heteranthy generally leads to the sterilization of the adaxial (upper) stamens into feeding stamens (accessible by the pollinator), whereas the abaxial (lower) stamens serve pollination (Koethe, Bloemer and Lunau, 2017).

Flower shape and symmetry are directly linked to pollination. The perception of symmetry is extremely important from the point of the pollinator, as a reflection of a coordinated evolution (Citerne et al., 2010; Specht et al., 2012; Van der Niet, Peakall and Johnson, 2014). Monosymmetry is strongly linked with synorganization of the flower and allows for different pollination strategies, which may be much more precise. Monosymmetric flowers can be subdivided in two kinds: flag-flowers and lip-flowers (Endress, 1994, 2012). Flag-flowers are sternotribic (pollination by the underside of the visitor through stamens that are exposed and curved up: e.g. Capparaceae, Caesalpinoid Leguminosae); lip-flowers have a lower lip functioning as a landing platform and are nototribic (pollination by the back of the insect; the stamens are usually concealed by the fused upper petals: e.g. Lamiales). The initial number and orientation of flower parts constrains the way a flower is differentiated into adaxial and abaxial parts (Donoghue, Ree and Baum, 1998). The almost ubiquitous presence of sepal two in adaxial position restricts the way two-lipped flowers are orientated (e.g. Lamiales, *Pelargonium* in Geraniaceae). Petals are generally arranged as an upper and lower lip in Lamiales, with three possible orientations of the petals: a 3:2 orientation has three abaxial and two adaxial petals; a 1:4 orientation has one abaxial and four abaxial petals; a 5:0 orientation has all petals moved to the abaxial side.[1] Such orientation appears rigid and is guided by the position of the sepals. Only in cases where sepal two is in abaxial position is an inverted arrangement possible (Donoghue, Ree and Baum, 1998; Endress, 1999). The orientation of the flower is also responsible for the reduction or loss of the adaxial stamen in many Lamiales. This is linked with the curvature of anthers and style in the upper part of the flower where they connect with the back of the pollinators and where an adaxial stamen would be in the way (Endress, 1994, 1999). In a few groups of plants (e.g. Cannaceae, Marantaceae, Caprifoliaceae, Vochysiaceae, some Leguminosae), flowers can be asymmetric. These flowers have no clear symmetry line, but are derived from monosymmetry (Endress, 1999, 2012). Their rarity may indicate that the pollinators do not appreciate such flowers, or that there are too many developmental constraints. However, asymmetric flowers appear often grouped in pairs with mirror-image flowers (enantiostyly: Marantaceae, *Wachendorfia* in

[1] I use the numerical symbols in the opposite way from Donoghue, Ree and Baum (1998), as it is more logical to read these from the abaxial (anterior) side corresponding with the perception of the pollinator (cf. Ronse De Craene et al., 2014).

Haemodoraceae, *Senna* in Leguminosae), or are small and arranged in compound inflorescences that appear regular (e.g. *Centhranthus* in Caprifoliaceae).

Although most evolutionary trends indicate that monosymmetric flowers are derived from polysymmetric flowers which represent the ancestral condition in angiosperms, there are several cases of polysymmetric flowers within mainly monosymmetric groups as evidence for a reversed evolution (e.g. Lamiales, Caprifoliaceae). Structural evidence indicates that fusion of posterior petals is linked with the loss of the adaxial stamen, leading to tetramerous flowers, and it appears to have arisen at least nine times in the asterids (e.g. Ronse De Craene and Smets, 1994; Donoghue, Ree and Baum, 1998; Endress, 1999; Bello et al., 2004; Soltis et al., 2005; Ronse De Craene, 2016). Alternatively, changes in *Cycloidea*-like genes are responsible for this shift (see p. 59).

1.7 Phyllotaxis

1.7.1 *Spirals and Whorls*

Flowers can have organs in spirals or in whorls similar to leaves. Spirals and whorls are regulated by phyllotaxis, which is the pattern of initiation of organs on vegetative stems and flowers. Guiding principles in floral phyllotaxis are the divergence angle (the angle between two organs arising in succession) and the plastochron (the time interval between the initiation of two successive organs). In spiral flowers organs arise in regular plastochrons with an equal divergence angle. In whorled flowers organs arise in pulses with unequal plastochron and unequal divergence angles. Within a whorl the plastochron is very short or approaching zero, while it is longer between different whorls (Endress, 1987; Endress and Doyle, 2007; Leins and Erbar, 2010). The sequence of initiation can be detected from contact spirals (parastichies) and contact lines (orthostichies). Parastichies and orthostichies can be mathematically described by Fibonacci series, reflecting the number of organs formed within a sequence. In spiral flowers there are several sets of parastichies with a particular number of organs (five, eight, thirteen, twenty-one, . . .) and no orthostichies, as is clearly visible on a pine cone or sunflower head. Whorled flowers have two equal parastichies running in opposite directions and several orthostichies. As a result change from one organ category to another (e.g. outer to inner tepals) is progressive in spiral flowers and different floral organ categories can be bridged by intermediate organs. In whorled flowers the transition between different organ categories tends to be abrupt. There has been much debate about what regulates organ development on the floral apex, either as the result of an inhibition zone, an auxin sink, or physico-mechanical causes (Ronse De

Craene, 2018). For reviews on phyllotaxis in flowers the reader is referred to Endress (1987, 2006), Endress and Doyle (2007) and Ronse De Craene (2018).

Changes in phyllotaxis are very common in flowering plants and happen at the transition of vegetative organs to flowers (e.g. bracteoles to sepals), but also between different organ categories (perianth and androecium, sepals and petals) and are dependent on a change in the size of the floral organs. In most Pentapetalae sepals follow a spiral phyllotaxis, which shifts to a whorled phyllotaxis at the level of the petals. As shown in this book, these shifts have important consequences for the arrangement of organs in flowers.

Phyllotaxis is best discussed in a phylogenetic context. A major evolutionary step in flowers is the transition from spiral to whorled flowers. There has been an extensive discussion of whether spiral or whorled phyllotaxis is ancestral in the angiosperms, with several authors favouring the latter option (see Endress and Doyle, 2007, 2015; Doyle and Endress, 2011; Sauquet et al., 2017; Soltis et al., 2018), although several basal angiosperms including *Amborella* have spiral flowers (Ronse De Craene, Soltis and Soltis, 2003). Erbar and Leins (e.g. Erbar and Leins, 1981, 1983; Leins and Erbar, 2010) demonstrated that in spiral flowers such as Magnoliaceae a switch to a whorled arrangement can result from an increase in size of the perianth parts. The transition between large tepals and much smaller stamens leads to an interruption in the continuous plastochron and a shift from a spiral to a whorled arrangement. Outer stamens tend to arise simultaneously as pairs in alternation with the inner perianth parts (see Figure 1.3A). Inner stamens and carpels may retain a spiral sequence, or organs tend to be arranged in alternating whorls repeating a sequence of pairs and non-paired organs (Ronse De Craene and Smets, 1993, 1994, 1998a). The paired arrangement of outer stamens can be retained in flowers while the number of parts is considerably reduced (Ronse De Craene and Smets, 1993; Iwamoto et al., 2018). Instead, Endress (1987, 1994) argued that stamens have been doubled as the result of a shortened plastochron and narrow organ shape. An increase in number of much smaller stamens in flowers can considerably disturb the regular whorled phyllotaxis (e.g. Annonaceae, some Papaveraceae; see Ronse De Craene, 2018 for an overview).

The acquisition of stable whorled flowers was considered an essential breakthrough in floral evolution (e.g. Endress, 1987, 1990; Ronse De Craene, Soltis and Soltis, 2003; Zanis et al., 2003; Endress and Doyle, 2007, 2015). Spiral flowers have a weak synorganization; the floral parts are loosely arranged with fluctuating organ number and little interaction between neighbouring organs. Whorled flowers have strong synorganization with close interaction between neighbouring organs; fusions and changes in symmetry are only possible with a concerted interaction between different organs in the flower (see

Endress, 1994, 2006, 2016). As a rule of thumb, phyllotaxis is labile in basal angiosperms and early diverging eudicots, contrary to Pentapetalae and monocots. Whorled flowers may retain a spiral initiation in one or several whorls. In most eudicots spiral phyllotaxis is retained in the calyx (with four to five sepals arranged in a 1/2 (decussate) or 2/5 (quincuncial) phyllotaxis. The corolla often arises in a rapid helical sequence but a spiral phyllotaxis is not clear at maturity. In some early diverging eudicots (e.g. Nelumbonaceae) and Pentapetalae (e.g. Paeoniaceae, Clusiaceae, Theaceae, Pentaphyllaceae) the whole perianth has occasionally reverted to a spiral sequence as a result of a secondary stamen and carpel increase (Ronse De Craene, 2007, 2016). Whorled phyllotaxis can revert to a spiral if constraints of synorganization disappear. The retardation or loss of petals may influence the phyllotaxis of stamens as in Apiaceae and Brassicaceae (Erbar and Leins, 1997) or Caryophyllaceae (Wei and Ronse De Craene, 2020). If constraints disappear (e.g. by loss of a perianth), whorled flowers may revert to a spiral or unordered initiation (e.g. *Cercidiphyllum*, *Trochodendron*, *Achlys*, *Theligonum*: Endress, 1986, 1989; Rutishauser et al., 1998).

1.7.2 Merism

Merism (or merosity) refers to the number of parts per whorl in the flower (usually based on petals or perianth, taken as reference). Merism is either indefinite, but more commonly trimerous or dimerous (basal angiosperms, basal eudicots, monocots), or pentamerous and tetramerous (Pentapetalae) (Figure 1.6). Merism is a major feature affecting the Bauplan of flowers besides symmetry and phyllotaxis. Origins of different merisms were discussed in Endress (1987), Kubitzki (1987) and Ronse De Craene and Smets (1994), and were recently reviewed by Ronse De Craene (2016).

Trimerous flowers (and by extension dimerous flowers) are widespread in basal angiosperms and monocots and represent a very stable Bauplan (Figure 1.6B, see Kubitzki, 1987). A derivation from a spiral condition is the most plausible interpretation for their origin. Dimery has arisen repeatedly by loss of a sector in each whorl of a trimerous flower, sometimes resembling a tetramerous flower, although the flower tends to be disymmetric (Figure 1.6D). Origins of pentamery are not well understood, although pentamerous flowers are found in the majority of angiosperms (Figure 1.6C). There are at least five origins of pentamery, the major one affecting the Pentapetalae (Ronse De Craene, Soltis and Soltis, 2003). The origin of pentamery has been postulated to be as a derivation from trimerous flowers (by amalgamation of two whorls), from dimerous flowers (through meristem expansion) or from spiral flowers (Figure 1.6A). Ronse De Craene (2007, 2017a) explains the origin of the pentamerous flower as a transition from

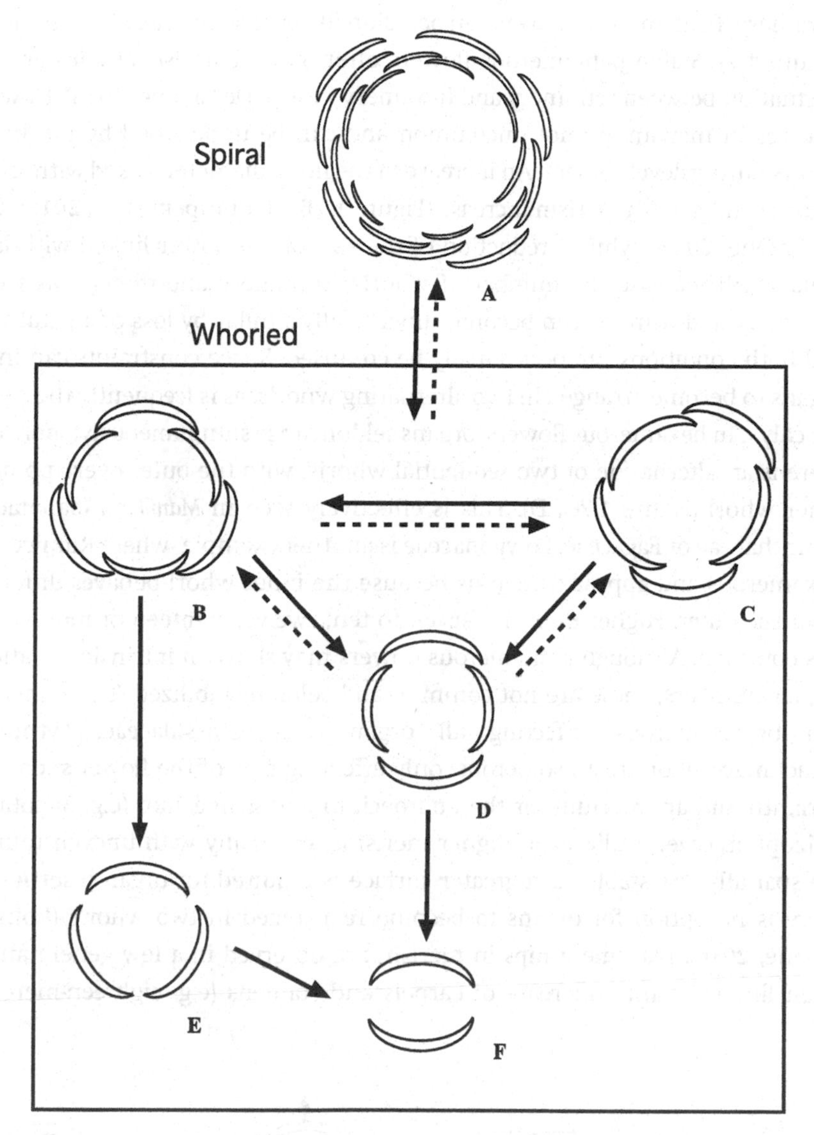

Figure 1.6 Relationships of merism in the perianth of angiosperms; undifferentiated perianth shown. A, spiral with variable number of tepals; B. trimery/hexamery; C. pentamery; D. dimery/tetramery; E. trimery, single whorl; F, dimery, single whorl

a spiral flower with undifferentiated perianth, reflected in extant Berberidopsidacae. Tetramerous flowers are further derived, either from trimerous flowers by an increase in size (e.g. Nymphaeaceae, Monimiaceae, Annonaceae), or by reduction in size from pentamerous flowers, occurring sporadically in pentamerous groups or characterizing a whole family. It is understood that a single mutation or proportionate size differences can trigger a change to tetramery. Pentamerous flowers can occasionally evolve into hexamerous flowers (Figure 1.7). Stable pentamerous flowers often have a statistically insignificant fluctuation between tetramery and hexamery (Ronse De Craene, 2016). However, changes in merism are not uncommon and can be understood by mechanical factors during development. An increase in the floral diameter linked with smaller organs can lead to a merism increase (Figure 1.7B; cf. Kümpers et al., 2016; Ronse De Craene, 2016), while a reduction of the size of the flower linked with larger organs will increase the number of whorls. Tetramery and dimery, as well as hexamery and trimery, can become superficially similar by loss of a petal whorl and both conditions can occasionally be confused. Space constraints can induce organs to become arranged in two alternating whorls, as is frequently the case for the calyx. In hexamerous flowers, organs seldom arise simultaneously; more often there is an alternation of two sequential whorls, with the outer overlapping the inner whorl (Figure 1.7C, D). This is effectively seen in *Manilkara* (Sapotaceae), Loranthaceae or Fagaceae. Polygonaceae is another example, where Rumiceae are hexamerous and appear trimerous because the inner whorl behaves differently from the outer. Higher merisms (seven to ten, twelve, eighteen or more) are far less common. Although pentamerous flowers may show an intrinsic variation to higher numbers, these are not common and seldom stabilized. These increases can be isomerous, affecting all organs (e.g. Crassulaceae, Lythraceae, Dirachmaceae) or are anisomerous, only affecting part of the flower such as the perianth and androecium, or the androecium and gynoecium (e.g. Sapotaceae, Rhizophoraceae, Araliaceae). Higher merisms, especially with uneven numbers are spatially less stable, as a greater surface is required for organ insertion and there is no option for organs to become rearranged in two whorls (Ronse De Craene, 2016). Extreme jumps in merism are observed in a few genera and are often linked with an increase of carpels and stamens (e.g. eighteen-merous to

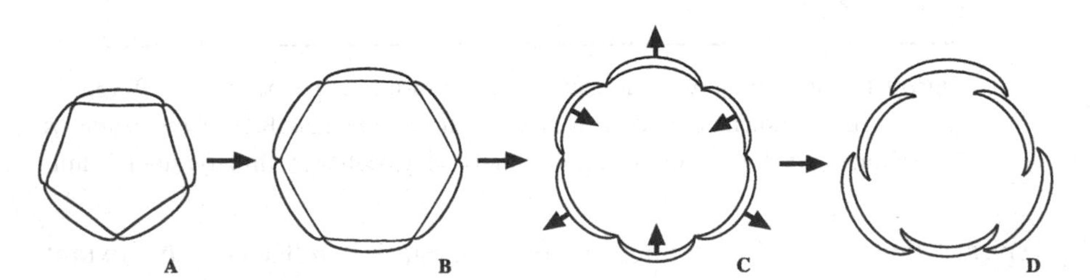

Figure 1.7 Merism change from pentamerous flower to hexamerous flower resembling a trimerous flower. A. pentamerous flower, perianth with valvate aestivation; B. meristic increase to hexamery by increase of the diameter of the floral apex; C. unequal development of the perianth and shift in two whorls; D. apparently trimerous flower with two alternating perianth whorls

thirty-two-merous *Sempervivum* in Crassulaceae). When the increase is anisomerous this can lead to distorted flowers by the lack of space for an expansion of organs (e.g. Araliaceae: Nuraliev et al., 2010, 2014).

Flowers are generally conservative in the presence of the same merism in most whorls. However, in some groups merism can fluctuate between different whorls (e.g. Nymphaeaceae, Winteraceae, Monimiaceae, Ranunculaceae, Sapotaceae). In basal angiosperms shifts in merism have been interpreted as the result of organ duplication (Staedler and Endress, 2009; Endress and Doyle, 2015). In several cases this jump can be related to smaller organ sizes, which I interpret as a result of a shift of a spiral phyllotaxis to a whorled phyllotaxis (see p. 35). Merism is intricately linked to phyllotaxis, as whorled phyllotaxis is a prerequisite for a stable merism. The combination of a whorled phyllotaxis with a stable merism has led to the clear-cut differentiation of the perianth into sepals and petals.

1.7.3 *Aestivation Patterns*

Aestivation is the relative arrangement of neighbouring perianth parts in a bud and is usually applied to the petals, which are often the most obvious organs at maturity. Three types are generally recognized: imbricate (with margins overlapping: Figure 1.8A–E), valvate (with margins touching: Figure 1.8F), and apert (or open, with margins not touching: Figure 1.8G). Aestivation is often – but not always – dependent on the initiation pattern of the perianth (Ronse De Craene, 2008). A sequential (spiral) initiation of organs will give rise to an imbricate aestivation (quincuncial or 2/5 pattern in pentamerous flowers: Figure 1.8B; decussate or 1/2 pattern in dimerous or tetramerous flowers: Figure 1.8D), in cases where some organs are completely outside and some completely within. Spiral flowers generally have an imbricate perianth arrangement. Simultaneous initiation pattern of the petals or sepals leads to a valvate or imbricate (contorted or convolute) aestivation (Figure 1.8E). Unidirectional initiation gives rise to an imbricate-cochleate pattern that can be ascending or descending (Figure 1.8C, e.g. Leguminosae). Apert aestivation is associated with retarded organs. In trimerous or hexamerous flowers, the outer perianth whorl can be imbricate, while the inner is apert (e.g. *Manilkara* in Sapotaceae), or the opposite occurs (e.g. *Monanthotaxis*, *Artabotrys* in Annonaceae). Aestivation is also correlated with the symmetry and the variable function of sepals and petals in flowers (Endress, 1994). Merism increase in flowers is generally associated with a valvate aestivation of the perianth (Ronse De Craene, 2016; Figure 1.7).

Despite having different origins, aestivation has systematic significance in several families of flowering plants and can be emphasized in floral diagrams.

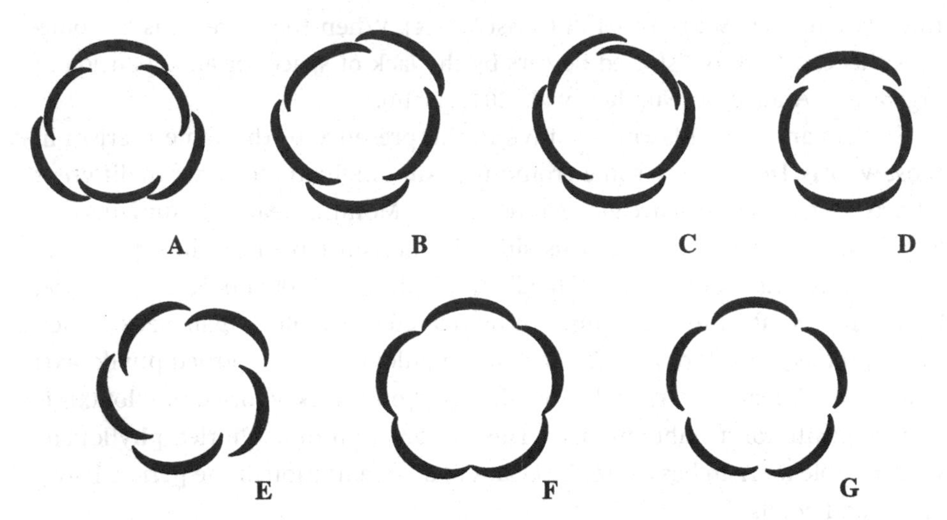

Figure 1.8 Aestivation patterns of the perianth. A–E. imbricate. A. imbricate-trimerous; B. quincuncial; C. cochleate; D. decussate-dimerous; E. contorted; F. valvate; G. apert

Aestivation of sepals is generally imbricate (2/5 aestivation), reflecting a spiral initiation sequence, especially in rosids and caryophyllids. Contorted sepal aestivation is rare. A valvate calyx is rare in pentamerous flowers, but when present is often accompanied by an epicalyx (e.g. Rosaceae, Malvaceae). For petals, aestivation patterns can be variable, but are often consistently contorted or valvate in some clades (e.g. Santalales, Malvales, Gentianales; Ronse De Craene, 2008).

1.8 Fusion of Floral Parts

Whorled phyllotaxis is a major requirement for fusion between organs in flowers. Therefore, basal angiosperms with spiral flowers rarely have fused organs. Fusion may occur independently in different organ whorls, such as sepals (synsepaly), tepals (syntepaly), petals (sympetaly), stamens (monadelphy), or carpels (syncarpy). In other cases different organ categories may be fused. Anthers and styles can be fused into a gynostemium by congenital fusion (e.g. Orchidaceae, Aristolochiaceae) or a gynostegium by postgenital fusion (e.g. Apocynaceae). Fusions of organs are often confused with the development of a hypanthium (see p. 24) as the limits between fused organs and a floral cup may not be visible.

Congenital and Postgenital Fusion

In the literature a distinction is commonly made between *congenital fusion* and *postgenital fusion*. However, this distinction is confusing and often difficult to make without evidence of floral development (Sokoloff et al., 2018a).

Postgenital fusion is a real process of fusion and arises by marginal adhesion of organs that were obviously separate at the start of development (false or pseudo-sympetaly: e.g. petals of *Napoleonaea*, *Correa*, *Oxalis*, *Pittosporum*; anthers of Asteraceae; stigmatic heads in Sapindales). There are different degrees in the process of fusion of surfaces leading to almost perfectly connected structures (Endress, 2006; Sokoloff et al., 2018a). This can fool us into believing that petals are truly fused. In some cases holes appear between the congenitally fused parts and the postgenitally fused section (fenestrations; e.g. *Paederia*: Puff and Igersheim, 1991). The development of a calyptra also belongs to this category: postgenital fusion affects the upper part of the corolla, which is dropped as a whole (e.g. some *Eucalyptus*, Myrtaceae; *Vitis*, Vitaceae). Similar developmental processes are involved in fusions of the upper parts of the calyx, androecium and gynoecium. However, a calyptra can also be formed by basal congenital fusion of sepal lobes, that are torn off at the base at their weakest point (Vasconcelos et al., 2020). This process is comparable to the dehiscence of a certain kind of capsule called pyxidium.

What is described as congenital fusion is difficult to ascribe as a fusion as there is no ontogenetic evidence of a process happening (cf. Sokoloff et al., 2018a). Morphological surfaces cannot be detected at the site of congenital coherence (Endress, 2006). Sattler (1977, 1978) described congenital fusion as interprimordial growth or zonal growth. Whenever a meristematic zone situated below primordia starts to grow, a tube is formed, be it a sepal tube, a petal tube, a stamen tube or a stamen-petal tube (Figure 1.5). This interpretation avoids the difficult question of fusion but it makes it occasionally difficult to separate different organs, such as the limits between the hypanthium and inferior ovary. I consider all sorts of tubes as extensions of the hypanthium in those cases where zonal growth leads to the formation of a common ring. In the same way an inferior ovary may arise as the result of extensive hypanthial growth (Figure 1.5H, I; cf. Leins and Erbar, 2010).

Fusion should be restricted to observable developmental processes. However, the developmental process of fusion can be interpreted as an evolutionary progression, leading to a complete immersion of organs that were originally separate. This can be seen in the progressive fusion of carpels (syncarpy), starting with postgenital fusion of individual carpel primordia (e.g. most monocots: Remizowa, Sokoloff and Rudall, 2010), the initial development of free carpel lobes followed by the development of a common basal zone (e.g. Rutaceae, Apocynaceae), to the complete development of a ring without individual parts (e.g. Primulaceae, Vitaceae) (for more examples, see also Endress, 2006, Sokoloff et al., 2018a, El Ottra et al., 2022, and Ronse De Craene, 2021).

Sympetaly and Common Stamen-Petal Tubes

The development of sympetaly (or syntepaly in monocots) is related to pollination strategies restricting rewards to long-tongued pollinators, so that nectaries are hidden deep in the flower and ovaries are better protected. The development of a stamen-petal tube is the most common form of sympetaly, but has been often misinterpreted in the past. Two processes are clearly involved in the development of sympetaly: (1) growth of a common meristematic zone below the petal and stamen lobes, lifting petals and stamens on a common tube (stamen-petal tube), and (2) lateral coalescence of petal lobes, possibly linked with a postgenital fusion with stamen bases (corolla tube) (Erbar, 1991; Leins and Erbar, 2010). The two processes are independent phenomena that occur in combination, although both are often indiscriminately described as a 'petal tube', and stamens as 'epipetalous stamens'. However, the common stamen-petal tube is to be interpreted as equivalent to a hypanthium, arising as a result of growth of an intercalary meristem, and the variable combination of these two growth processes gives highly different results (see also evidence in Sattler, 1977, 1978; Erbar and Leins, 2011). The petal tube proper can only exist above the insertion of the stamens. Erbar (1991) considered the two processes as strictly independent by concentrating on the corolla tube *sensu stricto*. She made a distinction between *early sympetaly* (in cases where the corolla tube arises before or together with the petal lobes) and *late sympetaly* (in cases where the corolla lobes are initially free but become connected by meristem fusion) (see also Erbar and Leins, 2011). In my opinion a strict distinction between the two processes cannot be made. Early sympetaly tends to be correlated with the formation of an early depression in the floral bud; stamens arise on the inner slope of the depression with petals on top, and petals tend to grow rapidly. Late sympetaly is linked with the formation of a convex or flat plateau with stamens on top and petals on the outer slope; petals tend to grow more slowly (e.g. Roels and Smets, 1994, 1996; Ronse De Craene and Smets, 2000). Erbar also recognized transitional stages (e.g. Acanthaceae, Apocynaceae – Asclepiadoideae), intermediate between early and late sympetaly. There is an obvious correlation between the position of the ovary and early or late sympetaly. For example, most Gentianales have late sympetaly and superior ovaries, while Rubiaceae have early sympetaly and inferior ovaries. Secondary loss of sympetaly occurs in several ‘sympetalous’ families by the failure of a common tube to develop (see Suessenguth, 1938; Montiniaceae: Ronse De Craene, Linder and Smets, 2000; Rubiaceae: Ronse De Craene and Smets, 2000; Plantaginaceae: Hufford, 1995). In some cases the corolla tube can be partially split (e.g. Goodeniaceae, Campanulaceae – Lobelioideae, Haemodoraceae).

2

The Significance of Floral Diagrams

2.1 Definition and Significance of Floral Diagrams

A floral diagram is a schematic cross section preferably through a flower bud in which all the individual organs or elements of the flower are projected into one plane. A floral diagram is the best way to show the number and topological properties of floral organs, their differentiation and their orientation. Different parts of the flower are represented by clear symbols and the spaces between organs are an approximate reflection of the distance between the organs. Fusions between different parts are shown by connecting lines, and the orientation of the flower is shown by placing reference points in relation to the axis of the inflorescence and the subtending bracts. Floral diagrams have two major attributes: (1) the information that can be retrieved from a good floral diagram is immense and replaces extensive descriptions or even drawings, and (2) they facilitate a whole-scale comparison of floral structures across the angiosperms. An inconvenience is that it is concentrated on a limited number of characters at the expense of others that do not fit on the diagrams. Endress (2008a) discussed the benefits of accurate drawings over detailed descriptions. Indeed, a floral diagram is a clearer way to convey information as a synthesis of the details of flowers.

There are different ways to create floral diagrams. This ranges from a simplified drawing of the position of organs in the flower in relation to each other, to a very complex representation of the differentiation of floral elements that cannot always be seen without a lens. Complexity and simplicity of flowers will undoubtedly be reflected in their floral diagrams, but a large amount of additional information can be presented in a clear way on the drawing. This information can include developmental evidence (sequence of initiation), details of organ morphology (such as the insertion of anthers, major vascular bundles, obvious hairs) or significant appendages that characterize certain flowers and have much significance for comparative morphology (e.g. nectaries, spurs, coronas). It can be difficult to represent organs that are small and occur in large numbers in the flower (such as many stamens – in some flowers ranging from one hundred to one thousand), and often no attempt is made to accurately represent these numbers. Information on the development of organs can facilitate their representation on floral diagrams, as a multitude of stamens can be derived from a few primary primordia. Floral diagrams can be used to represent the typical Bauplan of the flower of a certain family, order, genus or species, and this has much significance for didactic and research purposes. Because flowers are inherently conserved in their Bauplan, diagrams can be used as a clear reflection of affinity as floral diagrams generally represent patterns, not processes (Bukhari et al., 2017). However, it is possible to combine information obtained by the observation of different stages of the floral development into diagrams as a reflection of progressive changes within the flowers.

There is no rule as to how floral diagrams should be built. All floral diagrams imply a certain interpretation, even if efforts are made to depict the flower as accurately as possible. In this book a specific code of colours and shapes is used to depict different organs although this remains arbitrary (3, Table 3.1).

2.2 Types of Floral Diagrams

Depending on what needs to be emphasized, different kinds of floral diagrams can be constructed. *Empirical floral diagrams* represent the spatial relationships of organs in flowers without any interpretation of intervening changes or evolution. The representations can be simple, depicting the organs and their topological relationship in the flower, or more complex, showing other less obvious structures that a botanist wants to emphasize (e.g. presence of appendages on stamens or petals, nectaries, the orientation of organs). For a floral diagram to be useful special attention needs to be paid to the orientation of the flower relative to the inflorescence. Orientation of flowers in relation to subtending bracts is often overlooked.

Developmental floral diagrams are mainly based on information obtained through microscopic studies on the development of flowers. Examples of developmental diagrams are given in Leins and Erbar (2010). They can be combined with empirical floral diagrams. Developmental floral diagrams can represent specific stages of the development of a flower that need to be emphasized and often necessitate more than one diagram in some cases. The sequence of initiation of whorls or organs can be shown by numbers and can be more easily combined with an empirical diagram. Mature flowers with many stamens often look deceptively similar and it is difficult or sometimes impossible to know the basic androecium configuration. Developmental diagrams can depict the group-wise arrangement and sequence of initiation of stamens, something that cannot be seen in adult flowers. Figure 2.1 illustrates how floral diagrams can reproduce patterns of development in the flower of *Capparis cynophallophora* (Capparaceae). Figure 2.1A shows an early stage in the flower development with the numbered sequence of development of the stamens. This can be reproduced in an equivalent floral diagram (Figure 2.1B), or additional details of the flower can be added, including the limits of original stamen fascicles (Figure 2.1C). *Tropaeolum* is another example which shows a change in flower orientation during development of the flower (Ronse De Craene and Smets, 2001b; p. 219). A comparable change in orientation is caused by resupination of flowers (e.g. Orchidaceae, Figure 6.14; Leguminosae: Figure 9.55; Balsaminaceae, Figure 11.5). By a torsion of the pedicel, flowers can turn around for about 180 degrees or less, as shown by drawing a curved arrow on the floral diagram. Floral diagrams are useful in describing differences in patterns of development of the perianth and androecium. In this book details of the sequence of development of stamens are shown for polyandrous taxa. Phyllotaxis can be shown either by a sequence of numbers or by the aestivation of sepals and petals. Several examples illustrate this in this book (Figures 6.4, 7.2B, 7.7, 8.2, 9.56B, 9.63, 10.3, 10.18, 11.14).

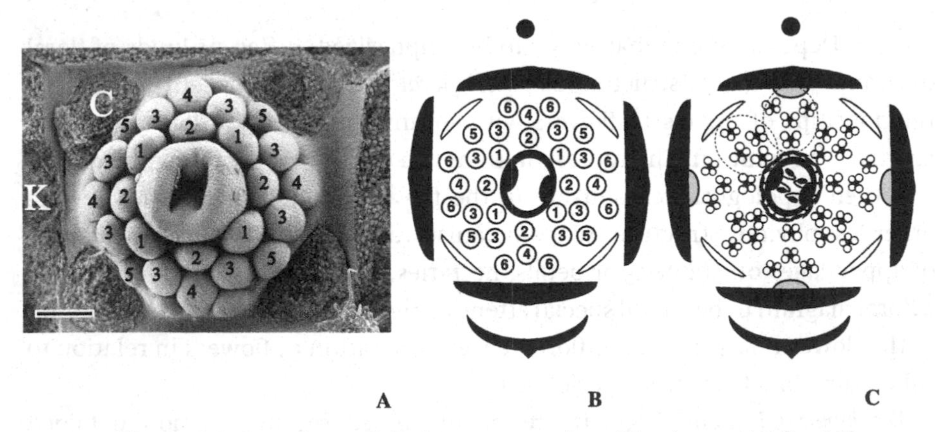

Figure 2.1 *Capparis cynophallophora*: presentation of developmental floral diagram. A. Early stage of development of the flower; sepals (K) and petals (C) removed. Numbers indicate sequence of stamen initiation. Bar = 100 microns. Photographer: L. Ronse De Craene; B. Corresponding floral diagram; C. Corresponding floral diagrams with more details including peltate hairs covering the ovary; broken lines represent initial stamen fascicles

Theoretical (or hypothetical) floral diagrams usually contain information that is not visible externally, certainly not in mature flowers. These include organs that have been reduced or lost but for which there is evidence that they were present in an ancestral flower. Clear examples where this can be used are Poaceae or Orchidaceae (Figures 6.14, 6.23), depicting missing perianth parts or stamens with a star. In some cases hypothetical and developmental diagrams can be combined. An example of this is *Melianthus* (Melianthaceae). Mature flowers have four petals only, but during development a fifth petal is initiated but growth aborts at mid-development and no trace is visible at maturity (Ronse De Craene et al., 2001). The sister genus *Bersama* has five petals, indicating that five petals were present in ancestral flowers of *Melianthus*. Evidence of the fifth petal can be shown by an asterisk on the floral diagram (Figure 9.11B). Even without developmental evidence floral diagrams can be used to present hypotheses in a clear way.

2.3 Floral Diagrams and Floral Formulae

Another widely used method to succinctly describe flowers is the use of floral formulae (Table 2.1). This provides information on the kind and number of organs, type of symmetry, presence of fusions and level of insertion of the ovary. However, it lacks detail of position and orientation of organs. The way to use floral formulae differs between the European and American traditions, but they tend to convey the same information. More extensive detail can be given for fluctuations within a whorl and recent attempts in improving the use of floral formulae are ongoing (e.g. Prenner and Klitgaard, 2008; Prenner, Bateman and Rudall, 2010; Ronse De Craene et al., 2014) and will be used in a similar way in this book in addition to floral diagrams. A floral formula can be given for a whole family or for a given species. In the first case numbers shown will give an approximate figure of variation, which is unhelpful in cases of high variation. For each individual species drawn, a floral formula is given following details presented in Table 2.1. If the family is variable in merism and number of organs, a general formula is also given, only referring to the number of floral organs.

2.4 Problems of Three-Dimensional Complexity

Floral diagrams are in essence two-dimensional representations of flowers. The structural information provided limits the description of differences that occur from the base to the top of the flower. In bilabiate flowers the lower part tends to be regular, while the upper part including the stamen-petal tube is highly monosymmetric. Inferior or half-inferior ovaries are taxonomically important and can be problematic to represent on a floral diagram.

Table 2.1 *Floral formulae: symbols used in this book*

Symmetry	
∗	polysymmetric (actinomorphic)
↓↑←→↖↗↙↘	monosymmetric (zygomorphic; orientation of arrow corresponds to orientation of the flower)
↺	spiral
⟷	disymmetric
↯	asymmetric
Floral organs	
P	perigon (number of tepals; no differentiation between calyx and corolla)
K	calyx (number of sepals)
C	corolla (number of petals)
A	androecium (number of stamens)
G	gynoecium (number of carpels): superior ($\underline{G}$); half-inferior (-G-); inferior (Ğ)
A°	staminode (sterile stamen)
G°	pistillode (sterile carpel)
(...)	fusion within a whorl
[...]	fusion involving two different whorls
+	more than one whorl can be distinguished
:	there is a clear morphological difference within a whorl
∞	number or organs is numerous or indefinite
–	refers to a variable number within a whorl

Examples

↓K(5) [C(3:2)A2:2°:1°] $\underline{G}$(2) *Streptocarpus rexii* (Gesneriaceae)

∗ K4 C4 A4+4 $\underline{G}$(4) *Ruta graveolens* (Rutaceae)

↯ K3 [C3 A1°+ $1^{½:½°}$:2°] Ğ(3) *Canna edulis* (Cannaceae)

A hypanthium, lifting stamens, petals and sepals, is found in several major groups and presents the same problems as inferior ovaries (Figure 1.5).

However, there are ways to solve the problem of three-dimensional complexity, either by superposing two diagrams from different levels of the flower (as a sequence of sections: Figure 2.2A), or by the addition of simple symbols. If upper and lower parts of the flower are markedly different (e.g. *Lonicera*, *Cuphea*), this can be shown by connecting two diagrams at different levels, similar to transverse sections in the flower (Figure 2.2B). An arrow can also show the orientation of symmetry in the upper part of the flower, especially when monosymmetry is weak and caused by the orientation of a single organ (e.g. the style in *Exacum*, Gentianaceae). Presence of a hypanthium can be shown by

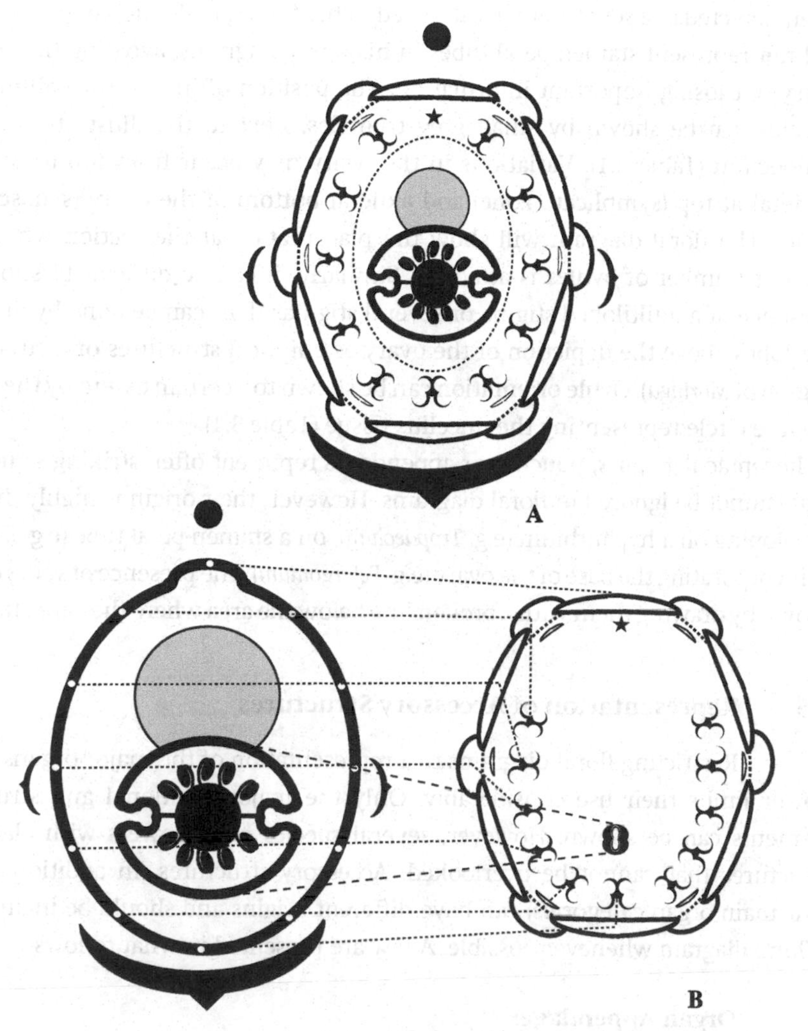

Figure 2.2 Two ways of accommodating three dimensions in a bi-dimensional floral diagram for *Cuphea micropetala* (Lythraceae): A. Compression of flower in one plane with the presentation of the basal ovary at the same level as the stamens inserted at the top of the hypanthium (broken circles); B. Presentations of two diagrams at different levels, the figure on the left representing the level of the ovary and nectary; the figure on the right shows the attachment of the stamens (broken line) and their position in the flower. White dots represent the vasculature of perianth and stamens

depicting two broken lines to delimit the area covered by the hypanthium. The same could be applied for petal-stamen tubes, which are connected by common growth, although this is not practical because one often cannot differentiate between different sections of the tube. In this book stamen-petal tubes have been depicted as a set of stamens attached to the fused petals. Eichler (1875, 1878) did not represent stamen-petal tubes in his floral diagrams, avoiding the controversy but losing important information. The position of inferior or half-inferior ovaries can be shown by small grey triangles, next to the illustration of the gynoecium (Table 3.1). Variations in the ovary may occur from top to bottom: parietal at top (symplicate zone) and axile at bottom of the ovary (synascidiate zone). The floral diagram will show the placentation at the section where the highest number of ovules is attached. Similarly, it can be difficult to show the presence of a multilobed stigma or several stigmas. This can be done by drawing the lobes above the depiction of the ovary as elliptical structures or a circle (e.g. stigma of *Moringa*). Ovule orientation can be shown to a certain extent by the use of a white circle representing the nucellus tissue (Table 3.1).

Receptacular spurs, pouches or appendages represent often striking structures that cannot be ignored in floral diagrams. However, their origin is highly diverse, developing on a hypanthium (e.g. *Tropaeolum*), on a stamen-petal tube (e.g. *Diascia*), or incorporating the base of the ovary (e.g. *Pelargonium*). The presence of spurs can be shown by drawing them with a broken line below the area where they are attached.

2.5 Representation of Accessory Structures

Restricting floral diagrams to a representation of the major organs of the flower limits their use considerably. Only the major positional and structural elements can be shown. However, several species have flowers with elaborate structures that cannot be overlooked. Accessory structures (in addition to the four main organ categories) can have different origins and should be included in a floral diagram whenever possible. A few are presented in what follows.

Organ Appendages

Petals and stamens, or more rarely sepals and carpels, can have obvious appendages in the form of awns, tufts of hairs, flaps, wings etc. These appendages are usually secondary elaborations on the floral organs, arising at a late stage of development, although they may occasionally represent whole organs such as staminodes. Representing these obvious features in floral diagrams helps to convey potentially significant information.

Petal appendages are tissue proliferations of the ventral surface of the petal or are the result of lateral divisions of petals. They have been variously interpreted as ligules, stipules, a corona or scales, but their interpretation often remains unclear or contradictory in the absence of developmental evidence. Appendages on petals have a function in pollination (e.g. nectar production, storage and cover, attraction, landing platforms) and are correlated with the architecture of flowers (Endress and Matthews, 2006b). If the appendages are clearly linked to the ventral surface of the petals, they are usually referred to as part of the petal. However, interpretations can be variable if the appendage is only loosely linked to the petal or not at all (e.g. the corona of *Narcissus*, Amaryllidaceae or *Passiflora*, Passifloraceae). A hollow space may be present between petal blade and appendage that can act as a nectary (e.g. Ranunculaceae). Leins and Erbar (2010) considered ventral appendages as evidence of sterile ventral pollen sacs and as an indication of a staminodial origin, at least in Caryophyllaceae and Ranunculaceae.

A true spur arises as a ventral invagination leaving an appendage to the dorsal side of a transformed petal (e.g. *Aquilegia*) or hypanthium (e.g. *Tropaeolum*). Spurs arise late in the development of flowers. They either combine storage and production of nectar (e.g. *Tropaeolum*) or only storage (*Corydalis, Delphinium, Aconitum, Viola*). Some taxa have spurs that are not visible externally (internal spurs: e.g. *Pelargonium*, Chrysobalanaceae). Bernardello (2007) gave an overview of nectar spurs.

By invagination of the petal blade an abaxial cavity can form on the back of the petal, similar to the development of an inverted spur (e.g. Jeiter, Langecker and Weigend, 2020). The invagination restricts entry to the flowers and is common in some lamiids (e.g. Boraginaceae, Plantaginaceae, Orobanchaceae, Gentianaceae).

Corona

A corona is a highly obvious outgrowth, usually of receptacular origin, but also arising as part of the perianth or the stamens. The corona can be in one or several series, is often showy and resembles the petals (e.g. paracorolla of Amaryllidaceae: *Narcissus*), or can be highly elaborate (e.g. Passifloraceae, Apocynaceae). In some exceptional cases the corona is staminodial in nature, such as in *Napoleonaea* (Lecythidaceae: Ronse De Craene, 2011) and *Pachynema* (Dilleniaceae: Endress and Matthews, 2006b). In *Narcissus*, the corona is tubular and is connected to the perianth tube; in other Amaryllidaceae, such as *Pancratium* or *Hymenocallis*, the corona is an outgrowth of the filaments (Guédès, 1979).

Pseudostaminodes

In some cases it is difficult to separate true staminodes (as derived from stamens) from receptacular or petal appendages that can have a similar morphology and function (pseudostaminodes: Ronse De Craene and Smets, 2001a). Examples are the stalked glands at the base of the stamens in several Laurales, glandular emergences between the stamens in *Francoa* or *Greyia* (Francoaceae: Ronse De Craene and Smets, 1999b), antepetalous appendages in *Brexia*

(Celastraceae: Edgell, 2004), the interstaminal appendages between the fused stamen bases of Amaranthaceae (Eliasson, 1988) or scales on the petals of several asterids (Ronse De Craene and Smets, 2001a). These appendages may be shown on floral diagrams, but they need to be distinguished from staminodes.

Nectaries

The presence of nectary tissue is often ignored in floral diagrams. However, it represents an important element of the flower that can occupy a considerable space (e.g. Rhamnaceae, Celastraceae) and cannot easily be ignored. The nectaries as well as any area of glandular tissue have been presented on the floral diagrams in all cases where they are discernable (grey colour). Eichler (1875, 1878) did not include nectaries in his floral diagrams, except in cases where they are very prominent.

2.6 Floral Heteromorphism and Unisexual Flowers

Flowers of a given species tend to be homogenous in most cases. However, there are instances for flowers of the same species to develop in different shapes and forms. Heteromorphic species usually have two morphologically distinct kinds of flowers (dimorphism), which is different from intrinsic variation found within a species. Heteromorphism can be occasional, depending on external factors such as the availability of pollinators, nutrients or the time of the season, or it can be inherent to a species. An occasional heteromorphism is dependent on nutrients, which affect the number of parts in certain flowers. This phenomenon is frequent in annual species with variable organ numbers (e.g. *Nigella* in Ranunculaceae; *Papaver* in Papaveraceae), and was experimentally studied in the beginning of the twentieth century (e.g. Murbeck, 1912).

Within an inflorescence, merism of flowers may fluctuate between a terminal tetramerous flower and lateral pentamerous flowers (e.g. *Ruta*, *Adoxa*), or lateral trimerous flowers and a terminal pentamerous flower (e.g. *Berberis*) (Eichler, 1875; Rudall and Bateman, 2003). In families with umbel-shaped or capitate inflorescences such as Apiaceae, Hydrangeaceae and Asteraceae, marginal flowers may differ from central flowers, as they gain bigger petals at the expense of fertility. The occurrence of cleistogamous flowers (flowers that do not open at anthesis, against chasmogamous flowers that do open) is a temporary phenomenon that may also be cause for an important heteromorphism, especially in cases where pollinators affect the structure of the flower. Cleistogamous flowers often have a different number of floral parts, with reduced petals, nectaries, stamens, pollen sac numbers per anther, carpels and ovules (Endress, 1994). This heteromorphism must be taken into account when producing floral diagrams.

The most important cause for dimorphism is unisexuality, which can be expressed to different degrees (Mitchell and Diggle, 2005). Heterostyly is a first step in the transition to unisexual flowers. Heterostyly can be complex and include two or more morphs, but flowers remain similar except for the length of stamens and styles. Mitchell and Diggle (2005) distinguished between two types of unisexual flowers. Type I represents unisexual flowers with rudiments of the other sex. In this case, unisexuality is a late-developmental phenomenon in cases where anthers or ovules fail to develop or reach maturity (e.g. Restionaceae, Sapindaceae). Very often, staminodes or carpellodes remain present in the mature flower. In other cases, sterility occurs much earlier in the floral development. Type II represents flowers that are unisexual from inception. The earlier abortion occurs in the development through a process called heterochrony (see p. 56); the earlier it occurs the greater the difference between staminate and pistillate morphs (e.g. Euphorbiaceae, Caricaceae, Begoniaceae). Unisexual flowers can be derived in different ways from bisexual flowers, by organ abortion, loss of organs or homeosis. The degree of reduction of one or the other gender in flowers affects the outcome of floral diagrams to a great extent. In some cases the reduced organ takes over different functions, such as the pistillodial nectary of *Carica* (Ronse De Craene and Smets, 1999a) or *Buxus* (Von Balthazar and Endress, 2002b), or carpels are replaced by organs of the other sex through homeosis (such as the three inner stamens in *Jatropha* of Euphorbiaceae).

In wind-pollinated flowers there is a whole range of adaptations including presence of unisexuality (see p. 72). The loss of the other gender is accompanied by an absence or reduction of the perianth, and various associations with bracts. This leads to highly simplified (reduced) flowers, often limited to the androecium and carpels, with an important dimorphism between the genders (e.g. Cercidiphyllaceae, Betulaceae, Casuarinaceae).

In this book two floral diagrams are presented where differences between genders are more pronounced (type II flowers). If unisexual flowers are only slightly different, this is mentioned in the text.

2.7 Floral Development and Floral Diagrams

2.7.1 *Meristems and Developmental Constraints*

It is clear that the floral diagram based on a study of the mature flower is the result of the developmental process and any fluctuations and shifts during development will affect the outcome in the diagram (see Ronse De Craene, 2016, 2018; Bakhari et al., 2017). Floral development (as well as vegetative development) is determined by a growing apical meristem. The molecular basis for floral development is being increasingly studied and different models have been proposed (see p. 58). It is mainly a combination of genetic control of the floral apex and mechanical forces that are responsible for the nature of organs as well as their position in the flower (Ronse De Craene, 2018). Phyllotaxis of flowers is influenced by the size of organ primordia and an apical inhibition zone, which determines the position of subsequent organs. The result is a clear pattern of organ distribution.

An important concept in floral morphology is Hofmeister's rule. The rule formulated by Wilhelm Hofmeister in 1868 insists that a subsequent primordium arises as far as possible from a primordium already formed on the apex. In flowers a whorl should always alternate with the previous one. While this rule holds for the majority of flowering plants, there are several exceptions. The main reason is that organs invariably arise where there is sufficient space for their initiation, and this can be variable during the initiation of the flower because of diverse influences such as shape of the apex, the position of previously formed organs and pressure exercised by them (Kirchoff, 2000; Wei and Ronse De Craene, 2020). In regular core eudicot flowers, whorls are usually alternate, and the antepetalous whorl is most prone to be lost. However, obhaplostemonous flowers are not that rare. They are usually linked with retarded petals, often arising from common stamen-petal primordia. Stamen and petal function as a single unit in that case, adapting to Hofmeister's rule. Interestingly, carpel number and position also have an influence on the number and position of stamens. Trimerous carpels may affect the number of stamens by reducing space for their initiation (e.g. *Hibbertia*, *Hypericum*, *Mollugo*, *Macarthuria*). Alternatively, carpel proliferations (e.g. *Tupidanthus*, Araliaceae) are linked to an increase of stamen numbers, even as carpels are initiated after the stamens. This means that spatial pattern formation of carpels is initiated before the onset of stamen primordia (see Ronse De Craene, 2018).

The developmental process will strongly determine the outcome of flower structure and arrangement of organs in mature flowers. Taxonomic groups that are largely monosymmetric will have a much earlier onset of monosymmetry, already starting during the development of the perianth. In groups where monosymmetry is a less current phenomenon, organs arise as in polysymmetric flowers and the onset of zygomorphy happens shortly before maturity (see Tucker, 1997). However, there are several exceptions to this generalized process (see Naghiloo, 2020). As discussed elsewhere, different factors are responsible for the high diversity of mature forms; these can be summarized as shifts in size, pressure and time operating during the development of flowers (Ronse De Craene, 2018).

Position of Organs

Floral organs develop in a specific sequence and position characteristic for a species, genus or family and this remains highly conserved throughout the phylogeny. An important aspect of floral diagrams is the orientation of the flower relative to the floral axis. There is an interesting analogy between Pentapetalae and monocots in the frequency with which an abnormal inversion of orientation occurs (see also Bukhari et al., 2017).

In most pentamerous Pentapetalae, sepals one and three are in an abaxial position and sepal two is in adaxial position; it is the reverse for the orientation of petals. This position is mainly regulated by transversally placed bracteoles, forcing the first floral organs in a median position. Loss of bracteoles may lead to displacement of the first formed sepals in a lateral position (see p. 30). In tetramerous flowers, sepals are generally positioned in median and transversal position, and petals in a diagonal position. There are few exceptions, which are not the result of a resupination of the flower enumerated by Eichler (1878), of which the Leguminosae is the most important representative. This abaxial-adaxial duality is very stable and is usually maintained whenever flowers have divergent petal sizes in monosymmetric or bilabiate flowers. The two adaxial petals may converge and fuse into one unit (this can be the origin of tetramerous flowers in some cases), while the abaxial petal can become much larger or become associated with the laterals in a large unit (lower lip). In Papilionoideae of Leguminosae, the two abaxial petals become associated as the keel.

Most monocots have their median outer tepal in abaxial position and the inner median in adaxial position. This position is reversed in a limited number of cases as a result of an addition of external bracts (e.g. *Tofieldia*), shifts of the bracteole to a median position (Iridaceae) or its loss (e.g. *Dioscorea*, *Allium*, *Agave*, *Lilium*), or an uncertain amalgamation of bract and abaxial tepal (e.g. *Acorus*, Juncaginaceae, Potamegetonaceae) (Buzgo, 2001; Remizowa and Sokoloff, 2003; Remizowa et al., 2012). Loss of the flower subtending bract (as in *Acorus*, *Triglochin* or *Potamogeton*) does not influence the orientation of the flower, implying a possible overlap in the development of bract and outer adaxial tepal organ resulting in a hybrid organ (Remizowa et al., 2012; see also under Acoraceae).

Stamens and carpels usually have a stable position. This has been discussed extensively for the androecium where loss of whorls can result in highly different configurations (e.g. Ronse De Craene and Smets, 1987, 1993, 1995b, 1998a). Position of carpels in eudicots is regulated by available space and number of stamen whorls. In cases where carpels are isomerous with the other whorls, they can either be in antepetalous or antesepalous position, and this may occasionally fluctuate in a family (e.g. Malvaceae, Rosaceae, Rutaceae, Caryophyllaceae). When lower than five, the carpels can have specific orientations. With three carpels, two are often abaxial and the odd adaxial in Pentapetalae, rarely the opposite (Ronse De Craene and Smets, 1998b), while it is the reverse in most monocots. With two carpels, position is mostly median, more rarely transversal. An oblique position is rare and is found in Solanaceae (Eichler, 1875), and occasionally in Capparaceae (Ronse De Craene and Smets, 1997a).

The Principle of Variable Proportions and Heterochrony

The principle of variable proportions was formulated by Troll (1956) and implies that the shape of structures can be modulated along a gradation; an organ can progressively be transformed into a different structure by a change of size and differential growth processes (cf. Ronse De Craene and Smets, 1991e; Ronse De Craene, De Laet and Smets, 1996). More recently the principle has become expressed in the concept of morphospace, investigating theoretical changes of space and applying these in an evolutionary and ecological context (Chartier et al., 2014; Ronse De Craene, 2018). In fact, the principle of variable proportions is linked to heterochrony, which describes the process by which the onset of initiation or differentiation of organs is shifted in time (see also Li and Johnston, 2000; Naghiloo and Claßen-Bockhoff, 2017; Buendía-Monreal and Gillmor, 2018; Ronse De Craene, 2018; Chinga, Pérez and Claßen-Bockhoff, 2021). The principle is mainly applied to the outcome of development, leading to juvenile shapes by a faster process of development (paedomorphosis) or more elaborate shapes by a longer development (peramorphosis) (Buendía-Monreal and Gillmor, 2018). However, I prefer to limit heterochrony to the process of development affecting the growth rate of development of different organs. For example, a delayed onset of petal development means that petal primordia are spatially taken over by the stamen primordia as to become absorbed in the stamen primordium as a common stamen-petal primordium and as a result undergo the genetic influence of the stamens. Flower development is largely guided by spatio-physical factors, such as an inverse relationship between the size of organs and the apical meristem. The larger the apical meristem the more organs can be produced, if they are smaller in size. The same applies for secondary primordia on primary primordia (see p. 13). Complex androecia can differentiate into a relatively large numbers of stamens (Figure 1.1), or much higher numbers if the size of individual stamens is decreased. In Annonaceae the number of stamens is indirectly related to the size of stamen primordia and the extent of development of the floral apex (Leins and Erbar, 1996; Xu and Ronse De Craene, 2010).

Heterochrony represents a fundamental aspect causing floral diversification. Although floral diagrams represent specific structures in time, they give an indication of the processes leading to a flexibility in shapes and novel structures.

2.7.2 *Reduction and Loss of Organs*

Flowers are the result of a long evolutionary process in which changes have progressively taken place. In many cases with reduced numbers of organs in a whorl, it is assumed that some organs were lost during evolution. Guédès (1979) refers to this condition as 'aborted' versus 'unborn', while Tucker (1988b, 1997) refers to 'suppressed' versus 'lost' organs. Sometimes this assumption can be verified by a study of floral development showing that a primordium is initiated, but is subsequently arrested in growth and has completely vanished at maturity (deletion or suppression). Such an organ can be described as ephemeral or as a 'phantom organ' that appears during a short period of development. Through the process of heterochrony, an organ can become progressively aborted at an earlier stage until it fails to initiate (e.g. Li and Johnston, 2000; Mitchell and Diggle, 2005; Ronse De Craene, 2018). Various examples are found in angiosperms, such as petals in Melianthaceae (*Melianthus*: Ronse De Craene et al., 2001), Leguminosae (*Amorpha*: McMahon and Hufford, 2005) and Polygalaceae (*Polygala*: Prenner, 2004b), the development of an outer staminode in the labellum of Zingiberaceae (*Gagnepainia*: Iwamoto et al., 2020), stamens in Leguminosae (*Bauhinia*, *Saraca*: Tucker, 1988a, 1997, 2000c), staminodes in pistillate Restionaceae (Ronse De Craene, Linder and Smets, 2002) and stamens and petals in Brassicaceae (*Lepidium*: Bowmann and Smyth, 1998). Organ abortion can occur at different stages of development. Loss of organs implies that certain organs were present in ancestral flowers but that they were lost in the descendant lineages. A clear example is the orchids with one to three functional stamens and one or two staminodes, derived from six stamens in two whorls (Rudall and Bateman, 2004; Stützel, 2006).

2.8 Evolutionary Developmental Genetics and Floral Diagrams

The approach to morphology has dramatically changed by increased emphasis on the genetic basis of organ and flower development. The discovery that several genes are responsible for the flower expression and the development of floral organs opened new areas in research that have strongly interacted with a classical approach to morphology. Flower development is the result of a cascade of expressions of different genes, transforming the inflorescence meristem into flowers and consequently leading to expression of different floral organs (e.g. Glover, 2007, Glover et al., 2015; Smyth, 2018). Different genes affect different factors responsible for a specific floral Bauplan, such as phyllotaxis, merism, organ boundaries and identity, symmetry and fusion (Smyth, 2018), although the set-up of a specific flower structure includes other mostly physico-mechanical factors (Claßen-Bockhoff, 2016; Ronse De Craene, 2016, 2018). Floral diagrams are ultimately an illustration of the morphological expression of genes responsible for flower induction and differentiation. At every level of flower differentiation, different genes are active in controlling developmental events, although they are not necessarily the cause, rather the consequence of self-organizing processes (see Claßen-Bockhoff, 2016; Ronse De Craene, 2018).

The ABC model of flower development was elaborated in the early 1990s (Coen and Meyerowitz, 1991). The identity of organs in each whorl is determined by a combination of three classes of identity genes: A genes are responsible for sepal expression, A+B function determines petal identity, B+C function specifies stamens, and C function determines carpel development. The model has been expanded with D and E genes with specific roles in the development of flowers; D genes are responsible for ovule and placenta development, while E genes tend to control the development of all organs except the sepals (e.g. Theißen et al., 2002; Glover, 2007). A and C genes are mutually antagonistic in their expression. The genetic model is based on the study of several mutants of model genera, such as *Arabidopsis*, *Antirrhinum* and *Oryza*, highly derived organisms. However, genetic evidence appears to be much more fluid in other groups of plants and cannot be applied to monocots or basal angiosperms in the same way as for Pentapetalae, with an easy transition of gene expressions across the organ boundaries and the evolution of numerous paralogous substitutions in the latter groups (e.g. Kim et al., 2005; Ronse De Craene, 2007; Litt and Kramer, 2010).

The definition of whorls is different from morphology, as describing a domain in which a single type of organ is produced in one or more concentric circles (Leyser and Day, 2003; Glover, 2007). Leins and Erbar (2010) rightly criticized the unfortunate use of 'organ whorls' in evo-devo where it is better to refer to 'organ categories'. The androecium or perianth can consist of more than one whorl and this is not accounted for. 'Asymmetry' has also been used in the sense of 'monosymmetry', clashing with the morphological definition of asymmetry (Endress, 1999).

The evolution of gene expression in the angiosperms is highly complex and can be applied to understand major evolutionary processes in flowers. Genes have undergone several duplications in their history, as is clear for the B-function genes *APETALA3* (*AP3*) and *PISTILATA* (*PI*) which are responsible for the expression of petal characters (Kramer, Di Stilio and Schlüter, 2003; Kramer et al., 2006, reviewed in Ronse De Craene, 2007). The origin of flowers has also been tested through an understanding of gene expressions (Theißen et al., 2002).

There are interesting parallelisms in the evolution of floral structures in the angiosperms and their gene counterparts. Basal angiosperms with spiral flowers and an undifferentiated perianth have a more diffuse expression pattern of A and B genes than the Pentapetalae, leading to the sliding boundaries (Kramer, Di Stilio and Schlüter, 2003; Kanno et al., 2007) or fading borders hypotheses (e.g. Buzgo, Soltis and Soltis, 2004; Kim et al., 2005). In the Pentapetalae there is a stricter compartmentalization of gene expression due to greater synorganization. Petal expression is strongly guided by *AP3* and *PI* genes, which have become confined within a specific region of the flower (Monniaux and Vandenbussche, 2018). The organ boundaries can occasionally break down by a reduction of the perianth, or a secondary stamen and carpel increase, and affect the development of the whole flower. Gene expression studies have also been applied to monocots to understand the homology of the perianth, especially Poaceae to understand the nature of lemma, palea and lodicules or the nature of the pappus in Asteraceae (Glover et al., 2015), but in most cases the answers are not straightforward or more complex than initially expected (Ronse De Craene, 2007).

Changes in the floral Bauplan are also regulated by genes. It is generally recognized that genes such as *CYCLOIDEA* (*CYC*) and *DIVARICATA* (*DIV*) are responsible for the change in symmetry from regular to bilateral flowers in *Antirrhinum* (e.g. Luo et al., 1996; Leyser and Day, 2003; Citerne, Pennington and Cronk, 2006; Busch and Zachgo, 2007; Citerne et al., 2010). The expression of *CYC* leads to the abortion of the adaxial stamen and the restriction of *DIV* to the abaxial side of the flower. Additional genes such as *DICHOTOMA* (*DICH*) and *RADIALIS* (*RAD*) affect the shape of abaxial petals or act as a repressor of *DIV* respectively (Reardon et al., 2014). Changes in expression patterns can lead to dramatic shifts in flower morphology, such as secondary polysymmetry in *Cadia purpurea* of

Leguminosae (Citerne, Pennington and Cronk, 2006) or abortion of lateral stamens in *Mohavea* of Plantaginaceae (Hileman, Kramer and Baum, 2003) by loss of expression of *CYC*. The pattern of genetic mutation is thought to cause a reversal to polysymmetric flowers in some asterids. Donoghue, Ree and Baum (1998) suggested that a return to polysymmetry could be triggered either by loss-of-function *CYC* mutants or by structural fusion of the posterior petals linked with a loss of stamens. The former possibility, based on cases of *Antirrhinum*, implies that the adaxial side of the flower develops like the abaxial side and leads to mutants with six petals (Luo et al., 1996). This evolution may be responsible for regular four- to six-merous flowers of *Sibthorpia* (Scrophulariaceae) or *Ramonda* (Gesneriaceae). Other cases, such as the tetramerous flowers of *Plantago*, are characterized by a complete loss of *CYC*, leading to a ventralization of the flower (Reardon et al., 2014).

Although evo-devo can give clearer indications of the nature of floral organs, an understanding of structural morphology remains essential. Pollinator co-evolution is probably the principal trigger to changes in flower morphology and this is not clarified by evo-devo studies. The fact that B genes are sometimes expressed in sepals is not an indication of a petal nature, but of petaloidy (Ronse De Craene, 2007). Gene expression can be extremely complex and future genetic studies will need to be closely linked with any morphological study. It is not always clear whether an altered gene function is the cause or a consequence of the morphological change, as other developmental factors can be instrumental for the occurring changes. Ronse De Craene (2018) suggested that the evolution of the flower is first regulated by spatial constraints during development that became genetically fixed. Subtle changes in development (in space and time) cause a diversification of flowers (see also Naghiloo and Claßen-Bockhoff, 2017; Chinga, Pérez and Claßen-Bockhoff, 2021). Floral diagrams are a way to convey morphological information that can be inspirational for further evo-devo research.

3

Floral Diagrams Used in This Book

Table 3.1 summarizes the symbols for the floral diagrams used in this book. When constructing floral diagrams it is essential to correctly orientate the flowers (Figure 3.1). Arrangement of individual organs and their orientation can only be accurately compared if a common reference point is being used. The main axis relative to the flower is shown by a black dot. The growing point or main axis of the inflorescence is shown by a crossed circle. Flowers are conventionally depicted with the axis on top and the main subtending bract at the bottom along the median line. Bracteoles are placed more or less in transversal position depending on their orientation relative to the main axis. To differentiate bracts and bracteoles from the perianth, a small triangle is placed on the abaxial (dorsal) side of the bract(eole). A bract or bracteole that is lost or early caducous is shown in white with a broken outline. A large straight arrow represents the main direction of monosymmetry, when present. If flowers are resupinate, this is shown by a curved arrow.

The distinction between sepals, petals and tepals is based on the presence versus absence of petaloidy, as the differentiation between sepals and petals is sometimes unclear (see p. 7). White curves represent pigmented (petaloid) organs, without distinction between sepals, petals or tepals as these represent homologous organs in most cases (cf. Ronse De Craene, 2007, 2008). Absence of petaloidy is shown by black arcs, which is mostly present in the green sepals but occasionally in the petals (e.g. *Callistemon citrinus*: Figure 9.15A). Stamens are represented as a cross section through the anther, with occasional variations reflecting differences of size and shape and dehiscence. In case many stamens are formed in a flower, stamens are represented by open circles, occasionally with the sequence of initiation shown by numbers. Staminodes are represented by a circle with a black dot, or occasionally by a blackened anther (representing a sterile anther). Staminodial petals are shown by an arc with a black central dot.

Table 3.1 *Abbreviations and symbols used in floral diagrams.*

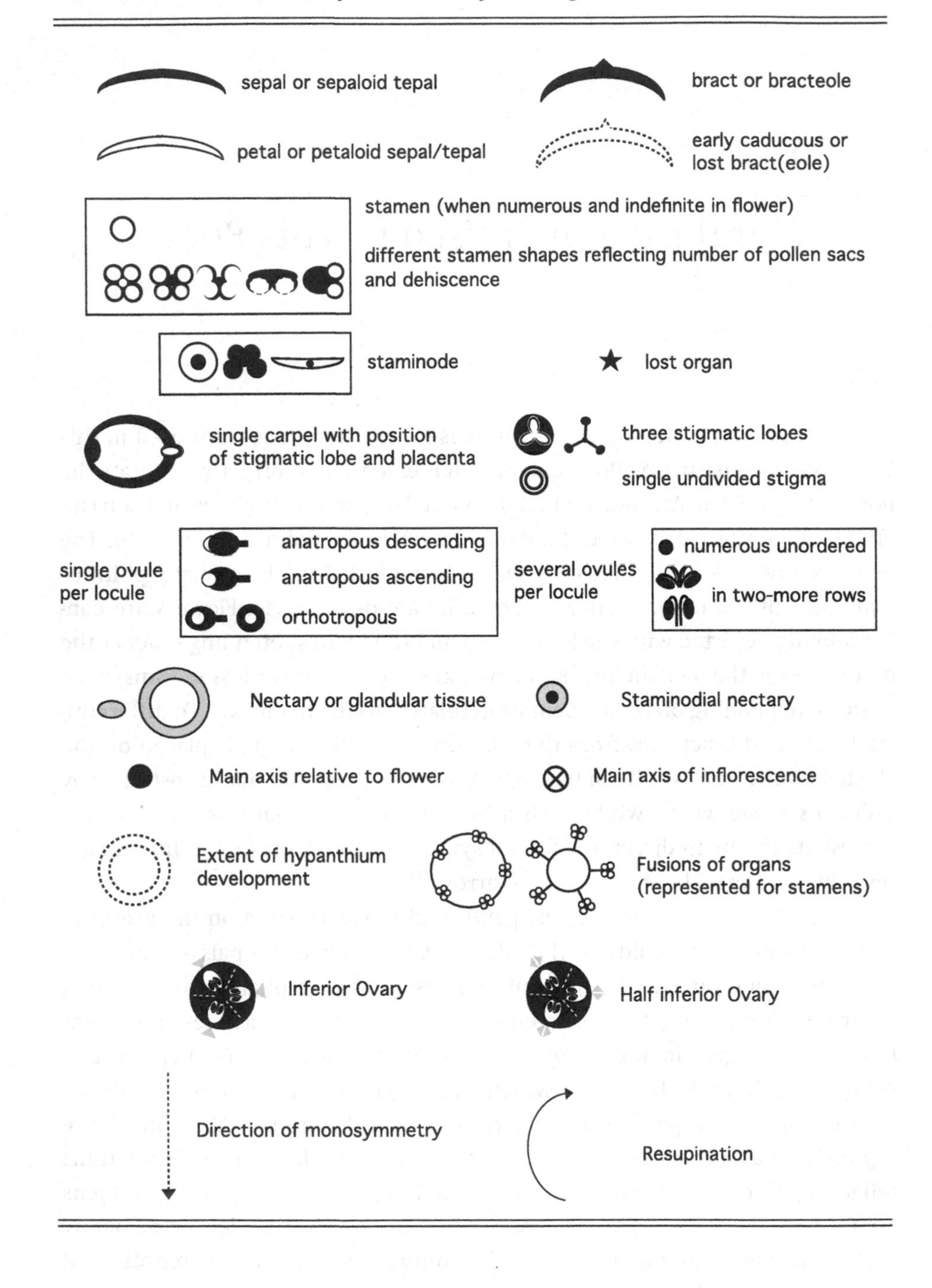

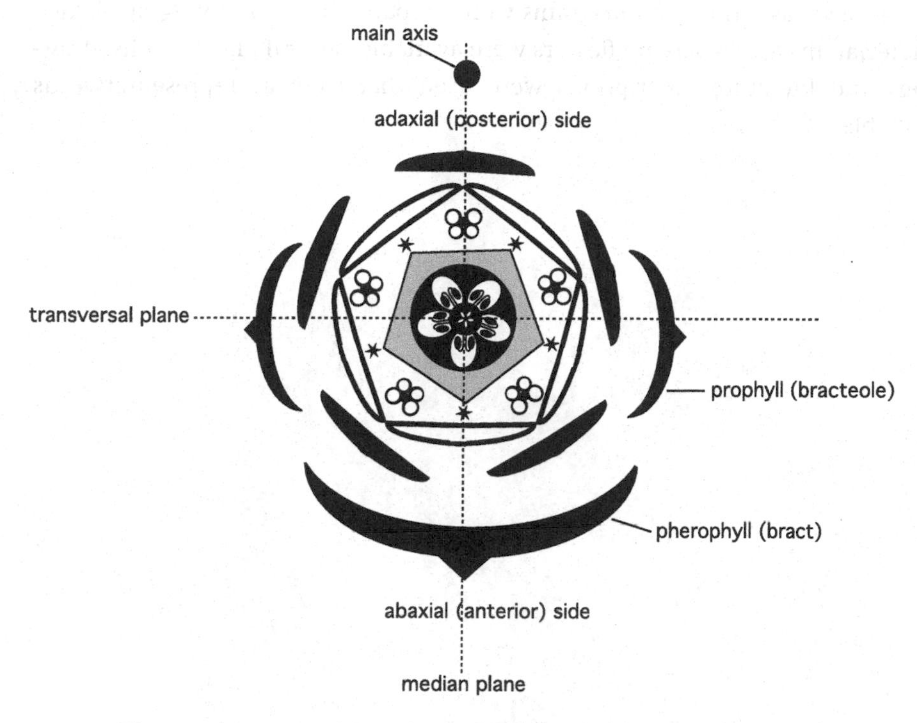

Figure 3.1 Representation of an idealized floral diagram

The ovary is represented as a cross section at the level of attachment of the ovules. When possible, the position of the styles or stigma lobes is shown, as these may be opposite the carpels (carinate) or alternate (commissural). Where the ovary is inferior or half-inferior this is shown by grey triangles inserted on the periphery of the ovary. In order to convey extra information, two drawings are occasionally combined, showing a diagram at the level of the style and a diagram of the inferior ovary separately. Ovules are depicted by different symbols - black dots when numerous and little differentiated, or ovules with funiculus. When there is a pair or a single ovule per carpel, it is possible to show the orientation of curvature as orthotropous, anatropous descending and anatropous ascending. Carpellodes are depicted by blackened or empty carpels. Secretory tissue, including nectaries, is represented in grey, wherever it occurs in the flower. Fusions between organs are shown by full connecting lines. The presence of a cup-like hypanthium is shown by two broken lines. Specific appendages of organs, such as conspicuous hairs or lobes, are shown on the diagram when they are conspicuous. In cases where there is abundant evidence to conclude that organs were present but were lost through abortion, their position is shown by an asterisk.

In most cases the floral diagrams were prepared based on living or pickled material. In cases where no flowers were available, several photographs, drawings and literature descriptions were used that were as representative as possible.

PART II FLORAL DIAGRAMS IN THE MAJOR CLADES OF FLOWERING PLANTS

4

Systematic Significance of Floral Diagrams

4.1 Floral Diagrams and the Molecular Phylogeny

4.1.1 *Molecular and Morphological Characters*

Since the early 1990s molecular systematics has dramatically changed the approach to studying relations of plants and led to major changes in the classification of plant groups. Pre-molecular classifications such as those of Cronquist (1981), Thorne (1992), Takhtajan (1997) and, to a lesser extent, Dahlgren (1975, 1983), were mostly intuitive with specifically selected characters considered more important than others. Especially for flowers, certain characters considered important in pre-molecular classifications were shown to be mere convergences by the molecular phylogenies. For example, families sharing three carpels with parietal placentation grouped in an order Violales or Parietales *sensu* Engler were shown to belong to three different lineages. Other comparable earlier associations of families including the 'Contortae' (based on contorted petal aestivation), 'Sympetalae' (taxa with stamen-petal tubes) or 'Rhoeadales' (Papaveraceae and Brassicaceae) were largely used in the book of Eichler (1875–78) and the Englerian systems.

Molecular phylogenies have proposed several shifts in relationships, some of them predicted by other characters (e.g. the link between Salicaceae and former Flacourtiaceae), others unexpected (e.g. the circumscription of Proteales and Rosales) or apparently questionable because of superficial morphological similarities (e.g. the separation of Oxalidaceae and Geraniaceae), and a few controversial cases (e.g. Anisophyllaceae in Cucurbitales). A number of major milestones, the Angiosperm Phylogeny Group (APG) classifications (APG I, 1998; APG II, 2003; APG III, 2009; APG IV, 2016: see Soltis et al., 2018) have increasingly brought stability to the system with most major groupings firmly supported by an increasing number of genes. More recently, a number of studies have improved understanding of phylogenetic relationships on a large scale and increased the predictive value of inferred relationships of major plant groups (e.g. Chase et al., 2005; Jansen et al., 2007; Brockington et al., 2009; Wang et al., 2009; Moore et al., 2010; Soltis et al., 2011; Xi et al., 2012; Zeng et al., 2017).

Nevertheless, morphology of flowers remains essential, even more important than ever before, in understanding affinities of plant families. Morphological characters belong to the most versatile category of phylogenetically useful information. Compared to molecular characters, morphology has far fewer characters, but most characters are phylogenetically informative. Morphology informs on phylogenetic relationships but also on function, and through function it is a reflection of natural selection (Bateman, Hilton and Rudall, 2006). A major drawback in understanding floral morphology is the occurrence of convergences towards adaptations to specific pollinators, or a secondary simplification due to changes towards wind pollination. The evolution of floral characters is shaped by ecology (flower–pollinator interactions) but also by developmental constraints and genetic mutations (see Ronse De Craene, 2018). Endress (2006, 2011) pointed to the importance of key innovations (e.g. monosymmetry, sympetaly, spurs, syncarpy), leading to bursts of diversification in flowers. Floral adaptations range from mechanisms to avoid selfing (dichogamy, herkogamy, dicliny) to specific adaptations to pollination (by wind, water or animals).

However, molecular data are equally riddled with problems and constraints (reviewed in Bateman, Hilton and Rudall, 2006). The study of morphological characters becomes increasingly important in the framework of a stable molecular phylogeny. Although morphological characters tend to improve the support for phylogenetic trees when used in combination with molecular characters, a combination of morphological characters also has great predictive value, giving indications about principal characteristics of taxa. Morphological characters are increasingly mapped on phylogenetic trees as an indication of evolution (e.g. Doyle and Endress, 2000; Ronse De Craene, Soltis and Soltis, 2003; Zanis et al., 2003; Endress and Doyle, 2009, 2015; Doyle and Endress, 2011; Soltis et al., 2018). As discussed in Ronse De Craene (2008) and Ochoterena et al. (2019), this is not without problems because of the uncertainty about the presence of characters in ancestral groups and the difficulty in accurately defining morphological characters.

The changes in the relationships of families proposed by molecular classifications have sometimes put the value of morphological characters in doubt, but they have also opened new challenges and emphasized the value of previously unrecognized characters. As flowers remain a prime tool in the identification of plants next to vegetative parts, the understanding of floral structures is essential, and this can be achieved through the use of floral diagrams. In this book efforts are made to demonstrate how floral diagrams can help in understanding relationships between or within families, but also to clarify the evolutionary shifts that have shaped floral structures. Floral diagrams are discussed within the system of APG IV (2016) and some more recent publications (e.g. Soltis et al., 2018). At the same time the existence of common characters or apomorphic tendencies for major groups is discussed in line with floral diagrams (see also p. 415).

4.1.2 *The Angiosperm Phylogeny Group Classification*

Figure 4.1 shows a phylogenetic tree with the major clades of angiosperms. Angiosperms are the sister group of the gymnosperms. There was much recent discussion about the nearest gymnosperm group, with contrary opinions based on molecular characters and fossil evidence. As it is not within the scope of this book to discuss the origin of angiosperms, the reader is referred to recent updates in Doyle (2008), Melzer, Wang and Theißen (2010), Doyle and Endress (2011), Endress and Doyle (2015), Sauquet et al. (2017), Sokoloff et al. (2018b) and Rümpler and Theißen (2019).

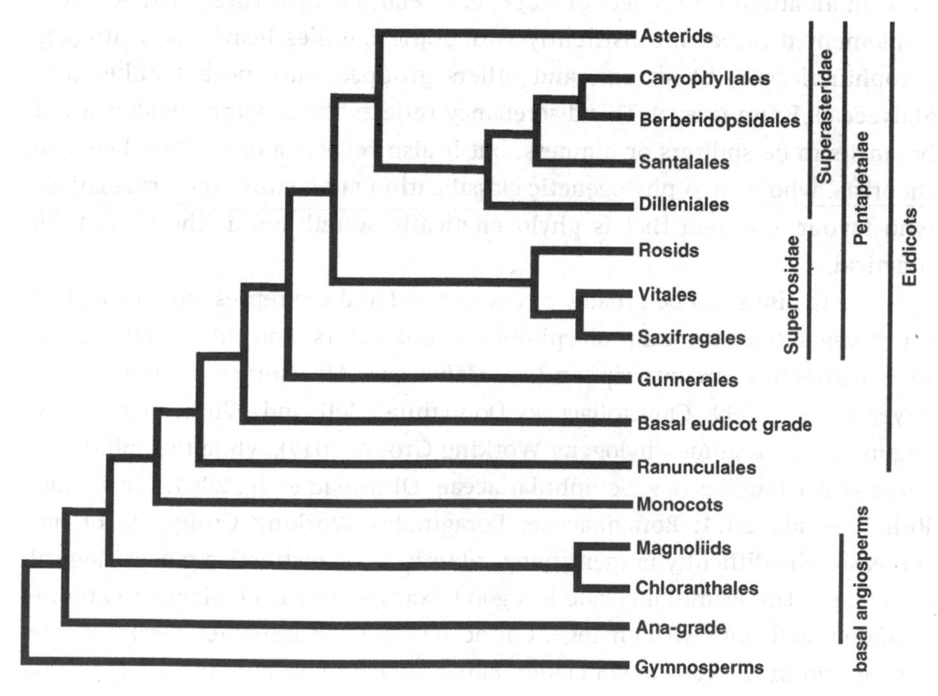

Figure 4.1 Phylogenetic tree of Angiosperms, based on Endress and Doyle (2015)

Although the relationships within major clades are not always fully supported, the foundations of the molecular classification are sound and are not expected to change radically in the future. The angiosperms consist of two major clades, the basal angiosperms, linked with the monocots, and the eudicots (consisting of a basal grade and core eudicots or Pentapetalae). Although a main morphological character is the presence of pollen with a single aperture in the former (monosulcate) and pollen with three apertures in the latter (tricolpate), differences of flowers are not clear-cut, with a gradual transition between basal groups and Pentapetalae through the basal eudicot grade. Pentapetalae, representing 75 per cent of the angiosperms (Soltis et al., 2003) consist of the well-supported larger Superrosidae and Superasteridae, but relationships among the major lineages are still partly unresolved. A lingering issue is the contrasting evidence of nuclear and chloroplast genes and the possibility of hybridization among different clades (see e.g. Petersen et al., 2016; Soltis et al., 2018)

Family Delimitation

For a number of large families (e.g. Scrophulariaceae, Malvaceae, Flacourtiaceae, Euphorbiaceae, Clusiaceae, Amaryllidaceae, Asparagaceae) delimitation has changed considerably compared with pre-molecular classifications in an attempt to reflect phylogenetic relationships. These changes were implemented rather inconsistently with some families heavily split up (e.g. Scrophulariaceae, Clusiaceae) and others grouped into mega families (e.g. Malvaceae, Primulaceae). This discrepancy reflects the inherent tendencies of botanists to be splitters or lumpers, but it also reflects a distinction between theorists, who want a phylogenetic classification at all costs, and pragmatists, who favour a system that is phylogenetically sound and at the same time practical.

Some families can be broadly circumscribed and are represented by a clear set of vegetative and floral morphological characters, and this is reflected in their taxonomic circumscription (e.g. Malvaceae: Alverson et al., 1998, 1999; Bayer et al., 1999; Caprifoliaceae: Donoghue, Bell and Winkworth, 2003; Leguminosae: Legume Phylogeny Working Group, 2017), while the split-up of some larger families (e.g. Scrophulariaceae: Olmstead et al., 2001; Clusiaceae: Ruhfel et al., 2011; Boraginaceae: Boraginales Working Group, 2016) has increased the difficulty in identifying subunits with distinctive morphological characters. The Primuloid clade is a good example of the challenge to accommodate practicality with phylogenetic accuracy. The paraphyletic Primulaceae can be split in several smaller clades. Källersjö, Bergqvist and Anderberg (2000) proposed the recognition of four families – Maesaceae, Theophrastaceae, Myrsinaceae and Primulaceae – although there are indications that these groupings lack obvious homogenous characters with transitional genera (e.g. *Samolus*, *Ardisiandra*, *Coris*) that can arguably be recognized as separate families. I agree with the statement of Källersjö, Bergqvist and Anderberg that 'The dilemma of phylogenetic analyses including molecular data is that although the results may be robust, and hence our best estimate of the evolutionary relationships, they may indicate a close relationship between taxa which may be very different morphologically' (p. 1339). However, I believe practical reasons need to be a major consideration in building classifications. A concept of a larger superfamily Primulaceae with a number of subfamilies makes more sense, as all taxa are characterized by a syndrome of common characters, including sympetaly, obhaplostemony and a free-central placentation. Another example of the difficulty in delimiting families is the relationship between Brassicaceae and Capparaceae. Capparaceae appears to be paraphyletic on molecular evidence, with subfamily Cleomoideae sister to Brassicaceae. Placing all Capparaceae in an expanded Brassicaceae would be phylogenetically correct but would obscure the many clear morphological characters. Therefore, Hall, Sytsma and Iltis (2002) recommended recognition of three well-supported monophyletic families – Capparaceae, Cleomaceae and Brassicaceae.

Several larger orders are phylogenetically well supported but remain unpractical with almost no sound morphological synapomorphies (e.g. Asparagales, Ericales, Brassicales, Malpighiales and Polygonales). There is a need to break up these large entities into workable units. This challenge is covered in this book by describing smaller clades within the orders whenever there is a possibility to do so. In addition it is important to consider other parameters to discuss affinities, such as apomorphic tendencies (see p. 415).

4.1.3 *Fossil Flowers and Floral Diagrams*

The past decades have seen a tremendous advance in the discovery of fossil angiosperms through improved technology (synchrotron radiation X-ray microtomography), and coupled with the building of molecular clocks it has been possible to identify the age of several angiosperm families and to use fossils in phylogenetic studies (e.g. Dilcher, 2000; Soltis and Soltis, 2004; Friis, Pedersen and Crane, 2006, 2016; Friis, Pedersen and Schönenberger, 2006; Crepet, 2008; Wang et al., 2009; Rothwell, Escapa and Tomescu, 2018; Schönenberger et al., 2020).

The earliest angiosperms date back to the Early Cretaceous (*c.*130 million years ago). The fossil record is remarkably concordant with floral evolution postulated for extant angiosperms. Earliest flowers tend to be perianthless or have a simple perianth resembling extant basal angiosperms and early diverging eudicots, while there is a broad-scale diversification of flower structures in mid-Cretaceous floras when the major clades become recognizable, including the core eudicot rosids and asterids. Radially symmetrical pentamerous flowers with calyx and corolla and diplostemony become well established in the early late Cretaceous and resemble extant rosids and early diverging asterids (Friis, Pedersen and Crane, 2016; Manchester et al., 2018). It is only in the Tertiary that large groups with floral tubes and monosymmetric flowers became abundant. Wind pollination arose several times and appears to be derived from perianth-bearing ancestors. The fossil diversity is huge and our understanding is increasing rapidly by constant new discoveries and by comparing fossil flowers with extant angiosperms (e.g. Friis, Pedersen and Crane, 2006; Endress, 2008b). The discovery of the early diverging *Archaefructus* and the recognition of several taxa with reduced floral structure (e.g. Ceratophyllaceae, Chloranthaceae, Hydatellaceae) have indicated an early diversification of angiosperms mainly in an aquatic environment (Doyle, 2008; Endress and Doyle, 2009, 2015). Inclusion of morphological data from extinct species has great potential in resolving deep node polytomies (Rothwell, Escapa and Tomescu, 2018).

Floral diagrams play an important part in representing our understanding of structures from fossil flowers, which have to be reconstructed from compressions or charcoalified remains. Fossil flowers become increasingly incorporated in phylogenetic studies and evidence from these flowers can be neatly summarized in floral diagrams. Several studies have incorporated floral diagrams in their analyses of fossil flowers (e.g. Friis, 1984; Crane, Friis and Pedersen, 1994; Keller, Herendeen and Crane, 1996; Gandolfo, Nixon and Crepet, 1998; Schönenberger and Friis, 2001; Schönenberger et al., 2001, 2012; Friis, Pedersen and Crane, 2006, 2016; Manchester et al., 2018; Crepet, Nixon and Weeks, 2018). These floral diagrams make comparisons with extant flowers much easier to interpret evolutionary trends.

4.2 Overview of Floral Diagrams in the Major Clades of Flowering Plants

The overview of families and their floral diagrams cannot be complete. Even Eichler did not cover all families, despite the fact that his overview was more extensive than mine. Compared to the first edition, twenty additional families are represented here and more genera are included for some families. Attempts were made to include as much diversity as possible as well as to include most major orders recognized by APG IV (2016). Missing from the list are some minor orders such as Canellales and Chloranthales (basal angiosperms), Petrosaviales (monocots), Huerteales, Picramniales, Zygophyllales (rosids), Icacinales, Vahliales, Garryales, Metteniusales, Escalloniales, Bruniales and Paracryphiales (asterids), as well as some families within larger orders.

5

Basal Angiosperms

The Ascent of Flowers

The basal angiosperms are a grade comprising the main part of Magnoliidae *sensu* Cronquist (1981) minus Ranunculales, which belong to the basal eudicots. The monocots are nested within the basal angiosperms. Figure 5.1 summarizes the phylogenetic relationships of the main families and orders according to Soltis et al. (2018). The position of Chloranthaceae and Ceratophyllaceae is debatable according to recent phylogenies, depending on emphasis on molecular or morphological analyses (see e.g. Moore et al., 1997; Jansen et al., 2007; Endress and Doyle, 2009, 2015). *Ceratophyllum* is highly distinctive as a submerged aquatic herb with an ancient fossil history. Its position has swung from being sister to the monocots, eudicots or Chloranthales. The latter appears to be better supported on a morphological basis (e.g. Iwamoto, Shimizu and Ohba, 2003; Endress and Doyle, 2009, 2015; Doyle and Endress, 2011; Iwamoto, Izumidate, and Ronse De Craene, 2016).

5.1 The Basalmost Angiosperms: Amborellales, Austrobaileyales, Nymphaeales

The basalmost angiosperms, also referred to as the ANITA-grade (Qiu et al., 1999) or ANA-grade (Soltis et al., 2005), groups the early-diverging lineages of the angiosperms. Among these taxa the monotypic *Amborella trichopoda* represents the basalmost extant angiosperm. Most taxa of the ANA-grade share spiral (or whorled) flowers with a morphological continuum between bracts and perianth, indefinite number of organs and a central receptacular residue (Endress, 2001; Ronse De Craene, Soltis and Soltis, 2003; Endress and Doyle, 2009, 2015). There is still some debate as to whether Nymphaeales or Amborellales are more primitive, and this is

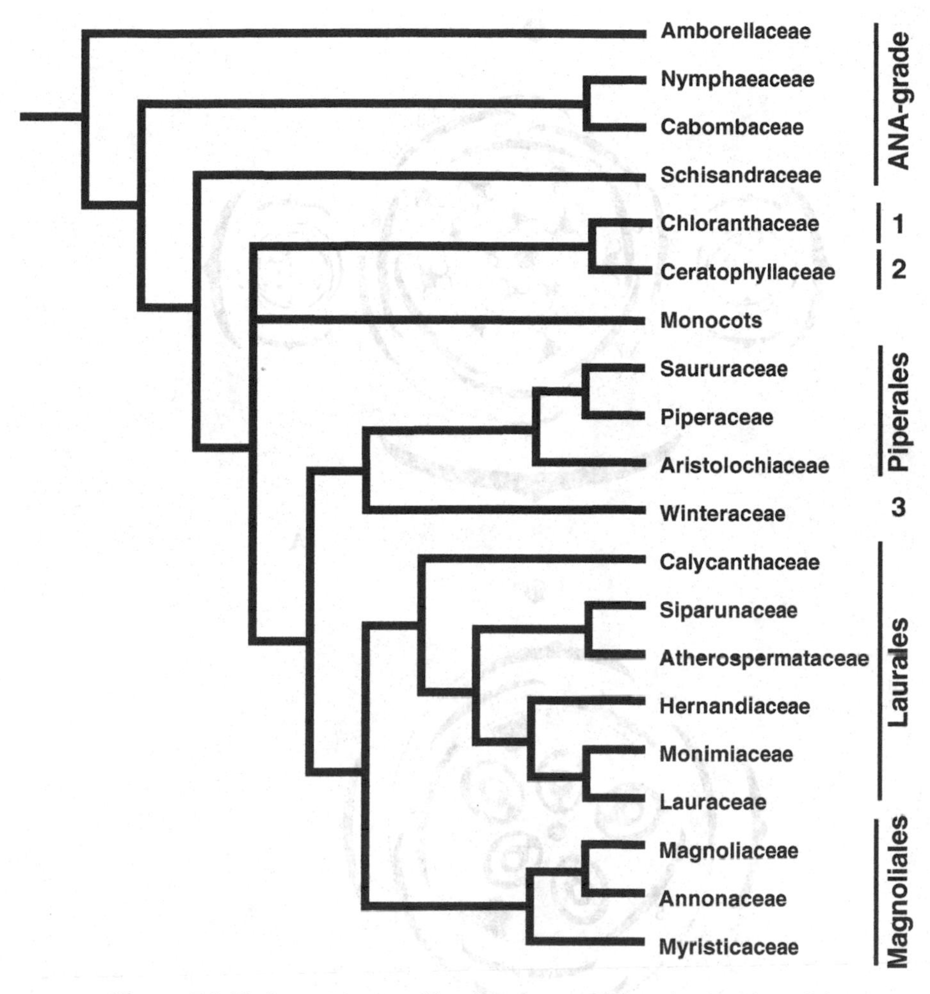

Figure 5.1 Phylogenetic tree of basal angiosperms, based on Soltis et al. (2005). 1. Chloranthales, 2. Ceratophyllales, 3. Canellales.

linked to the uncertainty whether spiral or whorled flowers are ancestral (Endress and Doyle, 2015). However, perianth initiation and differentiation are variable in the grade. Endress (2008c) argued that a multistep gradation from bracts to inner tepals may be secondary, as this is absent from *Amborella*. Carpels are mostly ascidiate and carpel closure is by secretion (Endress and Igersheim, 2000a). Some characters appear derived and have evolved independently, such as unisexual flowers and adaptations to specific pollinators.

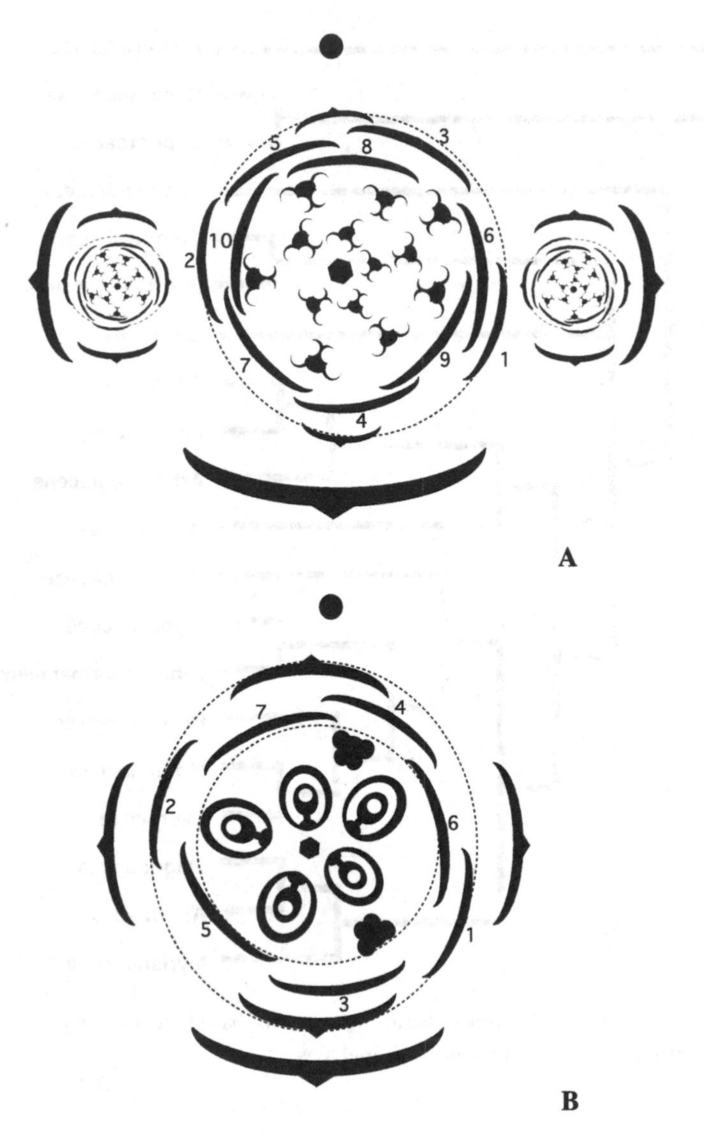

Figure 5.2 *Amborella trichopoda* (Amborellaceae) –staminate partial inflorescence (A) and pistillate flower (B). Numbers give sequence of initiation of the perianth.

Amborellales

Amborellaceae

Figure 5.2A–B *Amborella trichopoda* Baill., based on Endress and Igersheim (2000b) and Buzgo, Soltis and Soltis (2004)
Staminate: ↺ P(6)9–11(15) A(8)12–21(22) G0
Pistillate: ↺ P7–8 A(0)1°–2° $\underline{G}$(4)5(6)

Flowers have several characters considered ancestral in basal angiosperms: spiral flowers with undifferentiated perianth and no clear transition between bracts and perianth, variable number of floral parts and orthotropous ovules. However, unisexual flowers (with remnants of the other gender in pistillate flowers) and a floral cup (hypanthium) are derived (Ronse De Craene, Soltis and Soltis, 2003; Endress and Doyle, 2015).

The differentiation into vegetative shoot and flower is progressive with terminal flowers arranged in short cymose inflorescences. Two bracteoles are closely connected with the flower and occur at the transition of a decussate to a spiral phyllotaxis in the first two tepals (shown by numbers in Figure 5.2 A–B: Buzgo, Soltis and Soltis, 2004). In pistillate flowers staminodes resemble fertile stamens with sterile anthers. Both staminate and pistillate flowers have a central pyramidal extension representing a sterile apex (Endress and Igersheim, 2000b). At anthesis the inner tepals are larger and reflexed.

Nymphaeales

The order contains the three families Cabombaceae, Nymphaeaceae and Hydatellaceae. The latter was recently moved from an unsettled position in monocots to a basal position as sister to Nymphaeales, and this has important consequences for the interpretation of the earliest diverging flowers (Rudall et al., 2007). Flowers of Hydatellaceae (single genus *Trithuria* with twelve species) have simple, occasionally unisexual flowers that are highly different from other Nymphaeales and probably represent a reduction.

Compared to other early diverging angiosperms, flowers of Nymphaeaceae show important derived characters, such as a transition to whorled phyllotaxis, differentiation of sepals and petals with outer perianth parts larger than inner, development of a hypanthium, presence of nectaries and a secondary increase of floral parts (e.g. Ronse De Craene, Soltis and Soltis, 2003; Endress, 2008c).

Cabombaceae

Figure 5.3 *Brasenia schreberi* Gmel., based on Richardson (1969) and Ito (1986a)

∗ P3+3 A 6+6+3+3+6+6 $\underline{G}$3+6+3

General formula: ∗ P4–8[a] A3–36 G(1–)2–18

The trimerous flowers are axillary or extra-axillary and solitary without subtending bract. In *Brasenia* flowers are dull purple and wind pollinated without differentiation of sepals and petals, while *Cabomba* is insect pollinated with a differentiation of sepals and petals, and nectar secreted on the petal margins

[a] Or K(2–)3(–4) C(2–)3(–4) in *Cabomba*.

Figure 5.3 *Brasenia schreberi* (Cabombaceae), solitary flower. Numbers refer to the sequence of initiation of stamen and carpel whorls.

through localized glandular cells (nectarioles: Vogel, 1998a; Endress, 2008c). Petals are delayed in development compared to sepals, which is unusual for Nymphaeales (Endress, 2001; Schneider, Tucker and Williamson, 2003).

Stamens are numerous in *Brasenia* with a regular arrangement of trimerous whorls of stamens in double and single positions (Figure 5.3); carpels arise in one to three trimerous whorls (Richardson, 1969; Ronse De Craene and Smets, 1993). Only six stamens are found in *Cabomba* (rarely three), arising simultaneously and alternating with three carpels situated opposite the petals (Endress, 2001). The six stamens of *Cabomba* correspond to the three outer stamen pairs found in *Brasenia* and the lower stamen number is probably derived.

Carpels are strongly ascidiate and contain one to three ovules in variable position (laminar-diffuse placentation). In *Brasenia* one or two ovules are attached to the dorsal side of the carpel in a single row (Ito, 1986a). Floral diagrams of Cabombaceae are presented by Ito (1986a), but the position of carpels in *Cabomba* is erroneously placed as antesepalous.

Nymphaeaceae

Figure 5.4 *Nymphaea alba* L.
∗ K2+2 C4+8+4+4+8 A∞ $\underline{G}$14–18
General formula: ∗ K4–6(–12) C0-70 A14–200 G3–35

Only *Nuphar* has flowers associated with an abaxial bract, rarely with two basal phyllomes (El et al., 2020). In other genera a bract is lacking (Schneider,

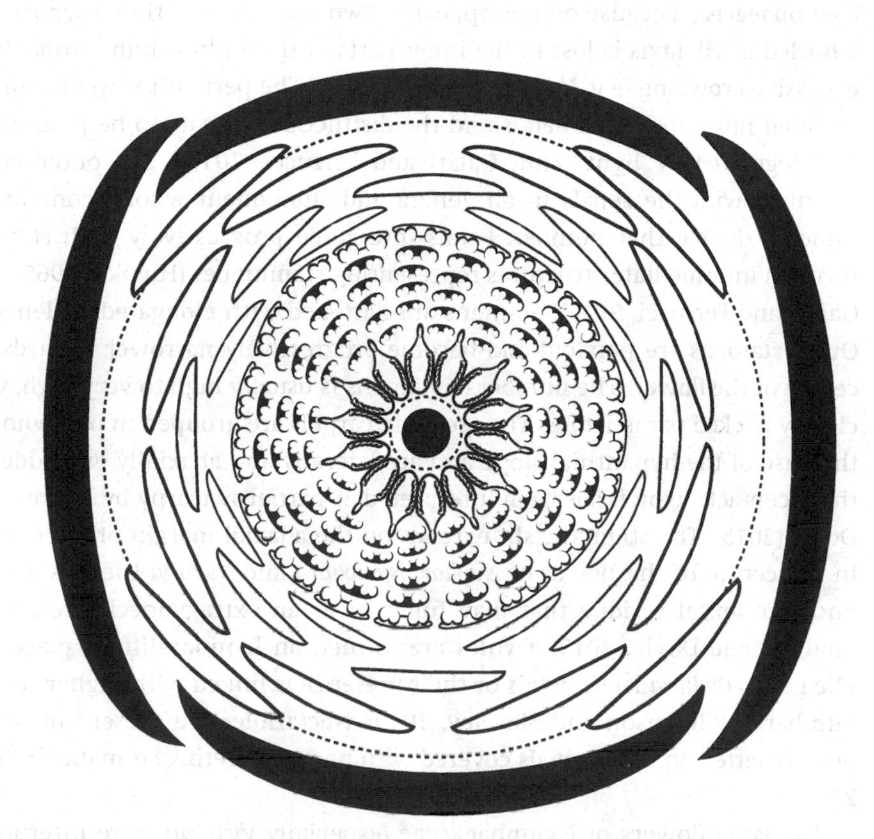

Figure 5.4 *Nymphaea alba* (Nymphaeaceae). Central receptacular residue shown by circle

Tucker and Williamson, 2003). Endress and Doyle (2009) discussed various possibilities that the bract has been lost or incorporated in the calyx of other genera.

Merism is variable in the family, ranging from di-, tri-, tetra-, to pentamery, but the ancestral phyllotaxis is unclear (Endress, 2001). According to El et al. (2020), the supposed pentamery appears to be problematic as evidence of an ancestral spiral phyllotaxis, as Nymphaeaceae appear to be basically whorled. The unstable merism appears to be linked to availability of space. Flowers are large to very large with a well-developed hypanthium in *Victoria* and *Nymphaea*. A depression arises by sinking of the apex during floral development (Schneider, Tucker and Williamson, 2003). Flower initiation is whorled with an easy transition between different merisms. Endress (2001) interpreted the flowers of *Nymphaea* and *Victoria* as tetramerous as a result of the extension of the diameter of a trimerous flower. Ronse De Craene, Soltis and Soltis (2003) suggested that tetramery is equivalent to dimery in Nymphaeaceae because organs appear in two sequences within a whorl. The whorled phyllotaxis is lost in the inner parts of the androecium, probably by excessive crowding (e.g. *Nuphar*: Endress, 2001). The perianth is weakly differentiated into sepals and petals and the distinction appears to be progressive and regulated by light (Zini, Galati and Ferrucci, 2017). The outer petals alternate with the sepals in all genera and subsequent whorls contain the same or double that number. Petals intergrade progressively with stamens through intermediate structures representing staminodes (Hiepko, 1965; Zini, Galati and Ferrucci, 2017). Stamens are flattened with elongated pollen sacs. Outer stamens are petaloid and become progressively narrower towards the centre of the flower. The number of stamens is usually high to very high, with closely packed parastichies. The ascidiate carpels are grouped in one whorl at the base of the hypanthial slope and are laterally and abaxially embedded in the receptacle, which has been interpreted as partial syncarpy by Endress and Doyle (2015). The stigmatic slit extends on the adaxial margin of the carpels. In the centre of the flower of *Nuphar*, *Nymphaea* and *Victoria* there is a small knoblike apical residue that may function as an extragynoecial compitum (Endress and Doyle, 2015). Ovules are formed on laminar-diffuse placentae. The genus *Ondinea* lacks petals or their presence is linked with higher stamen numbers (Williamson and Moseley, 1989). Nectarioles are present in *Nuphar* only, inserted on inner tepals covered with nectar-secreting stomata (Endress, 2008c).

The large flowers of Nymphaeaceae (especially *Victoria*) were interpreted as being secondarily increased in size, creating space for additional petals and stamens (Schneider, 1976; Schneider, Tucker and Williamson, 2003).

Austrobaileyales

The small order contains three families (Austrobaileyaceae, Schisandraceae and Trimeniaceae). All have spiral flowers with undifferentiated perianth and variable organ number in common (Endress, 2001). Bracts, outer and inner tepals overlap gradually on a spiral gradient, but outer tepals are smaller than inner ones in bud, as in *Amborella* (Endress, 2008c). Stamens and carpels are numerous, except for Trimeniaceae with a single carpel. In all taxa the transition between different organ categories is gradual, especially in Austrobaileyaceae with inner staminodes (Endress, 1980b, 2001).

Bisexual flowers are proterogynous, a common character in the basal angiosperms.

Schisandraceae

Figure 5.5 *Illicium simmonsii* Maxim.
↺ P14–19 A13–19 $\underline{G}$12–13
General formula: ↺ P7–33 A4–50 G5–21 (based on Endress, 2001)

Subterminal flowers are borne on short shoots enclosed by decussately arranged bracts. Transition between outer, inner tepals and stamens is progressive, but staminodes are lacking. Carpels are plicate with adaxial-median ovule. They appear to be inserted in a whorl but arise in helical sequence (Endress, 2001). *Kadsura* and *Schisandra* have unisexual flowers, while *Illicium* has bisexual flowers with carpels arranged in a whorl around a broad central vestigial apex. The other Schisandraceae have unisexual flowers appearing spiral throughout with a progressive transition from bracts to tepals; staminate and pistillate flowers lack any residue of the other gender (Endress, 2001: figs 5–6). However, intermediate bisexual flowers were reported in some taxa (Endress, 2001). A limited nectar secretion occurs at the base of tepals and stamens (Erbar, 2014).

Figure 5.5 *Illicium simmonsii* (Illiciaceae). Central receptacular residue shown by black circle

5.2 Magnoliales

The order is highly diverse and contains six families. Flowers range from indeterminate spiral flowers with many parts (e.g. Himantandraceae, Degeneriaceae, Eupomatiaceae) to trimerous whorled flowers (at least in the perianth: Magnoliaceae, Annonaceae) to reduced whorled flowers (e.g. Myristicaceae).

Eupomatiaceae have a single bract forming a circular calyptra and no perianth (Endress, 2003b). Corresponding calyptras are found in Magnoliaceae and Himantandraceae, but are less elaborate. The presence of a hypanthium with proterogynous flowers in some families and internal staminodes is similar to Calycanthaceae (Laurales), but outer staminodes are also found in other Magnoliales where they play a role in attraction (Himantandraceae, Eupomatiaceae). Nectaries are absent, except for inner tepals of some Annonaceae (Endress, 1994) and on the carpels of Magnoliaceae (Erbar, 2014).

Myristicaceae

Figure 5.6A–B *Myristica fragrans* Houtt., based on Armstrong and Tucker (1986)

Staminate flowers: ✶ P(3) A9–12[b] G0

Pistillate flowers: ✶ P(3) A0 $\underline{G}$1

General formula: staminate: ✶ P(2–)3(–5) A2–40 G0; pistillate ✶ P(2–)3(–5) G1

Myristicaceae differ from other Magnoliales by unisexual flowers with a single trimerous (rarely di-, tetra- or pentamerous) perianth. Pistillate and staminate flowers develop on an axis subtended by a pair of transversal bracts; a flower is terminal and enclosed by a single bract. In staminate flowers second order axes develop in the axil of the paired outer bracts and produce eighty to one hundred flowers in a helical sequence in the axil of bracts (Armstrong and Tucker, 1986). The single bract encloses the flower as a hooded structure.

The perianth is often partially to completely connate. Pistillate flowers have a single carpel with one basal anatropous ovule and are homogeneous throughout the family. Staminate flowers show much variation within the family. All stamens are fused into a synandrium of two to sixty anthers (Armstrong and Wilson, 1978; Sauquet, 2003; Xu and Ronse De Craene, 2010b). Anthers are extrorse and sessile, arranged as a crown on a column, sometimes resembling an inverted cone (*Knema*), or they are occasionally stalked (*Maloutchia*). Anthers develop four elongated pollen sacs with transversal septations. Each pair of bisporangiate lobes and a connecting vascular bundle are equivalent to

[b] Arranged as 3+3^3. Occasionally with an additional whorl.

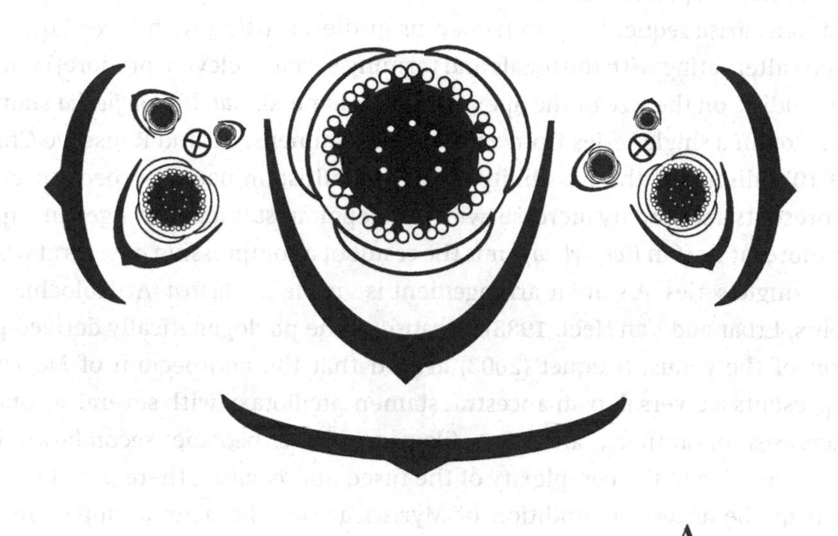

Figure 5.6 *Myristica fragrans* (Myristicaceae): A. staminate flower; B. pistillate flower. White dots represent the vascular bundles of the stamens.

a stamen (Armstrong and Wilson, 1978; Figure 5.6A). Some Myristicaceae have two or three anthers only. In *Myristica* staminate and pistillate flowers arise terminally on short shoots and are enclosed by a bract and a single abaxial bracteole (Figure 5.6A–B). Armstrong and Tucker (1986) and Xu and Ronse De Craene (2010b) demonstrated that the staminal column in *Myristica* is derived from the receptacle and not from fused filaments, as previously suggested. Stamens arise sequentially in trimerous girdles, starting with three larger stamens alternating with the tepals and forming seven to eleven (or more) stamens depending on the size of the apex in *Myristica* and *Knema*. In *Horsfieldia* stamens develop in a single series from antetepalous stamens. Xu and Ronse De Craene (2010b) discussed the possibility that the multistaminate androecium either represents a secondary increase (with antetepalous stamens arranged in triplets or more; at least in *Horsfieldia*), or is the result of a compression of several whorls in a single series. A similar arrangement is visible in *Thottea* (Aristolochiaceae: Leins, Erbar and Van Heel, 1988). Because of the phylogenetically derived position of the genus, Sauquet (2003) argued that the androecium of *Maloutchia* represents a reversal to an ancestral stamen phyllotaxy with several secondary increases of anthers, and that filaments have become secondarily free. However, due to the complexity of the fused androecium, there is little indication of the ancestral condition of Myristicaceae. The general appearance of staminate flowers of several Myristicaceae is very similar to staminate *Nepenthes* or Canellaceae and represents a remarkable convergence.

Magnoliaceae

Figure 5.7 *Magnolia paenetalauma* Dandy, based on Xu and Rudall (2006)

✶ K3C3+3 A∞ $\underline{G}$∞

General formula: ↺/✶ K3 C6–21 A(3–6)–∞ G(2–4)–∞

For a long time Magnoliaceae were considered the prototype for the ancestral flower type in the angiosperms (e.g. Eames, 1961). Nowadays the flower is considered relatively specialized (Nooteboom, 1993; Endress, 1994). The reason for the status of primitiveness is the often conically developed receptacle with many spirally inserted stamens and carpels. The generic circumscription ranges from two (*Magnolia, Liriodendron*) to 16 genera (Stevens, 2001 onwards).

Flowers are bisexual, rarely unisexual and dioecious (*Woonyoungia*: Fu et al., 2009). Flowers of Magnoliaceae are enclosed by one or more circular bracts, which are shed at anthesis (Nooteboom, 1993). The perianth is usually trimerous (very rarely pentamerous: Deroin, 2010), and differentiated in an outer whorl of three tepals (sepals), which is transitional in morphology and is genetically distinct (Wróblewska, Dolzblasz and Zagórska-Marek, 2016), and an inner part of two trimerous (petal) whorls, rarely in only two trimerous whorls or less (e.g. *Michelia figo*: Hiepko, 1965; *Woonyoungia*: Fu et al., 2009). The outer tepals have sometimes been interpreted as transformed bracts (e.g.

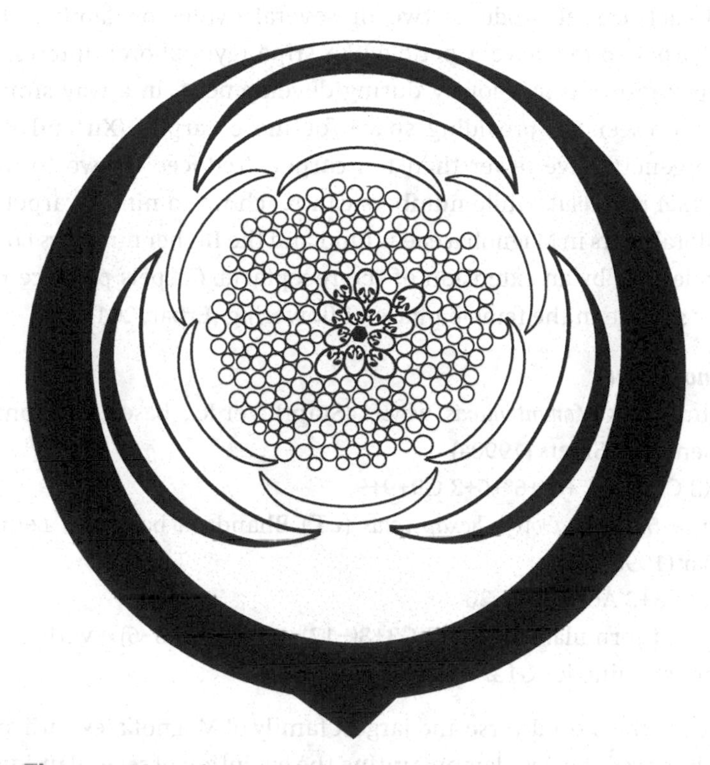

Figure 5.7 *Magnolia paenetalauma* (Magnoliaceae)

Baillon, 1868b; Eames, 1961), although ontogenetic evidence indicates that they belong to the flower and that they are differentiated into sepals and petals (Xu, 2006; Xu and Rudall, 2006). In some species the number of tepals is higher with spiral phyllotaxis (e.g. *Magnolia stellata, Michelia champaca*). In *M. stellata* additional tepals arise spirally within the trimerous perianth whorls and probably represent transformed stamens (Wróblewska, Dolzblasz and Zagórska-Marek, 2016). Erbar and Leins (1981, 1994) interpreted the flower as a transitional stage in the shift from spiral to whorled phyllotaxis. The transition between large perianth parts and androecium is abrupt and leads to a disruption of the spiral phyllotaxis. The initiation of the androecium does not follow a regular spiral, contrary to the claims of Erbar and Leins (1981, 1994), at least for some *Magnolia* species. Outer stamens tend to arise in flushes opposite the outer petal whorl with at least an indication of stamen pairs (Xu, 2006; Xu and Rudall, 2006), or clearly in pairs (see figures of *Liriodendron tulipifera* and *Magnolia denudata* in Erbar and Leins, 1981, 1994). The stamens have no distinct filament (except in *Liriodendron*), but consist of a laminate organ with embedded anther tissue.

Staminodes are absent, although outer stamens can be morphologically transitional with the perianth (e.g. *M. stellata*: Hiepko, 1965). The gynoecium is apocarpous and each carpel produces two or several ovules on laminar-diffuse placentae. Carpels often have a petiole-like stipe (gynophore). Interestingly, the floral apex grows continuously during development, in a way similar to racemose inflorescences, providing space for more carpels (Xu and Rudall, 2006). Some genera have fewer than ten carpels (reduced to two to four in pairs in *Michelia montana*), while numbers can reach up to ninety carpels. The number of floral parts in Magnoliaceae is fluid and the higher numbers could be secondarily derived by an extension of the receptacle. Carpels produce nectar on the stylar surface in the female phase of flowering (Erbar, 2014).

Annonaceae

Figure 5.8A *Monanthotaxis whytei* (Stapf) Verdc., based on Ronse De Craene and Smets (1990a)

∗ K3 C3+3 A6°+3°+6°+6+3 $\underline{G}$9+9+9

Figure 5.8B *Artabotrys hexapetalus* (L.f.) Bhandari, based on Leins and Erbar (1996)

∗ K3 C3+3 A6+∞ $\underline{G}$14–20

General formula: ∗ K(2)3(4) C3+3(−12) or C3+0 A(3–6)∞ with outer or inner staminodes G1,2–10,∞

Annonaceae are the most diverse and largest family of Magnoliales with around twenty-five hundred species, demonstrating the evolution of several unique and derived features (sympetaly, syncarpy and tetramery). Flowers are bisexual, rarely unisexual, and generally grouped in cymes, or they are solitary. A hypanthium is present in *Xylopia* (Steinecke, 1993). Flowers are mostly trimerous with three perianth whorls: three outer sepaloid or petaloid tepals (sepals) and two inner whorls of petaloid tepals (petals). Flowers are rarely tetramerous (e.g. *Tetrameranthus*: Saunders, 2010). The inner petal whorl may be reduced and apparently absent (e.g. *Annona*, *Dasymschalon*) or more rarely the outer (*Annickia*), or the calyx may be reduced to a rim (e.g. *Fenerivia*: Deroin, 2007; Saunders, 2010). More plasticity is found in *Ambavia* with supernumerary petals and variations between dimery and trimery (Deroin and Le Thomas, 1989). Petals are imbricate or valvate and are occasionally fused at the base or more extensively in a petal tube (e.g. *Isolona*). Inner petals may be erect and form a closed nuptial chamber linked to beetle pollination. The androecium consists either of three or six stamens (e.g. *Orophea*, *Mezzetiopsis*), or multiples of three (twelve to twenty-four) (e.g. *Bocagea*, *Monanthotaxis*: Baillon, 1868c; Steinecke, 1993; Ronse De Craene and Smets, 1990a; Xu and Ronse De Craene, 2010a). The six outer

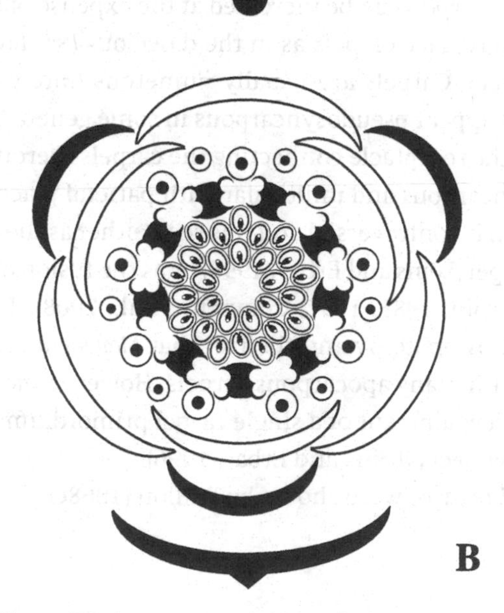

Figure 5.8 Annonaceae: A. *Monanthotaxis whytei*; B. *Artabotrys hexapetalus*

stamens (or staminodes) tend to be inserted as three pairs in alternation with the petals (Figure 5.8A). More often the androecium is highly polymerous (up to two thousand stamens in *Annona montana*) with a more or less whorled to irregular phyllotaxis, becoming spiral higher up on the receptacle (e.g. Leins and Erbar, 1996). Outer stamens always alternate with petals as in oligandrous flowers. More stamens appear in flushes, not in a regular spiral, and form parastichies. Leins and Erbar argued that the number of stamens has been increased at the expense of their size. In *Fenerivia*, there are six supernumerary petals corresponding in position and vasculature with outer stamen pairs (Deroin, 2007). The opposite condition with a replacement of inner petals by stamens is found in *Friesodielsia* (Guo, Thomas and Saunders, 2018), corresponding to a homeotic mutation comparable to some Papaveraceae (p. 152). In *Toussaintia*, there are also supernumerary petals arranged on an extended receptacle forming an androgynophore, although their relationship with stamens is less clear (Deroin, 2000). Several Annonaceae possess glandular tissue at the margin or the base of inner petals (Saunders, 2010). Staminodes are rarely inside (*Anaxogorea*, *Xylopia*), but mostly outside (Figure 5.8A; Saunders, 2010) fertile stamens. Only in *Xylopia* are they found in both positions. Anthers are usually extrorse, rarely latrorse and the connective is broadly developed, resembling a shield in some taxa. Pollen is released when stamen bases become detached from the receptacle. The transition between stamens and carpels is progressive or abrupt and the number of carpels can be increased at the expense of stamens (e.g. *Mezzetiopsis*) or stamens replace carpels as in the dioecious *Pseuduvaria* (Xu and Ronse De Craene, 2010a). Carpels are usually numerous (rarely a single carpel) and apocarpous, but appear pseudosyncarpous in some genera by development of a basal collar of the receptacle connecting the carpels (Deroin, 1997). *Monodora* and *Isolona* are syncarpous and unilocular with parietal placentation. The origin of this gynoecium is controversial and was seen either as the result of a subdivision of a single carpel (Leins and Erbar, 1996) or as the fusion of several carpels at their margin (Deroin, 1985, 1997; Couvreur et al., 2008). The latter interpretation is supported by anatomy and the fact that *Monodora* and *Isolona* are nested within a clade with many apocarpous carpels. However, early developmental stages show the development of a single carpel primordium comparable to other apocarpous gynoecia (Leins and Erbar, 1996).

Diagrams of *Popowia* and *Bocagea* were shown by Baillon (1868c).

5.3 Laurales

The order consists of six families in two clades, sister to a basal Calycanthaceae (Renner, 1999). Flowers are extremely variable with a floral phyllotaxis ranging from spiral to whorled or chaotic. Endress and Doyle (2007, 2015) assumed that spiral phyllotaxis found in Magnoliales and Laurales is derived, because both groups are nested deep within clades that are whorled. Phyllotactic patterns can jump easily between whorls and spirals in basal angiosperms, leading to repeated reversals but not unequivocally clarifying the origin of phyllotaxis, although basal Laurales are spiral. Endress and Doyle (2007) considered chaotic phyllotaxis to be derived from whorled phyllotaxis by doubling of the position of organs, leading to a loss of a regular phyllotaxis. Other factors probably lead to a chaotic initiation patterns, such as the distortion of the floral apex (cf. Winteraceae: Doust, 2002), as paired stamens are part of a cyclization event that appears several times within the basal angiosperms. While unicarpellate taxa have a whorled phyllotaxis throughout, pluricarpellate families appear to have a variable phyllotaxis (Staedler and Endress, 2009).

Merism is variable, and mainly trimerous only in Lauraceae. There is a tendency for building flowers with several superposed stamen whorls (e.g. Monimiaceae, Lauraceae). Most Laurales have stamens accompanied by a pair of nectariferous glands on the filament. These were taken for staminodes in the past although there is no evidence for this (Endress, 1980a; Ronse De Craene and Smets, 2001; Buzgo et al., 2007; Sajo et al., 2016). Staminodes occur frequently, either outside the stamens or as an inner whorl, and they are usually secretory. Dehiscence of anthers is mostly valvate by flaps and there are several trends leading to the reduction of ventral or dorsal pollen sacs (Endress and Hufford, 1989). Monimiaceae is the most variable family with evolutionary trends ranging from spiral bisexual flowers with well-developed tepals to highly specialized unisexual and dimerous flowers (Endress, 1980a; Staedler and Endress, 2009). Gynoecia are always apocarpous, with a tendency to reduction to monomery in several families and an occasional shift to an inferior ovary. A hypanthium is characteristic for all families (least developed in Lauraceae) and is linked with semi-inferior to inferior ovaries. The hypanthium can take huge proportions in some Monimiaceae enclosing the carpels as a pseudofruit and forming a hyperstigma. Each carpel contains a single ovule in adaxial median position, except for Calycanthaceae with two ovules (Endress and Igersheim, 1997).

Floral diagrams of several Atherospermataceae, Monimiaceae, Siparunaceae and Gomortegaceae are presented by Staedler and Endress (2009).

Calycanthaceae (incl. Idiospermaceae)

Figure 5.9 *Calycanthus floridus* L., partly based on Staedler, Weston and Endress (2007)

↺ P(26–)28(–29) A 13(–16) A°(16–)20(–22) $\underline{G}$(25–)34(–35)

General formula: ↺ P15–40 A5–30 A°∞ G(1–3)∞

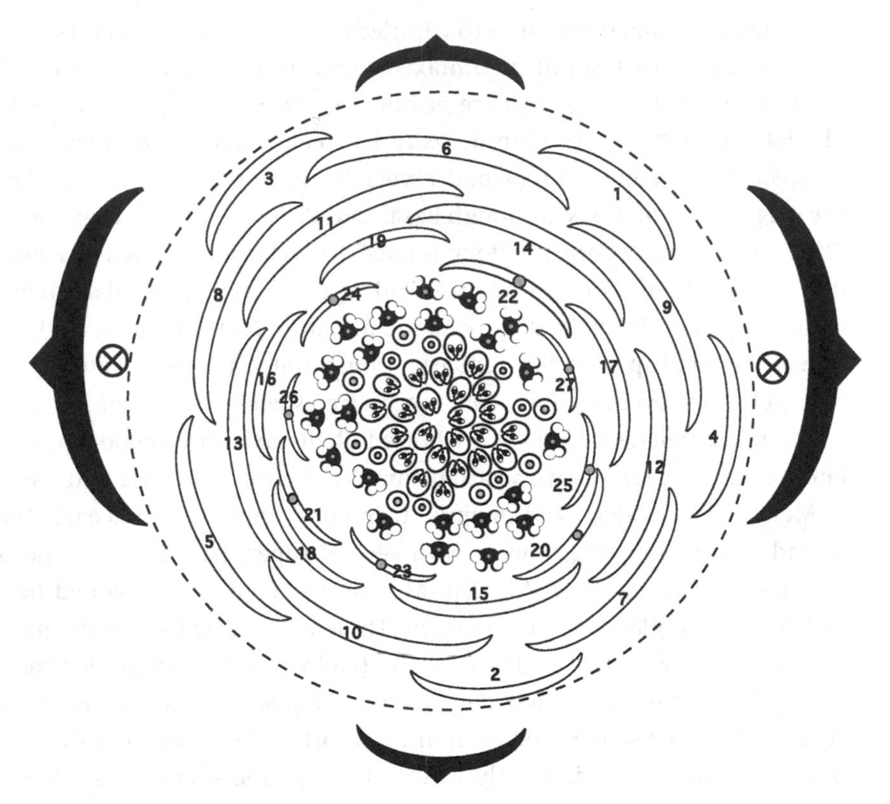

Figure 5.9 *Calycanthus floridus* (Calycanthaceae). Numbers refer to the initiation sequence of the perianth. Note the presence of food bodies on inner tepals and stamens.

Organ numbers are widely variable. In Figure 5.9 the tepal initiation sequence is numbered (one to twenty-seven), with several series of stamens and staminodes.

Phyllotaxis of the four genera of Calycanthaceae was studied by Staedler, Weston and Endress (2007). They found a regular transition from the decussate vegetative phyllotaxis to the spiral flower in all genera. The different organ categories are bridged by intermediate organs, and within a group of organs there are distinct Fibonacci series of eight, thirteen or twenty-one organs. This arrangement in distinct series can be seen as a preliminary phase in the transition to whorled phyllotaxis found in other Laurales. Stamens are extrorse. Inner stamens are reduced into staminodes that play a role in closing the floral bud. Carpels occupy the bottom and sides of a deep floral hypanthium and are variable in number. Long filiform stylar appendages reach the top of the depression between the packed staminodes. Each carpel contains two ovules, but the upper one degenerates (Endress and Igersheim, 1997).

Tepals of *Chimonanthus* secrete nectar through scattered glandular cells, while *Calycanthus* and *Sinocalycanthus* lack nectaries but offer food bodies on the inner staminodes, stamens and inner tepals (Vogel, 1998a).

Lauraceae
Figure 5.10 *Persea americana* Mill., based on Buzgo et al. (2007)
∗ P3+3 A3+3+3+3° $\underline{G}$1
General formula: ∗ P4–6 A3–30 G1

The perianth is made up of two inconspicuous, greenish-yellow to white tepal whorls held together in bud by trichomes. The outer whorl is much smaller in the parasitic genus *Cassytha* (Sastri, 1952). In a few species one or two outer stamen whorls are tepaloid in shape (e.g. *Eusideroxylon*), or the tepals are transformed into stamens (*Litsea*, *Lindera*: Rohwer, 1993a). The number of stamen whorls can be variable in the family with a distinction between outer and inner whorls. Reductions to three stamens occur in a few species (*Silvaea*, *Endiandra*: Eichler, 1878). Up to five whorls can be found in *Umbellularia*, or the number of stamens can be higher (up to thirty-two), probably in several whorls (Rohwer, 1993a). All stamens bear paired lateral appendages in only a few genera. In general, only the third whorl bears two lateral nectary appendages and tends to have a different anther orientation from the outer whorls. The inner whorl is

Figure 5.10 *Persea americana* (Lauraceae): partial inflorescence

reduced to small stub-like staminodes that can be secretory. Buzgo et al. (2007) interpreted the presence of the staminodes as the result of overlapping developmental signals between androecium and carpel. Anthers are either tetrasporangiate opening with four valves, or disporangiate through the reduction of two pollen sacs (Ecklund, 2000). The single carpel bears one anatropous, pendent ovule. The orientation of the carpel is variable; Buzgo et al. (2007) showed it transversal on their diagram, while in Eichler (1878) and Rohwer (1993a) it is median descending.

Laurus, among other genera, is unusual in having dimerous, unisexual flowers. These arise as a continuation of a subdecussate phyllotaxis found in several Lauraceae. The outer stamens alternate with the tepals (Eichler, 1878; Ronse De Craene and Smets, 1993). Lauraceae appear early in the fossil record and are well documented with a high variation of floral forms leading to extant taxa (e.g. Ecklund, 2000).

Hernandiaceae

Figure 5.11A–C *Hernandia nymphaeifolia* (C. Presl) Kubitzki, based on Endress and Lorence (2004)

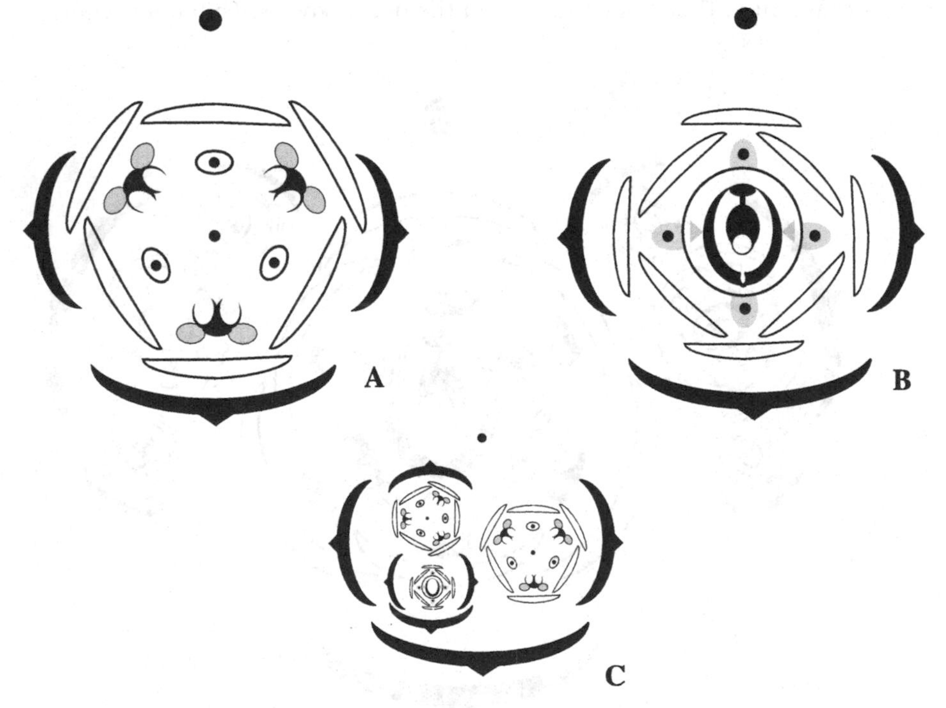

Figure 5.11 Staminate (A) and pistillate (B) flower and (C) inflorescence of *Hernandia nymphaeifolia* (Hernandiaceae)

Staminate: ✱ P3+3 A3+3° G0
Pistillate: ✱ P4+4 A4° Ğ1
General formula: ✱ P4–12 A3–5(-7) G1

The family is close to Lauraceae and shares similar flowers with undifferentiated perianth, anthers with flaps and basal nectaries, and an unicarpellate gynoecium. However, merism is more variable, ranging from tri- to tetra-, penta- or heptamery. Flowers are often unisexual and the ovary is inferior. The heptamerous perianth of *Gyrocarpus* is probably an amalgamation of two flowers with different merisms. Kubitzki (1969) provided several floral diagrams in his monograph of Hernandiaceae. In some species of *Hernandia* there is a tendency for the prophylls to fuse into a cupular structure enclosing the fruit (Kubitzki, 1993). In *Hernandia nymphaeifolia* flowers are grouped in triads derived from a monochasial cyme. Staminate flowers are trimerous and pistillate flowers tetramerous. Four nectaries are present in pistillate flowers and are probably staminodial in origin, corresponding to three appendages in staminate flowers (Endress and Lorence, 2004). The ovary is inferior with a single large, apical ovule (Endress and Igersheim, 1997).

Atherospermataceae

Figure 5.12A–C *Laurelia novae-zelandiae* A. Cunn., based on Sampson (1969)
Staminate: ↺ P4–9A7–13 G0
Pistillate and bisexual: ↺ P5–9 A0-6[c]A°13–24 $\underline{G}$8–12

Atherospermataceae used to be part of a traditionally circumscribed Monimiaceae (e.g. Philipson, 1993), which appears to be polyphyletic and has been subdivided into three families (Siparunaceae, Monimiaceae and Atherospermataceae) (Renner, 1999). Most inflorescences are thyrsoids consisting of lateral dichasial cymes, often with three flowers, or a single terminal flower. Flowers are bisexual or mostly unisexual with frequent transitions between flower types. There is much variation in the arrangement of organs, with spiral phyllotaxis (e.g. *Daphnandra*, *Laurelia*), often shifting towards a whorled dimerous arrangement with smaller tepals in double position (e.g. *Dryadodaphne*, *Atherosperma*: Staedler and Endress, 2009).

The perianth is undifferentiated and bract-like and the transition between bracteoles and tepals is often unclear. In *Laurelia* the two outer tepals continue the decussate arrangement of the bracteoles (Sampson, 1969). In bisexual and pistillate flowers the stamens or non-functional stamens are accompanied by

[c] Only in bisexual flowers.

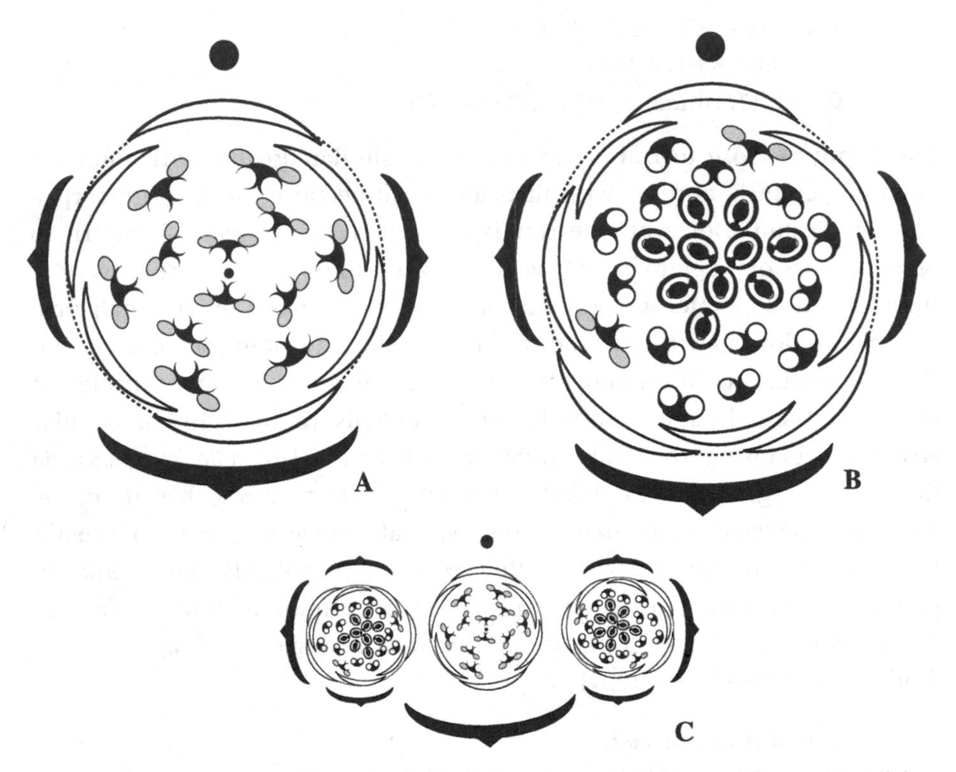

Figure 5.12 *Laurelia novae-zelandii*. A. Staminate and B. bisexual flower; C. partial inflorescence

smaller inner staminodes (*Atherosperma*, *Laurelia*). These are persistent and apparently assist in the closure of the floral cup after anthesis (Staedler and Endress, 2009). Staminate flowers have no carpels. All stamens bear glandular appendages. The number of microsporangia is often reduced to two with valvate dehiscence. Carpels bear an elongated stigmatic slit and have a single basal anatropous ovule.

Monimiaceae

Figure 5.13A–C *Ephippiandra myrtoidea* Decne, based on Lorence (1985) and Philipson (1993)
Staminate: ✱ P2+2 A2+2+2+2+2+2+2+2 G0
Pistillate: ↺ P2+2+2+2 A0 $\underline{G}$ 25–30
General formula: ✱ /↺ P(0)2–∞ A9–∞ $\underline{G}$ (1–)3–∞

Flowers are bisexual to unisexual, often arranged in monoecious or dioecious botryoids or dichasia. In *Ephippiandra* pistillate flowers are terminal and surrounded by axillary staminate flowers (Lorence, 1985; Figure 5.13C). Flowers are spiral (*Hortonia*, *Peumus*) or frequently dimerous, with decussate arrangement and paired position of outer tepals (*Tambourissa*, *Kibara*, *Ephippiandra* etc.). A hypanthium is usually well developed and petaloid, flattened or deeply urceolate with apical pore, and often takes over the function of the strongly reduced perianth (as in the related Siparunaceae). In *Peumus* the hypanthium forms a flattened platform exposing the spirally arranged tepals and stamens and paired nectaries are present at the base of the filaments.

Highly specialized flowers occur in *Tambourissa*, *Ephippiandra* or *Wilkiea*. In staminate flowers the number of stamens ranges from nine to eighteen hundred; carpel numbers range from a single carpel (e.g. *Xymalos*) and extend to two thousand in *Tambourissa* (Endress, 1980a; Endress and Igersheim, 1997). In some *Wilkiea* and *Kibara* the number of stamens is reduced to four to six (Staedler and Endress, 2009). Tiny tepals are arranged as several (generally five) decussate pairs around the apical pore or within the urceolate hypanthium. Dehiscence of the flower is by unequal fissuring of the cup forming a pseudoperianth (Endress, 1980a; Philipson, 1993). In *Epphippiandra* the fissuring is more regular, exposing decussate rows of anthers (Figure 5.13A). In pistillate flowers the hypanthium is either open and cupulate (e.g. *Ephippiandra*: Figure 5.13B) or urceolate. In the

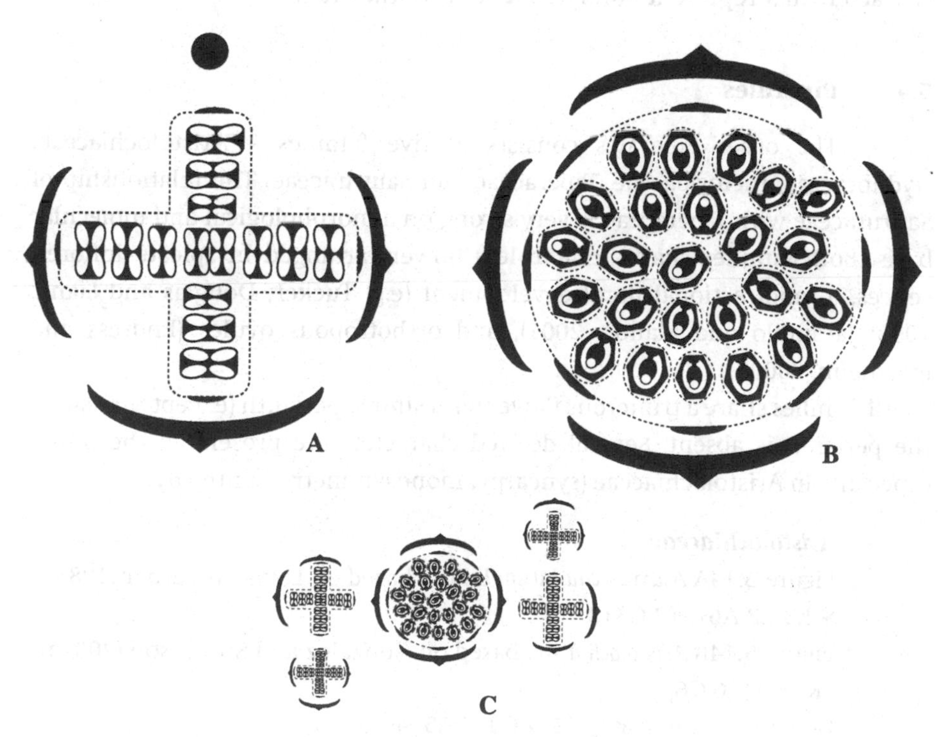

Figure 5.13 *Ephippiandra myrtoidea*. A. Staminate and B. pistillate flower; C, partial inflorescence

latter case the inner tepals are secretory and contribute to the formation of mucilage and a hyperstigma (a secretory pseudostigma). While in most other genera with a floral cup the styles protrude out of the apical pore (also in Siparunaceae), the pistillate flowers have the carpels enclosed in the hypanthium with the hyperstigma formed on the upper margin of the floral cup (Endress and Igersheim, 1997).

Tambourissa represents an extremely sophisticated evolution with a correlation between an increased floral size and increased stamen and carpel numbers. The perianth is reduced or lost and its protective and attractive role is taken over by the hypanthium. A hyperstigma is formed by the tepals and sterile carpels. In other genera there is a generalized trend to a reduction of the size and number of carpels, the reduction of the size of the entrance to the floral cup, linked with complete enclosure of carpels, stamens and tepals (Endress, 1980a).

Floral diagrams of several Monimiaceae and related families are presented in Staedler and Endress (2009). The distinction between Monimiaceae, Siparunaceae and Atherospermataceae as suggested by the molecular phylogeny appears less convincing on a morphological basis, with several recurrent characters that represent floral 'tendencies' of the order.

5.4 Piperales

The order Piperales consists of five families – Aristolochiaceae, Hydnoraceae, Lactoridaceae, Piperaceae and Saururaceae. The relationship of Saururaceae with Piperaceae is very strong on a morphological and molecular basis. Both families share perianthless flowers arranged in spicate inflorescences, a unidirectional floral development (e.g. Tucker, Douglas and Liang, 1993; Jaramillo and Manos, 2001), and orthotropous ovules (Endress and Igersheim, 2000b).

All families share a trimerous flower with simple perianth (except *Saruma*) or the perianth is absent. Several derived characters are present in the order, especially in Aristolochiaceae (syncarpy, monosymmetry, syntepaly).

Aristolochiaceae

Figure 5.14A *Asarum caudatum* Lindl., based on Leins and Erbar (1985)
✶ K3 C3 A6+3+3 Ğ3+3
Figure 5.14B *Aristolochia sp.*, based on González and Stevenson (2000a)
↓K3 C0 [A6 Ğ6]
General formula: ✶ /↓ K3–4 C0–3 A5–40 G4–6

Aristolochiaceae is a highly specialized family of basal angiosperms, with several evolutionary novelties in the flower culminating in the genus *Aristolochia* (gynostemium, monosymmetry, tubular perianth, inferior ovary and syncarpy). In *Asarum* and *Saruma*, the flowers arise at the end of shoots with distichous leaf arrangement and are enclosed by a bract that appears to arise on the adaxial side (Tucker and Douglas, 1996: Figure 7.2), although the flower could be terminal. In *Aristolochia* flowers arise successively in the axis of a bract (González and Stevenson, 2000a). *Saruma henryi* was considered the basal taxon, with well-developed petals. In other genera petals tend to become reduced or abort entirely. *Asarum* has either small petal appendages (e.g. *A. caudatum*: Figure 5.14A), or these are mostly absent (e.g. *A. europaeum*), while *Aristolochia* or *Thottea* have no petals at all (Leins and Erbar, 1985; Leins, Erbar and Van Heel, 1988; Kelly, 2001). The number of stamens ranges from six to forty-six. The most common arrangement of the androecium is whorled with six stamens arising opposite the sepals before the additional initiation of six more stamens in two whorls in *Asarum* (Leins and Erbar, 1985). The six additional stamens are shifted to the periphery by the pressure of the developing carpels. In *Saruma* a stamen pair is formed after the initiation of three petals and a larger stamen opposite the sepals. The order of development is more strictly centripetal (Leins and Erbar, 1995). *Thottea* is much more variable, with several stamen whorls of three, six, nine or twelve members, and inner staminodial structures in some species. The six stamens of *Aristolochia* correspond with the six first-formed stamens in other Aristolochiaceae. Some *Aristolochia* have a reduction to five stamens and five carpels (González and Stevenson, 2000b). Carpel number is also variable – three, six or nine – and there is often no correspondence between the number of carpels and stigmatic lobes. Carpels arise in two trimerous whorls in *Saruma* and *Asarum* when six in number, (Leins and Erbar, 1985, 1995). It is possible that a secondary increase of stamen and carpel numbers has affected some species of *Thottea*, linked with an expansion of the floral meristem. Leins, Erbar and Van Heel (1988: 369) attributed the increase of stamens in a whorl to 'a sudden phylogenetic change'. Stamens and carpels can occasionally proliferate up to twelve in subgenus *Pararistolochia* of *Aristolochia* (González and Stevenson, 2000b). Carpels are generally inferior but can be superior in subgenus *Heterotropa* of *Asarum* (Kelly, 2001). The ovary contains two rows of ovules per carpel, with little to variable fusion of the carpels in *Saruma* and *Asarum*, and strong fusion in *Aristolochia*.

Aristolochia is the most derived genus with a stable floral formula of P3 A6 G6. The perianth is highly diverse in structure, linked to a specific trap mechanism. Earlier theories, interpreting the perianth of *Aristolochia* as a single bract-like structure (such as a spathe of Araceae), were rejected by González and

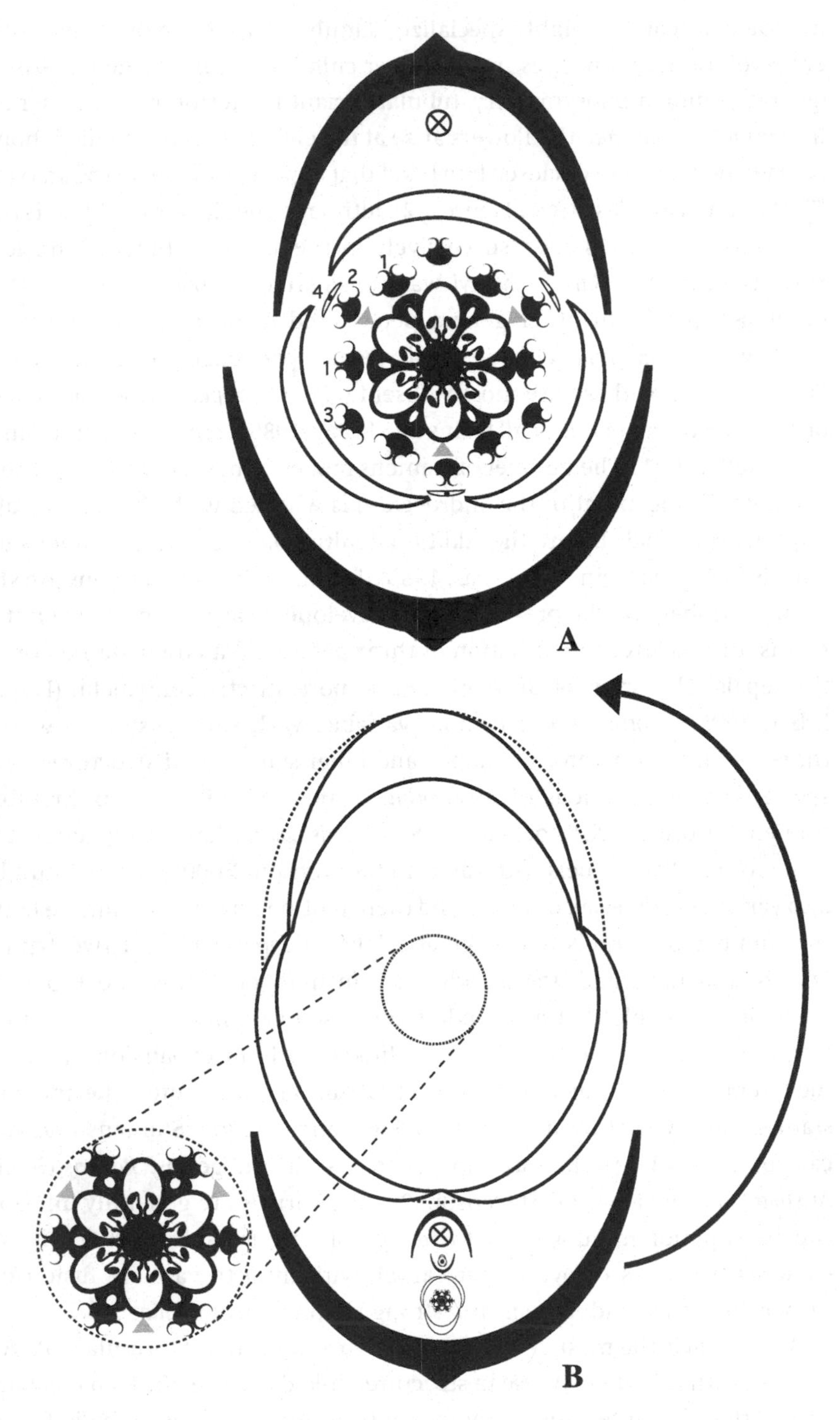

Figure 5.14 Aristolochiaceae. A. *Asarum caudatum* (Numbers give order of initiation of stamens and staminodes.) B. *Aristolochia* sp.

Stevenson (2000a), who showed that the perianth has a trimerous calyx-like nature. The development shows an early onset of monosymmetry in the development of the perianth linked to its curvature and the formation of a tube. Anthers are extrorse in Aristolochiaceae and the connectives become often fused with the upper part of the carpels in a gynostemium. In *Aristolochia* each gynostemium lobe is formed by extension of the commissural region of two carpels and their fusion with the stamens (González and Stevenson, 2000b); this corresponds with the alternating position of the stamens with the carpels.

Two possibilities exist for the evolution of the flower in Aristolochiaceae: the perianth is either basically biseriate, as in *Saruma*, and petals were progressively lost in other Aristolochiaceae, or the perianth is simple and trimerous, and petals were secondarily derived. The phylogeny does not support a basal position of *Saruma* (Kelly and González, 2003). A simple trimerous perianth is also found in other Piperales, such as Hydnoraceae and Lactoridaceae. González and Stevenson (2000a) discussed the difference between sepals and petals in Aristolochiaceae, implying that petals are different structures. Vestigial structures in *Asarum* were also interpreted as petals or staminodes. Kelly (2001) reported the occasional presence of vestigial anther sacs on the petals in *Asarum*. This indicates that petals in Aristolochiaceae may be structures derived from staminodes, which is rare in Angiosperms (see Ronse De Craene, Soltis and Soltis, 2003; Ronse De Craene, 2008). Further evidence comes from the genetic expression of B-class genes in Aristolochiaceae. Jaramillo and Kramer (2004) found the same gene expression in petals and stamens of *Saruma*, but a different expression in the perianth of *Aristolochia*. The expression of *AP3* and *PI* homologs in *Saruma* is also different from *Arabidopsis*, which may indicate a derivation from stamens.

Aristolochiaceae represent several plesiomorphic traits in the Piperales, despite their strong synorganization. The existence of three stamen pairs, trimerous simple perianth and multiple stamens link the family with other basal angiosperms. Lactoridaceae with trimerous flowers, a simple perianth and two whorls of stamens (Tucker and Douglas, 1996) has the same diagram as *Thottea tomentosa* (see Leins, Erbar and Van Heel, 1988) and may represent the ancestral condition.

Saururaceae

Figure 5.15 *Gymnotheca chinensis* Decne, based on Liang and Tucker (1989)

↔ P0 A6 Ğ(4)

General formula: ↔/✱/↓ P0 A3–6(8) G3–4

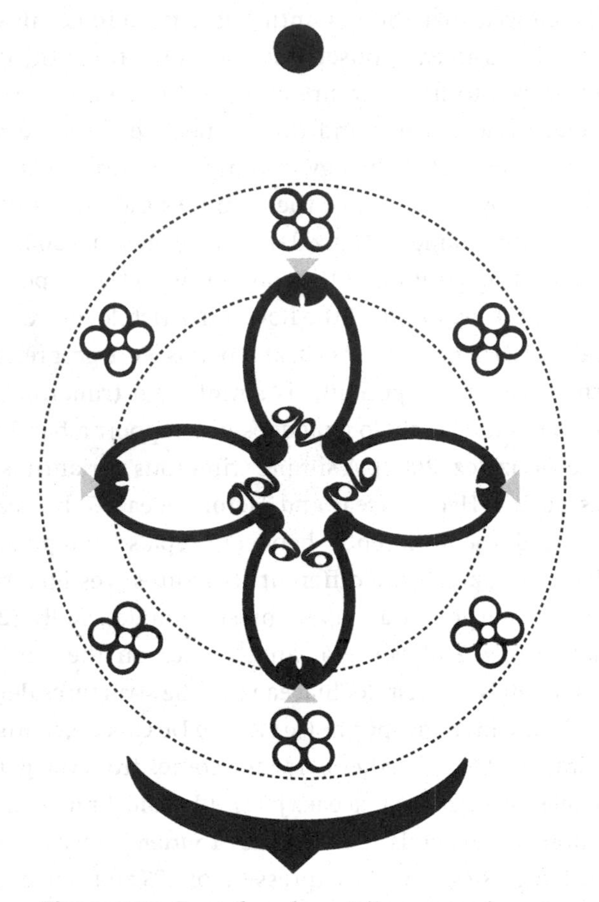

Figure 5.15 *Gymnotheca chinensis* (Saururaceae)

Piperaceae and Saururaceae share similar morphologies and are obviously closely related. The perianth is absent and its loss is thought to precede the evolution of both families (Dahlgren, 1983). Flowers are wind or insect pollinated with pollen as the main reward, although some *Peperomia* secrete nectar from glandular cells on their subtending bract (Vogel, 1998a).

Inflorescences are spicate racemes with each flower subtended by a single bract. In *Houttuynia* and *Gymnotheca* the basal involucral bracts are petaloid and the inflorescence mimics a flower. Flowers in Saururaceae are reduced, disymmetric (in cases where both sides are equally developed) or monosymmetric. The number of stamens in Saururaceae is six, except for *Houttuynia* with three stamens. Based on previous research Tucker, Douglas and Liang (1993) provided floral diagrams to illustrate the variation of floral initiation among the five genera of Saururaceae and some Piperaceae. In the absence of a perianth, floral development is usually unidirectional, either from the abaxial to the adaxial side (e.g. *Anemopsis, Houttuynia, Piper*) or the opposite (e.g. *Gymnotheca, Zippelia*). Six stamens probably represent the basal condition, but linked with a unidirectional initiation of stamens the lateral stamens can arise on common primordia. *Anemopsis* forms three stamen pairs arising from common primordia. Single stamens occupy the same position in *Houttuynia*, which Tucker (1985) interpreted as the result of splitting, although the opposite possibility cannot be excluded (Jaramillo, Manos and Zimmer, 2004). As stamen pairs are common in other basal angiosperms, including Aristolochiaceae, this could be evidence that the ancestral androecium of Piperales was hexamerous. In Piperaceae the stamens have a comparable unidirectional initiation but are further reduced to two in some genera (usually the first formed latero-anterior stamens) by terminal delation (Lei and Liang, 1998; Jaramillo and Manos, 2001). Jaramillo and Manos (2001) suggested that the loss of stamens in *Piper* is accompanied by closer packaging of flowers and larger anthers per flower.

The number of carpels ranges from four to three and the gynoecium is usually inferior and syncarpous through the development of a hypanthium (except in *Saururus*). Ovules are arranged in two rows on parietal placentae. In *Peperomia* (Piperaceae) the ovary is reduced to a single median carpel. Floral development is particularly labile in perianthless Piperales, leading to similar mature morphologies through different developmental pathways (see Tucker, Douglas and Liang, 1993; Jaramillo, Manos and Zimmer, 2004). The variability in stamen development is best considered the result of a gradual change in Piperales, not as having independent origins as suggested by Jaramillo, Manos and Zimmer (2004). This may be a reflection of the loss of the perianth inducing higher lability in floral development (Endress, 1987, 1994).

6

Monocots

Variation on a Trimerous Bauplan

Figure 6.1 shows the recent phylogeny of monocots based on Soltis et al. (2018). There is much uncertainty about the closest sister group of the monocots, which lies within the basal angiosperms. Options such as Ceratophyllaceae and Chloranthaceae were discussed in Soltis et al. (2005) but are not resolved. Moore et al. (1997) suggested a sister group relationship of monocots and eudicots, including Ceratophyllaceae, but this was questioned by Endress and Doyle (2009), who associated monocots with magnoliids. The trimerous monocot floral formula is found in a number of basal angiosperms, but there is no certainty of any morphological links and basal monocots have a more variable floral Bauplan. Within Alismatales several species have stamens arranged in pairs and Ronse De Craene (1995a) suggested that polycyclic androecia with paired stamens could be ancestral for monocots comparable to flowers of certain basal angiosperms (cf. Erbar and Leins, 1994). The presence of stamen pairs in the outer whorl of the androecium is probably plesiomorphic and is the result of a phylotactic shifting mediated by the trimerous perianth. Iwamoto et al. (2018) suggested that flowers with paired stamens, found in Butomaceae, Tofieldiaceae and Alismataceae, may form a link between monocots and basal angiosperms. The establishment of the typical monocot ground plan is not universal and is only firmly established at the node leading to Petrosaviales.

Remizowa Sokoloff and Rudall (2010) concluded that postgenitally fused carpels, associated with septal nectaries, represent the plesiomorphic condition in monocots; congenital fusion leads to a loss of the nectaries and has arisen several times independently, as well as fully apocarpous gynoecia.

Monocots present a high level of floral diversity with a strong convergence with Pentapetalae. While basal groups either have much reduced flowers or are polysymmetric with a variable degree of merism and apocarpy, most advanced orders are strictly guided by the pentacyclic trimerous floral Bauplan (P3+3A3+3G3) or

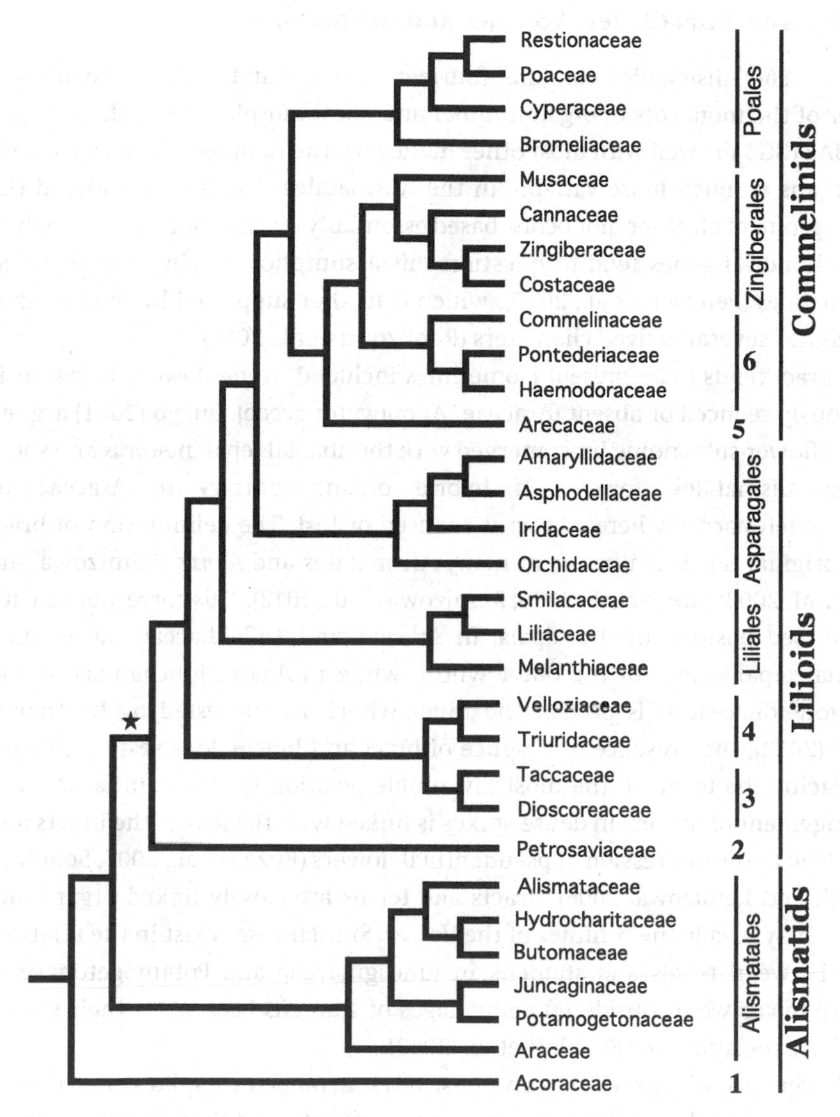

Figure 6.1 Phylogenetic tree of the monocots, based on Soltis et al. (2018). 1. Acorales; 2. Petrosaviaceae; 3. Dioscoreales; 4. Pandanales; 5. Arecales; 6. Commelinales. The asterisk refers to a stable trimerous pentacyclic Bauplan.

have undergone extensive reductions linked to wind pollination (Poales, Pandanales). Within this framework they demonstrate a high variation in floral morphs linked with monosymmetry and the development of hypanthia.

Monocots consist of three major units: a basal Acorales-Alismatales grade, a higher clade consisting of an asparagoid grade and a commelinid clade.

6.1 The Basal Clades: Acorales and Alismatales

The Alismatales contains fourteen families and is the most diverse order of the monocots in organ number and floral morphs. While the formula P3+3A3+3G3 fits well with most other monocots, the number and development of organs is much more variable in the Alismatales. *Acorus* is considered the sister group of all other monocots, based essentially on chloroplast genes, while mitochondrial genes tend to question this assumption placing the genus in Alismatales (Petersen et al., 2016), which is further supported by the fact that *Acorus* has several derived characters (Remizowa et al., 2010).

A bract tends to be present (sometimes included in the flower: *Acorus*) or is obviously reduced or absent (Araceae, Aponogetonaceae). Buzgo (2001) argued that a flower-subtending bract merged with the abaxial tepal in *Acorus* and some other Alismatales, forming a hybrid organ, contrary to Araceae or Potamogetonaceae where a bract is reduced or lost. The delimitation of bract and original tepals is unclear in many Alismatales and *Acorus* (Remizowa and Sokoloff, 2003; Buzgo et al., 2006; Remizowa et al., 2012). This corresponds with a reversed position of the tepals; in Araceae and Tofieldiaceae the median adaxial tepal is part of the outer whorl, while in *Acorus*, Juncaginaceae and Apogonetonaceae it is part of the inner whorl. As suggested by Remizowa et al. (2012), the presence or absence of bract and bracteole is responsible for displacing the tepals in the most favourable position for their initiation. The arrangement of flowers in dense spikes is linked with the loss of the bracts and may lead to the impression of pseudanthial flowers (Buzgo et al., 2006; Sokoloff, Rudall and Remizowa, 2006). Bracts and tepals are closely linked organs and bracts may invade the confines of the flower. Similar cases exist in the relationship between tepals and stamens in Juncaginaceae and Potamogetonaceae, where tepals were considered appendages of stamens because of their strong spatial association (see Sokoloff et al., 2013).

Flowers are dimerous or trimerous, often arranged in spikes or compact inflorescences. The perianth consists of two whorls, and there is a distinction between a petaloid clade (with marked distinction between a sepaloid outer whorl and petaloid inner whorl in Alismataceae) and a tepaloid clade (Figure 6.1). Families of the tepaloid clade have small flowers with a tendency for reduction of floral parts (tepals are undifferentiated and in two whorls, seldom three to one or even absent: e.g. Aponogetonaceae, Juncaginaceae, Potamogetonaceae), culminating in the highly adapted and reduced flowers of Posidoniaceae, Zosteraceae and Cymodoceaceae. The presence of nectaries is variable with the occurrence of septal nectaries without settled position (often infra-locular when present) or perigonal nectaries. The presence of septal nectaries is closely linked with the postgenital fusion of the carpels, which is most pronounced in the lower part of the ovary (e.g. Tofieldiaceae: Rudall, 2002; Remizowa and Sokoloff, 2003; Remizowa, Şokoloff and Rudall, 2006). It is assumed that apocarpous gynoecia are plesiomorphic in monocots because of significant differences between syncarpous gynoecia of early-diverging monocots (Remizowa, Sokoloff and Rudall, 2006; Remizowa et al., 2010). The gynoecium is variable in the order; it is often made up of numerous apocarpous carpels or two whorls of three (some Alismataceae, Aponogetonaceae, Butomaceae), or three to one carpel (e.g. Araceae, Tofieldiaceae, Acoraceae).

Stamen number is variable, ranging from three to very numerous. Paired outer stamens are common in Alismataceae. Some taxa have a secondary multiplication linked with a diminution of the size of stamen primordia (e.g. *Limnocharis* in Alismataceae: Sattler and Singh, 1973, 1977), or more commonly a reduction to the two-whorled androecium. Carpels can fluctuate in a similar fashion, with two whorls of three carpels found in all petaloid Alismatales, but with higher numbers associated with a smaller size (e.g. *Alisma*), resembling the carpel increase in some Rosaceae.

Alismatales make up four subclades if *Acorus* is not included (Figure 6.1; Buzgo et al., 2006). The petaloid Alismatales clade (three families) is characterized by relatively large pedicellate flowers subtended by bracts and a differentiation of calyx and corolla. In the tepaloid clade, there is a progressive reduction of flowers with undifferentiated perianth or without tepals and a diffuse distinction between flowers and inflorescence, and bracts and tepals.

Acoraceae

Figure 6.2 *Acorus calamus* L., based on Buzgo and Endress (2000)
↑P2–3?+3 A3+3 $\underline{G}(3)$

The monotypic family Acoraceae has become a major point of interest for systematists since it was placed at the base of monocots by molecular analyses. *Acorus* has typically trimerous monocot flowers and was interpreted as a prototype for all monocots (see Buzgo and Endress, 2000). However, other candidates for ancestral monocot flowers include members of Alismatales with polymerous flowers (Ronse De Craene and Smets, 1995a; Iwamoto et al., 2018). Buzgo and Endress (2000) studied the floral structure and development of *Acorus* and found similarities with Piperales.

The flower of *Acorus* is unusual, as the abaxial tepal cannot be distinguished from a flower-subtending bract. In the past this organ has either be interpreted as bract (e.g. Payer, 1857) or as part of the perianth (e.g. Sattler, 1973). The absence of a bract was the reason *Acorus* had been included in Araceae in the

Figure 6.2 *Acorus calamus* (Acoraceae). The asterisk refers to a lost abaxial outer tepal.

past (e.g. Eichler, 1875; Dahlgren, Clifford and Yeo, 1985). However, early initiation of the first floral organ is reminiscent of a bract and takes over tepal features later on (Buzgo and Endress, 2000). Three possibilities exist: the abaxial tepal is lost and replaced by the bract that became displaced close to the flower; the organ is a precocious tepal and there is no bract; the organ is a complex hybrid organ of bract and tepal.

The first interpretation is to be favoured (Figure 6.2), especially since the bract takes a larger size at maturity, covering the flower. The incursion of bracts in flowers is not rare in the angiosperms and could have led to the loss of the abaxial tepal. A comparison exists in some Saururaceae (e.g. *Saururus*, *Gymnotheca*) with a showy bract displaced close to the flower (Buzgo and Endress, 2000).

Flowers are sessile and arranged in spikes. The flower of *Acorus* is monosymmetric with unidirectional initiation. The syncarpous ovary contains orthotropous ovules on apical placentae. There is no style and the stigma is minute. Septal nectaries are absent, although non-secretory septal slits are present

(Buzgo and Endress, 2000; Ighersheim, Buzgo and Endress, 2001; Remizowa Sokoloff and Rudall, 2010).

Araceae
Figure 6.3 *Spathiphyllum patinii* (R. Hogg) N. E. Br.
∗ P3+3 A3+3 $\underline{G}$(3)
General formula: ∗ P0/6 A4–6 G2–3(–∞)

Figure 6.3 *Spathiphyllum patinii* (Araceae)

Flowers are grouped in condensed inflorescences on an inflated axis (a spadix) enclosed in a generally showy bract (spathe). The entire inflorescence functions as a pseudanthium and the apex is often enlarged and functions as an osmophore. Flowers are closely packed, forming regular parastichies, and lack a subtending bract. The orientation of flowers is unusual for monocots, with the median outer tepal in an adaxial position (Remizowa et al., 2012). Flowers are often unisexual on the same or different spadices. When monoecious staminate flowers develop distally and pistillate flowers proximally, with an intermediate zone arising on a gradient in between (e.g. *Philodendron*: Barabé and Lacroix, 2000). Tepals are not petaloid and are often reduced or absent in derived groups. Lehmann and Sattler (1992) showed that in *Calla* the perianth was replaced by stamens through a process of homeosis (cf. *Macleaya* in Papaveraceae). Stamens emerge at different levels and irregular intervals between tepals and gynoecium in *Spathiphyllum*. No distinct style is formed and stigmatic trichomes are formed on a triangular apical slit. Flowers generally lack nectaries. The number of carpels varies between one and several, and unicarpellate gynoecia may be pseudomonomerous. In *Spathiphyllum* three carpels bear two basally inserted collateral ovules each embedded in mucilage.

A distinction is made between basal Araceae (proto-Araceae) and other groups. *Gymnostachys* and Orontioideae are both dimerous with extrorse anthers, with transitions from trimery to dimery in *Orontium* (Buzgo, 2001).

Alismataceae (incl. Limnocharitaceae)

Figure 6.4A, B *Luronium natans* raf. based on Charlton (1999a)

∗ K3 C3 A6 $\underline{G}$3+3+3

Figure 6.4C, D *Sagittaria latifolia* Willd., based on Singh and Sattler (1973)

Staminate: ∗ K3 C3 A6+6+3+3+6+6+3+3+6 $\underline{G}^{o}$3+3

Pistillate: ∗ K3 C3 A^{o}6+3+3 $\underline{G}\infty$

General formula: ∗ K3 C3 A6–9–∞ G6–∞

Inflorescences often develop a succession of pseudowhorls interrupted by longer internodes (cf. Singh and Sattler, 1973; Charlton, 1999a, 2004). The pseudowhorls consist of a succession of three flowers with bracts and bracteoles (Figure 6.4B, E). *Ranalisma* is exceptional in forming pseudo-terminal flowers on a sympodial system (Charlton, 1991). Flowers are generally bisexual to unisexual, and always trimerous. The perianth is differentiated into outer greenish sepals and inner caducous petals. Stamen number is variable, ranging from three (*Wiesneria, Caldesia*) to many in several whorls, originating centripetally. Six stamens in three pairs are found in *Alisma, Damasonium, Baldellia* and *Luronium*, while additional whorls can be found in sequences of pairs and single stamens (Figure 6.4C). Carpels are six to numerous, in one to several whorls. There is often a discrepancy between stamen and carpel numbers within a flower, with fewer stamens and higher carpel numbers (e.g. *Ranalisma, Alisma, Echinodorus*: Sattler and Singh, 1978; Charlton, 1991, 2004). The stamens and carpels are characteristically whorled with the outer stamens arranged as three pairs next to the petals, which have a delayed development. *Wiesneria triandra* and *Caldesia parnassifolia* are exceptions with single antesepalous stamens (Charlton, 1999b, 2004). Sattler and Singh (1978) interpreted the association of the stamen pair and

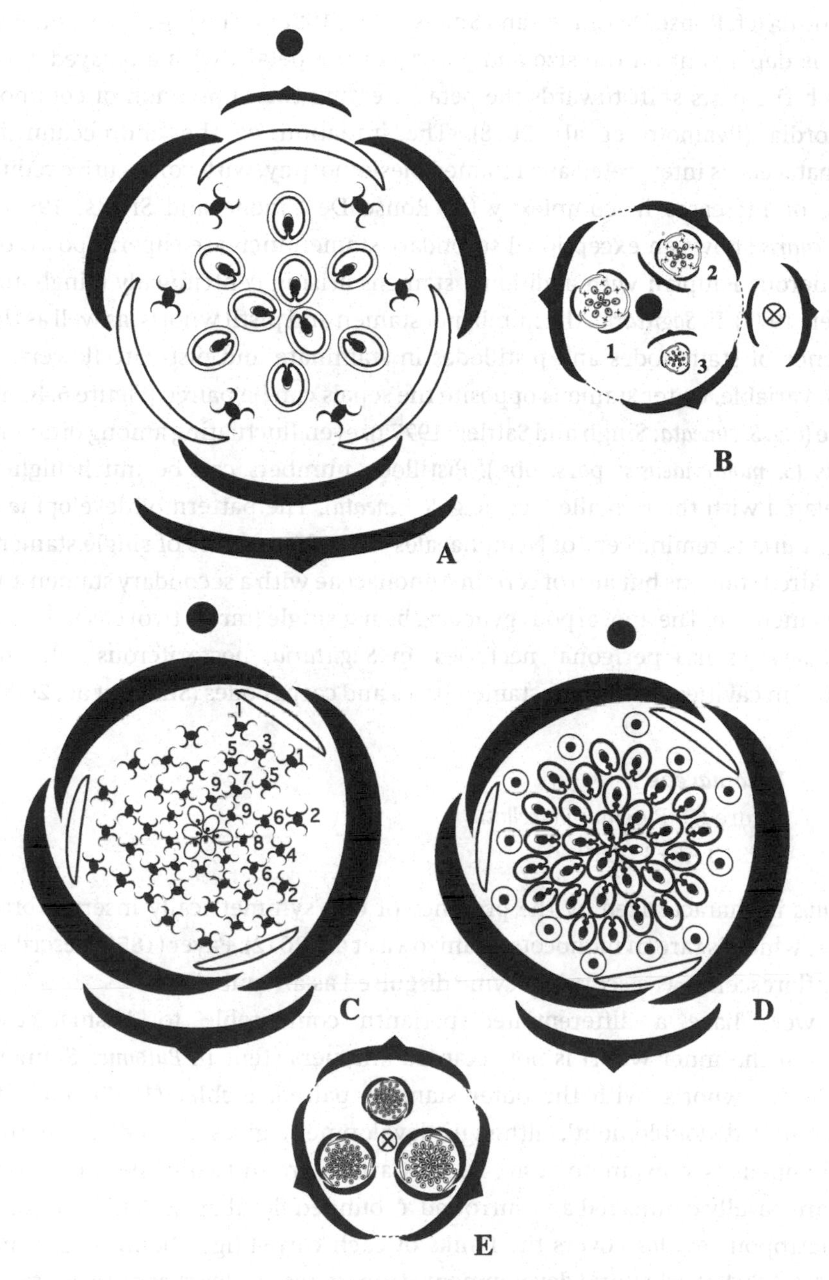

Figure 6.4 Alismataceae: flower (A) and inflorescence (B) of *Luronium natans*; (C) staminate, (D) pistillate flower and (E) inflorescence of *Sagittaria lancifolia*. In C numbers give order of initiation of whorls.

petal into common primordia as fundamental in the family and as the origin of a secondary stamen increase. Developmental evidence has shown that stamen pairs can be positionally associated with the petals but do not arise on common primordia (cf. Ronse De Craene and Smets, 1993, 1998a). The position of stamen pairs is dependent on the size and growth of the petals; with a delayed petal growth the pairs shift towards the petal, creating the impression of common primordia (Iwamoto et al., 2018). The initiation of the androecium in Alismataceae is interpreted as a retained plesiomorphy, with consecutive reductions, or increases in complexity (cf. Ronse De Craene and Smets, 1995a). *Limnocharis* shows an exceptional secondary stamen increase superimposed on a trimerous Bauplan with additional stamens arising centrifugally (Singh and Sattler, 1977). In *Sagittaria*, the number of stamen and pistil whorls, as well as the presence of staminodes and pistillodes in staminate and pistillate flowers, is much variable. Outer stamens opposite the sepals can be paired (Figure 6.4C) or single (e.g. *S. cuneata*: Singh and Sattler, 1977a), even fluctuating among different sepals (*S. montevidensis*: pers. obs.). Pistillode numbers can be much higher, correlated with their smaller size (e.g. *S. cuneata*). The pattern of development in *Sagittaria* is reminiscent of Nymphaeales with alternations of single stamens and paired stamens but also of certain Annonaceae with a secondary stamen and carpel increase. The apocarpous gynoecia bear a single (rarely two) basal ovules.

Echinodorus has perigonal nectaries. In *Sagittaria*, nectariferous cells are centred in cavities around the stamen bases and carpellodes (Smets et al., 2000).

Butomaceae

Figure 6.5 *Butomus umbellatus* L.
✶ P3+3 A6+3 $\underline{G}$3+3

Butomus is characterized by the presence of two symmetrically inserted bracteoles, which is rare in monocots (Remizowa et al., 2012). Payer (1857) described the inflorescence as a scorpioid cyme disguised as an umbel.

Flowers have a differentiated perianth comparable to Alismataceae, although the inner whorl is not retarded and persistent in *Butomus*. Stamens are in two whorls, with the outer stamens paired. Eichler (1875) refers to 'congenital dédoublement', although development gives no indication that this happens (see Iwamoto et al., 2018). Carpels are in two trimerous whorls that are basally connected and surround a rounded floral apex. A high number of anatropous ovules covers the flanks of each carpel (Igersheim, Buzgo and Endress, 2001). Early floral development strongly resembles that of *Damasonium* (Alismataceae), except for the smaller petals and absence of the inner stamen whorl in the latter (Charlton, 2004).

Compared to other Alismatales, *Butomus* presents a number of characteristics that appear plesiomorphic: a basically undifferentiated perianth, two carpel whorls, laminar placentation and paired outer stamens. The two bracteoles

Figure 6.5 *Butomus umbellatus* (Butomaceae)

could be remnant of the calyculus found in *Tofieldia* (see Remizowa and Sokoloff, 2003), in which the third abaxial member has been lost.

Hydrocharitaceae

Figure 6.6A–B *Hydrocharis morsus-ranae* L., based on Scribailo and Posluszny (1985) and Eichler (1875)

Staminate: ∗ K3 C3 A3+3+3+3°(+3° +3°)[a] G0

Pistillate: ∗ K3 C3 A°3[b] Ğ6

General formula: male ∗ K(0)2–3 C0–3 A1–18 G0; female ∗ K(0)2–3 C0–3 A1°–18° G(2–)3–6(–20)

Contrary to other Alismatales, flowers are epigynous and mostly unisexual. Inflorescences are complex monochasia with well-developed bracts enclosing several flowers. Flowers are variable, being either well developed (subf. Hydrocharitoideae), or strongly reduced as an adaptation to water pollination (e.g. Cox, 1998; Leins and Erbar, 2010). The perianth is either differentiated, rarely reduced to three or two tepals, or absent. Staminate flowers lack all traces of a pistillode, while pistillate flowers often possess staminodes. Pistillate flowers have carpels with carinal styles (Kaul, 1968; Cox, 1998).

[a] Developing as a nectary. [b] Occasionally as a pair.

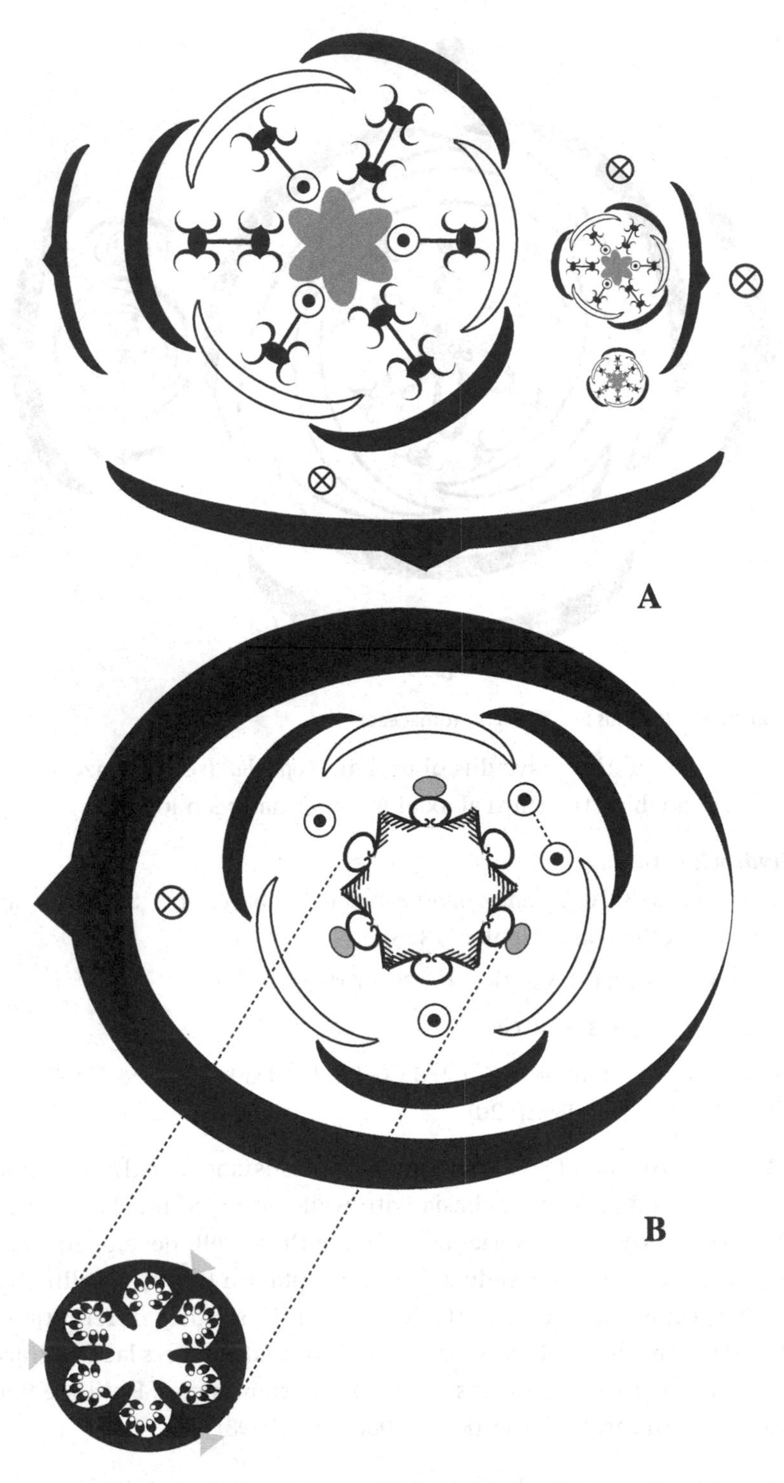

Figure 6.6 *Hydrocharis morsus-ranae* (Hydrocharitaceae): A. partial staminate inflorescence; B. pistillate flower

The androecium often contains paired stamens (as an outer pair opposite sepals: *Stratiotes*, *Ottelia*: Eichler, 1875; Kaul, 1968) and inner stamens are often staminodial. Paired staminodes are often found in pistillate flowers while staminodes may be absent in staminate flowers (e.g. *Limnobium*). In other species single stamens occupy the position of paired stamens. Iwamoto et al. (2018) argued that larger stamen primordia occupy the position of stamen pairs during development, due to space constrictions. Staminate flowers of *Hydrocharis* have a central nectary (Figure 6.6A), while pistillate flowers have nectaries as stylar appendages (Figure 6.6B). Although the nectary of staminate flowers was interpreted as a pistillode, Scribailo and Posluszny (1985) argued that it is best interpreted as fused staminodes, based on developmental evidence, and the same applies to the nectaries of pistillate flowers. In *Stratiotes* a whorl of small appendages surrounds fertile stamens in staminate flowers, and carpels in pistillate flowers. These were interpreted as staminodes (Kaul, 1968; Cox, 1998), although Eichler (1875) argued against this in favour of receptacular emergences. In small staminate flowers the number of stamens is often reduced to three in antesepalous position (e.g. *Vallisneria*), rarely one (*Maidenia*). In cases where there are only three perianth parts, stamens alternate with these (e.g. *Halophila*). In small pistillate flowers there are usually two or a single whorl of three carpels, mostly with staminodes.

The inferior ovary bears a protruding parietal placentation, with lobes corresponding to the number of carpels. Carpels can be multi-whorled (often as a multiple of three), and they are occasionally apocarpous (e.g. *Stratiotes*) despite being inferior, or carpels are united by their flanks to the receptacle. Kaul (1968) discussed the evolution of the gynoecium in the family as a progressive fusion of lateral carpel margins and their progressive reduction. Ovules range from one to three to numerous (circa fifty in *Hydrocharis*), and are anatropous to orthotropous (Igersheim, Buzgo and Endress, 2008). Compared to the related family Butomaceae with laminar placentation, the parietal placentation of Hydrocharitaceae can be seen as the result of a contraction of the placental areas to the carpel abaxial wall.

Juncaginaceae (incl. Lilaeaceae)

Figure 6.7 *Triglochin maritima* L., based on Dahlgren, Clifford and Yeo (1985) and Buzgo et al. (2006)

∗ P3+3 A3+3 G̲3+3

General formula: ∗ P(0)1–6 A1–6 G1–6

Inflorescences are dense spikes and flowers are ebracteate, although limits between bract and first tepal are unclear (Buzgo et al., 2006). Flowers have an extreme variation in perianth and stamen number, ranging from one to six (tri-, di-, or monomerous), and tepals are generally small and attached as an appendage to the extrorse stamens. Tepals are occasionally interpreted as an

Figure 6.7 *Triglochin maritima* (Juncaginaceae)

appendicular extension of the stamen, as the inner tepal whorl is distinctly inserted inside the outer stamen whorl (see Dahlgren, Clifford and Yeo, 1985; Endress, 1995a), but a true tepal nature appears to be more accurate (Sokoloff et al., 2013). Carpel numbers range from one to six (in two whorls). When in two whorls, the outer is sterile in *Triglochin* (Figure 6.7). Carpels are either free or basally fused to a variable extent via the floral apex without forming a style, with a single basal ovule. No nectaries are present, as flowers are wind pollinated. An extreme reduction occurs in *Lilaea* with unisexual flowers, a single stamen and carpel, and the perianth is reduced to a bract-like organ. Related families such as Scheuchzeriaceae and Aponogetonaceae have six stamens and three carpels, but the perianth is often incomplete, ranging from six to one tepals. Flower development of these three families is highly similar (Singh and Sattler, 1977b; Buzgo et al. 2006). In Potamogetonaceae flowers are dimerous, but they have the same arrangement of tepals attached to extrorse stamens. The tepaloid clade shares a wind pollination syndrome (extending to water pollination in derived families). This explains the reduction of flowers and the confusion of flowers with inflorescences.

6.2 The Lilioid Clades: Dioscoreales, Pandanales, Liliales and Asparagales

Floral synapomorphies are rare as most flowers conform to the generalized monocot floral diagram. In cases where monosymmetry is present, this is rarely structural (except in Orchidaceae).

Dioscoreales

The order contains a heterogeneous assemblage of three families, including the mycoheterotrophic Thismiaceae and Burmanniaceae, which have always been a problem to place due to lack of chloroplast DNA (Merckx et al., 2006) but are putatively placed in Dioscoreaceae (Soltis et al., 2018).

The perianth is undifferentiated in Nartheciaceae and Dioscoreaceae, but well differentiated into sepals and petals in the other families. Within Nartheciaceae the orientation of flowers is either with the outer median tepal abaxial (*Narthecium*), or adaxial (other genera: Tobe et al., 2018). The androecium is reduced to three inner stamens in Burmanniaceae and some Thismiaceae, or to three outer in some *Dioscorea* (Caddick, Rudall and Wilkin, 2000). Septal nectaries are generally present and associated with shifts in the ovary position in Nartheciaceae, but not in *Tacca* of Dioscoreaceae, which lacks nectaries (Tobe et al., 2018). Dioscoreales share a number of similar characteristics (Caddick, Rudall and Wilkin, 2000): reflexed stamens with a prolonged connective (in genera such as *Stenomeris* and *Trichopus* of Dioscoreaceae connectives are long and lie over the ovary), an umbrella-like stigma and urceolate floral chambers formed by the growth of a hypanthium. However, these characteristics are not found in Nartheciaceae (Remizowa, Sokoloff and Kondo, 2008; Tobe et al., 2018). The orientation of flowers in *Tacca* is inversed compared to most monocots; this may be linked with the absence of bracts.

Dioscoreaceae (incl. Taccaceae, Trichopodaceae, Thismiaceae)
Figure 6.8A–B *Tacca palmatifida* Baker
✶ K3 C3 A3+3 Ğ(3)

The inflorescence of *Tacca* consists of paired distichous cincinni surrounded by two pairs of bracts covering the hanging flowers. As most Dioscoreales, *Tacca* has the monocot floral diagram. Flowers and inflorescences are well adapted to a specific pollination syndrome of gnaths. Sepals and petals are differentiated; sepals are broad and reflexed; petals are much smaller and erect. Stamens are characteristically coiled with broad connective, providing shelter and enclosing the anthers. Young flowers of *Tacca* show the thecae strongly curved over the filaments as in other Dioscoreaceae and resembling *Saruma* (Aristolochiaceae). This similarity is later lost by curvature of the connective into a hood. The inferior ovary has three intrusive parietal placentae. Trichomes are situated in arcs on a hypanthium, but are reported not to represent perigonal nectaries (Dahlgren, Clifford and Yeo, 1985), as septal nectaries appear to be absent (cf. Caddick, Rudall and Wilkin, 2000 for other species), although their presence is reported in some species.

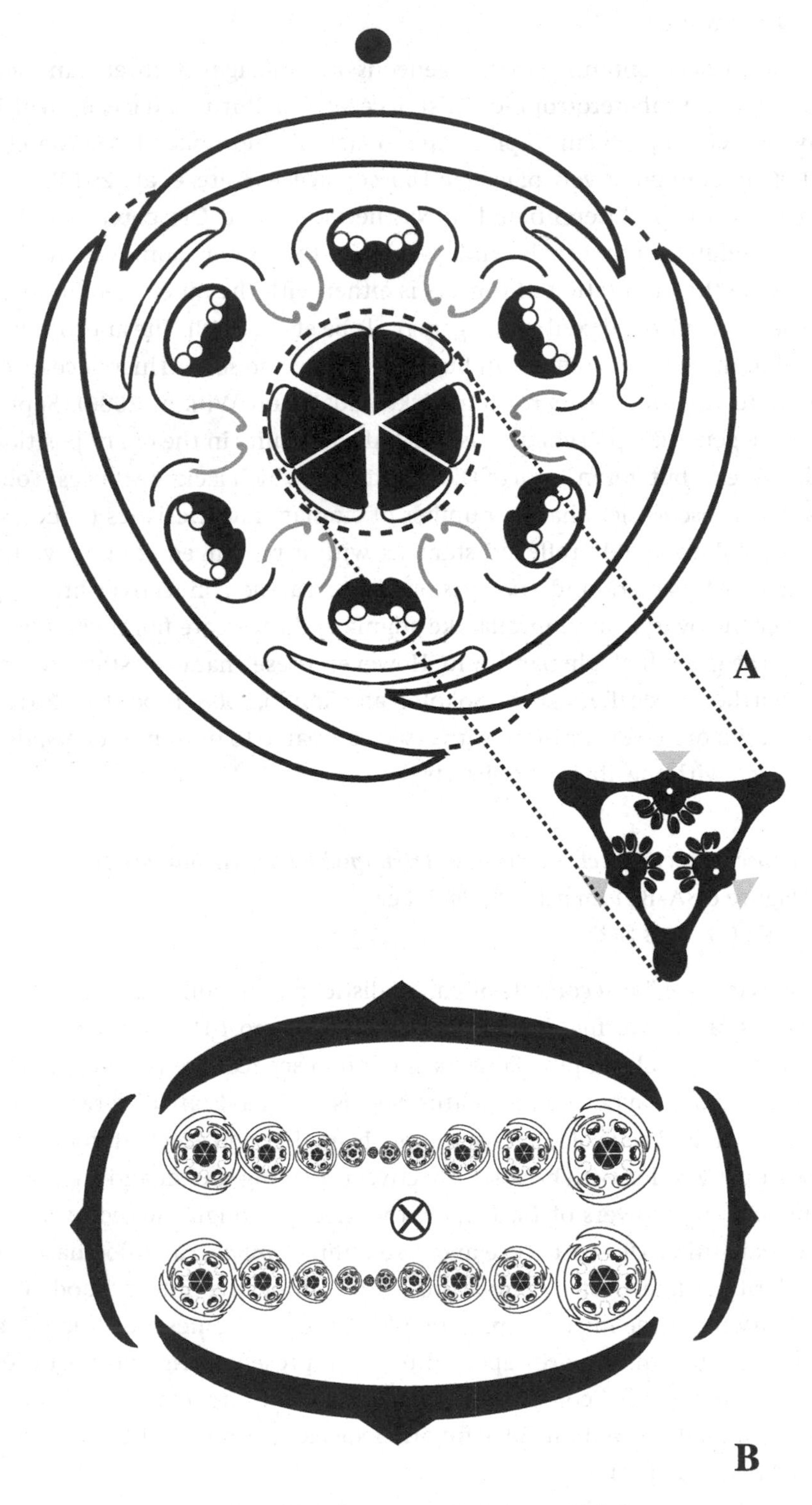

Figure 6.8 *Tacca palmatifida* (Dioscoreaceae): A. flower; B. inflorescence. Note the secretory (?) folds on the receptacle.

Other Dioscoreaceae have much more simple unisexual flowers with inferior ovary and two superposed ovules per carpel, but fruits are generally three-angled capsules. Septal nectaries are generally present (Igersheim, Buzgo and Endress, 2008).

Pandanales

The order consists of five families (Triuridaceae, Stemonaceae, Cyclanthaceae, Pandanaceae and Velloziaceae) with highly diverging habits and floral morphologies as adaptations to different ecological niches. Triuridaceae are mycoheterotrophic herbs with small, specialized flowers (Maas and Rübsamen, 1986). Cyclanthaceae and Pandanaceae are small palm-like trees producing pseudanthial inflorescences with reduced flowers (Rudall, 2003). Staminate and pistillate flowers of Pandanaceae and Cyclanthaceae are reduced with vestigial perianth and variable numbers of stamens or carpels. Stemonaceae are herbs with variable merism either two- or five-merous (*Pentastemona*). Ronse De Craene and Smets (1995a) hypothesized that *Pentastemona* with a stable pentamery represents a transitional condition between trimery and dimery by loss of an outer tepal and stamen. Velloziaceae appears to be the only family with 'classical' monocot characteristics.

Triuridaceae (incl. Lacandoniaceae)

Figure 6.9A–B *Lacandonia schismatica* E. Martínez and Ramos. A. flower; B. inflorescence, based on Ambrose et al. (2006)
∗ P(3+3) $\underline{G}$40-80 A3 (2–4)
General formula: ∗ P3–10 A(2)3–6 G6–80

Triuridaceae are mycoheterotrophic herbs characterized by small, mostly unisexual flowers, valvate tepals with pointed tips and many carpels. Flowers are arranged in terminal racemes. The perianth consists of equal tepals connate at the base in one series; they are generally reflexed resembling a star (Dahlgren, Clifford and Yeo, 1985). Appendages are frequently formed on the perianth and may function as osmophores (Rudall, 2003). In *Sciaphila* and *Triuris* staminate flowers have six tepals and three to six extrorse stamens on short filaments or fused into a column. *S. rubra* is dimerous with hermaphrodite flowers and two stamens (Rudall, 2003). Stamens can be split in half-anthers, embedded in a staminal column (*Triuris hexophtalma*). In *Seychellaria*, three staminodes alternate with the stamens. Pistillate units have the same arrangement of tepals with a generally high number of carpels. Each carpel contains a single basal ovule and gynobasic stylode.

Lacandonia is unique among angiosperms because of its inward-out flowers with three distal stamens and several proximal carpels. Flowers are cleistogamous and

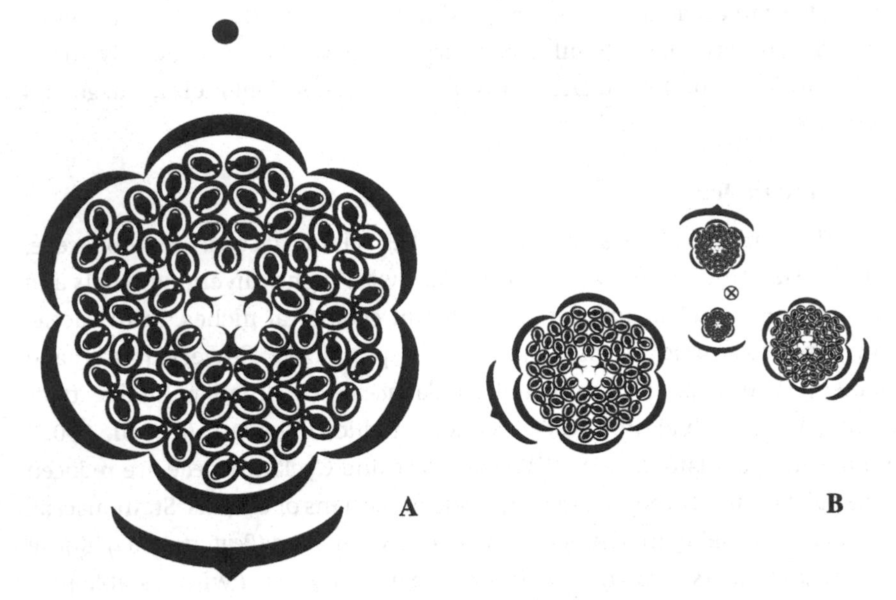

Figure 6.9 *Lacandonia schismatica* (Triuridaceae): A. flower, based on Vergara-Silva et al. (2003) and Ambrose et al. (2006); B. inflorescence

are grouped in racemes; they are enclosed by six valvate tepals opening as a star. Because of its singular arrangement the species was placed in its own family, Lacandoniaceae. Since its discovery in the late 1980s there was much speculation about the origin and homology of these flowers, interpreted either as the result of a homeotic mutation (Ronse De Craene, 2003; Vergara-Silva et al., 2003; Rudall, Alves and Sajo, 2016) or a pseudanthium (Rudall, 2003). Developmental, as well as genetic evidence support the former hypothesis (Ambrose et al., 2006; Álvarez-Buylla et al., 2010). Álvarez-Buylla et al. (2010) demonstrated a mutational change of C-genes into B-activity on the distal part of the flower.

Lacandonia develops two sets of common primordia; the upper one initiates three apical stamens and two rows of carpels in centrifugal sequence, while the lower one develops only carpels (Figure 6.9A; Ambrose et al., 2006). A similar development of fascicles was shown for pistillate *Triuris brevistylis*, while the staminate flowers develop only three extrorse stamens on an androphore. Some flowers also produce central stamens (Vergara-Silva et al., 2003). This indicates that *Lacandonia* belongs to Triuridaceae and that the upper stamens may not be the result of homeotic swapping of carpels and stamens (suggested by Ronse De Craene, 2003), because organs arise centrifugally on common primordia. The centrifugal development of several carpels in Triuridaceae on ridges in radial rows is unique in monocots and is comparable to the increase of stamens on common primordia. In contrast to other Triuridaceae the upper primordia develop as the three stamens in *Lacandonia*. The ridges were interpreted as the result of folding of an extended fasciated floral apex, similar to some Araliaceae (see Sokoloff et al., 2007; Endress, 2014; Rudall, Alves and Sajo, 2016). Multiplication of uniovulate carpels to increase seed set is an alternative to increase of number of ovules within the ovary.

Velloziaceae

Figure 6.10 *Xerophyta splendens* (Rendle) N. L. Menezes
✶ [P3+3 A3+3] Ğ(3)
General formula: ✶P3+3 A6–(18) Ğ 3

Figure 6.10 *Xerophyta splendens* (Velloziaceae). Note the adaxial appendages on tepals and septal nectaries.

Flowers are showy and trimerous, inserted solitary or with few in terminal position. Tepals are undifferentiated and petaloid, and often bear ventral appendages (described as a corona) enclosing the short filaments and sometimes adherent to them. The appendages were variously interpreted as stipules of the stamens, staminodes or perianth appendages, but are probably hypanthial in nature (see also Sajo, de Mello-Silva and Rudall, 2010). In *Barbacenia* two basal appendages enclose the base of the anthers as a sheath before expansion of the filaments. Although the basic number of stamens is six in two whorls, the stamens are often laterally increased by division, forming triplets (affecting inner and/or outer whorls) up to eighteen stamens (*Vellozia*: de Menezes, 1980; Sajo, de Mello-Silva and Rudall, 2010). Triplets may be free or fused into fascicles; the outer stamen whorl may be simple and the inner developed as triplets (*V. jolyi*). There is variation between species with a hypanthium and fewer stamens (e.g. *Barbacenia*, *Xerophyta*) and those with higher stamen numbers and without clear hypanthium (*Vellozia*). There is also a tendency for two lateral pollen sacs to become confluent by reduction of the middle partitions, leading to two-locular anthers. The inferior ovary has three large septal nectaries ending in holes at the bottom of the flower.

Liliales

The order contains ten families. Liliales share a number of synapomorphies, such as absence of septal nectaries reflecting congenitally fused carpels (Remizowa et al., 2010), absence of a hypanthium (free tepals) and the generalization of perigonal nectaries (Rudall, 2002). Flowers are mostly polysymmetric, more rarely monosymmetric (Corsiaceae, *Alstroemeria* in Alstroemeriaceae). Tepals are often spotted and stamens are often extrorse. Ovaries are variously superior or inferior. Stigmatic lobes tend to be long and separate, in contrast to most Asparagales. The Liliales respect the common floral diagram of monocots without loss of stamen whorls (except *Scoliopus*), although merism can be variable (Melanthiaceae).

Liliaceae

Figure 6.11A–B *Tricyrtis puberula* Nakai and Kitag
✶ K3 C3 A3+3 $\underline{G}$(3)
General formula: ✶ K3 C3 A3+(0–)3 $\underline{G}$(1–)3

Remizowa et al. (2012) describe the presence of two bracteoles in *Tricyrtis*, but bracts and bracteoles were reduced in the species investigated here. The perianth in *Tricyrtis* is differentiated in an outer whorl with basal pouches secreting nectar and inner tepals that are erect and have a dorsal crest. Both whorls are covered with spots, which is characteristic for the family. In *Scoliopus* the petals are small and linear and distinctive of the broader sepals, while the petals are broader in *Calochortus*. Most other Liliaceae lack the differentiation of outer and inner perianth parts. The androecium is two-whorled, except in *Scoliopus* where

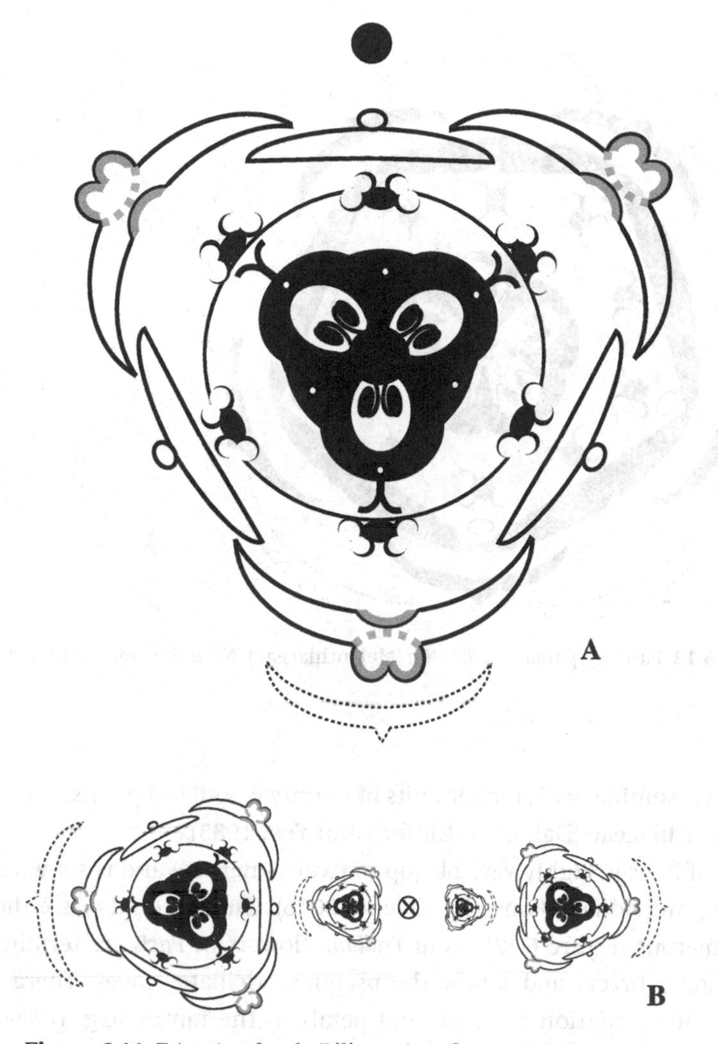

Figure 6.11 *Tricyrtis puberula* (Liliaceae): A. flower; B. inflorescence. Note the position of branched styles and nectariferous pouches on outer tepals.

the androecium is reduced to three stamens opposite the sepals. Gynoecia are superior with generally numerous ovules stacked in two rows on axile placentation. The family has been variously circumscribed in the past (e.g. Dahlgren, Clifford and Yeo, 1985; Stevens, 2001 onwards).

Melanthiaceae (incl. Trilliaceae)

Figure 6.12 *Paris polyphylla* Sm. var. *thibetica* H. Hara

∗ K 4–5 C4–5 A4–5+4–5 $\underline{G}$(4–5)

General formula: ∗ K 3(9–12) C3(–9) A6–18 $\underline{G}$3–9

Figure 6.12 *Paris polyphylla* var. *thibetica* (Melanthiaceae). Note the small rounded petals.

The family is an assemblage of smaller units of mainly woodland plants. *Paris* is often placed in Trilliaceae (Dahlgren, Clifford and Yeo, 1985).

The merism of *Paris* is highly variable (up to twelve-merous) and is shown to fluctuate among individuals (Ronse De Craene, 2016). Our material was either tetra- or pentamerous (Figure 6.12). As in *Trillium*, flowers of *Paris* are terminal without subtending bracts and follow the pseudoverticillate leaves. There is a tendency for differentiation of sepals and petals in the family (e.g. *Trillium*) and the sepals are often persistent. Petals are occasionally threadlike (Figure 6.12) or sometimes absent through suppression during development (*Paris tetraphylla*; also in *Trillium apetalon*: Narita and Takahashi, 2008). Takahashi (1994) argued that the loss of petals in *T. apetalon* is caused by homeosis, with replacement of petals by stamens. Takahashi (1994) and Narita and Takahashi (2008) interpreted the petals of *Paris* and *Trillium* as 'andropetals' derived from stamens. This would imply that the androecium was originally three-whorled, although no evidence exists for this in relatives of Liliales. An alternative explanation is that vacant space has become occupied by stamens with a subsequent shift in position (Ronse De Craene, 2003). The distinction between sepals and petals in *Paris* is derived from an original undifferentiated perianth and is a consequence of the retardation in development of the petals (stamen-petal primordia are currently found in

Paris). Stamens are in two whorls and in *Paris* the connective is apically extended. The gynoecium is superior with either parietal or axile placentation and free stylodes. In *Trillium* and *Veratrum* nectar is secreted from the tepal bases. Dahlgren, Clifford and Yeo (1985) mentioned that septal nectaries are reported in *Trillium*, although this is unlikely. No nectaries were seen in *Paris*.

Smilacaceae

Figure 6.13A–B *Smilax aspera* L.

Staminate: ✶ P3+3 A3+3 G0

Pistillate: ✶ P3+3 A3°+0 G(3)

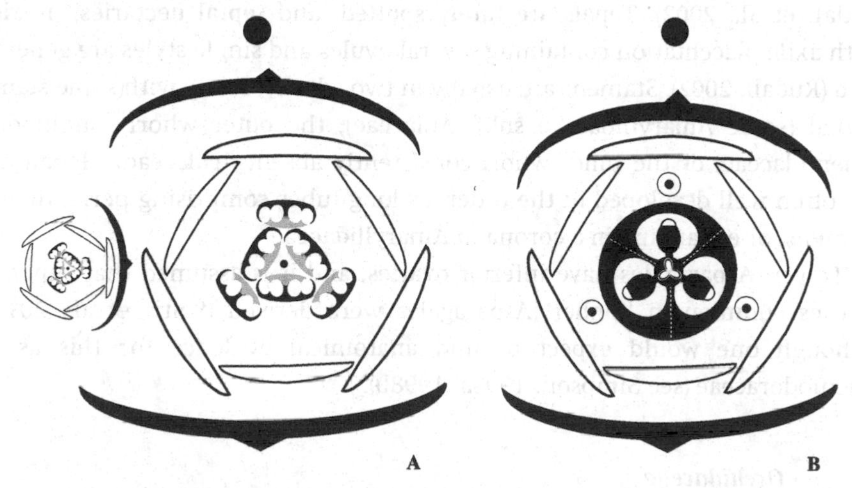

Figure 6.13 *Smilax aspera* (Smilacaceae): A. staminate partial inflorescence; B. pistillate flower

Smilax is dioecious with flowers grouped in clusters along the stem. Staminate flowers tend to be grouped in pairs in the axil of a bract, with one adaxial bracteole each. Pistillate flowers tend to be solitary in the axil of the bract and with fewer flowers per cluster. In staminate flowers the basal section of the filament bears nectariferous trichomes. In pistillate flowers filiform staminodes occur opposite the outer tepals. The three-carpellate ovary bears one apical ovule per locule (rarely two). There is no trace of nectaries. The style is inexistent.

Tepals are sometimes fused into a short to long tube. Stamens are in two whorls of three, rarely a single whorl, or three whorls of three (up to eighteen) (Dahlgren, Clifford and Yeo, 1985). Smilacaceae resemble Dioscoreaceae in habit and flower structure, except that the ovary is inferior in the latter.

Dahlgren, Clifford and Yeo, (1985) reported septal nectaries in the family, but this is erroneous.

Asparagales

The order comprises approximately fourteen families. Family delimitations are difficult as there are no obvious morphological characters, except for seed characters. In all Asparagales the median inner tepal is adaxial and carpels are opposite the outer tepals (Kocyan and Endress, 2001a, 2001b). Floral diversity is high, with a 'traditional' lily-like appearance, or with elaborate derivations. Ovaries are inferior to superior with several independent shifts and reversals. Flowers are either polysymmetric to weakly or highly monosymmetric (Orchidaceae: Rudall and Bateman, 2004; *Gilliesia* in Amaryllidaceae: Rudall et al., 2002). Tepals are rarely spotted, and septal nectaries, ovaries with axile placentation containing several ovules and single styles are generalized (Rudall, 2002). Stamens are usually in two whorls, rarely with some staminodial (some Amaryllidaceae subf. Alliaceae), the outer whorl staminodial (Themidaceae) or the inner whorl consistently absent (Iridaceae). Hypanthia are often well developed in the order, as long tubes comprising perianth and stamens, or expanding in a corona in Amaryllidaceae.

'Lower' Asparagales have inferior ovaries, and it is assumed that superior ovaries common in 'higher' Asparagales were derived (Soltis et al., 2018), although one would expect to find anatomical evidence for this as in Haemodoraceae (see Simpson, 1998a, 1998b).

Orchidaceae

Figure 6.14 *Oncidium altissimum* (Jacq.) Sw.
↑ P3+2:1 [A1+2° Ğ(3)]

Orchidaceae represents the most successful family of angiosperms with species estimates ranging between twenty-five and thirty-five thousand. Despite this the general floral Bauplan is remarkably conservative. The Orchidaceae are an early diverging clade of Asparagales with strictly monosymmetric flowers, an inferior ovary, lack of septal nectaries, a strong differentiation of tepals with development of an adaxial labellum and various reductions of the androecium to three, two or a single stamen, and one-two staminodes, all connected with the style in a monosymmetric gynostemium (Endress, 1994, 2016; Kocyan and Endress, 2001a, 2001b; Rudall and Bateman, 2004; Pabón-Mora and González, 2008). Basal Orchidaceae (Apostasioideae) have three adaxial fertile stamens (e.g. *Nieuwiedia*), occasionally with the middle one staminodial or missing (*Apostasia*) and flowers are weakly monosymmetric (Kocyan and Endress, 2001b). In other Orchidaceae the

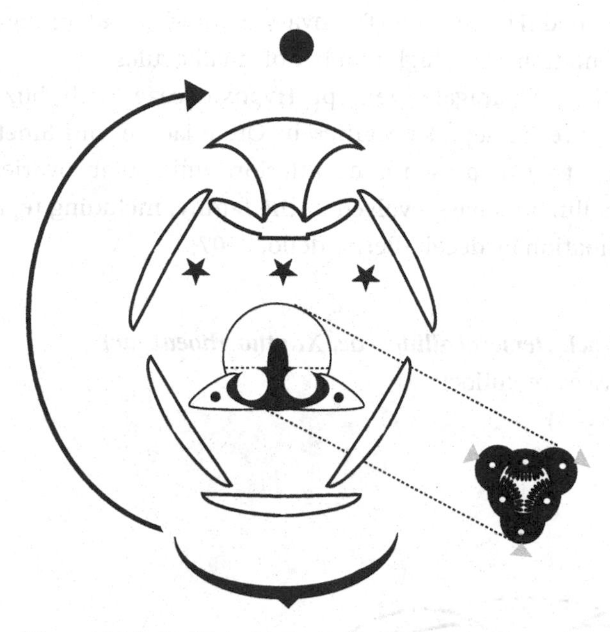

Figure 6.14 *Oncidium altissimum* (Orchidaceae). Note the appendage on the labellum. The fertile stamen and stigma are covered by a hood (broken line). The asterisks refer to lost adaxial stamens belonging to two whorls.

lateral stamens are staminodial (Figure 6.14; subf. Orchidoideae, Epidendroideae), or the median stamen is staminodial (subf. Cypripedioideae) (e.g. Graf, 1975; Dahlgren, Clifford and Yeo, 1985; Endress, 1994; Rudall and Batemann, 2004; Stützel, 2006). Basal Apostasioideae have a more or less axile placentation by postgenital fusion of carpels, and primordia of abaxial stamens are initiated but become soon repressed (Kocyan and Endress, 2001b). These characteristics are lost in other Orchidaceae. Flowers are generally grouped in racemes and subtended by a single bract. Flowers are resupinate at maturity, placing the labellum in abaxial position and stamens and stigma in an adaxial position.

The labellum (adaxial tepal of the inner whorl) is mostly elaborate and different from the other tepals, which may be identical, or differentiated in an outer and inner whorl. In *Oncidium* the labellum bears a conspicuous ridged protuberance (callus). In some Orchidaceae, the labellum may revert to a structure identical to the other inner tepals (e.g. *Gymnadenia*, *Telipogon*: Pabón-Mora and González, 2008). Except for Apostasioideae, the androecium is intimately fused with the style in a gynostemium, with the anther overtopping the stigma. Three commissural stigmatic lobes can be initiated and the adaxial lobe may develop as a protuberance (rostellum). Anthers have their pollen sacs confluent into two (basally) connected pollinia (Dahlgren, Clifford and Yeo, 1985). Pollinia are either exposed or covered by a hood (Figure 6.14). The ovary is inferior and strongly ribbed with parietal placentation and a high number of small ovules.

Contrary to most other Asparagales (except Hypoxidaceae with buzz-pollinated flowers) there are no septal nectaries in Orchidaceae and Smets et al. (2000) linked this to the presence of inferior, unilocular ovaries. Adaptations to various pollinators have evolved in the family, including tepal nectaries, spurs and pollination by deceit (Bernardello, 2007).

Asphodelaceae (incl. Hemerocallidaceae, Xanthorrhoeaceae)
Figure 6.15 *Aloe elgonica* Bullock
(↓)/∗ P(3)+3 A3+3 $\underline{G}$(3)

Figure 6.15 *Aloe elgonica* (Asphodellaceae)

Flowers are grouped in terminal racemes or spikes and are subtended by a single bract. Several Asphodelaceae have monosymmetric flowers enhanced by curvature of the pedicel and fertile organs, and compression of the lateral tepals. In *Aloe* outer tepals are fused up to the middle and variously connate to the free inner tepals, while certain genera have free tepals (*Asphodelus*). Fusion is linked with the development of a hypanthial tube which can be extensive (e.g. *Kniphofia*). In the species studied stamens develop sequentially and are curved to the abaxial side; the inner stamens are longer and mature first. Filaments are glabrous or often hairy (e.g. *Bulbine*). In some Australian genera inner stamens are staminodial (e.g. *Hodgsoniola*) or lost (e.g. *Stawellia*). The trimerous ovary has a simple style and abundant nectar is produced by septal nectaries. Morphological characters distinguishing the family from other Asparagales are unclear (Dahlgren, Clifford and Yeo, 1985).

Iridaceae

Figure 6.16A *Gladiolus communis* L. ssp. *byzantinus* (Miller) A. P. Ham.
↓ [P3+3 A3+0] Ğ(3)

Figure 6.16B–C *Sisyrinchium striatum* Sm.
∗ P3+3 A(3+0) Ğ(3)

Flowers are arranged in distichous spikes with flowers subtended by two unequal sheathlike bracts in a median position (Remizowa et al., 2012). All Iridaceae share the reduction of the inner stamen whorl, but have a variable ovary position. Stamens and carpels are opposite the outer perianth (carpels rarely alternating?). The flower of *Iris* is regular but functions as three floral units through the strong development of the three styles dividing the flower in three separate sections (see Stützel, 2006). The lower part of the flower is tubular (stamen-tepal tube) and stamens are distinct or monadelphous. Nectaries are either septal or are perigonal with nectar produced at the base of the tepals (subf. Iridoideae) (Smets et al., 2000). The ovary is mostly inferior and has three double rows of ovules. The fruit is mostly a loculicidal capsule.

Zygomorphic flowers of *Gladiolus* have a two-lipped perianth, with the outer lip often with a different pattern. The three stamens and stigma reach out at the adaxial side of the flower. Eichler (1875) wrongly depicted the zygomorphy as transversal in the flower. The basic construction of the flower is median monosymmetric and is a late developmental event. In *Diplarrhena* stamens are reduced to two by the loss of the abaxial stamen.

Amaryllidaceae (incl. Alliaceae and Agapanthaceae)

Figure 6.17 *Tulbaghia fragrans* Verdoorn
∗ [P3+3 A3+3] $\underline{G}$(3)

Flowers of Amaryllidaceae are in umbellate inflorescences enclosed by papery bracts. A variable number of smaller bracts can be found basal to individual flowers. In some genera (e.g. *Galanthus*, *Narcissus*) the number of flowers is

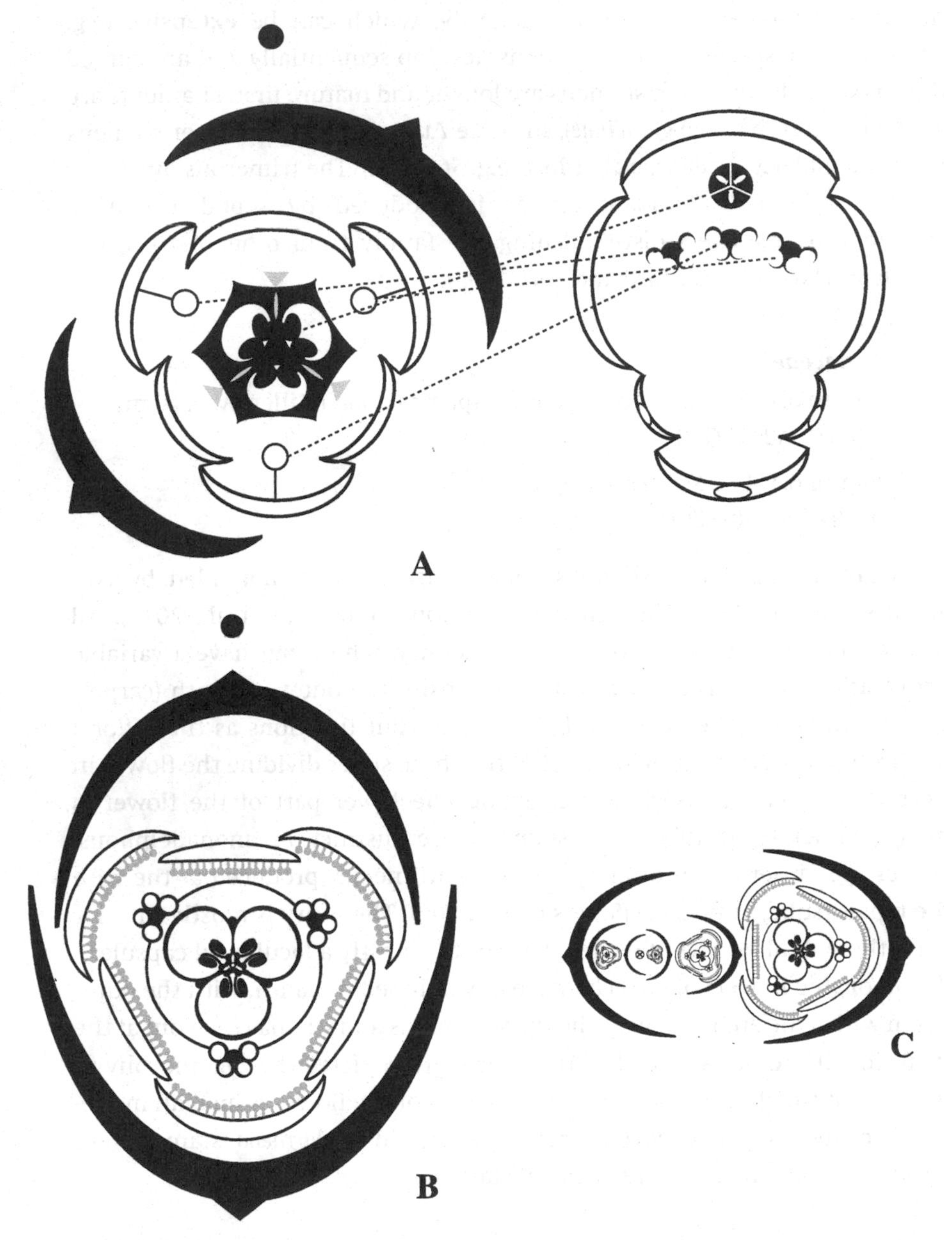

Figure 6.16 Iridaceae: A. *Gladiolus communis* var. *byzantinus*; left, lower section of flower; right, upper section; B. *Sisyrinchium striatum*, flower; C. inflorescence. In B, note the trichome nectaries on the tepals.

reduced to a single one. Flowers are generally trimerous with the typical monocot formula and diagram. In *Gethyllis* stamen number ranges from six up to eighteen (Dahlgren, Clifford and Yeo, 1985). Tepals are free or connected with

Figure 6.17 *Tulbaghia fragrans* (Amaryllidaceae). Note the corona with appendages surrounding the stamens.

the stamens in a hypanthial tube. The hypanthial tube can be extensive (e.g. *Crinum*), often forming appendages or a corona associated with the tepal lobes and/or filament bases (*Hymenocallis*, Figure 6.17). The nature of the corona has been hotly debated but appears to be hypanthial, comparable to any tubes in flowers. Waters et al. (2013) interpreted the corona of *Narcissus* as partly staminodial because of a similar gene expression, but this is unlikely. Contrary to most monocots, styles are solid in the family. The ovary is superior to inferior and ovules are in two rows. In *Allium* the number is reduced to two collateral ovules and the style is gynobasic. Septal nectaries are mostly present and well developed. Flowers are polysymmetric or weakly monosymmetric, except in *Gilliesia* and *Gethyum* (Rudall et al., 2002) with a strong structural monosymmetry (reduction of the three adaxial stamens and inner adaxial tepal).

6.3 The Commelinid Clade: Arecales, Commelinales, Poales and Zingiberales

Arecales

The Arecales consist of a single family, Arecaceae, which contains five subfamilies (Stevens, 2001 onwards). The order has a basal position in the commelinids.

Arecaceae

Figure 6.18A–C *Ptychosperma mooreanum* F. B. Essig, based on Uhl (1976a, 1976b)

Staminate: ✶ K3 C3 $A3^{6}+3^{3}$ $\underline{G}^{\circ}(3)$

Pistillate: ✶ K3 C3 A°3+3 $\underline{G}(1:2^{\circ})$

General formula: ✶ K3 C3 A(3-)6(-9-∞ G1–3(–4)

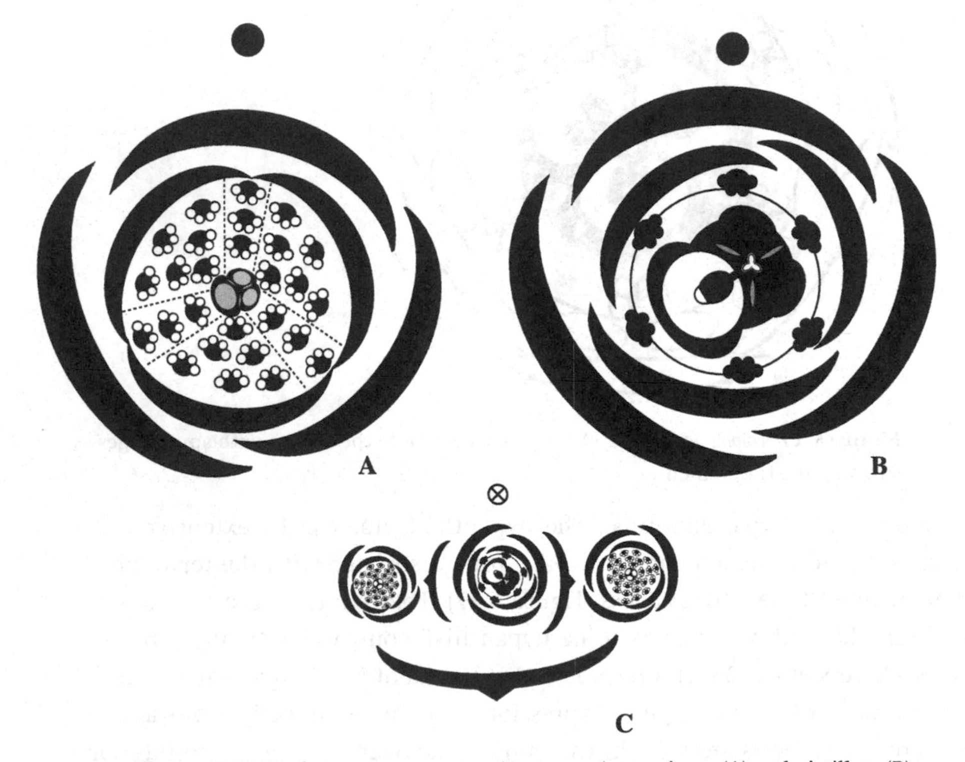

Figure 6.18 *Ptychosperma mooreanum* (Arecaceae): staminate (A) and pistillate (B) flower; C. partial inflorescence. In A the broken line delimits groups of stamens opposite a petal; note the septal nectary.

The basic floral pattern is trimerous with three imbricate sepals and petals, six stamens and three distinct uniovulate carpels (Moore and Uhl, 1982). Flowers are either bisexual or unisexual. Inflorescences range from solitary flowers to monopodial axes (Moore and Uhl, 1982). In the majority of palms, including *Ptychosperma*, flowers are arranged in triads of two staminate flowers and an upper pistillate flower (Figure 6.18C; Uhl, 1976a; Stauffer, Rutishauser and Endress, 2002). Flowers are mostly trimerous, but dimerous and tetramerous flowers occur in species of *Chelyocarpus* or *Palandra*. The perianth is imbricate but petals often are valvate in bisexual and staminate flowers. Extreme dimorphism occurs in Phytelephantoid palms (e.g. *Palandra*) with minute perianth segments in staminate flowers against very large petals in pistillate flowers (Uhl and Moore, 1977; Moore and Uhl, 1982).

The majority of taxa have six stamens in two whorls. However, about seventy genera have more than six stamens, with polyandry superimposed on a basic trimery (Moore, 1973; Uhl and Moore, 1980). *Nypa* has an undifferentiated reduced perianth; staminate flowers have only three outer (fused) stamens and no carpels; pistillate flowers lack staminodes.

Stamen development is either in centripetal (Caryotoid palms) or centrifugal direction (Phytelephantoid palms). In Caryotoid palms variation in stamen number depends on the size of the stamen primordia, the floral sector and the floral apex. The antepetalous sector is usually wider with space for more stamens, while the antesepalous sector generally has one row of stamens or more (Uhl and Moore, 1980). Up to two hundred stamens are formed in this way. In Phytelephantoid palms flowers are tetramerous with up to one thousand stamens in *Palandra* (Uhl and Moore, 1977).

The gynoecium is tricarpellate, apocarpous to syncarpous; in several genera the ovary is pseudomonomerous (Figure 6.18B). Carpels are increased to four up to ten, especially in the Phytelephantoid palms. Septal nectaries are variously developed, as palms can be occasionally wind pollinated, but are mainly insect pollinated.

In *Ptychosperma* (Figure 6.18A) staminate flowers have a row of antesepalous stamens and groups of antepetalous stamens, with a pistillode exuding nectar. The pistillate flower has two trimerous whorls of staminodes (occasionally more) that are often fused. The ovary is pseudomonomerous with a single pendent ovule. Septal nectaries are formed above the locule (Uhl, 1976b). Stauffer, Rutishauser and Endress (2002) and Stauffer and Endress (2003) described a comparable pistillate flower for *Geonoma* and other Geonomeae. In *Caryota* no trace of a pistillode is visible (apparently homeotically replaced by a stamen: Uhl and Moore, 1980).

Commelinales

The order as circumscribed by molecular data contains five families and is highly different from previous classifications (e.g. Dahlgren, Clifford and Yeo, 1985). Oblique or transversal monosymmetry is widespread in Commelinaceae, Pontederiaceae, Phylidraceae and Haemodoraceae (Stevens, 2001 onwards; Rudall and Bateman, 2004). Monosymmetry is often expressed in the androecium as there is a strong tendency for heteranthy or the androecium is variously reduced with or without staminodes, culminating in the single stamen of Phylidraceae and some Pontederiaceae. The order shows much variation in floral structure and adaptations to various pollination syndromes.

Haemodoraceae

Figure 6.19A *Anigozanthos flavidus* DC

↙ [P3+3 A3+3] G(3)

Fig. 6.19B–C *Xiphidium coeruleum* Aubl.

✶ P3+3 A0+3 G(3)

General formula: P3+3 A(1–)3–6 G(1–)3

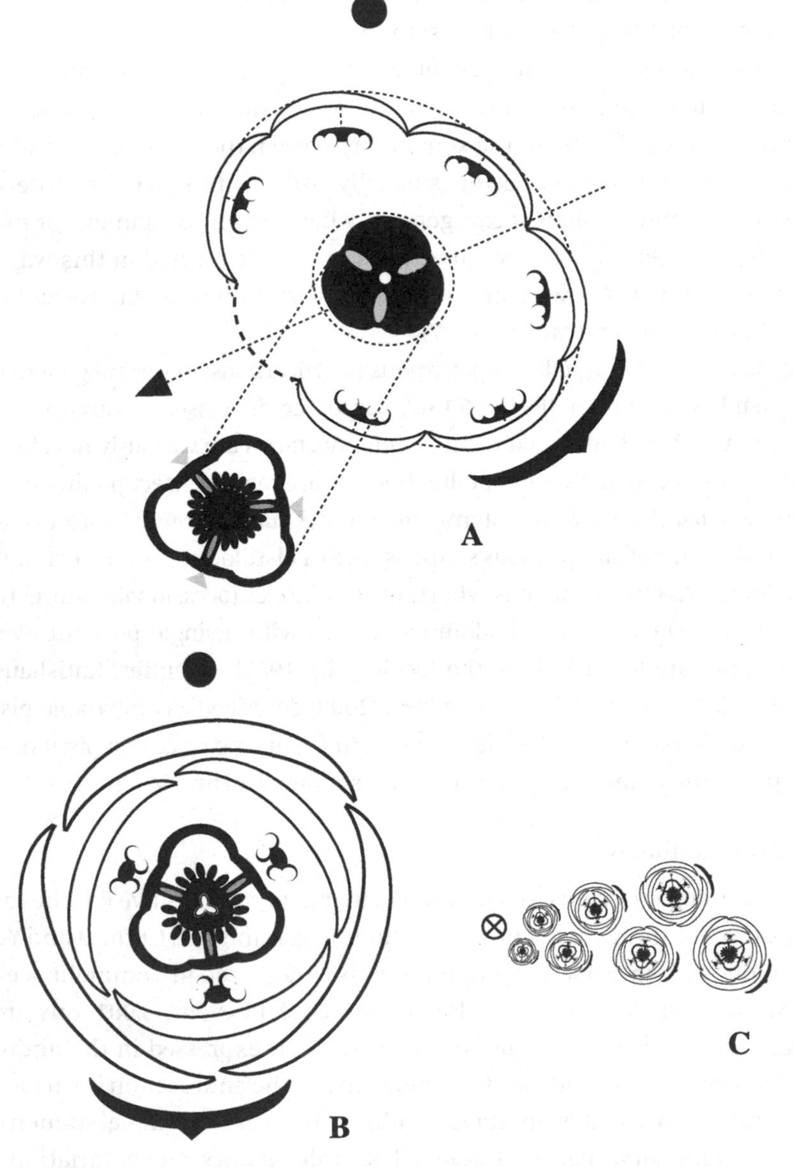

Figure 6.19 Haemodoraceae: A. *Anigozanthos flavidus*. *Xyphidium coeruleum*; B. flower; C. inflorescence. In A the thick broken line refers to the slit in the perianth tube.

The family is highly variable with inflorescences ranging from monochasial cymes (a bifurcate cincinnus in *Anigozanthos*) to racemes. Flowers are bisexual and range from nearly polysymmetric to strongly monosymmetric. The perianth is undifferentiated, imbricate in bud, but becomes valvate in *Anigozanthos* (Simpson, 1990). The outer median tepal is always in a posterior position and tepals are imbricately arranged (Simpson, 1990, 1998a, 1998b). In *Anigozanthos*, flowers have slightly oblique monosymmetry, which is linked with a slit formed on one side of the hypanthial tube connected to the tepals.

Several Haemodoraceae show a reduction to three stamens opposite the inner tepal whorl (e.g. *Xiphidium*, *Wachendorfia*, *Haemodora*: Simpson, 1998a, 1998b). In *Pyrrorhiza* the androecium is reduced to a single abaxial stamen with two staminodes, similar to *Hydrothryx* of Pontederiaceae. *Shiekia* has two supplementary staminodes opposite the abaxial outer tepals. In some genera with three stamens, including *Xiphidium*, the adaxial anther is larger (Simpson, 1990, 1998a). The gynoecium of three carpels has axile placentation or ovules are basal in *Phlebocarya*. The ovary is pseudomonomerous in *Barberetta* with abortion of the antero-lateral carpels (Simpson, 1998b).

Septal nectaries are usually present and sit on top of the ovary in epigynous flowers. *Xiphidium* is reported to lack septal nectaries, but I observed three slits just below the insertion of the style. The anterior septal nectary is reduced in *Wachendorfia* linked with a secondarily superior ovary in *Wachendorfia* (Simpson, 1998a). Several Haemodoraceae produce enantiostylous flowers with left and right flowers differing in the orientation of the largest of three stamens and style (Vogel, 1998b). The style of several Haemodoraceae, including *Xiphidium*, is also obliquely inserted.

Pontederiaceae

Figure 6.20 *Eichhornia crassipes* (Mart.) Solms: A. young flower bud; B. flower at anthesis

↓(⟷) P3+3 A3+3 $\underline{G}$(3)

General formula: P3+3 A(1–)3–6 G(1–)3

The family is closest sister to Haemodoracae and consists of freshwater, aquatic monocots. Flowers or *Eichhornia* are grouped in terminal racemes and lack bracts. All organs are basally connected by a hypanthium. Young floral buds of *Eichhornia* show a tendency for disymmetry, resulting in two trimerous tepal whorls that appear as three dimerous whorls (Figure 6.20A). Mature flowers are characterized by median monosymmetry, which is expressed early in the floral development (Strange, Rudall and Prychid, 2004). Stamens are arranged in three longer abaxial and three shorter adaxial stamens (also present in other Pontederiaceae), but

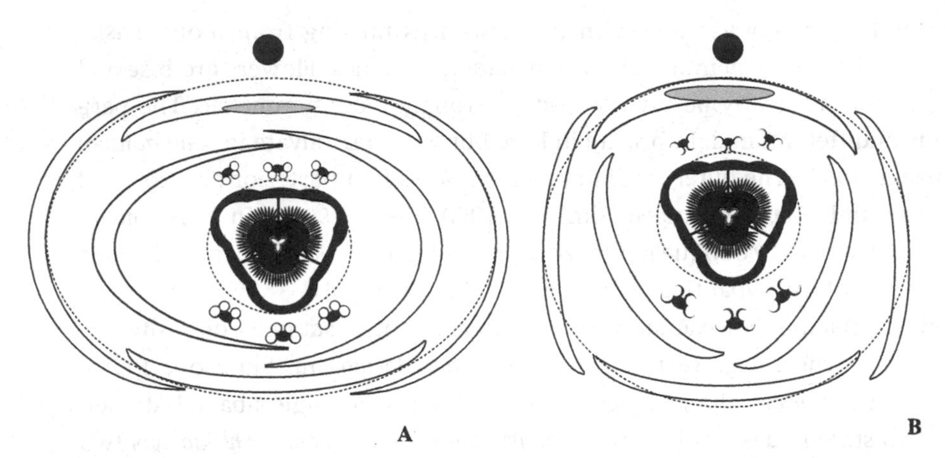

Figure 6.20 *Eichhornia crassipes* (Pontederiaceae): A. Young bud; the two trimerous tepal whorls are strongly disymmetrically flattened; B. Open flower. Note the larger anterior stamens.

reduced to the three abaxial stamens in *Heteranthera*, or to one stamen and two staminodes in *Hydrothryx* (Strange, Rudall and Prychid, 2004). The ovary is tricarpellate with axile or parietal placentation and a simple style with trilobate stigma. In *Pontederia* the ovary is pseudomonomerous by reduction of the two adaxial locules. Nectaries were not seen by me, although they are reported to be present in *E. crassipes*, but not in all species (Strange, Rudall and Prychid, 2004). Septal nectaries were lost in relation to heteranthy in the family.

Commelinaceae

Figure 6.21 *Callisia warscewicziana* (Kunth and Bouché) Hunt. A. flower; B. inflorescence

∗K3 C3 A3+3 $\underline{G}$(3)

General formula: ∗/↘ K3 C3 A(1−)2−6 G(3)

Flowers are readily recognizable by the formation of scorpioid cymes (cincinni), subtended by a laterally inserted bract, and differentiation of sepals and petals. The number of stamens is basically six in two equal whorls but there is much variation in some genera with variable numbers of staminodes. Either the outer (e.g. *Palisota*: A3°+3) or inner whorl (e.g. *Aploleia*, *Tripogandra*: A3+3°) is staminodial, or staminodes are spread over two whorls (A1:2°+2:1°) (Faden in Dahgren, Clifford and Yeo, 1985).

In *Plowmanianthus* and *Cochliostema* flowers are obliquely monosymmetric by suppression of three stamens (remaining staminodial in *Cochliostema* and *P. perforans*), enhanced by the larger size of the posterior sepal and occasionally

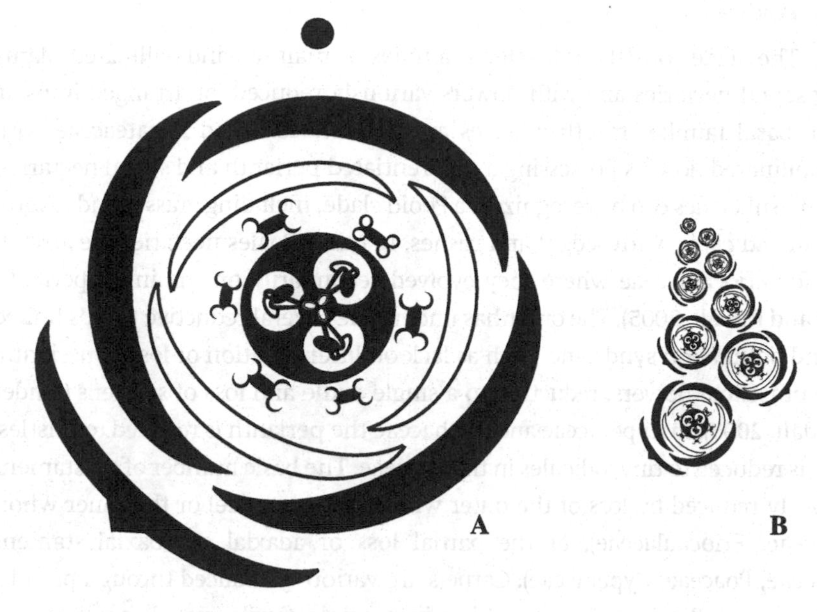

Figure 6.21 *Callisia warscewicziana* (Commelinaceae): A. flower; B. inflorescence. Anthers open sequentially.

a larger anterior petal and fusion of the fertile stamens (Hardy and Stevenson, 2000a; Hardy, Davis and Stevenson, 2004). As flowers are inserted sideways, orientation of flowers appears median, although it is oblique relative to the subtending bract. Other genera tend to be polysymmetric or slightly monosymmetric with differences in maturation of the stamens (e.g. *Callisia*: Figure 6.21A). Staminodes can have variable forms, being stamen-like without anther, or with anthers and functioning as fodder stamens (e.g. *Tripogandra*). Flowers offer pollen as reward and lack nectaries, although mechanisms have developed in the family to restrict access to pollen and its availability (Faden, 2000). Floral deception is common by the development of yellow moniliform filaments luring insects looking for pollen (*Cochliostema*). Pollen dimorphism is usually related to heteranthy between different anther sets, which can be subtle to strongly developed (Faden, 2000). The ovary is generally trilocular or bilocular with simple style and terminal stigma.

Hardy and Stevenson (2000b) analysed the floral development of another species of *Callisia*, *C. navicularis* and *Tradescantia*. They reported an unusual centrifugal initiation of the outer stamen whorl. This may be linked to the smaller petal primordia, providing more space for an earlier initiation of the antepetalous stamens, comparable to some obdiplostemonous eudicots. Clear floral diagrams were shown in Hardy and Stevenson (2000a) and Hardy, Davis and Stevenson (2004).

Poales

The clade consists of fourteen families of mainly wind-pollinated plants lacking septal nectaries and with flowers variously reduced, or arranged in pseudanthia. Basal families to other Poales are Bromeliaceae and Rapateaceae with insect-pollinated flowers possessing a differentiated perianth and septal nectaries. Two main subclades can be recognized: a Pooid clade, including grasses and restios, and a juncoid clade, with sedges and rushes. In other Poales nectaries are absent, except for Eriocaulaceae where they evolved secondarily on the inner perianth (Linder and Rudall, 2005). The order has undergone several reductive trends linked to a wind pollination syndrome, such as lack of differentiation or loss of perianth, shift to unisexual flowers, reduction to a single ovule and loss of stamens (Linder and Rudall, 2005). In Cyperaceae and Typhaceae the perianth is reduced to bristles, while it is reduced to tiny lodicules in the Poaceae. The basic number of six stamens is variously reduced by loss of the outer whorl (Restionaceae) or the inner whorl (Xyridaceae, Eriocaulaceae), or the partial loss of adaxial or abaxial stamens (Xyridaceae, Poaceae, Cyperaceae). Carpels are variously reduced through pseudomonomery (e.g. Restionaceae), resulting in a single fertile carpel (e.g. Poaceae, Cyperaceae). What is currently missing is a clear mapping of floral structural characters on a phylogeny to determine the reductive trends in different floral whorls.

Restionaceae (incl. Anarthriaceae, Centrolepidaceae)

Figure 6.22A–B *Hypodiscus aristatus* Nees

Staminate: ⟷ K3 C3 A0+3 G0

Pistillate: ⟷ K3 A0+3° $\underline{G}$(1:1°)

Figure 6.22C–D *Elegia cuspidata* Mast

Staminate: ⟷ K3 C3 A0+3 G0

Pistillate: ⟷ K3 A0+3° $\underline{G}$(1:2°)

Restionaceae are dioecious plants, often with a strong dimorphism between staminate and pistillate inflorescences. Flowers are arranged in spikelets and are subtended by a single scarious bract. Staminate and pistillate reproductive structures are variously enclosed and protected by tepals, floral bracts and inflorescence bracts (Linder, 1991). The perianth is either undifferentiated, or outer tepals differ strongly from inner tepals, with the outer lateral tepals strongly keeled (Linder, 1991, 1992a). The basic floral diagram of Restionaceae is trimerous, less often dimerous, and consists of two tepal whorls, a single whorl of stamens (inner) and three carpels. While this floral arrangement is found in some genera (e.g. *Dovea*, *Askidiosperma*), most genera have undergone various reductions of carpels (Linder, 1991, 1992a, 1992b; Ronse De Craene, Linder and Smets, 2001, 2002). The ovary is often

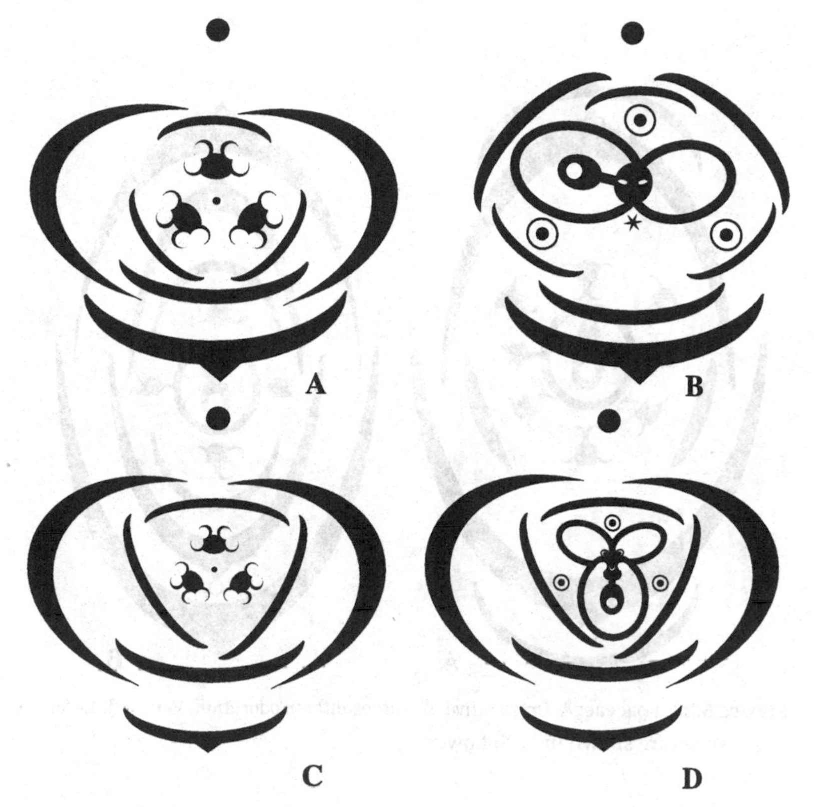

Figure 6.22 Restionaceae: staminate (A) and pistillate (B) flower of *Hypodiscus aristatus*; staminate (C) and pistillate (D) flower of *Elegia cuspidata*. The asterisk refers to the lost carpel.

pseudomonomerous, with reductions affecting any of the three carpels (Figure 6.22B, D; Ronse De Craene, Linder and Smets, 2002). This reduction is gradual, first affecting the fertility of locules, followed by loss of the sterile locules, which may still be represented by styles or vascular bundles (Linder, 1992a, 1992b). Staminodes may be variously developed or are absent in pistillate flowers, similar to pistillodes in staminate flowers. A single pendulous, orthotropous ovule is present per locule, with one to three carinal styles.

Poaceae

Figure 6.23A *Oryza sativa* L.

∗ P2 A3+3 G̲(1/2°)

Figure 6.23B *Anthoxanthum odoratum* L.

⟷P2 A2 G̲(1/2°)[c]

General formula: ∗/⟷/↓ P0–3 A1–3(6–∞) G(1/2°)

[c] Arber (1934: figure 71) does not mention lodicules (overlooked?).

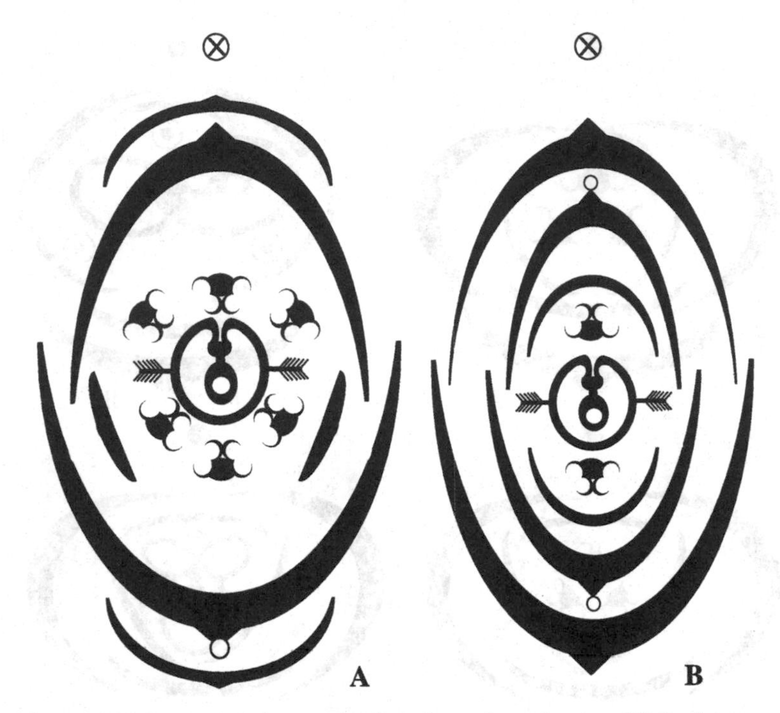

Figure 6.23 Poaceae: A. *Oryza sativa*; B. *Anthoxanthum odoratum*. White dots represent awns; styles are shown in both flowers.

Flowers of Poaceae are highly reduced with derived morphology and there has been considerable speculation about their floral structure (see Rudall et al., 2005; Sajo, Longhi-Wagner and Rudall, 2007, 2008). Grasses are adapted to wind pollination, although insects were observed gathering pollen (Dahlgren, Clifford and Yeo, 1985).

Flowers are grouped in spikelets, short shoots with transformed bracts, which make up the basic unit of the inflorescence. There is tremendous variation in the construction of spikelets. Two outer bracts (glumes) are usually present, enclosing a shoot with two rows of flowers in a distichous arrangement, or a single flower, and each individual flower (floret) is enclosed by two (lemma and palea) or more bracts. The palea is often a two-parted structure that is interpreted as the fusion product of two bracts (or tepals: see later in this chapter) and can be seen as developing from two primordia (e.g. Reinheimer, Pozner and Vegetti, 2005). The palea may be missing (e.g. *Alopecurus*) or is much reduced (e.g. *Zea*). Glumes and inner bracts may bear appendages such as hairs, bristles or awns. The flowers are mostly trimerous, rarely dimerous (e.g. *Anthoxanthum*: Figure 6.23B). A single whorl of two or three green organs called lodicules, rarely more or none, is interpreted as representing the inner tepal whorl (Dahlgren, Clifford and Yeo, 1985; Cocucci and Anton, 1988; Sajo, Longhi-Wagner and Rudall, 2008). Lodicules play a role in the opening of the flower at maturity as swelling organs. The majority of Poaceae have two lodicules and three alternating stamens, as the adaxial lodicule is absent, leading to a monosymmetric flower. Rudall and Bateman (2004) illustrate the variation in stamen position within Poaceae, but more floral morphological investigations are required to understand the evolution of stamen positions in the family. Bambusoideae represent the basal group with three to six stamens and three lodicules. *Oryza sativa* (rice) belongs to this subfamily. It has six stamens but the perianth is reduced (Figure 6.23A). The number of stamens is mostly reduced to three, two, or one, rarely increased up to 170 stamens (*Ochlandra*). *Anthoxanthum* with two opposite stamens was interpreted as the result of a retention of one stamen of the outer whorl with one of the inner whorl (Rudall and Bateman, 2004). Suppression of either stamens or pistil leads to unisexual florets (Le Roux and Kellogg, 1999; Reinheimer, Pozner and Vegetti, 2005; Sajo, Longhi-Wagner and Rudall, 2007), in the same (e.g. *Panicum* group) or in distinct spikelets (e.g. *Zea mays*). The ovary is tricarpellate, but strongly pseudomonomerous with various degrees of reduction of two carpels comparable to Restionaceae (Philipson, 1985; Ronse De Craene, Linder and Smets, 2002; Sajo, Longhi-Wagner and Rudall, 2007, 2008). Two styles are formed on the sterile carpels, and the ovary rarely has three styles (*Streptochaeta*). The ovary with single ovule develops as a caryopsis (one-seeded, indehiscent fruit with fused testa and pericarp) and has a unique embryo development.

The interpretation of the grass flower remains controversial (Figure 6.24). Some authors interpreted the palea as part of an outer perianth (the palea is occasionally two-keeled), consisting of two fused outer adaxial tepals with the abaxial outer tepal missing (Figure 6.24A; e.g. Eichler, 1875; Stützel, 2006; Sajo, Longhi-Wagner and Rudall, 2008). The inner tepal whorl consists of two abaxial lodicules and the adaxial one is missing (probably through pressure of the floret against the axis). The androecium of three stamens is usually interpreted as an outer stamen whorl alternating with the lodicules (e.g. Eichler, 1875; Stützel, 2006). Another interpretation regards lodicules and palea as bracts and the flower as naked (Figure 6.24B; Dahlgren, Clifford and Yeo, 1985). Cocucci and Anton (1988) interpreted the androecium as basically two-whorled and the flower as monosymmetric with adaxial stamens and lodicule reduced through the inhibitory influence of the palea (Figure 6.24C). Rudall et al. (2005) favoured the first interpretation on the basis of the floral morphology of *Ecdeiocolea* in Ecdeiocoleaceae, which are considered the nearest sister group of Poaceae. Staminate flowers have six tepals and four stamens arranged in two whorls in

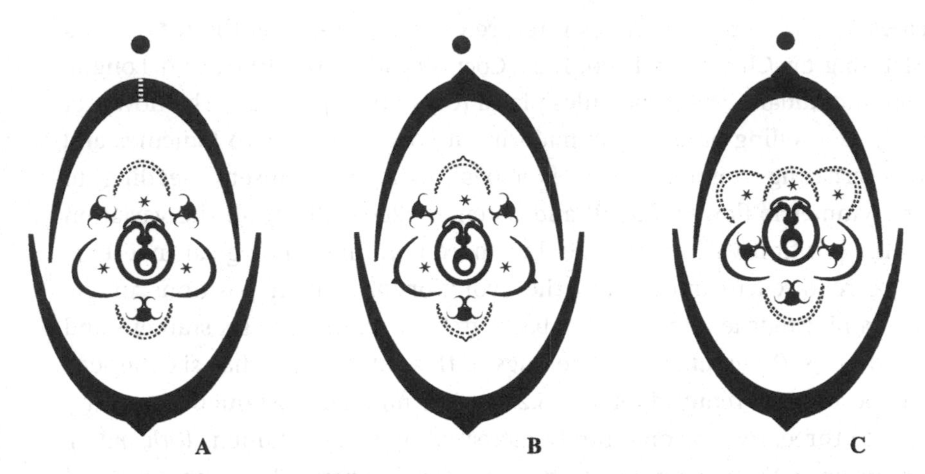

Figure 6.24 Different interpretations of the floral evolution of Poaceae: A. Eichler (1875); B. Dahlgren, Clifford and Yeo (1985); C. Cocucci and Anton (1988). Broken lines refer to lost tepals or bracts; asterisks refer to lost stamens.

a monosymmetric arrangement. The early diverging grass *Anomochloa* also has four stamens, arranged as three outer stamens and an adaxial inner stamen comparable to *Ecdeiocolea*. Among the sister groups of Poaceae, Flagellariaceae, Joinvilleaceae and *Georgeantha* of Ecdeiocoleaceae have a regular monocot floral formula (Rudall et al., 2005). Sajo, Longhi-Wagner and Rudall (2008) demonstrated that the grass spikelet could be derived from an intermediate structure as in the basal grass genus *Streptochaeta*. Single flowers arise on top of (pseudo) spikelets surrounded by seven or eight basal bracts. Lower bracts (one to five) are interpreted as glumes; bract six (with an awn) bears a modified flower in its axis, or the flower arises in the continuation of the main axis. Bracts seven and eight represent outer tepals, one of which is lost, and the three inner tepals are comparable to the lodicules. Differentiation of glumes, palea and lemma arises later in the evolution of the grasses.

Cyperaceae

Figure 6.25 *Carex brasiliensis* A. St.-Hil., based on M. Monteiro (unpubl. data)

Staminate: ∗A4 (6–8) G0

Pistillate: ∗A0 <u>G(3)</u>

General formula: ∗/↘ P0–6 A(1–)3(6–20) G2–3(–8)

Sedges belong to a subclade with Juncaceae and Thurniaceae. While the two latter families are characterized by the pentacyclic trimerous Bauplan,

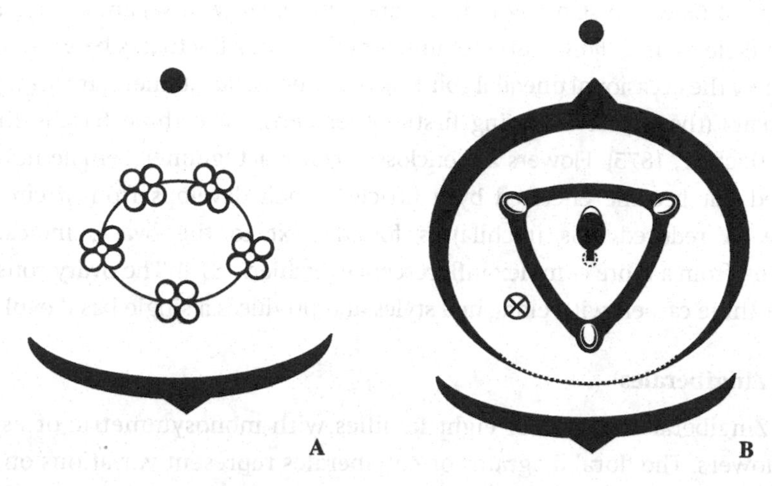

Figure 6.25 *Carex brasiliensis* (Cyperaceae): A. staminate flower; B. pistillate flower

flowers are variously reduced in the larger family Cyperaceae, with a shift to dimery, presence or absence of a perianth and the loss of at least one stamen whorl.

Inflorescences (and flowers) are bisexual to unisexual, characteristically arranged in dense spikelets. Flowers are generally small and strongly reduced. The strong reduction of flowers blurs the limits between flowers and pseudanthia, especially in subfamily Mapanioideae. When present the perianth consists of bristles, scales or hairs arranged in two trimerous whorls. The inner whorl is generally delayed and occasionally absent. Stamens range from one to three and are occasionally numerous in *Evandra* (twelve to twenty: Dahlgren, Clifford and Yeo, 1985). Eichler (1875) explains the high number as the result of a compression of several male flowers around a terminal female flower in a pseudanthium. When three in number stamens are arranged opposite the outer tepals and it can be assumed that an inner stamen whorl has been lost (Vrijdaghs et al., 2005). The ovary consists of two or three carpels with a single anatropous basal ovule. Developmentally the ovary wall develops independently of the ovule, which emerges by differentiation of the floral apex. A similar development occurs in certain core Caryophyllales (see p. 317). This dislocation of ovule and carpel wall allows for the occasional shift of styles from a lateral to a median position in bicarpellate ovaries (Reynders et al., 2012). Flowers are generally strongly compressed, leading to the loss of the abaxial tepal and the abaxial carpel (e.g. *Pleurostachys*).

Flowers of *Carex* are strongly reduced and unisexual with separate male and female spikelets. In *C. brasiliensis* the number of stamens fluctuates between four and eight by the occasional unequal splitting of stamens, but stamens are arranged as two pairs (the lateral emerging first). Other *Carex* have three basally fused stamens (Eichler, 1875). Flowers are enclosed by a bract (glume). Female flowers are naked but become enclosed by a utricle, which develops from a circular bracteole. A reduced axis (rachila) is found next to the ovary, indicating a reduction from a more complex inflorescence (Eichler, 1875). The ovary consists of two or three carpels with elongated styles and produces a single basal ovule.

Zingiberales

Zingiberales consist of eight families with monosymmetric or asymmetric flowers. The floral diagrams of Zingiberales represent variations on the basic monocot diagram, with reductions and transformations affecting mainly the androecium.

In the order there is a remarkable variation in floral forms related with intricate pollination mechanisms. Two groups can be recognized: a more basal 'banana clade', including Musaceae, Lowiaceae and Strelitziaceae, that often has an odd non-functional (?) staminode in the outer whorl, and the 'ginger group', including Heliconiaceae, where the pattern is reversed with a reduction of the abaxial stamen(s) leading to highly specialized constructs, including a further heterotopic transformation of stamens into petaloid staminodes and only a single (half-) stamen (Walker-Larsen and Harder, 2000; Rudall and Bateman, 2004; Specht et al., 2012). A hypanthial tube is usually developed connecting stamens and petals to various degrees and there is a gradual transformation into a dimorphic perianth. The labellum is a major morphological structure in Lowiaceae, Cannaceae, Costaceae and Zingiberaceae, but is not homologous between families (Kirchoff, 1992), consisting of variable parts of the androecium in the three latter families. In Lowiaceae the labellum is formed by the inner adaxial petal but is positioned abaxially by resupination (Kirchoff, Liu and Liao, 2020). Highly specialized asymmetric flowers have evolved in Marantaceae and Cannaceae, leading to the development of half-stamens. The ovary is always inferior and trimerous (or dimerous through the reduction of the abaxial carpel). Septal nectaries are present in all families of the order except in Lowiaceae where they are aborted, and Zingiberaceae where they are highly transformed (cf. Rao, Karnik and Gupte, 1954; Pai and Tilak, 1965; Kirchoff, 1997). Flowers tend to be arranged in various inflorescences built on the same Bauplan of partial inflorescences, which is a cincinnus

(Dahlgren, Clifford and Yeo, 1985). A series of floral diagrams of Zingiberales were first presented by Kress (1990).

Musaceae
Figure 6.26A–B *Musa campestris* Becc.
↑P(5):1 A3+2 Ğ(3)

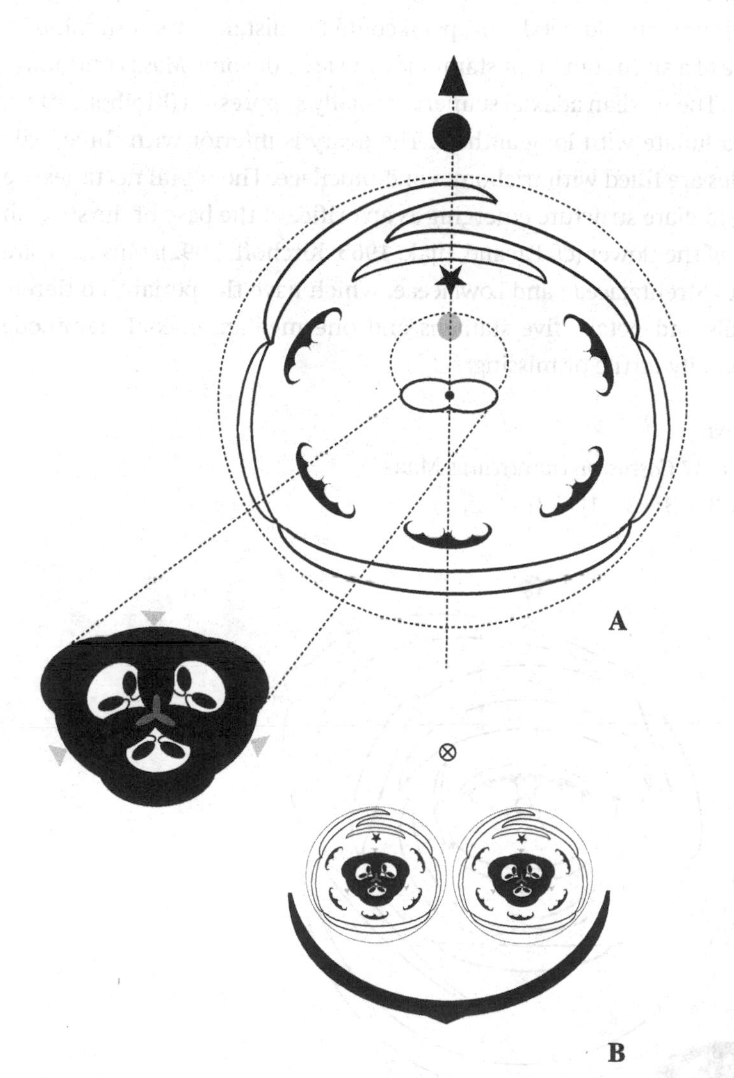

Figure 6.26 *Musa campestris* (Musaceae): A. flower; B. partial inflorescence. Note the single aperture for the septal nectary.

The family contains two genera: *Musa* and *Ensete*. Flowers are arranged in tiers in the axil of large caducous bracts. The number varies between a pair and up to forty flowers (modified cincinni: Dahlgren, Clifford and Yeo, 1985). Individual flowers may be subtended by bracts, but these were absent in *M. campestris*. Flowers tend to be staminate in the distal portion of the inflorescence and pistillate in the proximal part. All floral organs are basally connected by hypanthial growth. The perianth is undifferentiated and completely fused except for the adaxial side where the folds overlap and surround an odd petal. This petal could be mistaken for a staminode, but the presence of a sixth stamen or staminode in *Ensete* or some *Musa* contradicts this assumption. The median adaxial stamen is usually suppressed (Kirchoff, 1992). Stamens are spathulate with long anthers. The ovary is inferior with three axile placentae; locules are filled with trichomes and mucilage. The septal nectaries are merged into a triradiate structure emerging as an orifice at the base of the style on the adaxial side of the flower (cf. Pai and Tilak, 1965; Kirchoff, 1992). Musaceae are closely related to Strelitziaceae and Lowiaceae, which have the perianth differentiated into sepals and petals, five stamens and one median adaxial staminode which is occasionally fertile or missing.

Costaceae

Figure 6.27 *Costus curvibracteatus* Maas

↓K(3) [C3 A(3°+2°):1] Ğ(3)

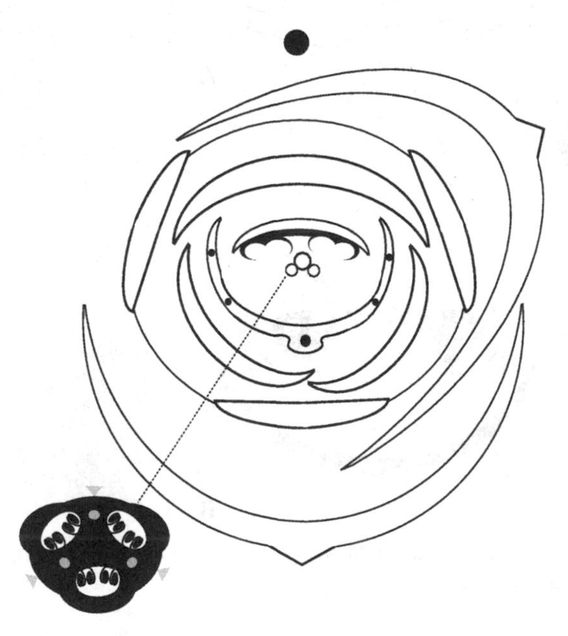

Figure 6.27 *Costus curvibracteatus* (Costacaceae). Note the labellum consisting of five fused staminodes (black dots).

Flowers are arranged in terminal racemes. A coloured bract and bracteole are present, but are indistinguishable in colour from the perianth in *C. curvibracteatus*. The calyx is tubular. The large petaloid stamen is enclosed by a median abaxial petaloid labellum, which is similar in texture and shape. There are no lateral staminodes and the labellum is interpreted by Kirchoff (1988: figure 2) based on floral developmental evidence as consisting of five staminodes. The style fits between the two anther lobes as in Zingiberaceae. The inferior ovary has two or four rows of ovules on each placenta with fewer ovules in the abaxial locule towards the bract. Contrary to Zingiberaceae three glands are present in the locular zone of the ovary but do not have a slit-like shape characteristic of septal nectaries (Rao, Karnik and Gupte, 1954; Newman and Kirchoff, 1992; Endress, 1994). In some members of the family the anterior locule is reduced and the two antero-lateral septa are fused (Newman and Kirchoff, 1992).

Zingiberaceae
Figure 6.28A–B *Roscoea cautleoides* Gagnep
↓ K(2)[d] [C3 A2°+2°:1] Ğ(3)

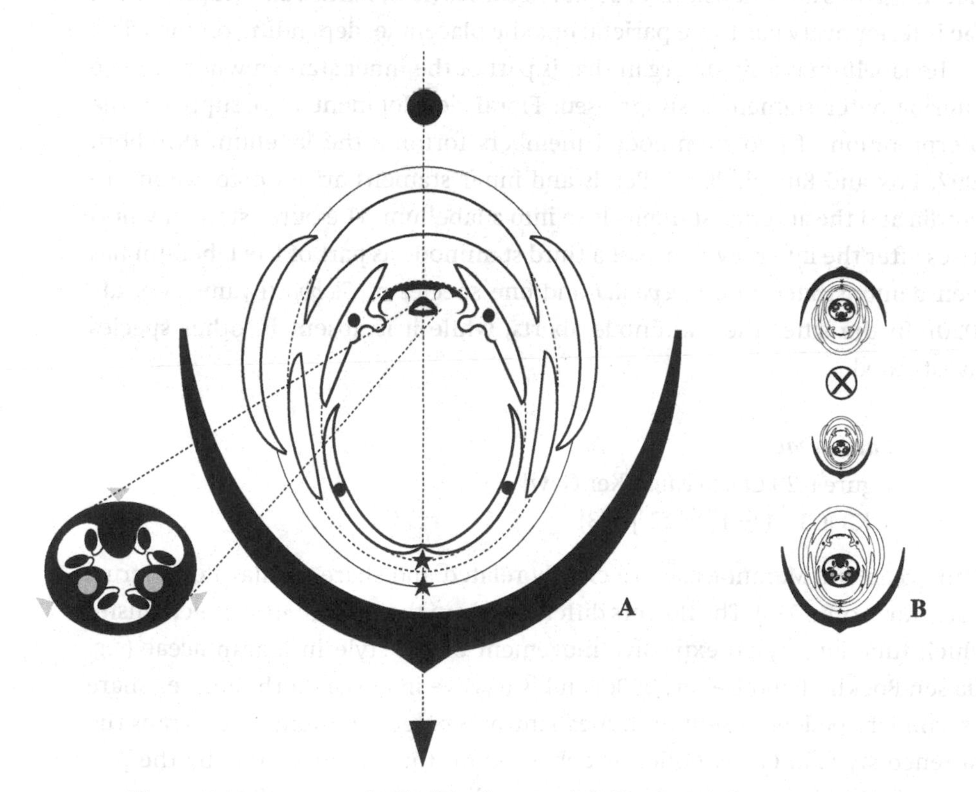

Figure 6.28 *Roscoea cautleoides* (Zingiberaceae): A. flower; B. inflorescence. The lower asterisk refers to a lost outer sepal; the upper asterisk refers to a lost staminode.

[d] K(3) in other Zingiberaceae; a third staminode of the outer whorl is sometimes present and included in the labellum (see Iwamoto et al., 2020).

There is a single bract, occasionally with a second adaxial bracteole. Bracts can be strongly coloured, contrasting with the perianth (e.g. *Globba*). In *Roscoea* only two sepals are present; the adaxial median sepal is missing. Petals and stamens are fused into a tubular structure with apical appendages. Petals develop in a median-adaxial hood and two lateral appendages that connect with the androecium. There is only a single massive fertile stamen enclosing the single style positioned on the adaxial side. This stamen is flanked by two lateral petaloid staminodes in *Roscoea*, as in other members of tribe Zingibereae. In other tribes lateral staminodes are free, and in tribe Alpinieae they are very small to absent. A large labellum characteristic for Zingiberaceae is situated in a median abaxial position. The anther can develop lateral outgrowths (*Globba*: Box and Rudall, 2006; Cao et al., 2019). In other species of *Roscoea* (e.g. *R. purpurea*, *R. scillifolia*), the lower part of the anther is transformed in a lever mechanism resembling *Salvia* (Endress, 1994; pers. obs.). The style is flanked at the base by two glandular appendages that are variously developed as crescentic or filiform structures around the base of the style. Anatomical evidence suggests that the glands are extensions of septal nectaries (Rao, Karnik and Gupte, 1954). The inferior ovary has three parietal or axile placentae, depending on the tribe.

The labellum is a single organ that is part of the inner stamen whorl, as one anterior outer stamen is suppressed. Floral development also supports the interpretation of two staminodial members forming the labellum (Kirchoff, 1997; Box and Rudall, 2006). Petals and inner stamens arise on common primordia and the anterior stamens fuse into a labellum. The outer stamen whorl arises after the inner. Evidence of a third staminode as part of the labellum has been demonstrated in *Gagnepainia* and one species of *Globba* (Iwamoto et al., 2020). In the latter the staminode aborts, while it is absent in other species investigated.

Cannaceae

Figure 6.29 *Canna edulis* Ker Gawl

↯ K3 [C3 A1°+ $1^{½:½°}$:2°] Ğ(3)

Cannaceae and Marantaceae are closely related and share similar floral structures (Kirchoff, 1983). The flowers differ mainly in their pollination mechanism, which functions by an explosive movement of the style in Marantaceae (e.g. Claßen-Bockhoff and Heller, 2006) and is passive in *Canna*. Both families share a secondary pollen presentation mechanism whereby the stamen enwraps the flattened style. In *Canna* pollen is deposited on the flattened style by the presence of small hairs on one of its margins (Glinos and Cocucci, 2011).

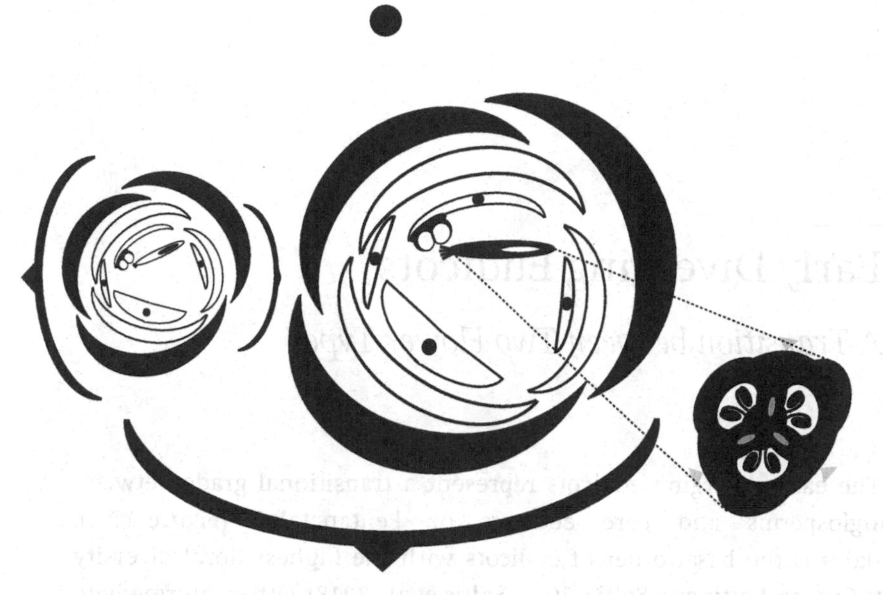

Figure 6.29 *Canna edulis* (Cannaceae): partial inflorescence

Flowers are arranged in pairs, equal in Maranthaceae, but one of which is smaller and subtended by a second bract in Cannaceae. The second flower is rarely fully developed (Kirchoff, 1983). Contrary to Zingiberaceae, the median adaxial inner petal is smaller and enclosed by lateral petals. Only four to five members of the androecium are present (cf. Pai, 1965; Kirchoff, 1983), and the single fertile inner adaxial stamen is half-fertile, while the other half is petaloid. Pai (1965) interpreted half of the anther as reduced and the crest as a prolongation of the connective, but this is refuted by ontogenetic data showing an early division of two equal primordia, one developing into a theca and the other into the crest (Kirchoff, 1983). The other staminodes as well as the style are petaloid. One of the inner staminodes develops into a reflexed labellum while the third is erect and petaloid. Contrary to Zingiberaceae and Costaceae, the labellum is a single organ (Pai, 1965; Kirchoff, 1983, 1997). The outer stamens are variously reduced or absent. In *Canna edulis* only one staminode is present (mostly the outer abaxial as in Zingiberaceae and Marantaceae). In *C. indica* two staminodes occur in the outer whorl. The inferior ovary has axile placentation with ovules borne in two rows. Septal nectaries are present on the septal arms.

7

Early Diverging Eudicots

A Transition between Two Flower Types

The early diverging eudicots represent a transitional grade between basal angiosperms and core eudicots or Pentapetalae (Figure 7.1). Ranunculales is the basal order of eudicots with the highest floral diversity (Ronse De Craene, Soltis and Soltis, 2003; Soltis et al., 2018). Other intermediate orders (e.g. Proteales, Buxales, Trochodendrales) generally have much reduced, dimerous flowers and the link between Ranunculales and core eudicots remains unclear on a floral morphological basis (e.g. Ronse De Craene, 2004; Wanntorp and Ronse De Craene, 2005). As Gunnerales are morphologically much closer to early diverging eudicots than to Pentapetalae, I prefer to treat them as part of the early diverging eudicots.

7.1 Ranunculales

The order is highly diverse in terms of floral structure. As such it occupies a transitional position between basal angiosperms and core eudicots (Ronse De Craene, Soltis and Soltis, 2003). Several derived characters tend to be concentrated in Menispermaceae, Ranunculaceae and Papaveraceae, such as median or transversal monosymmetry, sepal and petal differentiation and fusion, petal appendages in the shape of nectariferous structures or spurs, syncarpy and (pentamerous) cyclic flowers. Unisexual flowers with synandry have evolved in Menispermaceae and Lardizabalaceae (Endress, 1995b).

Merism is highly variable in Ranunculales, ranging from dimery to trimery, pentamery or indefinite. Ancestral flowers were probably trimerous with several perianth whorls (Ronse De Craene, Soltis and Soltis, 2003; Carrive et al., 2020).

The androecium is highly variable, ranging from numerous spirally arranged stamens to a single stamen. Most core Ranunculales share nectariferous petals

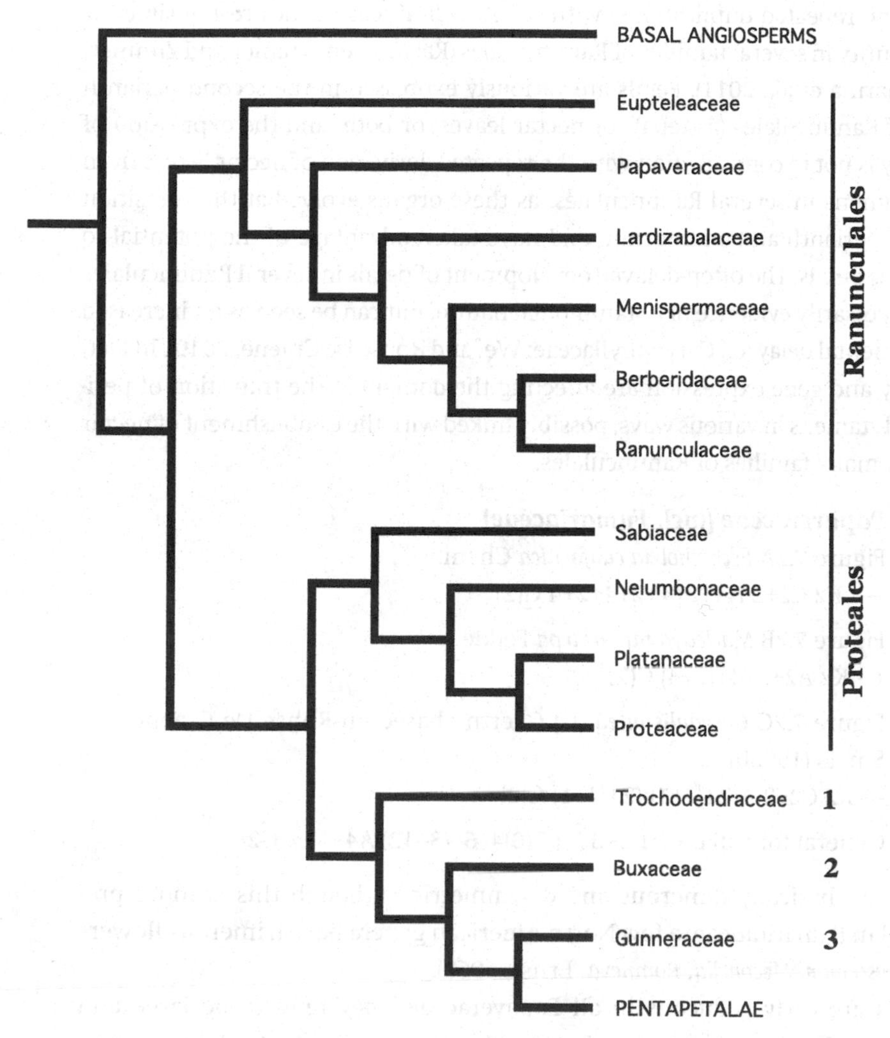

Figure 7.1 Phylogenetic tree of the early diverging eudicots, based on Soltis et al. (2018). 1. Trochodendrales; 2. Buxales; 3. Gunnerales

that are of probably staminodial origin (Erbar, Kusma and Leins, 1998; Walker-Larsen and Harder, 2000; Ronse De Craene, Soltis and Soltis, 2003). The perianthless Eupteleaceae is sister group of Ranunculales, but flowers are strongly derived with an increase of stamens and carpels as an adaptation to wind pollination (Ren et al., 2007).

Genetic studies have supported the idea that there is only a single origin for petals in Ranunculales, in contrast to earlier interpretations of several derivations of petals from outer stamens (e.g. Kosuge, 1994; Ronse De Craene and Smets, 1995c; Ronse De Craene, 2007). In Ranunculales *AP3*–like genes have undergone repeated duplications, with the *AP3–III* lineage genes responsible for petal identity in several families of Ranunculales (Rasmussen, Kramer and Zimmer, 2009; Sharma et al., 2011). Petals are variously expressed in the second perianth whorl of Ranunculales (as tepals or nectar leaves, or both) and the expression of petaloidy is not in contradiction with the repeated derivation of nectar leaves from outer stamens in several Ranunculales, as these organs evolved at the transition between perianth and androecium and have taken advantage of the potential to develop as petals. The often-delayed development of petals in several Ranunculales is not necessarily evidence of a staminodial nature, but can be seen as an increased developmental delay (cf. Caryophyllaceae: Wei and Ronse De Craene, 2019). In fact, petaloidy and gene expression are affecting the domain at the transition of perianth and stamens in various ways, possibly linked with the establishment of nectar leaves in many families of Ranunculales.

Papaveraceae (incl. Fumariaceae)

Figure 7.2A *Eschscholzia californica* Cham.
↔ K2 C2+2 A4+2+4+2+4+2+4 $\underline{G}$(2)

Figure 7.2B *Macleaya microcarpa* Fedde
↔K2 A2+2+4+(2+4) $\underline{G}$(2)

Figure 7.2C *Corydalis lutea* (L.) Gaertn., based on Ronse De Craene and Smets (1992b)
→ K2 C2+2 A(21/2+1):(21/2+1) $\underline{G}$(2)

General formula: ↔K2–3 (4) C(0)4–6–(8–12) A4–6–∞ G2–∞

Flowers are basically dimerous and disymmetric, although this is more pronounced in Fumarioideae; a few North American genera have trimerous flowers (e.g. *Platystemon*, *Meconella*, *Romneya*: Ernst, 1967).

Sepals are early caducous in all Papaveraceae; they tend to be large and saccate or calyptrate, or small and reduced in Fumarioideae. Petals are clearly arranged in two imbricate whorls and are often crumpled in bud. Stamens arise in regular whorls of four or two in dimerous flowers (forming pseudowhorls of six), and outer stamens are always inserted in two pairs or alternate with the petals (e.g. Murbeck, 1912; Ronse De Craene and Smets, 1990b; Karrer, 1991). The number of stamens can be highly fluctuating; in the specimens of *Eschscholzia* studied stamens ranged from four to twenty-two, although the number can be higher. The regular arrangement of pairs and single stamens is interpreted as the result of the compression of an ancestral helical initiation, although others interpret the pairs as the result of doubling (e.g. Endress, 1987) with oscillations between dimerous and tetramerous whorls. In cases where stamen numbers are low, the four or six outer stamens are always present (e.g.

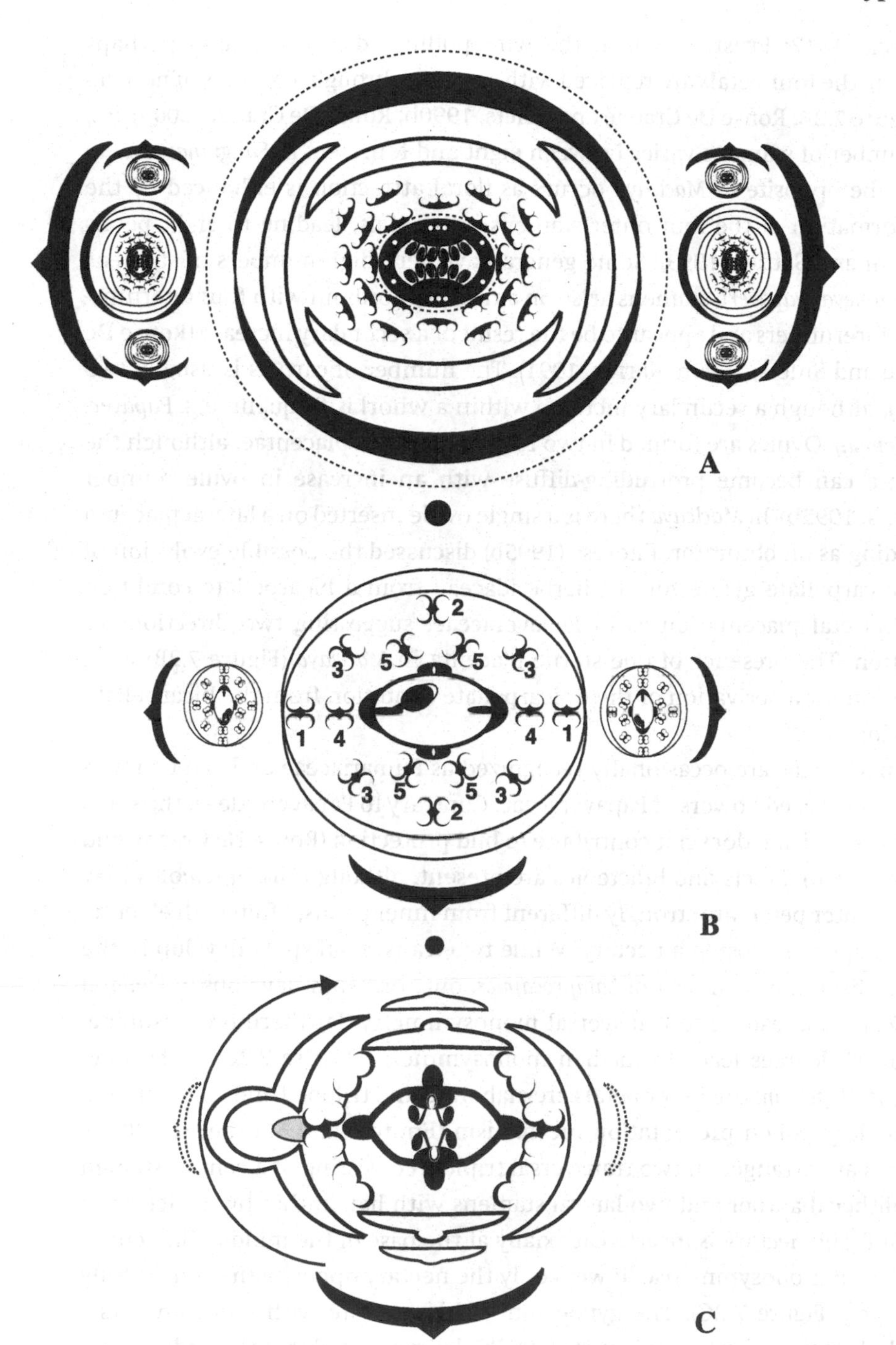

Figure 7.2 Papaveraceae: A. *Eschscholzia californica*, partial inflorescence; B. *Macleaya microcarpa*, partial inflorescence; C. *Corydalis lutea*. In B numbers refer to sequence of stamen initiation.

Murbeck, 1912; Ernst, 1967). In the wind-pollinated *Macleaya* (and perhaps *Bocconia*), the four petals are replaced with stamens through a process of homeosis (Figure 7.2B; Ronse De Craene and Smets, 1990b; Ronse De Craene, 2003) and the number of stamens varies between eight and fourteen. In *Sanguinaria canadensis*, the opposite of *Macleaya* occurs as floral attraction is enhanced by the transformation of the four outer stamens into petals, leading to eight petals (Lehman and Sattler, 1993). Some genera have very high numbers of stamens (e.g. *Romneya*, *Papaver*); stamens arise on a ring primordium with four alternipetalous forerunners and appear to be the result of a secondary increase (Ronse De Craene and Smets, 1990b; Karrer, 1991). The number of carpels is usually two (three), although a secondary increase within a whorl is frequent (e.g. *Papaver*, *Platystemon*). Ovules are formed in two rows on parietal placentae, although the placenta can become protruding-diffuse with an increase in ovule number (Endress, 1995b). In *Macleaya* there is a single ovule inserted on a lateral placenta extending as an obturator. Endress (1995b) discussed the possible evolution of the unicarpellate gynoecium of Berberidaceae from a bicarpellate condition with parietal placentation as in Papaveraceae, suggesting two directions of evolution. The presence of one sterile placenta in *Macleaya* (Figure 7.2B) is an indication for a derivation of the unicarpellate condition from the bicarpellate condition.

Fumarioideae are occasionally recognized as Fumariaceae and can be interpreted as reduced flowers of Papaveraceae. Contrary to Papaveroideae, the sepal whorl is small and does not contribute to bud protection (Ronse De Craene and Smets, 1992b). Bracts and bracteoles are present, although the bracteoles may be lost. Outer petals are strongly different from inner petals, often with elaborations or spurs enclosing a nectary. While two transversal spurs develop in the regular disymmetric flower of *Lamprocapnos*, only one spur develops in *Fumaria* and *Capnoides*, leading to transversal monosymmetry. Furthermore, resupination of 90 degrees leads to median monosymmetry (Figure 7.2C; Weberling, 1989). In *Hypecoum* the inner petals are elaborate and trilobed and contribute to a secondary pollen presentation mechanism (Endress and Matthews, 2006b). Stamens are arranged in two transversal triplets consisting of a central stamen with dithecal anther and two lateral stamens with half-anther by reduction of one half. The nectary is inserted abaxially at the base of the middle filament of a triplet; in monosymmetric flowers only the nectary opposite the spur is fully developed (Figure 7.2C). The gynoecium is bicarpellate with two transversal carpels. Ronse De Craene and Smets (1992b) demonstrated that the androecium of Fumarioideae is derived from two whorls consisting of four outer half-stamens alternating with the petals and two transversal inner stamens. The more external position of the second stamen whorl is linked with the strong median compression of the flower, and there is no indication that two additional median stamens have been lost as in Brassicaceae (p. 228). In *Hypecoum* with seemingly two median and two transversal stamens, the outer half-stamens fuse postgenitally into two larger median stamens (Ronse De Craene and Smets, 1992b). In *Pteridophyllum* the inner stamens are lost, resulting in four alternipetalous stamens.

Lardizabalaceae (incl. Sargentodoxaceae)

Figure 7.3A–C *Holboellia angustifolia* Wall.

Staminate: $\ast$ K3+3 [C3+3 A3+3] $\underline{G}3^{\circ}$

Pistillate: $\ast$ K3+3 [C3+3 A3°+3°] $\underline{G}3$

Lardizabalaceae share several floral characters with Menispermaceae, including trimerous, unisexual flowers, synandry, nectar leaves and a comparable floral formula. In *Holboellia* the sepal whorl is petaloid and the nectar leaves are small and attached as scales to the stamens (Figure 7.3A–B). In *Decaisnea* and *Akebia* no

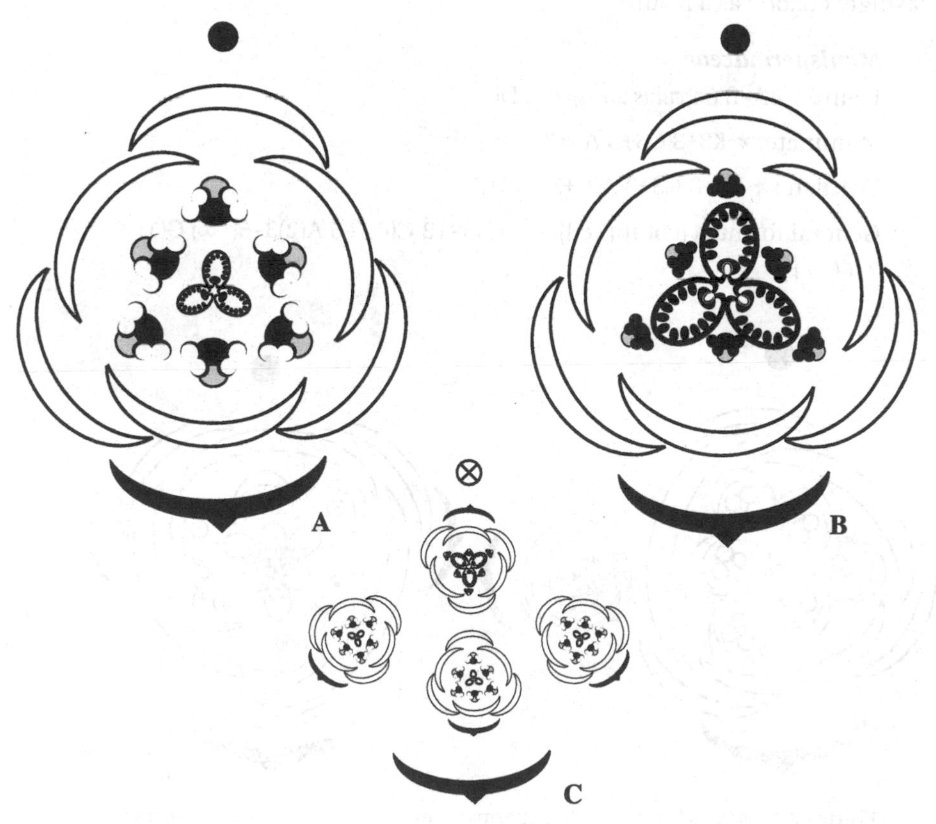

Figure 7.3 *Hollboelia angustifolia*. A. staminate flower; B. pistillate flower; C. partial inflorescence. Note the small nectar leaves attached to stamens and staminodes.

nectar leaves are present. Flowers of *Akebia* initiate six sepals, but one to three inner sepals do not develop further (Zhang and Ren, 2011). Flowers of *Akebia* show a strong floral dimorphism between staminate and pistillate flowers, but both are inserted on the same inflorescence. Stamens are either free or basally united in a synandrium and have extrorse anthers. Small nectar leaves are attached outside the stamens in *Holboelia* and are strongly retarded, while they have a more complex morphology in *Sinofranchetia* and *Sargentodoxa* (Zhang and Ren, 2008, 2011). Carpel number ranges from three to twelve (Eichler, 1878). The apocarpous gynoecium is plicate and has a laminar-diffuse placenta reminiscent of Nymphaeales (Endress, 1995b).

Sargentodoxa differs from other Lardizabalaceae in the higher carpel number in pistillate flowers (up to ninety), linked with ascidiate, uniovulate carpels (Zhang and Ren, 2008). Staminate flowers are trimerous with a floral diagram resembling *Holboellia*, while bisexual and pistillate flowers have a highly variable merism with fluctuating perianth and carpel numbers. It is likely that the carpel number has been secondarily multiplied and that the flower has become increasingly chaotic as a result.

Menispermaceae

Figure 7.4A–B *Cocculus laurifolius* DC.

Staminate: ✶ K3+3 C3+3 A3+3 $\underline{G}$3°

Pistillate: ✶ K3+3 C3+3 A3°+3° $\underline{G}$3+3

General formula (combined): ✶ K(1)3–12 C(0)1–6 A(2)3–6(–∞) G(1) 3–6(–∞)

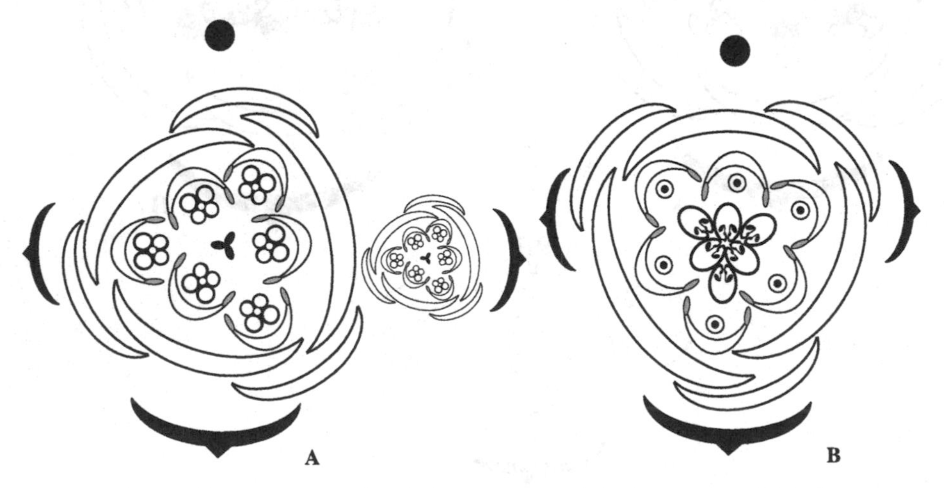

Figure 7.4 *Cocculus laurifolius* (Menispermaceae): A. staminate partial inflorescence; pistillate flower

Flowers of Menispermaceae are highly heterogeneous, ranging from spiral, multistaminate (e.g. *Hypserpa*) to trimerous (e.g. *Cocculus*) or dimerous (e.g. *Cissampelos*) flowers with one-two stamen whorls (Schaeppi, 1976; Endress, 1995b). Compared to other Ranunculales there is high diversity in floral forms with occasional fusions of sepals or petals and even zygomorphy through reduction of organs. Flowers are always unisexual (dioecious) with variable dimorphism between the genders, which is also expressed in the complexity of the inflorescences (Kessler, 1993; Endress, 1995b). The perianth is undifferentiated and spiral, or there is a distinction between sepals and petals. Sepals are inserted in several alternating whorls (e.g. *Sciadotenia*: diagram shown by Endress, 1995b) or more generally in two whorls. Petals or nectar leaves are much smaller and are homologized with staminodes as in Berberidaceae and Lardizabalaceae (Ronse De Craene and Smets, 2001a). They are occasionally absent (e.g. *Abuta*). Stamens range from many (e.g. *Hypserpa*: Endress, 1995b) to twelve, six, three, or exceptionally one (Kessler, 1993). Carpels are free and plicate, and range from twelve to three (rarely one). Two ovules are present but only one develops as a seed.

Cocculus resembles Berberidaceae in its floral diagram, although flowers are unisexual with rudiments of carpels and stamens. Staminate flowers have three carpellodes rather than six, while pistillate flowers have six rudiments of stamens (cf. Wang et al., 2006). The petals are sheathlike and enclose the filaments at maturity. Flowers are highly reduced in some dioecious genera (e.g. *Stephania*, *Cissampelos*) with strong dimorphism between staminate and pistillate flowers and no traces of lost organs (Eichler, 1878; Wang et al., 2006; Meng et al., 2012). Reduction of perianth parts differs between staminate and pistillate flowers, being often more pronounced in pistillate flowers and leading to monosymmetry (Meng et al., 2012). In staminate flowers with complete reduction of the gynoecium the androecium often develops as a synandrium consisting of three-four (e.g. *Stephania*: Wang et al., 2006) or six stamens (*Dioscoreophyllum*: Schaeppi, 1976). Dehiscence of the synandrium is horizontal (Meng et al., 2012). The gynoecium is reduced to a single carpel with a multilobed stigma.

Berberidaceae

Figure 7.5A *Epimedium x versicolor* E. Morren
⟷K2+2(+2+2+2+2) C2+2 A2+2 $\underline{G}$1

Figure 7.5B *Podophyllum peltatum* L., based on DeMaggio and Wilson (1986) and Schmidt (1928)
∗ K3/3+3 C3+3/3+3+3 A3/3^2 +3^3 $\underline{G}$1

General formula: ∗ K3–12 C4–6 A4–6(–∞) G1

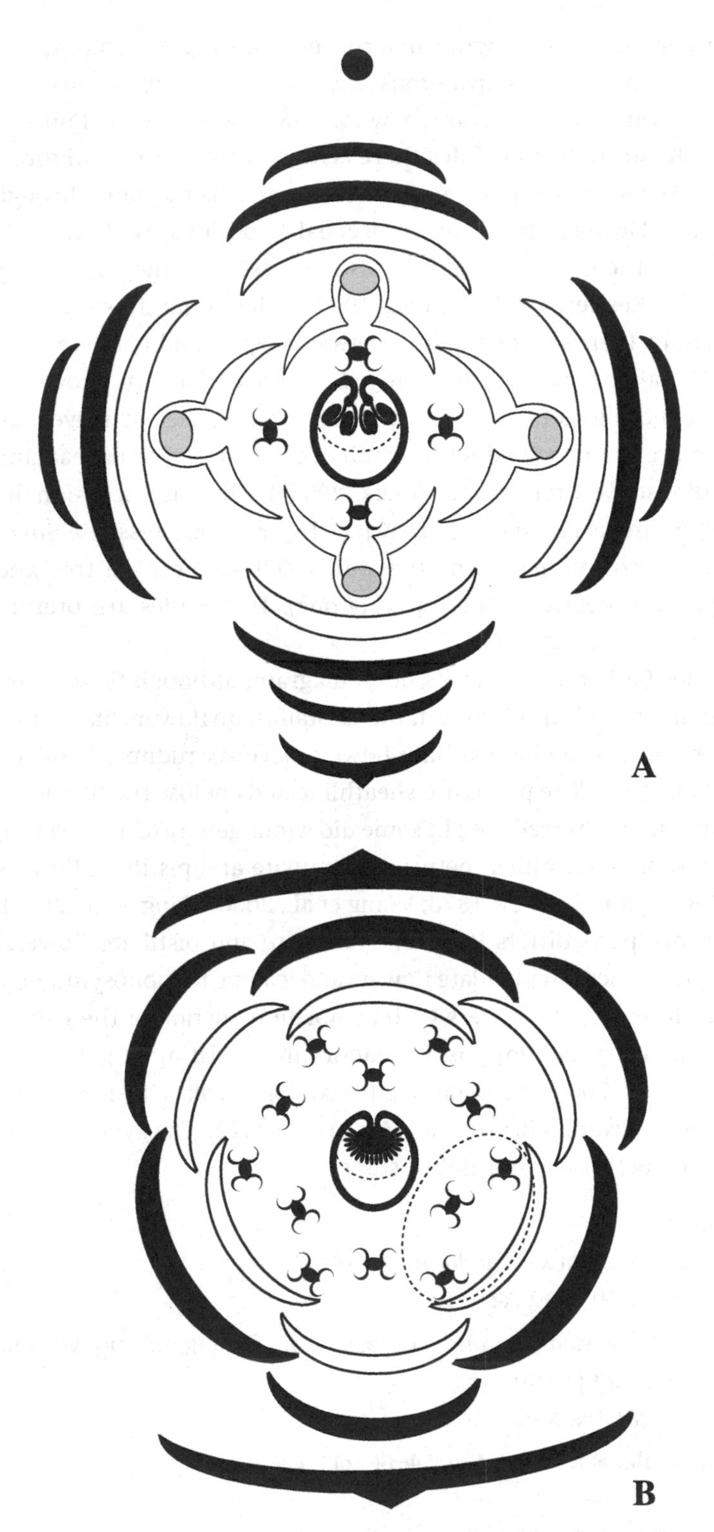

Figure 7.5 Berberidaceae: A. *Epimedium x versicolor*; B. *Podophyllum peltatum*. The broken line refers to a group of stamens and the delimitation of the stigma, respectively.

Flowers are arranged in complex cymes, racemes, or are solitary (*Podophyllum*). The flower of *Podophyllum peltatum* is terminal and inserted between the petioles of two leaves (DeMaggio and Wilson, 1986). Flowers are mostly trimerous, rarely variable or dimerous (e.g. *Epimedium*). The inflorescence of *Berberis* has one terminal pentamerous flower while lateral flowers are trimerous (Eichler, 1875, 1878; Endress, 1987). These were extensively discussed as to whether pentamerous flowers are derived from trimerous flowers or the opposite.

The perianth is variable but differentiated in sepals and petals in a similar way as in Menispermaceae (e.g. *Berberis*, *Mahonia*, *Epimedium*) with two to three (occasionally more) whorls of sepals and generally two whorls of nectar leaves (petals) and two whorls of three stamens. However, the perianth in Berberidaceae may have several origins from bracts and stamens (Zhang and Ren, 2011). *Vancouveria* and *Nandina* have several trimerous whorls of bracts below the six sepals. Petals of *Podophyllum* lack floral nectaries and are sometimes in double position. The trimerous flower of *Gymnospermium* is unusual in its development in that two lateral sepals precede four median sepals, which could indicate that these flowers are derived from dimerous progenitors (Xue et al., 2017). *Podophyllum peltatum* is unusual in having more than six stamens (ranging from six to twelve to eighteen), which were interpreted as a result of dédoublement (Payer, 1857; Schmidt, 1928; Terabayashi, 1983). DeMaggio and Wilson (1986) demonstrated that stamens arise in groups starting with three to six primordia arising opposite the petals, and that stamens may arise after initiation of the gynoecium, suggesting a secondary increase. A secondary increase is mostly linked to the inner whorl. A perianth is lost in *Achlys*, leading to a breakdown of the regular merism of the flowers (Endress, 1989).

Nectar leaves develop from common primordia with the stamens (e.g. Schmidt, 1928; Brett and Posluszny, 1982; Xue et al., 2017) and they lag behind the stamens in development. They are best interpreted as stamen derived (cf. Brett and Posluszny, 1982; Endress, 1995b; Ronse De Craene, Soltis and Soltis, 2003). While most taxa have two marginal nectaries at the base of the nectar leaf, in *Epimedium* the nectar is concealed in a spur similar to *Aquilegia* (Ranunculaceae). Anther dehiscence is longitudinal or valvate with two valves. The single carpel develops as a strictly ascidiate structure without ventral suture (Endress, 1995b). Placentation is marginal with two rows of ovules, sometimes protruding-diffuse with a high number of ovules (e.g. *Podophyllum*).

Ranunculaceae

Figure 7.6A–B *Ranunculus ficaria* L.
✱ K3–5 C8–11 A15–31 $\underline{G}$9–10

Figure 7.6C *Aconitum lycoctonum* L.
↓P5 A°2:6 A∞[a] $\underline{G}$3

Figure 7.6D–E *Aquilegia sp.*
✱ K5 C5 A5 (x13) + 5°+5° $\underline{G}$5

General formula: ✱/↻/↓/↔ P1–∞ or K3–8 C0–13 A5–∞ G1–∞

The family is probably the most diverse in flower structure among angiosperms. Inflorescences are variable and consist of single terminal flowers as part of monochasial units (e.g. *Ranunculus ficaria*), dichasia (e.g. *Clematis*) or racemes (e.g. *Actaea, Aconitum*). Bracts may hide lateral flowers enclosed by two bracteoles. In *Aquilegia*, terminal flowers are overtopped by the lateral branches. The flower of Ranunculaceae is in a flux, with several evolutionary novelties, such as a transition to pentamery, different forms of petaloid perianths including staminodial petals, transitions between whorled and spiral flowers, and monosymmetry. Perianth diversity ranges from an undifferentiated tepalar perianth to a biseriate perianth consisting of outer tepals and inner nectar leaves. The presence of a petaloid perianth is linked with the expression of *AP3–III* petal identity genes in most Ranunculales, and this appears to be ancestral in the order (Rasmussen, Kramer and Zimmer, 2009). However, in a number of genera petals have been secondarily lost (e.g. *Thalictrum*) accompanied by a deletion of the *AP3–III* genes (Zhang et al., 2013). In *Thalictrum*, this loss corresponds to a shift to wind pollination, making nectaries redundant.

Ranunculaceae were studied extensively for their phyllotaxis and the family was used as a model to interpret the evolution of petals in the angiosperms (e.g. Eichler, 1878; Schöffel, 1932). The perianth is either spiral and undifferentiated, or whorled (dimerous, trimerous, or pentamerous). Genera with a petaloid perianth (e.g. *Anemone*) tend to have a variable number of tepals (occasionally stabilized to three or five), and a high number of stamens and carpels in a spiral. An involucre of three bracts is usually present and can be close to the perianth, mimicking a calyx (e.g. *Barneoudia, Hepatica*). Clear transitions from spirals to whorls are present in *Clematis* with a dimerous perianth, a whorled arrangement of outer stamens, and a transition to a spiral arrangement in the inner stamens and gynoecium (Ronse De Craene and Smets, 1996a; Ronse De Craene, Soltis and Soltis, 2003). Flowers of Ranunculaceae often have transitional series of Fibonacci numbers between different organ categories. Sepals are often five in number (2/5), with eight petals or nectar leaves (3/8) (e.g. *Adonis, Aconitum, Delphinium, Nigella*: Figure 7.6C). Some

[a] Generally 26 to 27.

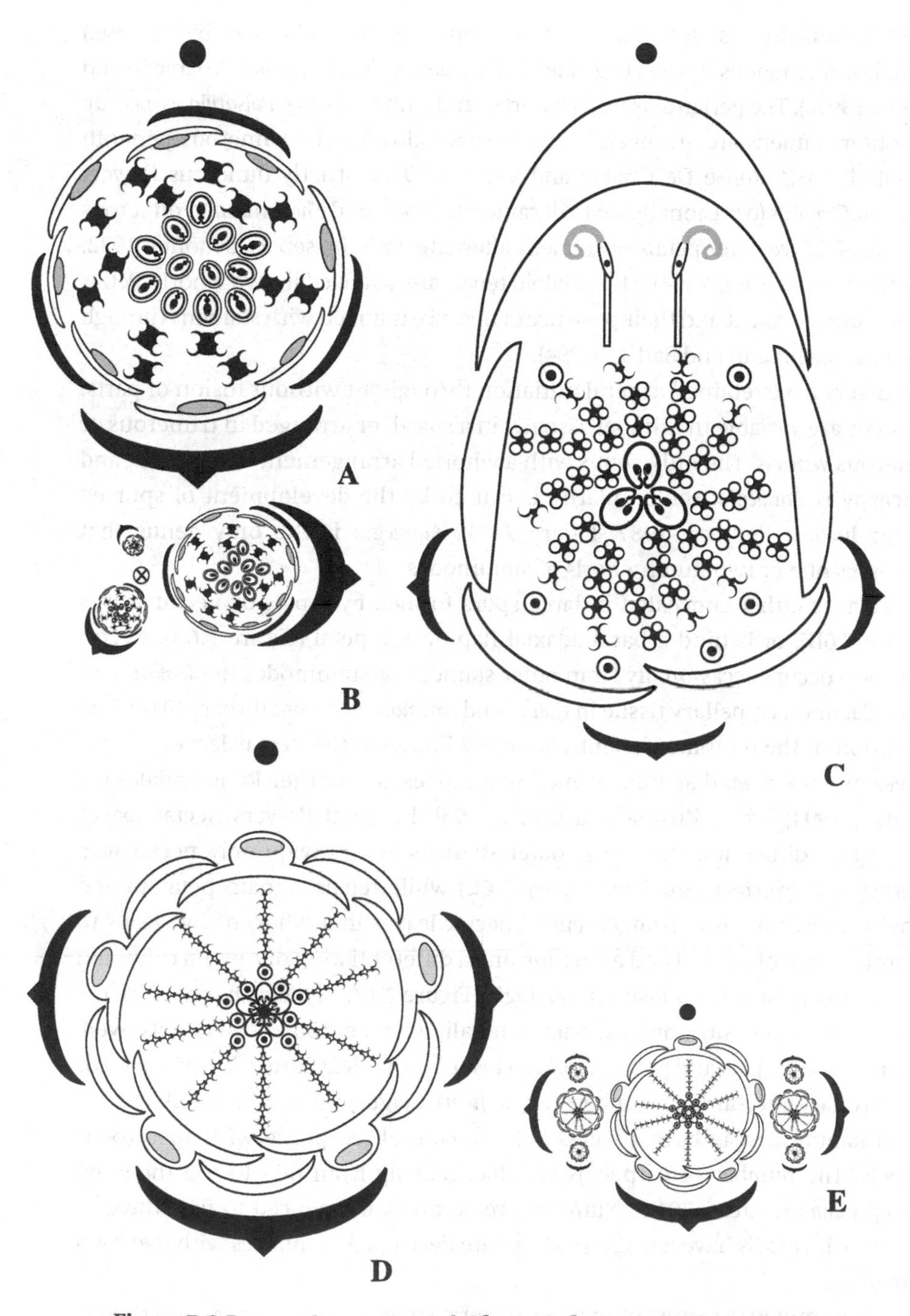

Figure 7.6 Ranunculaceae: *Ranunculus ficaria*: A. flower; B. partial inflorescence; C. *Aconitum lycoctonum*; *Aquilegia sp.*: D. flower; E. partial inflorescence. Note the different types of nectar leaves.

pentamerous flowers with reduced stamen and carpel number can be confused with diplostemonous flowers (e.g. *Xanthorhiza*, *Ranunculus sp.*: Ronse De Craene and Smets, 1995c). The perianth is often whorled and trimerous (e.g. *Pulsatilla*, *Hepatica*); the outer stamens are arranged in three pairs following the trimerous perianth (Schöffel, 1932; Ronse De Craene and Smets, 1995c). Strictly dimerous flowers occur in *Clematis* (occasionally up to decamery), *Actaea*, and *Thalictrum*. In dimerous flowers of *Actaea* outer petals or stamens alternate with the sepals (Schöffel, 1932; Lehmann and Sattler, 1994). The petaloid tepals are arranged in two whorls. Petals are variously present and their presence or absence is linked with stamens through homeosis (Lehmann and Sattler, 1994).

Flowers have retained a spiral initiation throughout without fusion of parts. Stamens are variable in number, usually in a spiral, or arranged in trimerous or dimerous whorls. The only genus with a whorled arrangement throughout and syncarpy is *Aquilegia*; the regularity is caused by the development of spurred nectar leaves: Endress, 1987; Figure 7.6D). *Aquilegia* is the only genus that possesses one or two inner whorls of staminodes.

Nectar is either concealed in large spurs formed by tepals or nectar leaves (Figure 7.6D), or behind a basal adaxial flap on the petal (Figure 7.6A). Nectar secretion occurs occasionally from outer stamens or staminodes in *Clematis* and *Pulsatilla*, or on carpellary tissue in *Caltha* and *Anemone nemerosa* (Erbar, 2014). The evolution of the biseriate perianth is closely linked to the nectar leaves. Nectar leaves are interpreted as transformed staminodes, as in other Ranunculales (cf. Kosuge, 1994; Erbar, Kusma and Leins, 1998). In spiral flowers nectar leaves develop by differentiation of the outer stamens and provide only nectar (e.g. *Helleborus*, *Delphinium*, *Aconitum*: Figure 7.6C) while tepals remain petaloid and provide attraction. An arrangement of nectar leaves in a whorl of five leads to a combination of reward and attraction and a differentiation of a green calyx and coloured corolla (e.g. *Ranunculus*, *Aquilegia*: Figure 7.6A, D). *Ranunculus* is highly variable in flower morphology, with generally pentamerous flowers with well-differentiated sepals and petals, a partial loss of nectar leaves (e.g. *R. auricomus*), or an increase of nectar leaf numbers (e.g. *R. ficaria*: Figure 7.6A, *R. chilensis*).

All Ranunculaceae have a well-developed conical receptacle with apocarpous carpels. The number of carpels is variable, ranging from one to ten thousand (*Laccopetalum*: Endress, 2014). Numbers are occasionally reduced to five, three or one carpel. Carpels have a single ovule, or are developed as follicles with two rows of ovules.

Monosymmetry is present in a few genera, superimposed on a regular, polymerous flower (Figure 7.6C). It appears late in the development of flowers of tribe Delphineae (e.g. *Delphinium*, *Aconitum*) and includes a pentamerous undifferentiated perianth and two abaxial nectar leaves (initially eight: Eichler, 1878;

Jabbour et al., 2009; Zalko et al., 2021). The remaining nectar leaves (staminodes) are variously developed or reduced along a gradient running from the adaxial to the abaxial side of the flower. The nectar leaves of *Aconitum* are hammer-like and fit in a depression formed by the posterior tepal (Figure 7.6C). In other Delphineae the two nectar leaves produce depressions and spurs that fit within spurs formed by the posterior tepal regulating access to pollinators. In *Staphisagria* and *Delphinium* the nectar leaves are postgenitally fused (Zalko et al., 2021). In some *Delphinium* one to four abaxial nectar leaves are developed, while the adaxial organs are barely visible at maturity.

The evolution of the flowers in Ranunculaceae appears remarkable, with several innovations that are reflected in Pentapetalae. However, the origins of pentamerous flowers and petals appear to be unique novelties not related to the evolution in Pentapetalae.

7.2 The Basal Eudicot Grade: Proteales, Trochodendrales, Buxales, Gunnerales

Proteales

The association of Proteaceae, Nelumbonaceae and Platanaceae in a well-supported Proteales is one of the greatest surprises of molecular systematics, as the three families used to be placed in widely diverging groups (e.g. Cronquist, 1981: Rosidae, Magnoliidae, Hamamelidae). However, Proteales are probably end branches of a long evolution dating back to about 120 million years, with a much higher fossil diversity, as recognized for Platanaceae (Stevens, 2001 onwards; Von Balthazar and Schönenberger, 2009). More recently, the small family Sabiaceae has been included at the base of Proteales.

Sabiaceae (Meliosmaceae)

Figure 7.7A *Sabia limoniacea* Wall., based on Ronse De Craene, Quandt and Wanntorp (2015a)

✶ K4–5 C5 A5 $\underline{G}$(2)

Figure 7.7B *Meliosma dilleniifolia* Wall., based on Ronse De Craene and Wanntorp (2008)

↘ K4–5 C2:3 A2:3°] $\underline{G}$(2)

Sabiaceae is a small family of three genera: *Sabia*, *Ophiocaryon* and *Meliosma*.[b] Based on a series of micromorphological and embryological characters,

[b] *Meliosma alba* is an unusual disjunct species that is sister to *Ophiocaryon*, but is morphologically more similar to other species of *Meliosma* (Thaowetsuwan, Honorio Coronado and Ronse De Craene, 2017). Zúñiga (2015) proposed the name *Kingsboroughia alba* to solve this dilemma, but further investigations are needed.

Sabiaceae appears close to Menispermaceae (data summarized in Ronse De Craene and Wanntorp, 2008 and Ronse De Craene, Quandt and Wanntorp, 2015a). The flowers are arranged in terminal panicles and include numerous to solitary flowers that do not open in a regular succession. Flowers are enclosed by two bracteoles with variable position on the pedicel, occasionally intergrading with sepals. A single caducous median bract subtending the flower is situated at the base of the pedicel. Bracteoles, sepals and petals are morphologically similar, especially visible early in their development. The pattern of insertion of the sepals appears more or less decussate.

Flowers of *Meliosma* and *Ophiocaryon* are deceivingly complex (see Wanntorp and Ronse De Craene, 2007; Ronse De Craene and Wanntorp, 2008; Thaowetsuwan, Coronado and Ronse De Craene, 2017). They consist of four to five small sepals, five petals of two different shapes and sizes, three staminodes, two fertile stamens, and a superior, bicarpellate ovary. Staminodes are inserted opposite the large petals and are basally adnate to them. Stamens are dorsally fused with a small petal.

Anthers in *Meliosma* and *Ophiocaryon* consist of a broad basal platform, in some species extending into a crenulate rim, and bear two globular pollen sacs in an apical-adaxial position. The platform goes over into a narrow, flattened filament and might function as a secondary pollen presentation mechanism (see Thaowetsuwan, Honorio Coronado and Ronse De Craene, 2017). In young flower buds of *Meliosma*, staminodes surround the young styles in a coherent unit and the young anthers fit in adaxial folds formed by the staminodes. At this stage filaments are bent inwards in the middle and the anthers are hidden from view. One staminode is symmetrically developed and encapsulates one pollen sac of each anther; the two other staminodes have only one lobe developed which enclose the other pollen sac (Figure 7.7B). When the flower expands, filaments bend abruptly outwards and anthers become detached from the tight grip of the staminodes, mostly in an explosive manner. The pollination mechanism might explain the heteromorphy of the petals; the broad petals enclose the bud and hold the staminodes erect, while the smaller often bifid petals do allow for the filaments to curve outwards. The genus *Ophiocarpon* can be considered a diminutive derivative of *Meliosma* that has arisen as a result of paedomorphosis (Thaowetsuwan, Honorio Coronado and Ronse De Craene, 2017). Flowers of *Sabia* are regular with superposed sepals, petals and stamens, and without a dimorphic androecium. Anthers are dithecal monosporangiate in all genera. The gynoecium consists of two fused carpels bearing two parallel ovules each with two connivent, occasionally twisted styles. The ovules become superposed at maturity due to space constrictions. The base of the ovary is surrounded by a conspicuous receptacular nectary with

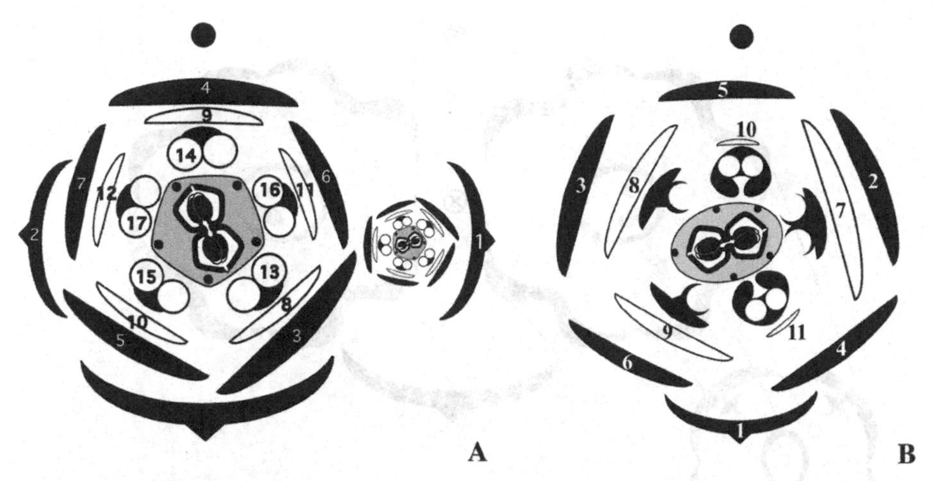

Figure 7.7 Sabiaceae. A. *Sabia limoniacea*; B. *Meliosma dilleniifolia*. Dots on nectary represent secretory appendages; numbers refer to order of initiation of the perianth (and stamens).

five prominent appendages alternating with the stamens (and staminodes). At maturity the nectary is either buoy shaped with weak crenellations, or bears five prominent appendages topped with one-three stomata.

Flowers of Sabiaceae are unusual in being pentamerous and having superposed sepals, petals and stamens. Based on developmental evidence, Ronse De Craene, Quandt and Wanntorp (2015b) suggested that the pentamerous flower of *Sabia* is derived from progenitors with a floral formula similar to trimerous Ranunculales. The superimposed arrangement of stamens with petals fits well in the general Bauplan found in the early diverging eudicots.

Proteaceae

Figure 7.8 *Lomatia tinctoria* (Labill.) R.Br.

↓ [P2+2 A2+2] $\underline{G}$1

Most genera are bisexual, rarely dioecious (e.g. *Aulax*, *Leucadendron*). Flowers share a similar floral Bauplan throughout the family, although inflorescence structure can be highly variable (occasional formation of pseudanthia as in *Protea*). In Grevilloideae, flowers are always paired in the axil of a bract (Figure 7.8), while in other subfamilies they are single. Douglas and Tucker (1996b) demonstrated that the two flowers are lateral branches of a short shoot. Individual flowers are generally subtended by a bract (not in *Lomatia*). The orientation of flowers relative to the main axis can be variable, ranging from straight to oblique (Douglas and Tucker, 1996b). The flowers are dimerous (apparently tetramerous) with two whorls of tepals (with valvate aestivation) and opposite adnate stamens. Stamens

Figure 7.8 *Lomatia tinctoria* (Proteaceae): partial inflorescence. The upper level with stamen attachment is shown on the left flower.

are rarely free from the tepals (e.g. *Symphyonema*). Interpretations of the perianth were controversial in the past, ranging from petals, sepals (with loss of petals), or undifferentiated tepals. Tepals appear to be postgenitally fused and are apically reflexed at anthesis exposing the attached anthers. Dehiscence of tepals is variable, either resulting in a polysymmetric flower by equal partitioning, or monosymmetric by a deeper slit on the adaxial side of the flower and the curving of all tepals towards the abaxial side (Figure 7.8). A few Proteaceae are structurally monosymmetric with a bilabiate perianth and abaxial (e.g. *Placospermum*) or adaxial (e.g. *Synaphea*) staminodes (Douglas and Tucker, 1996a, 1997). Half of the lateral anthers is sterile. In several *Protea* one stamen is sterile (Haber, 1966). The gynoecium is sessile or inserted on a gynophore and is always unicarpellate with two marginal rows of ovules or a single basal or apical mostly orthotropous ovule. Orientation of the single carpel tends to be variable in the family with eight possible orientations in Grevilloideae, depending on the space available for initiation (Douglas and Tucker, 1996c). The stigma is club shaped and functions as a receptor for pollen (secondary pollen presentation). Nectaries are present at the base of the ovary in alternation with the stamens (occasionally fewer: Figure 7.8). They have been interpreted as petals (e.g. Eames, 1961; Haber, 1966), although there is no evidence to support this (Douglas and Tucker, 1996a; Ronse De Craene and Smets, 2001a).

Proteaceae appear to be more closely related with Platanaceae and share some floral characters, such as interstaminal scales and fruiting structures. *Platanus* has a whorl of scales alternating with stamens that were interpreted as staminodes and could represent a second stamen whorl, as would be the inner glands of Proteaceae (Von Balthazar and Schönenberger, 2009).

Nelumbonaceae

Figure 7.9 *Nelumbo nucifera* L., based on Hayes, Schneider and Carlquist (2000)

∗/⟷ K2 C∞ A∞ $\underline{\text{G}}$∞

Nelumbo was placed in Nymphaeaceae in pre-molecular classifications based on strong convergent characteristics linked to a similar aquatic habitat (e.g. Ito, 1986b). However, a placement within Proteales comes as a surprise. The arrangement of solitary flowers on a submerged shoot is complex and little understood (Eichler, 1878; Stevens, 2001 onwards). Flowers are massive and have several characters in common with basal angiosperms (spiral petals, ascidiate carpels closed by secretion), although they are probably secondarily elaborated with an increase of petals, stamens and carpels (Hayes, Schneider and Carlquist, 2000). There is a hint of a dimerous flower in two outer sepals enclosing the flower. They resemble the outer perianth of Papaveraceae or the involucre of Montiaceae. Petals arise in a spiral sequence and enclose the androecium of a very high number of stamens. The fact that stamens arise on a ring primordium (Hayes, Schneider and Carlquist, 2000) and a fascicled vascular supply (Ito, 1986b) is an indication that they were secondarily increased. Carpels are scattered and embedded in a flattened expanded receptacle. Each carpel encloses a single apical anatropous ovule. There is no nectary tissue.

Trochodendrales

Trochodendraceae

Figure 7.10 *Tetracentron sinense* Oliv.

∗ P2+2 A2+2-G- (4)

Formula for *Trochodendron*: ∗/↺ P0 A∞- G- (4–)6–17

The family Trochodendraceae consists of two monospecific genera *Tetracentron* and *Trochodendron* with highly diverging morphologies. However, both share several less conspicuous characters (anther morphology, pollen, nectaries, carpel morphology: Endress, 1986).

Tetracentron is dimerous with flowers arranged spirally in catkin-like spikes (Chen et al., 2007). Young flowers look conspicuously like some Araceae, although the carpels are diagonally inserted, a condition found also in Potamogetonaceae (Alismatales). In *Trochodendron*, flowers are indefinite (variably spiral or whorled) with several stamens and carpels. The number of

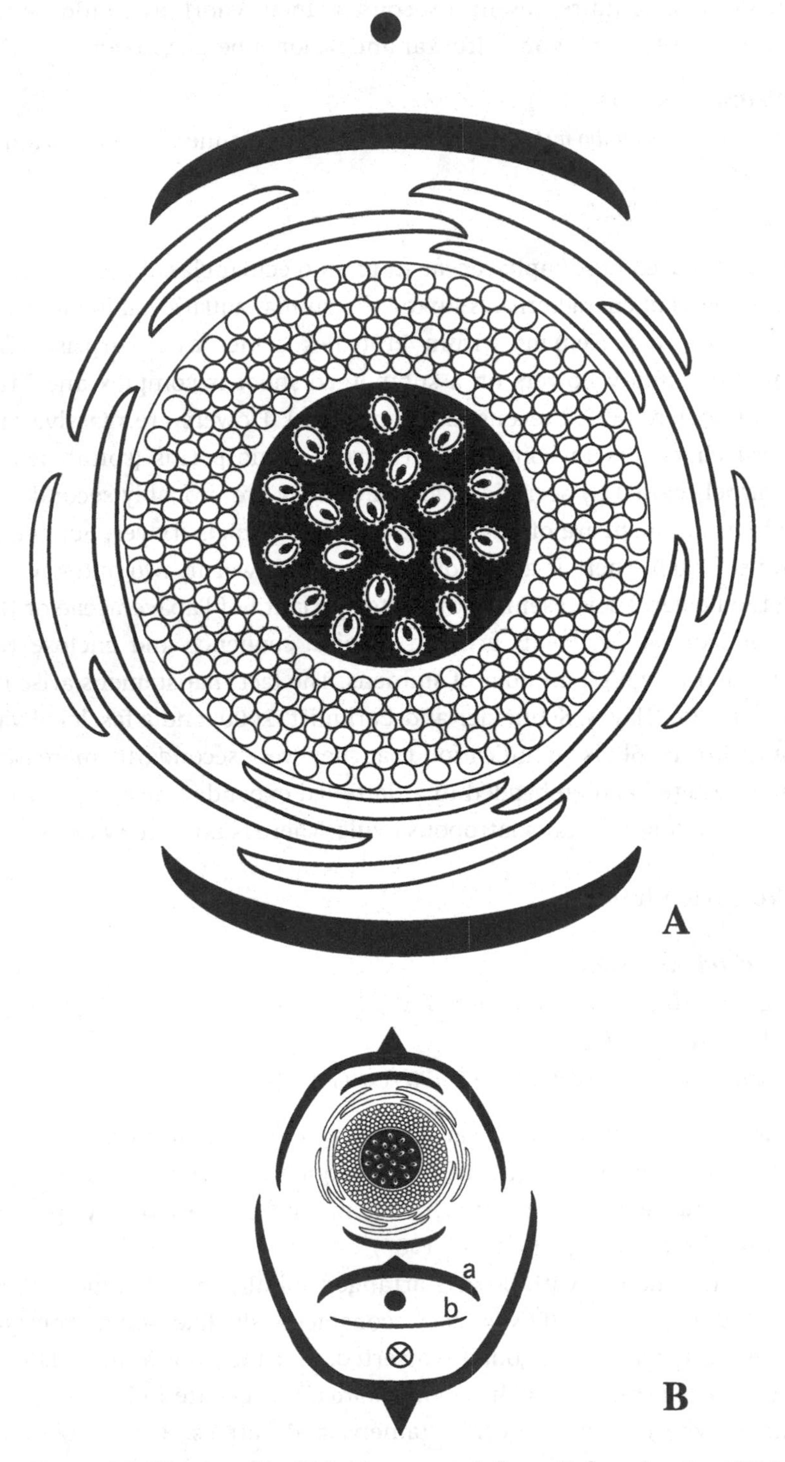

Figure 7.9 *Nelumbo nucifera* (Nelumbonaceae): A. flower; B. flowering shoot. White dots represent stamens arising centripetally on a ring primordium. a, leaf; b, stipule of leaf

Figure 7.10 *Tetracentron sinense* (Trochodendraceae). Nectary shown on the abaxial side of the carpels

stamens in *Trochodendron* ranges from thirty-nine to around seventy, and stamens are initiated in a helical sequence (Endress, 1986). Tepals are inconspicuous in *Tetracentron*, but reduced to small scales in *Trochodendron*, which has occasionally been interpreted as lacking tepals (Endress, 1986). Wu, Su and Hu (2007) recently confirmed the presence of tepals and suggested that the perianth was secondarily lost based on the presence of micromorphological characters of petals on several floral parts of *Trochodendron*. As in other early diverging eudicots (e.g. Buxaceae, Gunneraceae), the distinction between bracts and tepals is unclear (see Von Balthazar and Endress, 2002a, 2002b; Wanntorp and Ronse De Craene, 2005; Ronse De Craene, 2008). *Trochodendron* has two tiny lateral bracteoles that could be homologized with the lateral tepals of *Tetracentron*. In both genera the anthers open by two lateral valves.

The ovaries of *Tetracentron* and *Trochodendron* are half inferior with a small residual area and basally fused plicate carpels (Chen et al., 2007). In *Trochodendron* the carpel number is increased. Two rows of ovules are found on marginal placentae. The abaxial sloping side of the ovary develops as a nectary in both genera (Endress, 1986), and this is comparable to nectaries found in pistillate *Buxus*.

Trochodendraceae share a dimerous flower structure and absence of differentiation of the perianth with other early diverging eudicots. As for Buxaceae the existence of flowers with undifferentiated perianth and a nectary point to a secondary reversal from wind-pollinated ancestors to insect pollination.

Buxales

The order contains one family, Buxaceae. Flowers are simple, dimerous (or spiral) and apetalous, as in several other early diverging eudicots.

Buxaceae (incl. Didymelaceae and Haptanthaceae)

Figure 7.11A–C. *Buxus sempervirens* L., based on Von Balthazar and Endress (2002a, 2002b)

Staminate: ⟷ P2+2 A2+2 G0

Pistillate: ✱ P0 A0 $\underline{G}$(3)

General formula: staminate: ✱/⟷ P(0)2–4 A4–∞ G0; pistillate: ✱ P0 A0 G(2)3(4)

Floral morphology of the family has been extensively documented by Von Balthazar and Endress (2002a, 2002b). Flowers are unisexual and basically dimerous or spiral. Inflorescences are varied but in *Buxus* are mostly short compact spikes composed of lateral staminate flowers and a terminal pistillate flower (Figure 7.11C). Within inflorescences the arrangement of flowers can fluctuate between a decussate and spiral phyllotaxis (Von Balthazar and Endress, 2002b).

Staminate flowers invariably have a dimerous arrangement with four tepals and four stamens. In *Styloceras* a perianth is missing. In *Notobuxus* median stamens are arranged in a series of six or eight stamens, but it remains uncertain whether there are one, two or three whorls of stamens (Von Balthazar and Endress, 2002b: figure 10). In staminate flowers the gynoecium is replaced by a flat nectariferous pistillode or is absent (*Styloceras*).

Pistillate flowers invariably arise in a terminal position. The two to three carpels are surrounded by a variable number of bract-like phyllomes arising in a spiral sequence. In *Buxus*, pistillate flowers bear nectaries between the stylar lobes that appear convergent with the septal nectaries of monocots (Smets, 1988), and are probably homologous to the nectaries of Trochodendraceae. Each carpel is topped by a massive style with adaxial stigma and contains two parallel pendent ovules on an axile placentation.

Figure 7.11 *Buxus sempervirens* L. (Buxaceae): A. staminate flower; B. pistillate flower; C. inflorescence

In *Styloceras* and *Pachysandra* a false septum divides the ovary into four locules (Von Balthazar and Endress, 2002a). In *Pachysandra* and *Sarcoccoca* only staminate flowers are nectariferous and nectarless pistillate flowers take advantage of this (Vogel, 1998b).

Especially for pistillate flowers it is unclear whether a perianth is present. Bracts and tepals cannot be distinguished morphologically as they arise in a continuous spiral sequence, except for a slightly longer plastochron between presumed sepals and bracts, the occasional distinction between empty bracts and the bracts immediately surrounding the flowers, and fusions of the upper bracts (Von Balthazar and Endress, 2002b). On this basis Von Balthazar and Endress (2002b) interpreted the bracts immediately preceding the reproductive organs as a weakly differentiated perianth and as a progression in the elaboration of a perianth as found in core eudicots. However, it is also possible that a perianth is missing in pistillate flowers and perhaps in staminate flowers, as part of an evolutionary process of reduction affecting the early diverging eudicots and culminating in Gunnerales. A perianth is missing in the related *Didymeles*.

Gunnerales

The small order consists of two monogeneric families: Gunneraceae and Myrothamnaceae. Gunneraceae have traditionally been associated with Haloragaceae (e.g. Schindler, 1905), mainly because their reduced flower morphologies look similar. Molecular analyses (e.g. Soltis et al., 2003) have placed Gunnerales as the sister group of all Pentapetalae with strong support. The dimerous floral Bauplan of Gunnerales tends to be more similar to the basal eudicot grade, such as Buxales or Trochodendrales, than to Pentapetalae, and this has led to a questioning of the origin of pentamery from dimerous ancestors at the base of Pentapetalae (Ronse De Craene, 2004; Wanntorp and Ronse De Craene, 2005). Within Gunneraceae there is a general trend to floral reduction and unisexual flowers and it appears structurally difficult to derive pentamerous flowers from a prototype such as *Gunnera* (Wanntorp and Ronse De Craene, 2005; Ronse De Craene and Wanntorp, 2006). Absence of clear intermediates, both extant and fossil, makes the understanding of floral evolution difficult (cf. Ronse De Craene, Soltis and Soltis, 2003; Ronse De Craene, 2008) and the floral evolution of Gunnerales may well be a unique specialization linked to wind pollination, as suggested by Wanntorp and Ronse De Craene (2005). Flowers of Myrothamnaceae are even more reduced without clear differentiation of a perianth and a distinction between staminate and pistillate plants. (e.g. Jäger-Zurn, 1966). Flowers tend to be labile with variable number of stamens and carpels and a perianth of uncertain homology.

Gunneraceae

Figure 7.12 *Gunnera manicata* Linden ex André
⟷ K2 C2 A2 Ğ(2)

Flowers are grouped in dense compound inflorescences, with each partial inflorescence ending with a fully developed terminal flower. The basic floral condition is generally found in subgenus *Panke* to which *G. manicata* belongs (Ronse De Craene and Wanntorp, 2006; González and Bello, 2009). There is a progressive reduction from the distal side of the inflorescence to the proximal side with loss of petals and stamens. Stamens and petals tend to be genetically connected and arise from common primordia (González and Bello, 2009). Loss of stamens tends to be correlated with reduction of petals and corresponds with several independent derivations of unisexual flowers in the different subgenera of *Gunnera*. Ronse De Craene and Wanntorp (2006: figure 73) provided a diagram linking different flower types. They demonstrated a clear tendency for unisexuality linked with the loss of perianth parts. Flowers are wind pollinated and petals tend to play a role in anther protection and pollen release (González and Bello, 2009).

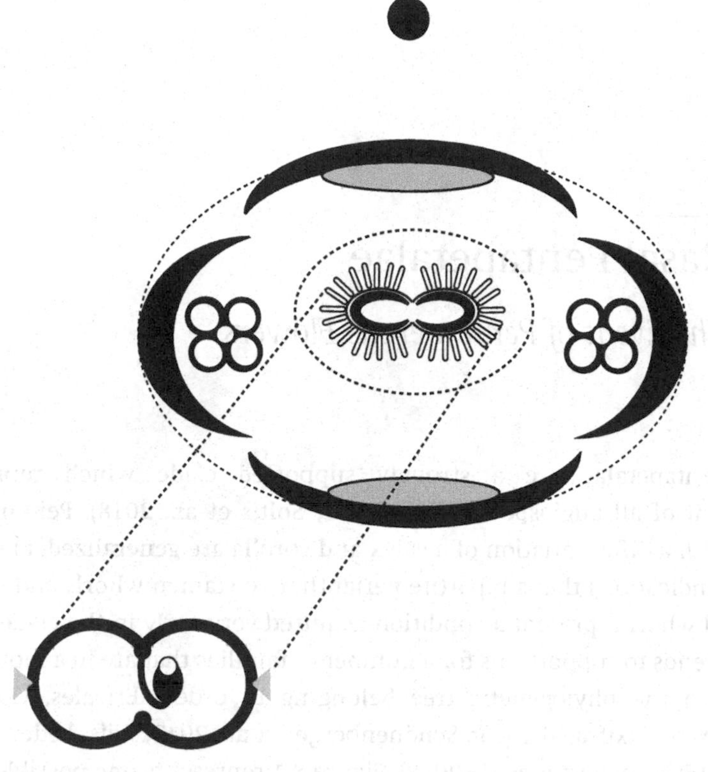

Figure 7.12 *Gunnera manicata* (Gunneraceae). Glandular tissue consists of hydathodes.

Flowers are unusual in the dimerous, almost distichous arrangement of petals, stamens and carpels. Sepals are median and persistent and were compared to bracteoles (e.g. Soltis et al., 2003); they have an inflated base by the presence of a hydathode while the apical part shrivels at maturity (Ronse De Craene and Wanntorp, 2006; González and Bello, 2009). Flowers have a half-inferior to inferior ovary with a single apical ovule; the ovary is considered pseudomonomerous (Eckardt, 1937). Two strongly developed styles are usually present, especially in unisexual flowers, where they take up the greatest area of the flower (Ronse De Craene and Wanntorp, 2006).

A possible interpretation for the unusual distichous arrangement of petals, stamens and carpels was proposed by Wanntorp and Ronse De Craene (2006) as the result of partial loss of organs from a dimerous, decussate ancestor similar to Buxaceae or *Tetracentron*.

8

Basal Pentapetalae

The Event of Pentamerous Flowers

Pentapetalae are a strongly supported clade which represents 70 per cent of all angiosperms (Figure 8.1; Soltis et al., 2018). Pentamerous flowers with a differentiation of a calyx and corolla are generalized, and there are good indications that a bipartite perianth, two stamen whorls and isomerous carpel whorl represent a condition acquired very early in the clade. Fossil evidence tends to support this for a number of families that are in a more basal position on the phylogenetic tree belonging to orders Ericales, Cornales, Cunoniales or Saxifragales (e.g. Schönenberger et al., 2012; Friis, Pedersen and Crane, 2016; Manchester et al., 2018). Figure 8.1 represents one possible topology out of twelve recent proposals for affinities between major lineages of Pentapetalae, although lack of resolution does not allow for a clear understanding of floral evolution (see Zeng et al., 2017). The positions of Berberidopsidales, Dilleniaceae, and Santalales remain largely unsettled, reflecting a rapid radiation at the base of Pentapetalae.

8.1 Berberidopsidales

The order was created by Soltis et al. (2003) in order to accommodate two families - Berberidopsidaceae and Aextoxicaceae - at the base of Pentapetalae. Berberidopsidaceae used to be placed in the heterogeneous Flacourtiaceae and Aextoxicaceae used to have an uncertain position somewhere close to Euphorbiaceae. Berberidopsidales have been variously placed in core eudicots at the base of the clade (Soltis et al., 2003), or in Superasteridae as sister to a clade consisting of asterids and caryophyllids (Wang et al., 2009; Figure 8.1), but its position remains uncertain (Zeng et al., 2017). Morphological evidence for a close affinity between the two families was given by Ronse De Craene (2004, 2007, 2017a) and Ronse De Craene and Stuppy (2010), who also

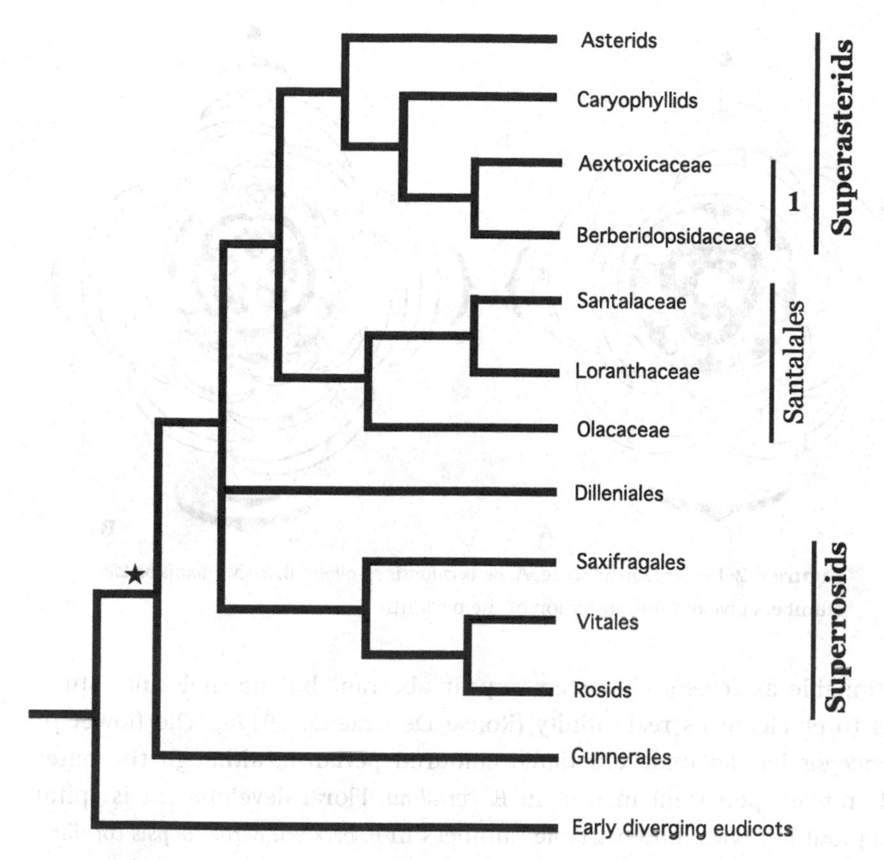

Figure 8.1 Phylogenetic tree of Pentapetalae, based on Soltis et al. (2018). 1. Berberidopsidales. Asterisk: limit of Pentapetalae floral Bauplan

demonstrated the basal position of *Berberidopsis* in Pentapetalae on the basis of floral developmental evidence. Both families represent a transitional stage in the evolution of early Pentapetalae, as they possess spiral flowers with an undifferentiated perianth or with petals of a transitional nature.

Berberidopsidaceae

Figure 8.2A *Berberidopsis corallina* Hook. f., based on Ronse De Craene (2004)
↻* P12 A8 G3

Figure 8.2B *Berberidopsis beckleri* (F. Muell.) Veldkamp, based on Ronse De Craene (2017a)
↻* P15–18 A(8–)10–11(–13) G$\underline{5}$(6)

Two species of *Berberidopsis* and the monotypic *Streptothamnus* make up Berberidopsidaceae. The position of *Streptothamnus* in Berberidopsidaceae is

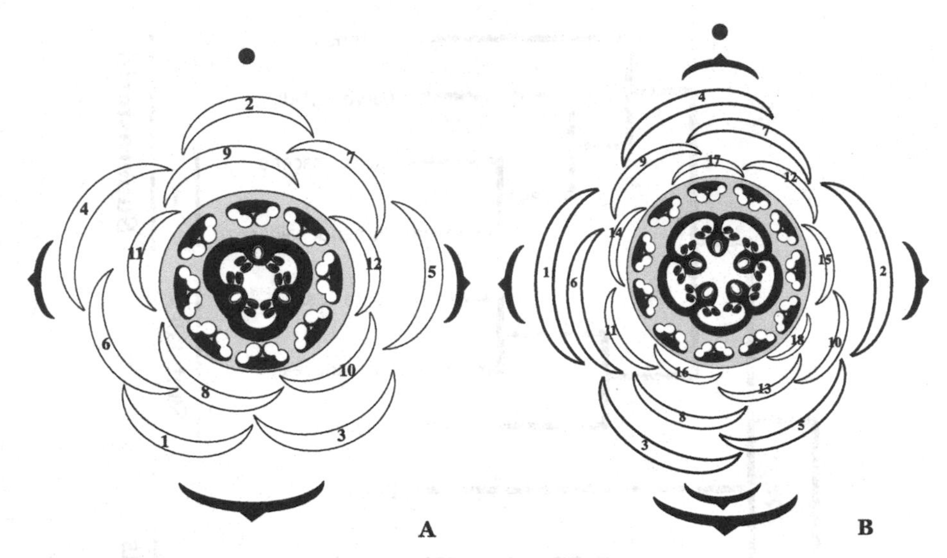

Figure 8.2 Berberidopsidaceae. A. *Berberidopsis corallina*; B. *Berberidopsis beckleri*. Numbers give order of initiation of the perianth.

questionable as several characters appear aberrant but no molecular study exists to elucidate its real affinity (Ronse De Craene, 2017a). The flower of *Berberidopsis* has an undifferentiated coloured perianth, although the outer tepals remain persistent in fruit in *B. corallina*. Floral development is spiral throughout but with more variable numbers in *B. beckleri*. *Berberidopsis corallina* has eight stamens in a single whorl, while the number is higher and more variable in *B. beckleri*. Anthers are broad with protruding connective and the filament is short. A nectary develops as an extrastaminal ring. The ovary is globular with tapering style and multilobed stigma developing by the extension of ridges alternating with the carpels. Ovules develop on parietal placentae in two unequal rows.

Ronse De Craene (2004, 2017a) hypothesized that the flower of *Berberidopsis* could act as a precursor to the biseriate perianth of Pentapetalae, by a progressive differentiation of an outer and inner whorl of tepals into sepals and petals and their arrangement in pentamerous whorls. Ronse De Craene (2017a) discussed the possibility of a direct derivation of the paracarpous gynoecium with parietal placentation from apocarpous precursors (cf. Doyle and Endress, 2009). Kubitzki (2007) criticized the assumption that *Berberidopsis* is a basal link in floral evolution, on the basis that the family has several derived characteristics, including the parietal ovary, differentiation of calyx and corolla and multiple androecium in *Streptothamnus*. However, it is unlikely that the floral morphology of *Berberidopsis* represents a reversal as both species have a unique morphology and there is no disruption in the developmental sequence of the flowers as found in other taxa with a reversal to a spiral flower (e.g. Paeoniaceae: p. 187). The flower probably represents an early core eudicot preceding the stable pentacyclic Pentapetalae Bauplan.

Aextoxicaceae

Figure 8.3A–B *Aextoxicon punctatum* Ruíz and Pav., based on Ronse De Craene and Stuppy (2010)
∗ K5–6 C5–6 A5–6 $\underline{G}$1

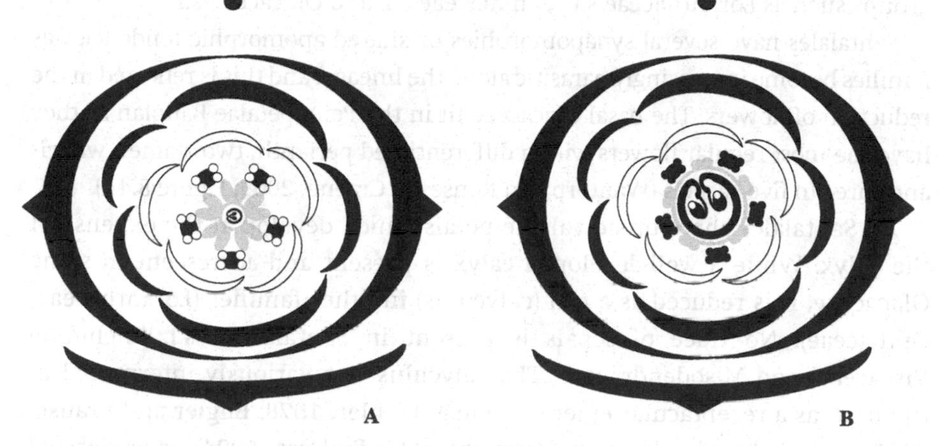

Figure 8.3 Staminate (A) and pistillate (B) flower of *Aextoxicon punctatum* (Aextoxicaceae). Note the difference in size between nectaries of staminate and pistillate flowers.

The monotypic genus *Aextoxicon* is morphologically highly different from *Berberidopsis*, although several characters tend to be common to both genera, including wood anatomy and micromorphological evidence (Ronse De Craene, 2004, 2017a). Flowers are functionally staminate or pistillate, with a differentiation of genders arising late in the development. Two bracteoles enclose the flower bud as a calyptra that is shed at anthesis. The same decussate arrangement as the bracteoles is found in the outer sepals where a transition to the spiral perianth occurs (Ronse De Craene and Stuppy, 2010). Transitions between sepals, petals and stamens appear to be progressive. The ovary is unicarpellate with variable position and with two apical marginal ovules. The style is very short and compressed against the globular ovary. The nectary consists of broad wing-like appendages inserted between the stamens; glandular tissue is more extensively developed in staminate flowers due to space restrictions in the pistillate flowers.

Aextoxicon compares well with *Berberidopsis* in the spiral flowers with clear homology of perianth parts. Both families represent a clear transitional stage in the evolution of Pentapetalae.

8.2 Santalales

The order contains about seventeen families with a basal paraphyletic Olacaceae and containing several small monophyletic entities (Nickrent et al., 2010). Resolution within the clade is poor and prevents rearrangement in more practical entities (Soltis et al., 2018), while I would prefer to consider larger groups such as Loranthaceae s.l., Santalaceae s.l. and Olacaceae s.l.

Santalales have several synapomorphies or shared apomorphic tendencies as families become increasingly parasitic along the lineage, and this is reflected in the reduction of flowers. The basal Olacaceae fit in the Pentapetalae Bauplan as they have the most regular flowers with a differentiated perianth, two stamen whorls and three to five carpels (Wantorp and Ronse De Craene, 2009; Figure 8.4A).

All Santalales share broad valvate petals which develop at the expense of the calyx. While a well-developed calyx is present and accrescent in some Olacaceae, it is reduced as a rim (calyculus) in other families (Loranthaceae, Opiliaceae). No trace of sepals is present in Santalaceae s.l. (including Viscaceae) and Misodendraceae. The calyculus was variously interpreted in the past, as a receptacular emergence (e.g. Eichler, 1878; Engler and Krause, 1935), as a reduced calyx (e.g. Sleumer, 1935; Endress, 1994) or as derived from bracteoles (Venkata Rao, 1963; Wanntorp and Ronse De Craene, 2009). Recent morphological and developmental investigations (Kuijt, 2013; Suaza-Gaviria et al., 2016, 2017) support the origin of the calyculus as a reduced calyx rim. Obhaplostemony is present in most taxa by loss of antesepalous stamens. A few genera of Olacaceae are diplostemonous (e.g. *Heisteria*, *Strombosia*). The gynoecium is rarely isomerous, mostly tricarpellate and superior to inferior with a single style. The placentation is free-central, usually with three ovule primordia. While more basal groups have ovules developing two integuments, there is a progressive reduction within the order to unitegmic and ategmic ovules, or even without ovule differentiation (Brown, Nickrent and Gasser, 2010). A nectary develops as a disc around or on top of the ovary. Several Santalales have tufts of hair between the petals and the stamens. Their function is currently unknown, but they may play a role in the retention of nectar (Endress and Matthews, 2006b).

The floral structure and development of several of the former Olacaceae is still unknown but can give important clues about the evolution of flower and perianth in the order (cf. Wanntorp and Ronse De Craene, 2009).

Olacaceae s.l. (incl. Coulaceae, Aptandraceae, Ximeniaceae, Strombosiaceae, Erythropalaceae)
Figure 8.4A *Heisteria parviflora* Sm.
∗ K(5) C5 A5+5 G(3)

Figure 8.4 Olacaceae: A. *Heisteria parviflora*, partial inflorescence; B. *Coula edulis*. Note the alternipetalous triplets of stamens.

Figure 8.4.B *Coula edulis* Baill.
∗ K(?) C5 A5^{3}+5 $\underline{G}$(3)
General formula: ∗ K(0)3–6 C3–6 A3–12(–18) G(2–)3(–5)

Olacaceae are recognized as paraphyletic, and several genera, including *Coula*, belong to different clades. *Olax* is more closely related to Loranthaceae and shares a similar calyculus (Wanntorp and Ronse De Craene, 2009). More basal Olacaceae have a well-differentiated calyx and corolla with diplostemonous flowers (e.g. *Strombosia*, *Heisteria*: Figure 8.4A) and a tendency for loss of antesepalous stamens (e.g. *Diogoa*), very rarely antepetalous stamens (e.g. *Heisteria pentandra*: Sleumer, 1935). *Coula* is exceptional in Santalales because it has twelve to twenty stamens in a diplostemonous arrangement, with antesepalous stamens arising as triplets (Figure 8.4B). The antepetalous stamens are occasionally missing. Previous authors interpreted the smaller lateral stamens of *Coula* as pairs. Sleumer (1935: 7) showed floral diagrams of several Olacaceae, including *Coula* with fifteen stamens.

The calyx, while being narrow and basally fused, is accrescent in some genera (e.g. *Heisteria*) and plays an important role in seed dispersal by a difference in colour with the ovary. In other genera, including *Coula*, the calyx is represented by a shallow rim with narrow teeth, comparable to a calyculus. Petals are usually thick and valvate, compressing the stamens against the ovary in bud. *Olax* has hexamerous flowers with petals fused into pairs and a variable number of stamens alternating with staminodes (Wanntorp and Ronse De Craene, 2009). The ovary is (two-) three- to five-carpellate and septa are often formed, leading to axile placentation. However, there is a tendency for septa to become reduced (e.g. *Cathedra*). Ovules in *Heisteria* are narrow without clear development of integuments.

Loranthaceae s.l. (incl. Misodendraceae, Schoepfiaceae)
Figure 8.5 *Phthirusa pyrifolia* Eichler
∗ K? [C6/3+3 A6/3+3] Ğ(3)

Several former Olacaceae appear to be closely related to Loranthaceae and Opiliaceae (including *Olax* and *Schoepfia*). All share the presence of a calyculus enwrapping the flower.

In most Loranthaceae flowers are arranged as triplets (basically a dichasium) and bracts and bracteoles (in cases where they are present) are basally fused (Engler and Krause, 1935). In *Struthanthus* there are no extra bracteoles (Wanntorp and Ronse De Craene, 2009). Flowers are bisexual or unisexual, with the laterals of the triplet staminate (*Nuytsia*: Narayana, 1958b). Flowers are tetramerous, pentamerous or hexamerous (Figure 8.5), or even occasionally heptamerous (e.g. *Phthirusa*, *Nuytsia*). Hexamerous flowers appear trimerous (cf.

Figure 8.5 *Phthirusa pyrifolia* (Loranthaceae): partial inflorescence

Polygonaceae) because petals and stamens are arranged in two whorls of different size (e.g. *Phthirusa, Loranthus*). The petals have valvate aestivation and can have a long basal tube (e.g. *Psittacanthus*) and stamens appear to be fused with the petals as in Proteaceae (cf. Venkata Rao, 1963). The flower may appear monosymmetric by unequal division of the corolla lobes. This is probably caused by space restrictions. The ovary is inferior and is often covered by a broad disc surrounding a central erect style. The ovary is rarely septate (*Lysania*: Narayana 1958a), usually unilocular with central placenta or is solid (e.g. *Phthirusa*). Ovules lack a nucellus and integuments. Only one ovule develops as a single seed. Narayana (1958b) presents a progressive reduction series in the family from a well-developed central placenta with ovules to a complete suppression of the central column.

Quinchamalium (Schoepfiaceae) lacks a calyculus but the bract and bracteoles fuse into a cuplike involucre that resembles a calyx and was described as a calyculus by Pilger (1935).

Santalaceae (incl. Thesiaceae, Cervantiesiaceae, Comandraceae, Nanodeaceae, Santalaceae, Balanophoraceae, Amphorogynaceae, Viscaceae)

Figure 8.6 *Thesium strictum* Berg

∗ K0 [C5 A5] Ğ(3)

General formula: ∗ K0 C3–5 A3–5 G (2–5)

Figure 8.6 *Thesium strictum* (Santalaceae), partial inflorescence

Santalaceae closely resemble Loranthaceae morphologically but they lack the typical calyculus of the latter. Flowers are bisexual or unisexual often with a residue of the other gender (e.g. *Osyris*, *Colpoon*). Inflorescences are varied but a dichasial arrangement of flowers is most common. Flowers are pentamerous, tetramerous or trimerous, mostly subtended by a bract and two bracteoles, which may be rarely absent (e.g. *Thesium ebracteatum*: Pilger, 1935). Santalaceae have relatively simple flowers with a valvate corolla and antepetalous anthers. Petals have occasionally been described as tepals, but a comparison with other Santalales supports the interpretation of a corolla. The odd petal is abaxial in pentamerous flowers as is to be expected for petals. Filaments are usually very short and anthers are broadly flattened. A tuft of hairs usually develops between petal and stamen. In *Thesium* petals and stamens develop from common primordia (Ronse De Craene, unpubl. data). The gynoecium is often inferior, rarely half-inferior or superior, and the style is embedded in a broad dish-like nectary that often extends as erect lobes between the stamens (e.g. *Santalum*, *Osyris*). The number of carpels is usually three, rarely less (two) or more (four or five). Stamens and petals are often lifted by a hypanthium enclosing the inferior ovary. No septa are formed and the central axis develops as a placental column bearing three long ovular protuberances without integuments (cf. Sattler, 1973 for *Comandra*). The placental column remains short or can grow into a coiled snake-like protuberance filling the ovarian cavity. Viscaceae have often been associated with Loranthaceae (e.g. Eichler, 1878; Engler and Krause, 1935).

However, their affinities are closer to Santalaceae with whom they share simplified flowers without calyculus (Malécot and Nickrent, 2008).

8.3 Dilleniales

Dilleniaceae

Figure 8.7 *Hibbertia cuneiformis* (Labill.) Smith
∗ K5 C5 A5^4 $\underline{\text{G}}$5

General diagram: ∗/↓ K(3)4–5(–18) C(2)3–5(–7) A(1–3)5–∞, rarely 5+5 G1–10(20)

Dilleniaceae were viewed as an archaic family at the base of Dilleniidae in the past (e.g. Cronquist, 1981). The position of Dilleniaceae is debatable, as the family was variously associated with caryophyllids and asterids, or Superrosidae (e.g. Stevens, 2001 onwards; Soltis et al., 2003, 2018; Zeng et al., 2017). Zeng et al. (2017) list a number of characters to link Dilleniaceae with Caryophyllales, such as the persistent calyx, centrifugally developing multistaminate androecium, variable carpel numbers and campylotrous ovules with thin outer integuments. A better understanding of floral structure and development of basal Caryophyllales, such as Rhabdodendraceae, might give clues about the relationships of Dilleniaceae (Horn, 2007).

Dilleniaceae are nectarless pollen flowers, principally bee pollinated and usually with many stamens (Endress, 2007). Flowers can be terminal and solitary, or are arranged in various cymose inflorescences. Characteristic features of the family are: imbricate perianth aestivation (quincuncial in pentamerous flowers), showy, (usually) polysymmetric and rapidly caducous petals crumpled in bud, secondary polyandry with common primordia or a ring primordium, and basifixed anthers (Judd and Olmstead, 2004; Horn, 2007).

Dilleniaceae have a spiral perianth with imbricate sepals and petals, but the flower is clearly whorled with several derived patterns, especially in the androecium. Merism can be highly variable, although pentamery is predominant and is probably the ancestral condition. The highest variation in merism is found in the calyx of *Tetracera*, ranging from two smaller and three larger sepals to seven to fifteen helically arranged sepals (Horn, 2007). The number of petals can be reduced to three in some species of *Hibbertia*. *Davilla* has three smaller outer and two larger inner sepals. The latter enclose the bud as two lids before and after anthesis.

The greatest variation is found in the androecium. Most Dilleniaceae are polyandric and there is a remarkable correlation between the number of carpels and the extent of development of the androecium. The most common

Figure 8.7 *Hibbertia cuneiformis* (Dilleniaceae). Note the protruding styles between the stamen fascicles.

development is five common stamen primordia alternating with five carpels (Figure 8.7). With three carpels, two stamens remain simple and the three alternating with the carpels are common primordia dividing into more stamens (e.g. *H. stellaris*: pers. obs.). Many more stamens (up to nine hundred in some *Dillenia*) arise by the development of a ring primordium, and this is often correlated with a lateral increase of carpels up to ten (twenty) and much larger flowers (Endress, 1997, 2014). The total number of stamens depends on the number of common primordia and the duration of meristematic activity as the outermost stamens often remain undeveloped. Tucker and Bernhardt (2000) showed a remarkable variation in stamen number and development in *Hibbertia*. The number of stamens is increased by the development from common primordia, but several trends have led to further increases or reductions. Heteranthy occurs in some genera with different sets of inner and outer stamens. Several authors have suggested that polyandry is ancestral in the family (e.g. Horn, 2007; Dickison, 1970), although this seems of little consequence in the light of the high variation in stamen development that can be linked to a single stamen whorl in most cases.

Some species of *Dillenia*, *Hibbertia* or *Didesmandra* are monosymmetric by the reduction of one side of the flower. Petal initiation is unidirectional instead of helical and the androecium is restricted to the adaxial side of the flower, occasionally with the presence of staminodes on the abaxial side (*H. empetrifolia*, *H. hypericoides*: Tucker and Bernhardt, 2000). Zygomorphy is correlated with a reduction of carpels to two and can result in a single stamen only.

Hibbertia (subg. *Adrastea*) *salicifolia* was described as an obdiplostemonous flower with ten stamens (Eichler, 1878). The floral development is highly unusual in the centrifugal initiation of the antepetalous stamens on a ring primordium and heteranthy between the two whorls of stamens (Tucker and Bernhardt, 2000). The occasional presence of more external stamens indicates that this has nothing to do with obdiplostemony but appears to be an interrupted centrifugal multiplication of the androecium, similar to an analogous development in *Triumfetta* (Malvaceae: Van Heel, 1966) or *Sericolea* (Elaeocarpaceae (Ronse De Craene and Bull-Hereñu, 2016).

The gynoecium is normally apocarpous. In some *Dillenia* with a specific pollination mechanism, carpels can be syncarpous by basal concrescence to the receptacle (Endress 1997). Higher carpel numbers are clearly derived, and arrangement is in one whorl often with unequal closure of carpels in the centre of the flower due to lack of space (Endress, 2014), rarely two whorls (*H. grossulariifolia*: Tucker and Bernhardt, 2000). The latter condition is questionable, as carpels may have become displaced in the flower. Carpels are always opposite the petals in cases where they are isomerous; with fewer carpels the position can vary. The orientation of monosymmetric flowers of *Hibbertia* is transverse and so are the two carpels, in contrast to *Hibbertia* subg. *Adrastea* with median orientation (see Eichler, 1978; Horn, 2007). Ovules are inserted in two, up to six rows per carpel, or reduced to one-two with basal placentation. A diagram of *H. cuneiformis* was shown by Baillon (1868a); diagrams of other *Hibbertia* were shown in Tucker and Bernhardt (2000: 1919).

9

Rosids

The Diplostemonous Alliance

The rosid clade is well supported, but it is the least resolved major clade of Pentapetalae, containing more than a quarter of all angiosperm species (Schönenberger and Von Balthazar, 2006). It is clear that Saxifragales represent an ancient early diverging lineage in the Pentapetalae, closely related to rosids (Soltis et al., 2005, 2018; Magallón, 2007). Recent analyses incorporating a high number of genes have clarified the internal relationships of rosids with the recognition of two main clades – fabids (or Fabidae) and malvids (or Malvidae) – with the inclusion of Saxifragales as a basal order and Vitaceae as sister to the rosid clade in a large clade called Superrosidae (Wang et al., 2009; Soltis et al., 2011). Figure 9.1 represents a phylogenetic tree of the rosids based on Wang et al. (2009).

9.1 The Basal Clades: Saxifragales and Vitales

Saxifragales

Saxifragales *sensu* APG (2003) comprises an assemblage of highly diverse families (about fifteen), including core Saxifrages and allies, part of the former Hamamelidae of Cronquist (1981), and strongly reduced flowers.

The general flower morphology fits well with the syndrome found in other Superrosidae. Flowers tend to be generally pentamerous or tetramerous and often share a hypanthium with a half-inferior ovary. Merism increase evolved independently in Penthoraceae and Crassulaceae. Obdiplostemony is a common feature in families with two whorls of stamens (e.g. Saxifragaceae, Haloragaceae, Crassulaceae: Ronse De Craene and Smets, 1995b; Ronse De Craene and Bull-Hereñu, 2016). Obdiplostemony may be linked to the general weakening of the petals, which are lost in several genera. There is a general

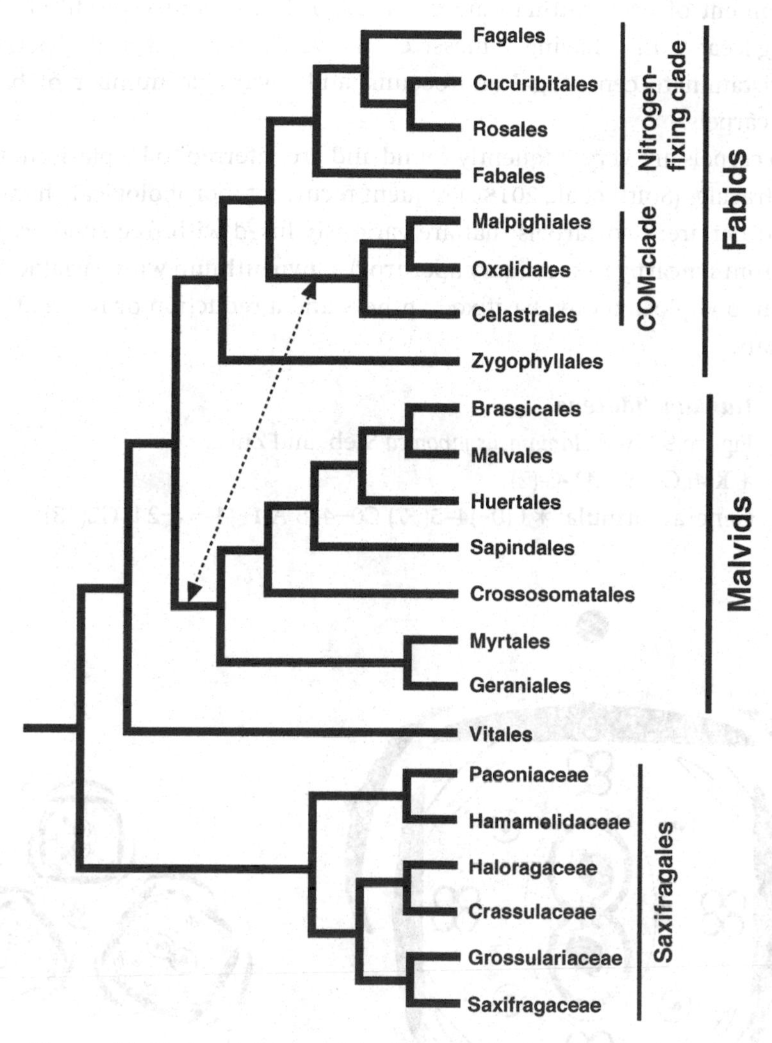

Figure 9.1 Phylogenetic tree of the Superrosidae, based on Soltis et al. (2018). The double-facing arrow points to the alternative position of the COM clade.

trend to haplostemony (e.g. Iteaceae, Grossulariaceae, Saxifragaceae), and this is often linked with a reduction of the size of the petals and petaloidy of the hypanthium and sepal lobes. Within Saxifragales is an entire range of transitions between flowers with well-developed petals to apetalous genera (e.g. Saxifragaceae, Pterostemonaceae, Grossulariaceae). The perianth is lost in a number of families (part of Hamamelidaceae, Cercidiphyllaceae, Daphniphyllaceae). Sympetaly is rare and has only been found in some Crassulaceae where it is linked with hypanthial growth. Hamamelidaceae show high variation in floral diagrams, linked with wind pollination and the development of pseudanthia (Endress 1967, 1976). Paeoniaceae differs from Saxifragaceae in having massive flowers with spiral perianth, a multistaminate centrifugal androecium and a variable number of basally united carpels.

Two carpels are very frequently found and are inferred to be plesiomorphic in Saxifragales (Soltis et al., 2018). Frequent recurrent morphological characters in the order are: two carpels that are variously fused with free stylodes (with shifts from syncarpy to secondary apocarpy), a hypanthium with variable ovary position, obdiplostemony, basifixed anthers and a reduction or retardation of the petals.

Hamamelidaceae

Figure 9.2A–B *Hamamelis japonica* Sieb. and Zucc.

∗ K(4) C4 A4+4° -G-(2)

General formula: ∗ K(0–)4–5(–7) C0–4–5 A(1–)4–5 (–24) G2(–3)

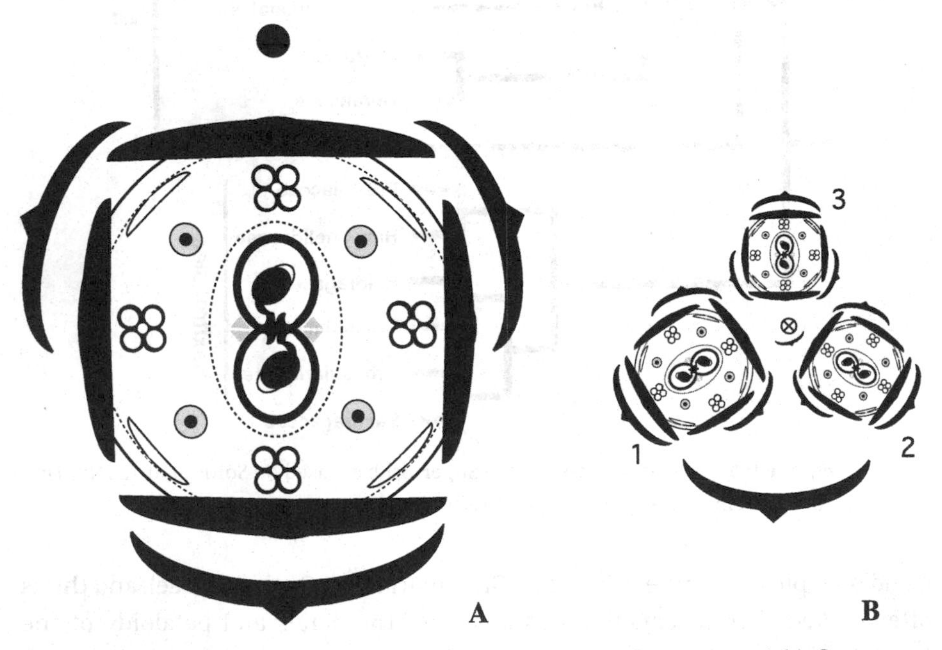

Figure 9.2 *Hamamelis japonica* (Hamamelidaceae): A. flower; B. partial inflorescence with three flowers

The family is highly diverse and the variability is often linked with the evolution of a wind pollination syndrome. The ovary is usually bicarpellate in the family, partly fused and with free stylodes. Flowers of *Hamamelis* are grouped in triplets or pairs on a short shoot bearing a terminal aborted apex. The genus *Hamamelis* is closest to other Saxifragales in having narrow petals, a hypanthium, a bicarpellate gynoecium and weak obdiplostemony. Petals, when present, are narrow and coiled and superposed to filament-shaped nectariferous staminodes in *Hamamelis*. In other genera such as *Corylopsis*, ten intrastaminal protuberances produce nectar besides the staminodes, or nectaries are present on the petal bases (*Disanthus*). The fertile antesepalous whorl bears dithecal monosporangiate anthers dehiscing by adaxially curved flaps (also in *Exbucklandia*, but disporangiate in other genera). The ovary is superior to (half) inferior and has two apical ovules (rarely more), of which one aborts in *Hamamelis* (Eichler, 1878). Diagrams of inflorescences and flowers of *H. virginiana* were shown in Mione and Bogle (1990).

In some genera the number and arrangement of floral parts become irregular with the reduction and loss of petals or the entire perianth and the aggregation of several reduced flowers in elaborate pseudanthial structures with large bracts functioning as attractive organs (e.g. *Parrotia, Distylium, Rhodoleia, Parrotiopsis*: Endress, 1978). In apetalous genera flowers are often unisexual (Endress, 1978). A few taxa have undergone a secondary increase in the androecium (*Fothergilla, Matudea*: Endress, 1976, 1978). Some taxa have an additional whorl of 'sterile phyllomes' or scales within the staminodial whorl (e.g. *Maingaya, Loropetalum, Corylopsis*) that were interpreted together with the petals as staminodial in origin (Mione and Bogle, 1990). This would make the androecium ancestrally multiwhorled. However, this interpretation is uncertain as scales arise very late in ontogeny (Endress, 1967) and the nearest sister groups of Hamamelidaceae are diplostemonous. The inner scales are best interpreted as receptacular emergences. Magallón (2007) analysed character evolution in the Hamamelidoideae relative to Saxifragales; she concluded that pentamerous and tetramerous flowers were derived three times, and that a bicarpellate gynoecium is plesiomorphic in the family.

Paeoniaceae

Figure 9.3 *Paeonia delavayi* Franch

↻ K2–5 C7–9 A5$^{\infty}$ $\underline{G}$3–4

General floral formula: ↻ K3–5(-7) C5–13 A5$^{\infty}$ G2–15

Flowers are usually solitary at the end of branches, often with a smaller flower just below. Flowers are large and have a variable number of bracts, sepals and petals, with a spiral initiation throughout. The flower bears a hypanthium that can be more or less deeply developed. Bracts are leaflike structures closely pressed to the flower and progressively transgressing into the sepals, followed by the petals. Sepals and petals intergrade and transitional organs are frequently found, although outer sepals and bracts are persistent. Sepals

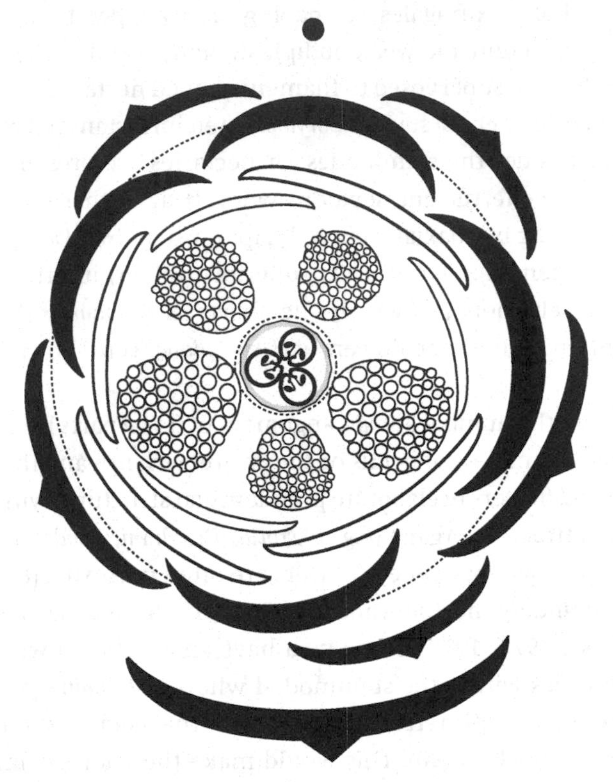

Figure 9.3 *Paeonia delavayi* (Paeoniaceae). Note the stamens arising on five common primordia.

occasionally show a 2/5 arrangement, but numbers may fluctuate with bracts invading the confines of the flower (e.g. *P. veitchii*: pers. obs.). Stamens are always numerous, arising centrifugally from five common primordia alternating with the petals (Hiepko, 1964; Leins and Erbar, 1991). The outermost stamens may be occasionally staminodial. A crenelated receptacular disc (nectary) develops between androecium and gynoecium. This was interpreted as of staminodial origin on the basis of vasculature (Melville, 1984), although not supported by floral ontogeny. According to Hiepko (1964, 1966) the disc should be interpreted as a receptacular emergence. Hiepko (1966) mentioned that the disc is not secretory, although secretion from crushed inner stamens occurs in some species. It is possible that the nectary has become defunct due to a transfer to pollen rewards through the many stamens. Erbar (2014) mentions the production of nectar in some species. The gynoecium is apocarpous with a variable number of follicular carpels fused at the base. Ovules develop in two rows on the folded margins. Paeoniaceae were associated with Ranunculaceae or

Dilleniaceae in the past because of superficial similarities in the flower construction. Its placement in Saxifragales is supported by the presence of a hypanthium and apocarpous gynoecium, although the multistaminate androecium and apocarpy are clearly derived (cf. Soltis et al., 2005). Paeoniaceae are anomalous in Pentapetalae in having massive flowers without clear boundaries between bracts, sepals and petals. It is postulated that the increase in stamens triggered a disturbance of the genetic boundaries of different organ categories and this has led to a breakdown of the whorled arrangement and a proliferation of perianth parts (Ronse De Craene, 2007). A simplified floral diagram of *Paeonia officinalis* was presented by Leins and Erbar (1991).

Haloragaceae

Figure 9.4 *Haloragis erecta* (Murr.) Oken

∗ K4 C4 A4+4 Ğ(4)

General formula: ∗ K(0)2–4 C(0)2–4 A(2–)4–8 Ğ 2–4

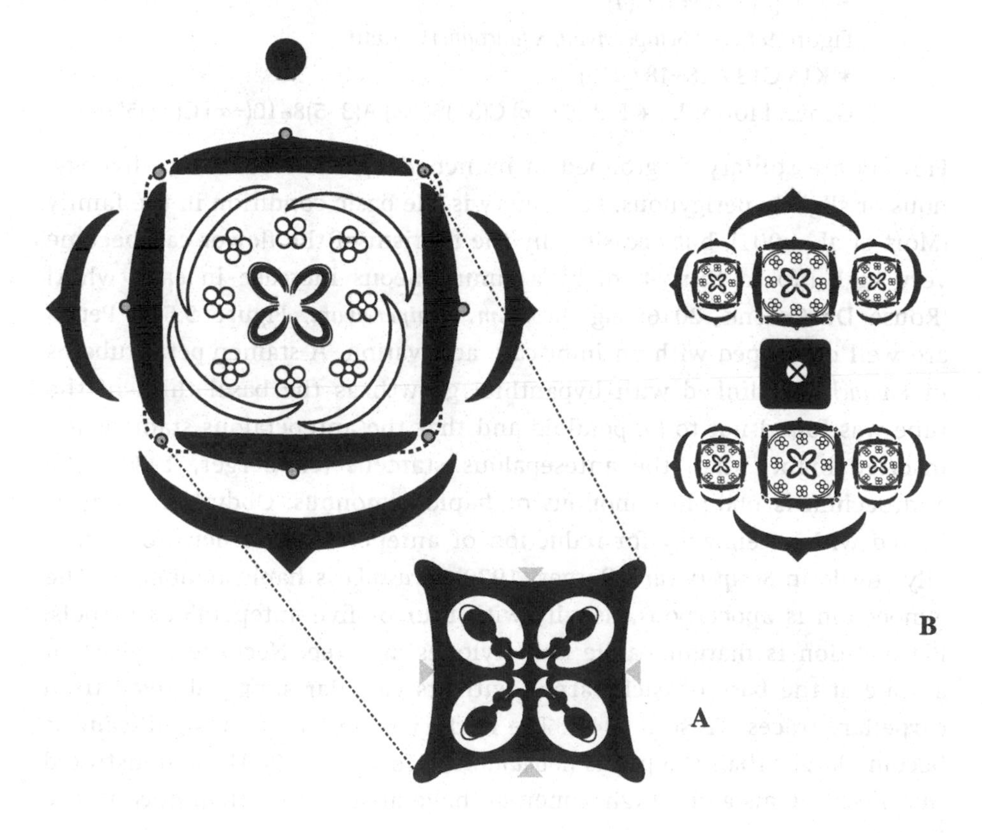

Figure 9.4 *Haloragis erecta* (Haloragaceae): A. flower; B. partial inflorescence

Flowers are bisexual or occasionally unisexual, arranged in dichasia with staminate or bisexual flowers distal to the basal pistillate flowers. Flowers are generally tetramerous (rarely dimerous) with persistent, valvate sepals. Petals are contorted (occasionally absent or rudimentary: pistillate *Myriophyllum*), and drop off with the anthers, leaving filiform filaments behind. The androecium is weakly obdiplostemonous with all stamens inserted at the same level. Antepetalous stamens are lost in some genera. In *Proserpinaca* the flowers are trimerous and haplostemonous with reduced petals (Eichler, 1878). The ovary is inferior, two- to four-carpellate with free stylodes on top of the inferior ovary. In *Haloragis* the ovary has four wings and is partitioned by four septa with one apical ovule per locule. There are occasionally two present but one aborts (Schindler, 1905).

Crassulaceae

Figure 9.5A–B *Kalanchoe fedtschenkoi* Raym.-Hamet and H. Perrier

∗ K(4) [C(4) A4+4] $\underline{G}$(4)

Figure 9.5C–D *Sempervivum x fauconnettii* Reut.

∗ K18 C18 A18+18 $\underline{G}$(18)

General formula: ∗ K(3–)5(–∞) C(3–)5(–∞) A(3–5)8–10(–∞) G(3–)5(–∞)

Flowers are solitary or grouped in branched cymes. Flowers are hypogynous or slightly perigynous. Pentamery is the basic condition in the family (Mort et al., 2001), but occasionally the merism of the flower can become very high (up to thirty-two) by a simultaneous increase in each whorl (Ronse De Craene, 2016; e.g. *Aeonium*, *Sempervivum*: Figure 9.5C). Petals are well developed with an imbricate aestivation. A stamen-petal tube as in *Kalanchoe* is linked with hypanthial growth as the basal third of the tube was found not to be petaloid and that the antepetalous stamens are inserted higher than the antesepalous stamens (cf. Berger, 1930). The androecium is obdiplostemonous or haplostemonous. Obdiplostemony is linked with a tendency for reduction of antepetalous stamens (occasionally sterile in *Sempervivum*: Berger, 1930). *Crassula* is haplostemonous. The gynoecium is apocarpous, usually with four or five antepetalous carpels. Placentation is marginal-axile and stylodes are free. Nectaries consist of a scale at the base of each carpel, with its vascular supply derived from carpellary traces (Tillson, 1940). The nectary scales can be insignificant or become larger than the petals (*Monanthes*). Mort et al. (2001) demonstrated that fused petals and a higher merism have arisen more than once in the family.

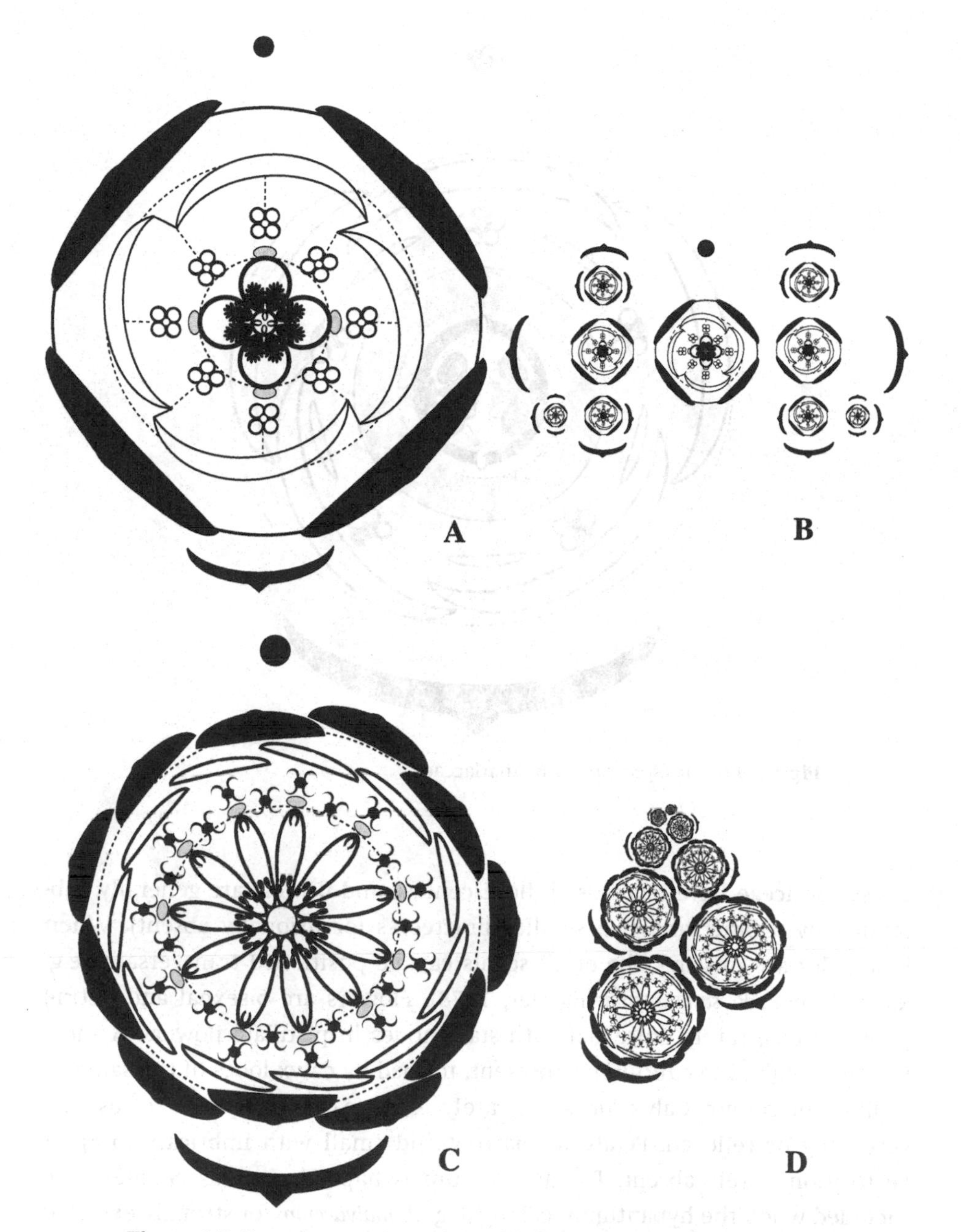

Figure 9.5 Crassulaceae: Flower (A) and inflorescence (B) of *Kalanchoe fedtschenkoi*; flower (C) and inflorescence (D) of *Sempervivum x fauconnettii*

Grossulariaceae

Figure 9.6 *Ribes speciosum* Pursh

∗ [K5 C5 A5] Ğ(2)

General formula: ∗ K(3–)5(–9) C(0–3)–5(–9) A(3–5) Ğ(2)

Figure 9.6 *Ribes speciosum* (Grossulariaceae)

Grossulariaceae have racemose inflorescences, and flowers are generally subtended by a bract and two smaller bracteoles (occasionally absent). When bracteoles are missing, the outer sepals can be positioned transversally (e.g. *Ribes alpinum*, *R. sanguineum*: Eichler, 1878). Flowers are bisexual and (tetra-) pentamerous, rarely unisexual with staminodes in pistillate flowers. A short to deep petaloid hypanthium is present, topped by calyx lobes of the same or a different colour. Calyx lobes are rarely erect (e.g. in *R. speciosum*), usually spreading or reflexed. Petals are narrow and small with imbricate to apert aestivation, rarely absent. The androecium is haplostemonous. Stamens are included when the hypanthium is long (e.g. *R. malvaceum*), or strongly exserted (e.g. *R. speciosum*). In some species the connective bears a distal gland. A well-developed nectary is present as a five-lobed disc, with lobes alternating with stamens (Weigend, 2007) or is present as an inconspicuous annular zone on the hypanthium. The ovary is inferior to half-inferior with two parietal placentae covered with numerous ovules. The ovary is topped with two carinal styles in median position.

Saxifragaceae

Figure 9.7A–B *Rodgersia aesculifolia* Batalin
∗ K5 C0 A5+5 -G-(2)
Figure 9.7C *Saxifraga fortunei* Hook.f
↘K5 C5 A5+5 -$\underline{\text{G}}$-2)

General formula: ∗ (↘/↓) K4–5 C(0–)4–5 A5–10 G2–(5)

Molecular phylogenies have greatly improved the delimitation of Saxifragaceae by removing several genera that were previously included as subfamilies (*Bauera* in Cunoniaceae, *Francoa* in Francoaceae, *Parnassia* and *Lepuropetalon* in Celastraceae, *Vahlia* in Vahliaceae) and whole subfamilies that were raised to the rank of families (Hydrangeoideae, Montinioideae and Escallonioideae: Morgan and Soltis, 1993). The closest relatives of Saxifragaceae are Grossulariaceae, Iteaceae and Pterostemonaceae, which share similarities in floral morphology and anatomy (e.g. Bensel and Palser, 1975b). Inflorescences are variable and built on a monotelic or polytelic pattern. Sepal lobes are large and are linked to the hypanthium, thus appearing basally fused. There is much variation in the extent of development of petals, ranging from well-developed imbricate petals (e.g. *Saxifraga, Lithophragma*) to small, weakly developed appendages (e.g. *Mitella, Tiarella*) or their complete loss (e.g. *Rodgersia*). Petals have a small insertion base and are often narrow and fimbriate (e.g. Endress and Matthews, 2006b). As in Grossulariaceae, hypanthium, bracts and sepals may be petaloid in case petals are missing (*Chrysosplenium, Rodgersia*: Figure 10.7A). Obdiplostemony is commonly found and tends to be correlated with a retardation of the petals and the development of the hypanthium (Gelius, 1967; Ronse De Craene and Smets, 1995b). In the genus *Mitella* the androecium can exceptionally be more variable, ranging from obdiplostemonous to (ob) haplostemonous arrangements (Cronquist, 1981).

A hypanthium is usually present, weakly to strongly developed, and there is a nectary at its base that may expand on the slopes of the ovary (Erbar, 2014). Ovary position can fluctuate from superior to inferior, even within a genus (e.g. *Lithophragma*: Kuzoff, Hufford and Soltis, 2001), and this shift is related to variable hypanthial growth. A characteristic feature is that the ovary is restricted to two carpels in a median position and is partially fused. There is generally a transition between axile placentation in the synascidiate part of the ovary to parietal placentation in the symplicate part leading to free carinal styles (Bensel and Palser, 1975b).

Zygomorphy is rare and is found in genera with predominantly polysymmetric flowers (e.g. *Saxifraga fortunei*: Figure 9.7C, *Heuchera richardsonii*); it is more rarely characteristic for a genus (e.g. *Tolmiea* with loss of one abaxial petal: Klopfer, 1973).

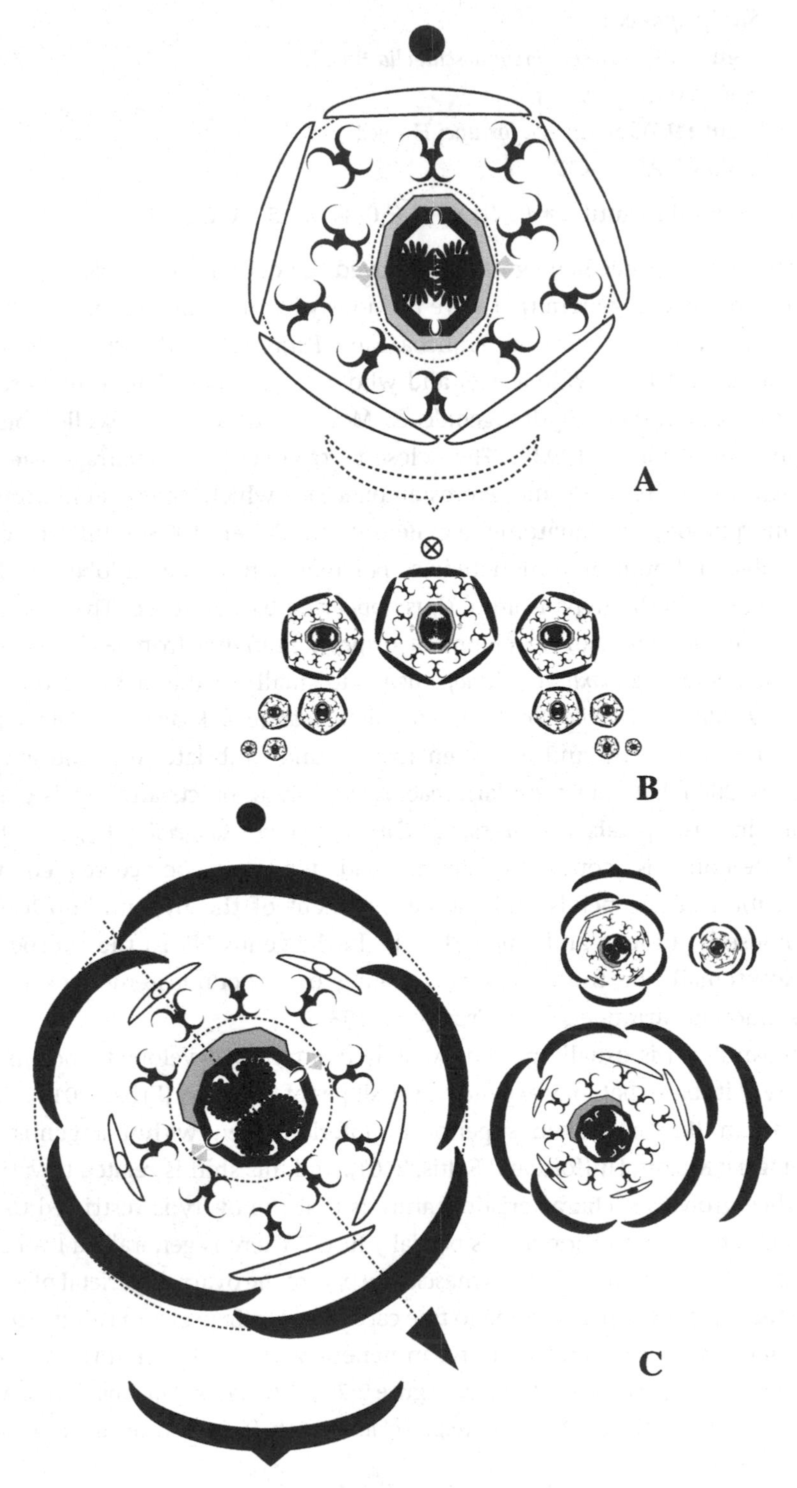

Figure 9.7 Saxifragaceae: Flower (A) and partial inflorescence (B) of *Rodgersia pinnata*; (C) partial inflorescence of *Saxifraga fortunei*

Vitales

Vitaceae (incl. Leeaceae)

Figure 9.8 *Leea guineensis* G. Don

∗ K(5) C5 A5 $\underline{G}$(3)

General formula: ∗K4–5 C4–5 A4–5 $\underline{G}$2–3

Figure 9.8 *Leea guineensis* (Vitaceae): partial inflorescence

APG IV (2016) includes *Leea* in Vitaceae, although the genus differs in a number of characters, such as the development of a specialized gynoecial disc without nectar production, three carpels instead of two, the presence of false septa and various vegetative differences (Ickert-Bond, Gerrath and Wen, 2014).

In *Leea* stamens alternate with bilobed structures that have been interpreted as staminodes with an absence of a disc (Ickert-Bond, Gerrath and Wen, 2014). However, as shown by Gerrath et al. (1990), they arise inside the stamens as a rim without indications of a staminodial nature. This is best interpreted as homologous with the expanded disc of other Vitaceae and is similar to pseudostaminodes (Ronse De Craene and Smets, 2001a). Anthers are initially introrse but shift between the disc lobes and become extrorse at maturity while the filaments break off. Each carpel bears two basal ovules, which become separated by the formation of a false septum, absent in other Vitaceae. Petals, stamens and disc are basally connected by hypanthial growth.

Flowers of Vitaceae are either pentamerous or tetramerous. Specific characters of Vitaceae are the small calyx (often reduced to a rim), obhaplostemony with stamens and petals developing from common primordia, a strongly developed disc nectary (reduced in *Parthenocissus*) and a superior ovary with septa interrupted in the middle and two lateral ovules in the synascidiate part. Petals are always valvate, free or fused with the stamen tube. In some cases petals are postgenitally fused in a calyptra that is shed at anthesis (*Vitis*).

These characters appear strongly apomorphic in the basal Superrosidae.

9.2 Malvids

Figure 9.9 shows the phylogenetic tree of malvids based on Soltis et al. (2018). Malvids contains eight orders with Geraniales and Myrtales at the base.

9.2.1 *Early Diverging Malvids: Geraniales and Myrtales*

Geraniales

Some authors (Palazzesi et al., 2012; Jeiter et al., 2017) recognize five families in this morphologically heterogenous order, although APG IV (2016) recognizes only two families: Francoaceae and Geraniaceae.

All Geraniales share a persistent calyx, isomerous antepetalous carpels with axile placentation and clear separation of the style from the ovary, and obdiplostemony with a clear trend towards haplostemony by sterilization or loss of the antepetalous stamens. Nectaries are extra- or interstaminal receptacular outgrowths (never an intrastaminal disc) and occur as separate entities. Different pollination mechanisms have evolved within the order linked with changes in the floral morphology. There is a strong tendency to evolve median monosymmetric flowers with displacement of the extrastaminal receptacular nectary to the adaxial side (*Pelargonium, Melianthus*). *Rhynchotheca* is wind pollinated and lacks petals and nectaries. Jeiter et al. (2017) present a coherent series of floral diagrams of the main genera of Geraniales.

Geraniaceae (including Hypseocharitaceae)

Figure 9.10A–B *Erodium leucanthemum* Boiss.

∗/↓ K5 C5 A(5+5°) $\underline{G}$(5)

Figure 9.10C –D *Pelargonium abrotanifolium* (L.f.) Jacq.

↓ K5 C3:2 A(5+3°:2) $\underline{G}$(5)

General formula: ∗/↓K(4–)5 C(2–)5 A5–15 G(4–)5

Inflorescences are cymose with flowers seemingly umbellate through contraction of internodes (Figure 9.10B, D). Flowers of Geraniaceae are basically

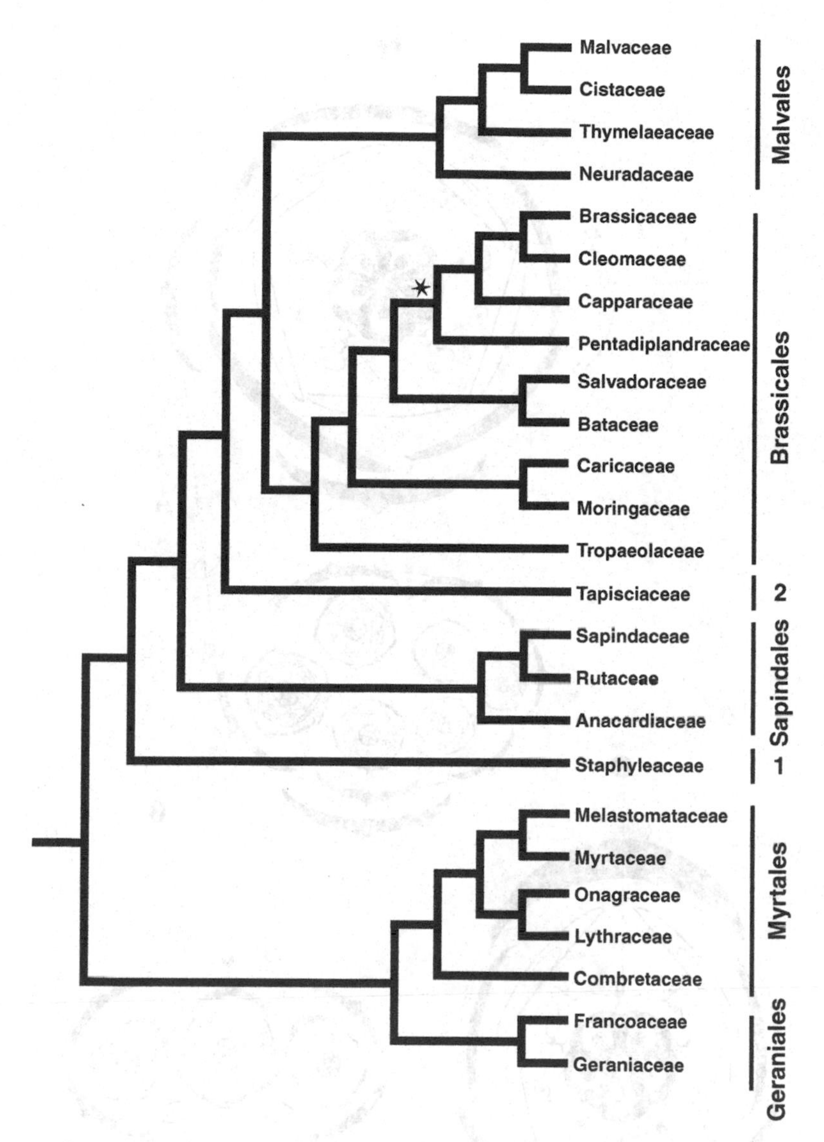

Figure 9.9 Phylogenetic tree of malvids, based on Soltis et al. (2018). 1. Crossomatales; 2. Huertales. The asterisk refers to core Brassicales.

polysymmetric and pentamerous, with isomery in all floral whorls. While *Geranium* or *Monsonia* are strictly polysymmetric, *Erodium* has a tendency for the adaxial petals to differ from the abaxial in size and texture. *Pelargonium* (Figure 9.10C) represents truly monosymmetric flowers with a reduced number of stamens. The two upper petals differ in shape, colour and size from the abaxial petals, with variably descending aestivation. In *P. mutans* the median

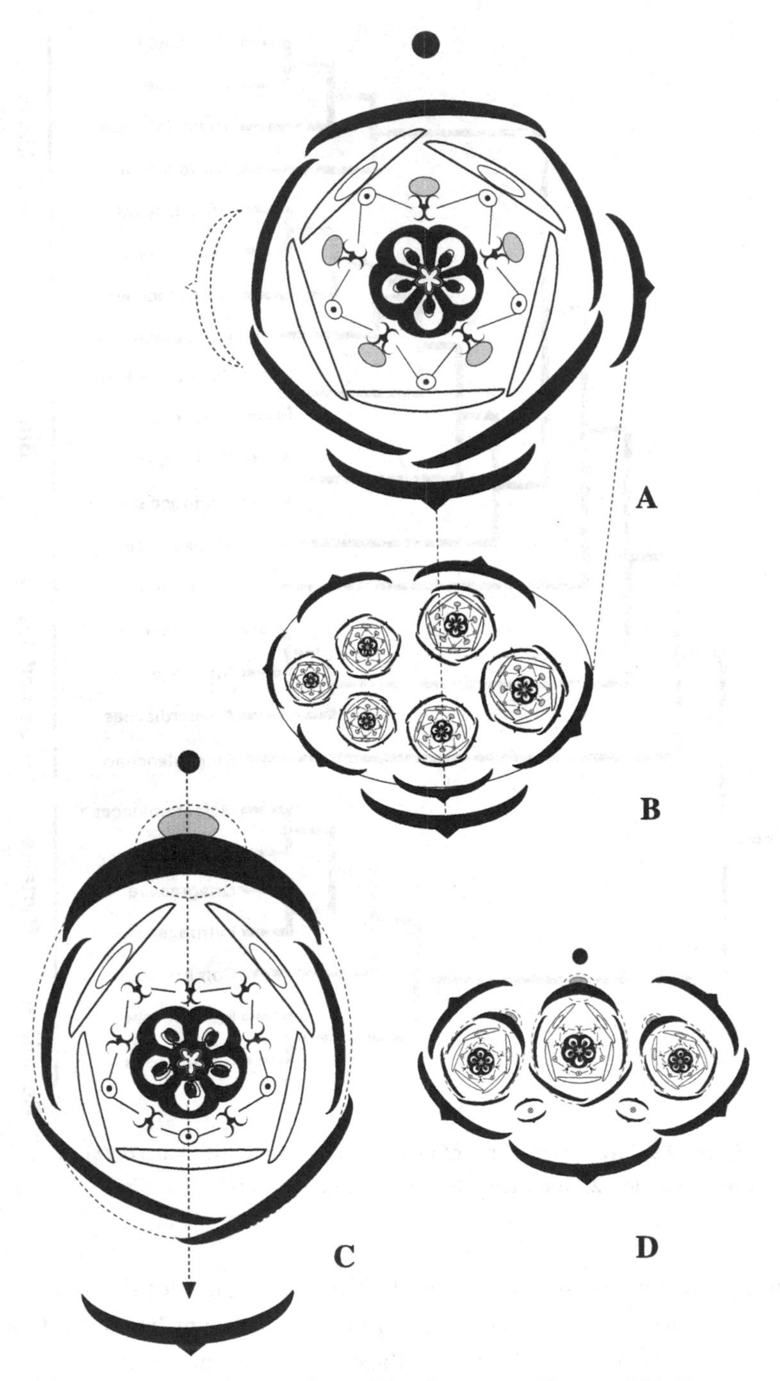

Figure 9.10 Geraniaceae: Flower (A) and corresponding partial inflorescence; (B) of *Erodium leucanthemum*; C. *Pelargonium abrotanifolium*. The dotted bracteole in A is lost.

abaxial petal is absent (pers. obs.). This trend is also expressed in the nectaries which are found as five equal dorsal bumps at the base of the filaments in *Geranium*, as unequally sized nectaries in *Erodium* and as a single adaxial nectary in *Pelargonium* located at the base of a spur (Vogel, 1998c; Jeiter et al., 2017). The spur is hypanthial in nature and produces nectar at the base of a tube extending below the adaxial sepal. Nectar is withheld in the flower through the erect adaxial sepal and posterior petals, while other sepals and petals may be reflexed in some species. Petals are retarded in their development and occasionally arise from common primordia (e.g. Payer, 1857; Ronse De Craene, Clinckemaillie and Smets, 1993; Erbar, 1998). The androecium of Geraniaceae is obdiplostemonous, and this is caused by a shift in position of the antepetalous stamens in connection with common primordia with the petals (Ronse De Craene and Smets, 1995b; Erbar, 1998; Endress, 2010a). In *Erodium* antepetalous stamens develop into filamentous staminodes. A secondary stamen increase is found in *Monsonia* and *Hypseocharis* leading to fifteen stamens, although it is not clear whether this is the result of a lateral division of antepetalous stamens or the development of antesepalous triplets. In *Monsonia* antesepalous stamens are connected with the filaments of neighbouring antepetalous stamens. In *Pelargonium* three adaxial staminodes are formed through unequal development (Sattler, 1973); they are opposite the anterior petals. Except for *Hypseocharis* the gynoecium has a stylar beak separated from the ovary. The close connection of antepetalous stamens and petals creates separate canals with entry to the nectar produced at the back of the antesepalous stamens (Endress, 2010a).

Carpels are superior, isomerous and antepetalous. Ovules are paired or single per locule. In *Pelargonium* ovules are superposed by constricted space.

Francoaceae (including Melianthaceae and Vivianiaceae)

Figure 9.11A *Francoa sonchifolia* Cav., based on living material and Ronse De Craene and Smets (1999b).

✶ K4 C4 A4+4 -G-(4)

Fig. 9.11B. *Melianthus major* L., based on Ronse De Craene et al. (2001)

↓ K5 C4 A4 G̲(4)

General formula: ✶/↓K4–5 C4–5 A4–5(+4–5) G4–5

Ronse De Craene and Smets (1999) and Ronse De Craene et al. (2001) discussed several morphological characters linking Francoaceae and Melianthaceae. There is high floral diversity in the small family Francoaceae with five genera: *Bersama*, *Melianthus*, *Francoa*, *Greyia* and *Tetilla*. *Greyia*, *Tetilla* and *Francoa* share the presence of conspicuous extrastaminal to interstaminal receptacular nectaries, which were interpreted as staminodes by several authors. Flowers are

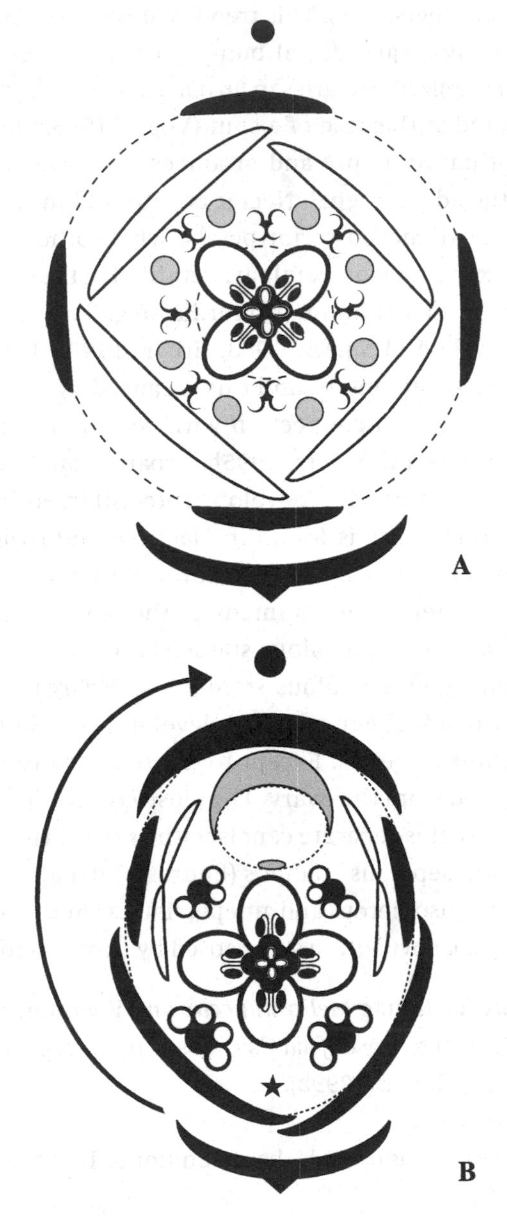

Figure 9.11 Melianthaceae. A. *Francoa sonchifolia*; B. *Melianthus major*. The asterisk refers to the missing petal.

pentamerous or tetramerous (*Francoa* is occasionally pentamerous), with obdiplostemonous (or haplostemonous) androecium, semi-inferior ovary and axile placentation. Ovules are numerous and arranged in two rows per locule. Petals are characteristically contorted to cochleate-ascending.

Bersama and *Melianthus* are closely related but are remarkable in their divergent evolution. *Melianthus* has monosymmetric flowers, which are exacerbated by unequal growth of a shallow hypanthium, comprising a large cuplike adaxial nectary, and shift of the petals to the adaxial side. Bilateral symmetry is initiated early in the development of the flower, disrupting the 2/5 sequence of initiation of the sepals. The abaxial petal is missing, although initiated but suppressed later in the development (Figure 9.11B; Payer, 1857; Ronse De Craene et al., 2001). The adaxial sepal encloses the nectary as a hoodlike appendage. The receptacular nectary develops as a rim partially surrounding a depression in which black nectar accumulates. On the side of the gynoecium a protrusion of the receptacle develops into a beak which has been interpreted as a posterior staminode (Eichler, 1878). As the flower resupinates at maturity, the petals and adaxial sepal withhold the nectar. Petals are marginally connected by trichomes. Principal pollinators of *Melianthus* are sunbirds attracted by abundant nectar. In *Bersama* flowers are nearly polysymmetric by the equal development of the five petals and the smaller adaxial nectary, which is sometimes circular. Early initiation of flowers of *Melianthus* and *Bersama* is monosymmetric, an unusual pattern in the rosids. The androecium and gynoecium are tetramerous. Two rows of two to four ovules develop within each carpel of *Melianthus*, while only a single basal ovule develops in *Bersama* by reduction of the other.

The original Vivianiaceae (with three genera) show the occasional reduction to three carpels and a variable presence of nectaries (absent in *Balbisia* and *Rhynchotheca*: Jeiter et al., 2017).

Members of Francoaceae share several similarities (Ronse De Craene and Smets, 1999b; Ronse De Craene et al., 2001), including racemose inflorescences, the absence of bracteoles and the presence of a sterile bract on top of the inflorescence, a tendency for developing tetramerous flowers, resupination of flowers (in *Greyia*, *Melianthus* and *Bersama*), a shallow hypanthium with sunken ovary, and stomata on the anthers. Some species of *Bersama* have five stamens, indicating a derivation from pentamerous polysymmetric flowers. The genus *Tetilla*, sister to *Francoa*, is monosymmetric by unequal differentiation of the petals (slightly so in *Greyia*), reflecting the fluctuating level of symmetry in the Geraniales.

Myrtales

Myrtales is a homogenous order of thirteen families which has not changed much since the introduction of molecular systematics, except for the inclusion of Vochysiaceae as a sister to Myrtaceae. The order shares a number of striking floral morphological characters that enable an easy recognition in the field (see

Dahlgren and Thorne, 1984; Judd and Olmstead, 2004). Flowers are generally bisexual and tetra- or pentamerous (occ. hexamerous). A common characteristic is the presence of a deep hypanthium lifting perianth and androecium high above the gynoecium. As a result stamens are generally incurved in the bud. The ovary consists of two to five carpels, is inferior or less often superior, with axile placentation, often with numerous ovules on protruding placentae and a single style. Petals are always free and the androecium is basically diplostemonous with a secondary stamen increase in some families (Dahlgren and Thorne, 1984). Petals are reduced or lost in many taxa (e.g. Combretaceae, Penaeaceae). Monosymmetry is rare and weakly developed, except in Vochysiaceae where it results from reduction (Litt and Stevenson, 2003a, 2003b).

There is a strong tendency for obhaplostemony in Myrtales, present in eight families. Some small families are exclusively obhaplostemonous (e.g. Rhynchocalycaceae, Oliniaceae, Penaeaceae, Alzateaceae) with a tendency for reduction or loss of the petals (Schönenberger and Conti, 2003). The single stamen of Vochysiaceae arises opposite a petal but is occasionally displaced towards a sepal in some species leading to asymmetric flowers (Litt and Stevenson, 2003b). Obhaplostemonous flowers of Myrtaceae have undergone a secondary stamen increase (Ronse De Craene and Smets, 1991a). Obdiplostemony, which is associated with a reduction of the antepetalous stamens, is present in some families (Ronse De Craene and Bull-Hereñu, 2016). In cases where a nectary is present, it is usually associated with the hypanthium as an inconspicuous disc or is epigynous.

Combretaceae

Figure 9.12 *Guiera senegalensis* J. F. Gmel

∗ K5 C5 A5+5 Ğ(5)

General formula: ∗ K4–5 C4–5/0 A4–5(+4–5) G2–3–5?

Inflorescences are capitate or spicate, rarely racemose. Flowers are pentamerous to tetramerous. In *Guiera* flowers are arranged in terminal capitula enclosed by four bracts and resembling a flower bud. Individual flowers of *Guiera* lack bracts, which are present in other genera (Stace, 2007). Flowers have a well-developed saucer- to tube-shaped hypanthium with valvate sepal lobes and stamens inserted on two different levels (obdiplostemony). Stace (2007) considered the hypanthium to be of two parts, a lower hypanthium (usually) fused to the ovary, and an upper hypanthium bearing the perianth and stamens. This would make the ovary virtually superior. The ovary is mostly inferior to semi-inferior in *Strephonema* (Stace, 2007), and is mostly topped by a broad nectary surrounding a single style. Petals are small and narrow, and often absent (e.g. *Terminalia*). One stamen whorl is occasionally

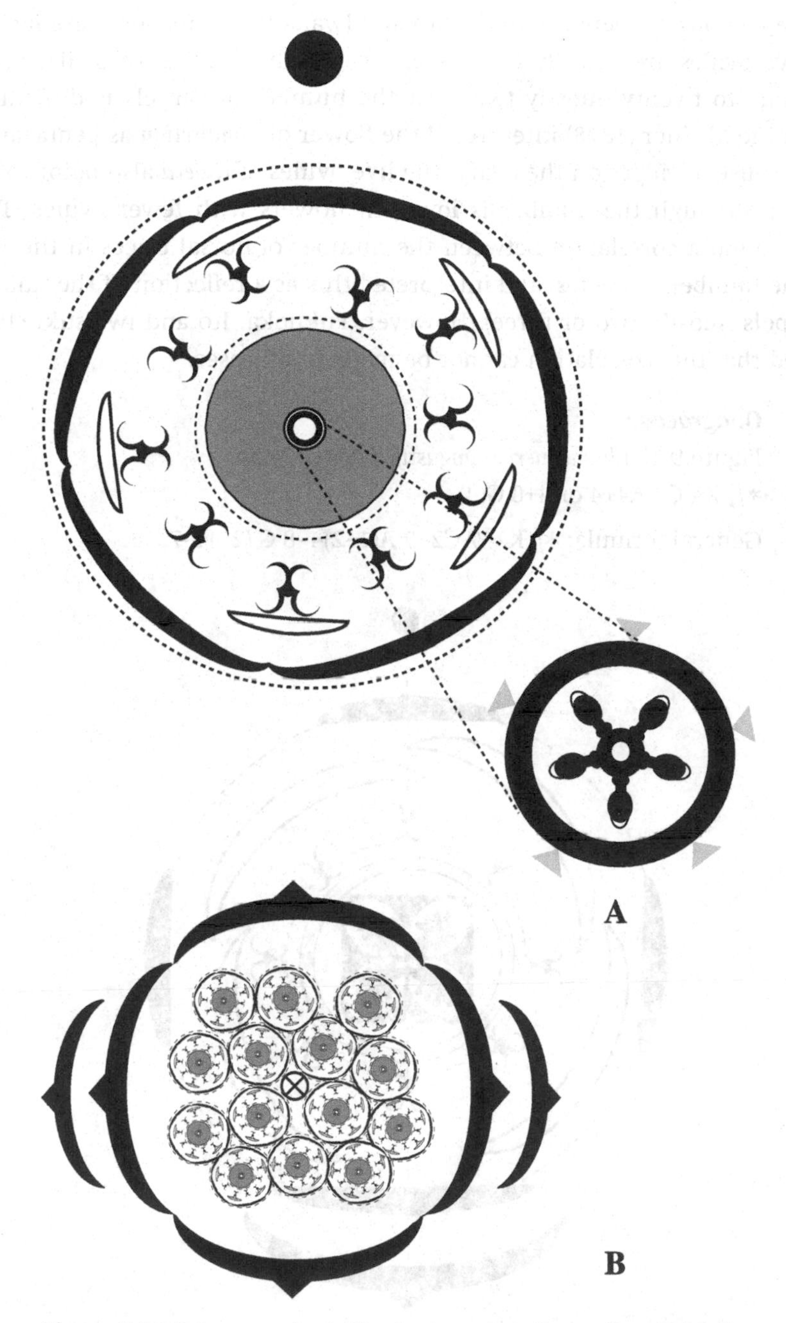

Figure 9.12 *Guiera senegalensis* (Combretaceae): A. flower; B. partial inflorescence

staminodial or absent (antepetalous stamens in some *Combretum sp.*, antesepalous stamens in other *Combretum* and *Terminalia tetrandra*). Eichler (1878) mentions an increase of stamens by dédoublement in some *Combretum*. In *Lumnizera* the number of stamens is often lower than ten by partial reduction of antepetalous stamens (Fukuoka, Ito and Iwatsuki, 1986). Septa are lacking and five ovules are apically inserted in *Guiera*. The number of ovules varies from one to twenty (mostly two), but the number of carpels is difficult to determine. Eichler (1878) interpreted the flower of *Combretum* as pentacarpellate because of ridges on the ovary; the five ovules of *Guiera* also point to five carpels, although the number is lower in flowers with fewer ovules. Tiagi (1969) found a correlation between the number of dorsal traces in the style and the number of ovules, and interpreted this as a reflection of the number of carpels (mostly two or three). However, Fukuoka, Ito and Iwatsuki (1986) showed that this correlation cannot be made in all cases.

Onagraceae

Figure 9.13 *Chamaenerion angustifolium* (L.) Scop.

∗/↓ K4 C4 A4+4 or 4+0 Ğ(4)

General formula: ∗/↓K2–7 C2–7 A(1–2)4–8 G (2–)4–7

Figure 9.13 *Chamaenerion angustifolium* (Onagraceae). Note the sequence of maturation of the stamens.

Onagraceae and Lythraceae are sister groups sharing a valvate calyx and a long tubular hypanthium (Judd and Olmstead, 2004).

Flower merism is mostly four, rarely two (*Circaea*) or a higher number (five to seven in some *Ludwigia*: Eyde, 1977). Flowers have a strongly developed hypanthium linking the perianth and androecium. The gynoecium is inferior and the style protrudes through the narrow hypanthium, which can be very long. There is an epigynous nectary. The flower is well adapted to long-tongue insects (hawkmoths) or birds. Flowers are regular to weakly monosymmetric. In *Chamaenerion*, monosymmetry is caused by a larger abaxial sepal and wider space between anterior petals. Stamens and style are curved to the anterior side (Figure 9.12). Petals are often clawed and usually present; they are lost in some species of *Fuchsia*. The androecium is diplostemonous to weakly obdiplostemonous (pers. obs.; Sattler, 1973), occasionally with antepetalous staminodes (*Clarkia sp.*), or haplostemonous (some *Ludwigia*, *Circaea*); there is no secondary increase of stamens. Only *Lopezia* is strongly monosymmetric with two median stamens, of which the adaxial one is a petaloid staminode (Eyde and Morgan, 1973).

The gynoecium is isomerous and carpels are inserted opposite the petals. The ovary has axile placentation with protruding placentae covered by small ovules or when fewer, ovules are uniseriate (Eyde, 1977).

Lythraceae (inl. Punicaceae and Sonneratiaceae)
Figure 9.14 *Cuphea micropetala* Kunth
↓ [K6 C6/0 A5+6] $\underline{G}$(2)
General formula: ∗/↓ K4–6 C(0–)4–6 A4+4/6+6 $\underline{G}$2–4–6(–12)[a]

Partial inflorescences develop as dichasia, which can be reduced to single flowers (e.g. *Cuphea*). Tobe, Graham and Raven (1998) described the floral morphology and presented a number of floral diagrams for Lythraceae but did not include *Cuphea*. Floral morphology is highly diverse, but all taxa share a long tubular hypanthium and superior ovary. Sepal and petal lobes, as well as stamens, are inserted at different levels on the tube. Hexamery is very common in the family (70 per cent), while the remainder is tetramerous or has a variable and higher merism (e.g. *Lafoensia*). Petals are well developed (crumpled in bud), sometimes reduced or missing in some genera. In *Cuphea* they are barely visible and are often overlooked. Petals are occasionally absent (e.g. *C. ignea*). An epicalyx is present at the top of the hypanthial tube in *Cuphea* and *Lythrum* and was interpreted as emergences of the congenitally fused sepals (Mayr, 1969),

[a] Flowers are rarely eight to sixteen merous.

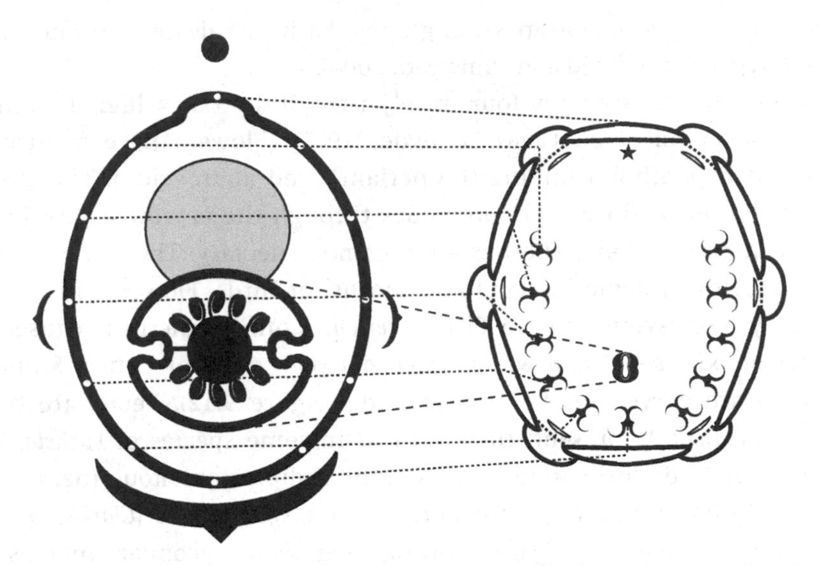

Figure 9.14 *Cuphea micropetala* (Lythraceae), showing level of ovary and level of insertion of the stamens and perianth

which is not satisfactory as explanation. Recent evidence points to the epicalyx being part of the calyx and resulting from folds of the valvate calyx lobes (Barroca, 2014; Sinjushin and Ploshinskaya, 2020). The androecium is basically diplostemonous with occasional reductions happening in both whorls, leading to haplostemony or more rarely obhaplostemony. Paired stamens occur occasionally opposite the sepals. When stamens are situated at the base of the hypanthium, a secondary increase is centrifugal from antepetalous primary primordia as in *Lagerstroemia* (Ronse De Craene and Smets, 1991a). In *Punica* a secondary stamen increase runs centripetally from antepetalous sectors (Leins, 1988).

The ovary can be occasionally (semi) inferior but is mostly superior. Apart from the highly aberrant ovary of *Punica* with one to more extra carpel whorls within the ovary (see e.g. Ronse De Craene and Smets, 1998b), the gynoecium is mostly bicarpellate with axile placentation (with septa interrupted at the top of the ovary). The number of carpels is occasionally isomerous (six) to higher (*Sonneratia*: up to twenty; Tomlinson, 1986).

A nectary is variously present on the inner slopes of the hypanthium, at the base of the ovary, or can be absent (Tobe, Graham and Raven, 1998). In *Cuphea* it arises on the lower part of the ovary, as in *Decodon* but shifts towards the junction between ovary and hypanthium. Zygomorphy is rare, and in *Cuphea* this is emphasized by the loss of a stamen, the unequal development of the carpels and the development of an abaxial nectary at the base of the gynoecium (Figure 9.14; Sinjushin and Ploshinskaya, 2020).

Myrtaceae

Figure 9.15A *Callistemon citrinus* (Curtis) Skeels
∗ K5 C5 A0+5$^{\infty}$ $\breve{G}$(3)

Figure 9.15B *Syzygium australe* (link) B. Hyland
∗ K4 C4 A∞ $\breve{G}$(2)

General formula: ∗ K4–5 C4–5 A4–∞ G(1–)2–3(–5–10)

Myrtaceae is a distinctive family characterized by an inferior ovary and a broad hypanthial cup lined with nectary tissue. Sepals, petals and stamens are inserted on the upper margins of the cup. Bracteoles are variously developed in *Callistemon*; they determine the position of the first sepals. In *C. citrinus* the outer sepals are lateral in the absence of bracteoles (Figure 9.15A), contrary to species with bracteoles (e.g. Orlovich, Drinnan and Ladiges, 1999). Sepals and petals usually have an imbricate quincuncial (in pentamerous flowers) or decussate aestivation (in tetramerous flowers), or are cochleate. Calyx and corolla can be both calyptrate, or only the calyx (*Calyptranthes*) or corolla (*Eucalyptus sp.*), although this does not necessarily involve fusion of the petal tips (pseudocalyptrate: Vasconcelos et al., 2020). The corolla is generally early caducous or short and attraction is taken over by stamens. Petals arise in a 2/5 sequence following sepal initiation and cannot be distinguished from the calyx in early stages. In *Callistemon citrinus* petals are greenish as the sepals. Other genera have been occasionally described as apetalous with two whorls of four sepals (*Osbornia*: Tomlinson, 1986), although this may be linked to the poor differentiation of petals from sepals. Stamen number in Myrtaceae is highly variable (four to more than one thousand per flower: Schmid, 1980), but can most often be described as complex obhaplostemonous. Primary primordia arise opposite the petals (occasionally from common stamen-petal primordia: *Eucalyptus*: Drinnan and Ladiges, 1989) and stamens have a centripetal or lateral development, sometimes covering the upper inner margins of the cup (Figure 9.15B). The stamens remain in clear fascicles (e.g. *Melaleuca, Lophostemon, Arillastrum*) or expand into a girdle (e.g. *Callistemon, Syzygium, Myrcia*: Figure 9.15A, B; Orlovich, Drinnan and Ladiges, 1996, 1999; Bohte and Drinnan, 2005; Vasconcelos et al., 2017). In *Syzygium* the first stamen primordia arise opposite the petals but extend in the intermediate areas finally covering a narrow margin of two or three rows around a broad sloping nectariferous hypanthium (Figure 9.15B; Ronse De Craene and Smets, 1991a; Belsham and Orlovich, 2003). The expansion and growth of the hypanthium are responsible for a lateral and centripetal stamen increase (Ronse De Craene and Smets, 1991a; Carrucan and Drinnan, 2000).

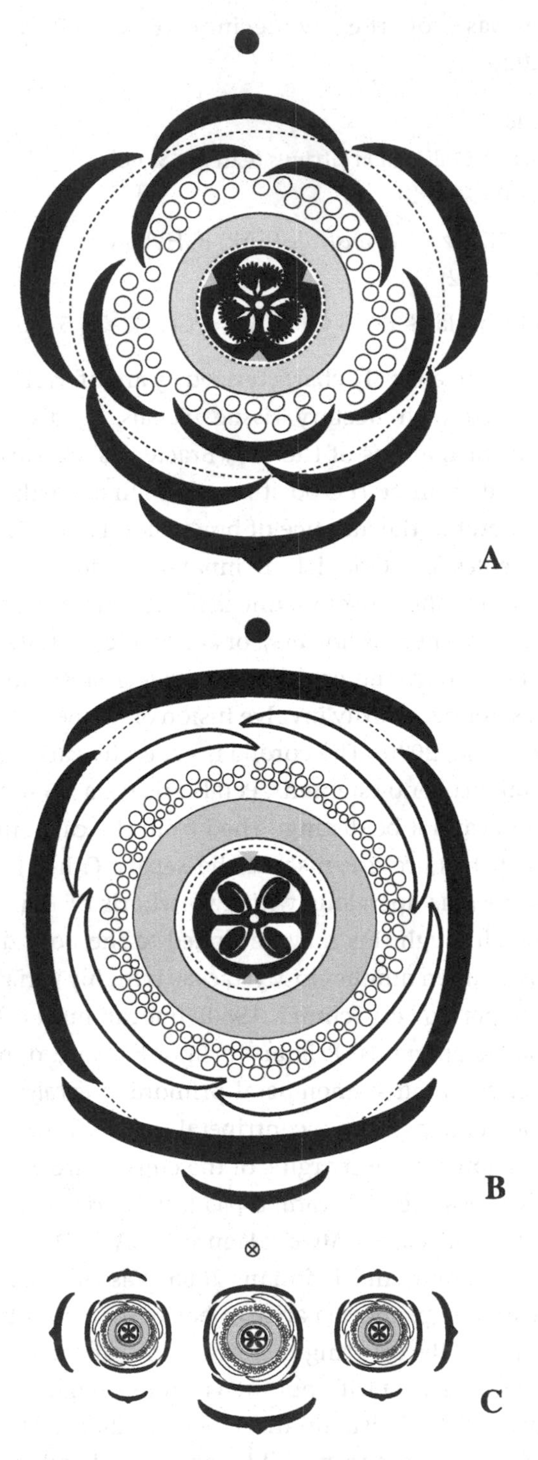

Figure 9.15 Myrtaceae: A. *Callistemon citrinus*; *Syzygium australe*; B. flower; C. partial inflorescence

Outer stamens may be staminodial, adding to the attractiveness of the flower as petals are insignificant (*Arillastrum*, *Chamaelaucium*). Stamens are typically coiled in bud, although the curvature of filaments is variable and strongly connected with the growth of the hypanthium. Vasconcelos et al. (2015) recognize three patterns of stamen orientation in Myrteae (straight, semi-curved and strongly incurved), corresponding with the depth of the hypanthium and as an adaptation to different pollination strategies. Parallel growth patterns involve an increased expansion of the hypanthium, leading to a a progressive reduction of the calyx and corolla, and replacement by a hypanthium (Bohte and Drinnan, 2005; Vasconcelos et al., 2017). In the latter case the hypanthium splits open at anthesis to expose the stamens and nectary, in a manner reminiscent of fruiting in Monimiaceae and Siparunaceae. Similarly in a number of Australian genera the hypanthium behaves as a floral structure, being abscised off at fruiting from the rest of the flower (Bohte and Drinnan, 2005).

Some Myrtaceae have less-developed hypanthia with an obhaplostemonous androecium (e.g. *Backea*, *Micromyrtus*) or with stamens grouped in pairs or triplets (e.g. *Thryptomene*, *Chamaelaucium*) (Carrucan and Drinnan 2000). The reports of (ob) diplostemony for *Psiloxylon*, *Heteropyxis* and other genera enumerated by Schmid (1980) are suspicious without the backing of floral developmental evidence.

Carpels are isomerous in *Eucalyptus* and *Leptospermum*, but reduced to two or three in other genera. Exceptionally a single carpel with two ovules is formed (*Calytrix*). The mostly inferior ovary has a single terminal style; placentation is axile and two rows of ovules are arranged on an U-shaped placenta in each locule. The number of ovules ranges from few to numerous (Figure 9.15A). A few Myrtaceae have a parietal, basal or apical placentation, which is clearly derived (e.g. *Chamaelaucium*: Eichler, 1878).

Myrtaceae can be confused with some Rosaceae in their general flower construction. The main differences are the arrangement of the stamens and single terminal style.

Melastomataceae

Figure 9.16 *Medinilla magnifica* Lindl

∗ [K5 C5 A5+5] Ğ(5)

General formula: ∗/↓ K(3)4–5(–8) C(3)4–5(–8) A5–10(–∞) G(2–)3–6

Flowers are mosty tetra- or pentamerous (occasionally from tri- up to decamerous). Flowers have a deep hypanthium with superior to inferior ovary. Sepals are free lobes or are fused into a calyptra (*Conostegia*). Petals are generally strongly contorted and inserted with the stamens on top of the hypanthial rim. The androecium is mostly diplostemonous and the filaments are characteristically bent over in bud, with the anthers fitting in holes formed in the hypanthial tissue

Figure 9.16 *Medinilla magnifica* (Melastomataceae): partial inflorescence. Note the depressions around the ovary with fitting anthers.

surrounding the ovary (Figure 9.16). Appendages of the connective, which are often elaborate and different in colour from the anthers, are a main feature of the family. Anthers are incurved and poricidal, hanging over on the adaxial side of the flower. Antepetalous stamens are rarely staminodial or suppressed (*Poteranthera*, *Dissochaeta*: Eichler, 1878). Polyandry is found in a few genera (e.g. *Conostegia*, *Clidemia*, *Plethiandra*). In *Conostegia* the stamen number is laterally increased in a single whorl, often in correlation with a concordant carpel increase (Wanntorp et al., 2011). Merism increase and lateral stamen increase operate independently in *Conostegia*, leading to a greater diversity in number of parts. The ovary is mostly isomerous and antepetalous (not antesepalous as suggested by Eichler, 1878), or reduced to two. Placentation is mostly axile on protruding placental columns.

Most Melastomataceae are nectarless as they produce pollen or secretions on the connective as reward. However, in some genera, which reverted to nectar production, nectar is exuded from a cleft at the site where the filaments are sharply bent in bud (e.g. *Myconia*: Vogel, 1997). *Medinilla* exudes nectar from its petal tips in bud.

9.2.2 *Remaining Malvids: Crossosomatales, Sapindales, Brassicales and Malvales*

Crossosomatales

The order as recognized by APG IV (2016) comprises seven morphologically distinct, relatively small families with typical rosid characteristics. The clade appears to be homogeneous, sharing tetramerous or pentamerous flowers, although it is in need of further study. Matthews and Endress (2005) recognize the presence of a hypanthium, imbricate sepals and petals (occasionally missing) and a shortly stalked gynoecium with free postgenitally united styles as characters shared by the order. In Strasburgeriaceae sepals range from nine to eleven and intergrade spirally with the petals. The androecium is mostly (ob) diplostemonous or haplostemonous. There is generally an intrastaminal nectary, developed as a disc or free lobes.

Staphyleaceae

Figure 9.17 *Staphyllea pinnata* L.

∗ K5 C5 A5 $\underline{G}(2)$

Figure 9.17 *Staphylea pinnata* L. (Staphyleaceae)

Two genera formerly included (*Tapiscia* and *Huertea*) have been moved to Huerteales into separate families. *Tapiscia* differs from *Staphylea* in the fused calyx, absence of a nectary disc and a unilocular ovary with a single ovule; *Huertea* has two ovules on a basal septum and separate nectaries alternating with the stamens (Dickison, 1986). Inflorescences are terminal consisting of two or three flowered racemes. Perianth and stamens are connected by a shallow hypanthium connecting with a narrow disc. The ovary is (half-)inferior and has two appressed styles. Ovules are in two rows on axile placentae within each locule. Floral anatomical sections are presented by Dickison (1986) and Matthews and Endress (2005).

Sapindales

The order represents a clade of nine families forming three subclades: a broadly defined Sapindaceae, Kirkiaceae-Anacardiaceae-Burseraceae and Rutaceae-Meliaceae-Simaroubaceae (Ronse De Craene and Haston, 2006; Soltis et al., 2018). Sapindales share a syndrome of characters or interesting features that are not synapomorphic. Flowers are usually small and grouped in cymose inflorescences. They bear the typical characteristics of rosids: flowers are usually tetra- or pentamerous, polysymmetric and rarely monosymmetric (generally oblique and as a late developmental event) with free, imbricate petals and a diplostemonous androecium. The gynoecium is superior, basically isomerous or reduced to two or three carpels with axile placentation. There is a remarkable fluctuation in the degree of development of the synascidiate, symplicate and plicate zones of Sapindalean gynoecia, ranging from fully syncarpous to apocarpous gynoecia (El Ottra et al., 2022). A residual floral apex is generally integrated in the ovary of most taxa. A synapomorphy for the order appears to be a stigmatic head formed by the postgenital fusion of carpel tips (El Ottra, Demarco and Pirani, 2019). A prominent disc nectary is present and is either intrastaminal or extrastaminal. The superior ovary contains few large ovules. Functionally unisexual flowers occur frequently and arise through late abortion during development.

Sapindaceae (incl. Aceraceae and Hippocastanaceae)

Figure 9.18A, B *Acer griseum* (Franch.) Pax

⟷ K5–6 C5–6 A10–12 $\underline{G}$(2)

Figure 9.18C, D *Serjania glabrata* HBK

↘ K5 C2:2 A5+3 $\underline{G}$(3)[b]

General formula: ⟷/✱/↘ K(4)5 C0/(4)5 A4–10(–∞) G(1–)2–3

Sapindaceae is a highly diverse family containing four subfamilies. A frequent combination of characters in Sapindaceae is: flowers functionally unisexual,

[b] Flowers are unisexual with aborted stamens and ovary.

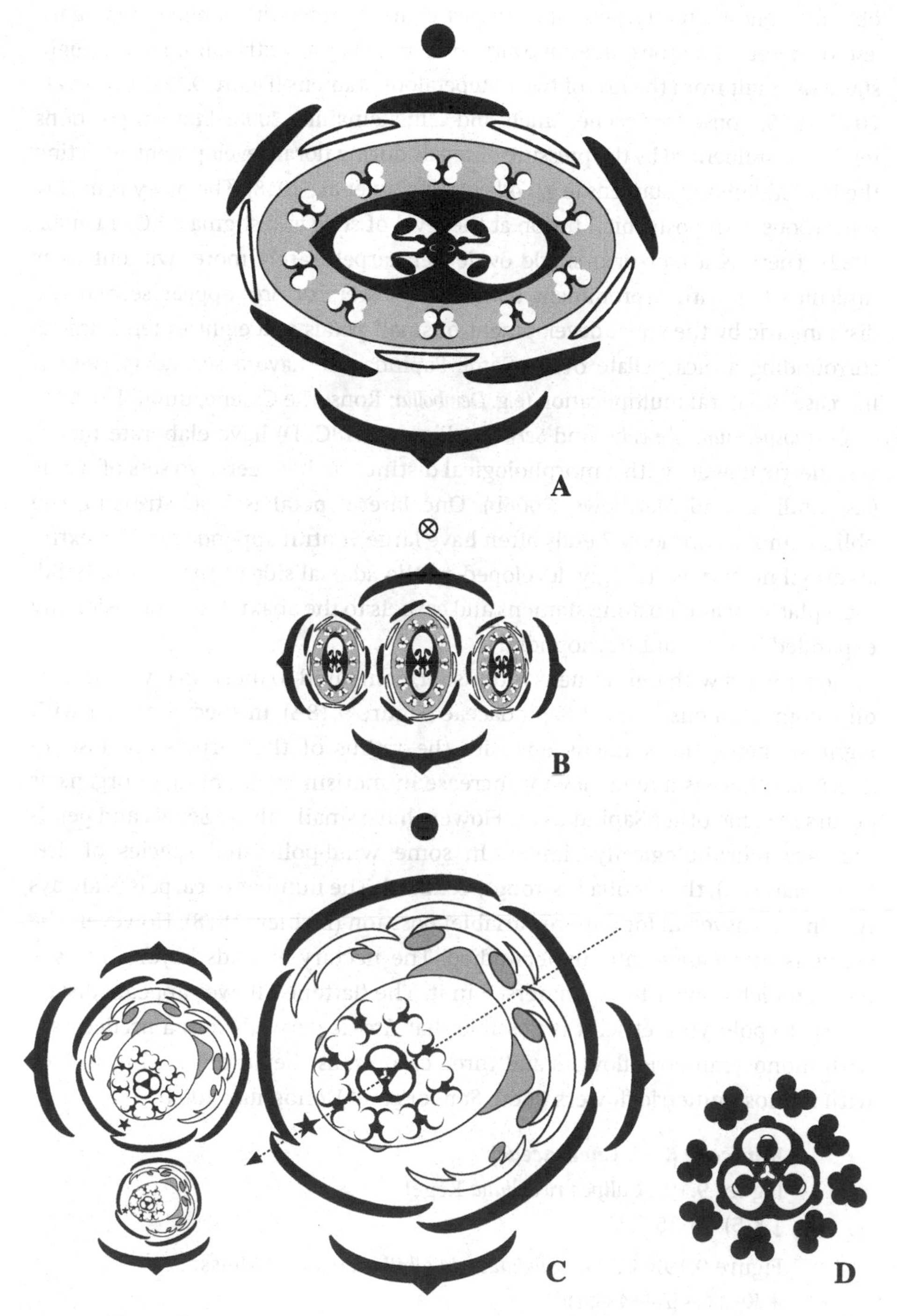

Figure 9.18 Sapindaceae: *Acer griseum*. A. flower; B. inflorescence; *Serjania glabrata*; C. partial inflorescence with staminate flowers; D. reproductive organs of pistillate flower

oblique monosymmetry, petal appendages, eight stamens with papillose filaments, and trimerous-dimerous superior ovary surrounded by an extrastaminal disc. Eight stamens result from the loss of two antepetalous stamens (Figure 9.18C; Cao et al., 2017, 2018; Ronse De Craene, Smets and Clinckemaillie, 2000). Stamen positions tend to be influenced by the pressure of sepals during floral development, affecting the loss of different stamens (e.g. *Koelreuteria*: Cao et al., 2018). The ovary is mostly syncarpous with postgenital fusion at the level of style and stigma (El Ottra et al., 2022). There is a pair or a single ovule per carpel (rarely more) without clear funiculus but with a prominent obturator. Flowers of *Acer* appear secondarily disymmetric by the equal development of small petals and eight to ten stamens surrounding a bicarpellate ovary. Some Sapindaceae have a secondary stamen increase by lateral multiplication (e.g. *Deinbollia*: Ronse De Craene, unpubl. data).

Cardiospermum, *Aesculus* and *Serjania* (Figure 9.18C, D) have elaborate monosymmetric flowers with a morphological distinction between two sets of petals (see Endress and Matthews, 2006b). One lateral petal is lost, stressing the oblique monosymmetry. Petals often have large ventral appendages. The extrastaminal nectary is strongly developed on the adaxial side of the flower, building a platform and pushing stamens and carpels to the abaxial side, occasionally expanded into an androgynophore.

Acer griseum with ten stamens is unusual, compared to most *Acer*, which have only eight stamens as most Sapindaceae (Figure 9.18A). In species of *Acer* with eight stamens, the stamens opposite the radius of the carpels are lost. In *A. griseum* there is a tendency for increase in merism by doubling of organs as occurs in some other Sapindaceae. Flowers have small valvate sepals and petals that are morphologically similar. In some wind-pollinated species of *Acer* (*A. saccharinum*), the corolla is strongly reduced. The number of carpels is always two in a transversal (or a more variable) position (Eichler, 1878). However, the genus is morphologically understudied. The nectary extends beyond the stamens, which appear to be immersed in it. The flattened flower appears disymmetric to polysymmetric. It is possible that *Acer* is derived from a predecessor with monosymmetric flowers and three carpels, as they are nested in a clade with monosymmetric flowers (Judd, Sanders and Donoghue, 2004).

Rutaceae (incl. Cneoraceae)

Figure 9.19A *Galipea riedeliana* Regel
↓ K(5) C5 A5 $\underline{G}(5)$

Figure 9.19B–C. *Boenninghausenia albiflora* (Hook.) Meissn.
∗ K(4) C4 [A4+4 $\underline{G}(4)$]

General diagram: ∗/↓/↙ K(3) 4–5 C(3) 4–5 A(2–)4–5–10(–∞) G(1–2)4–5 (–10)

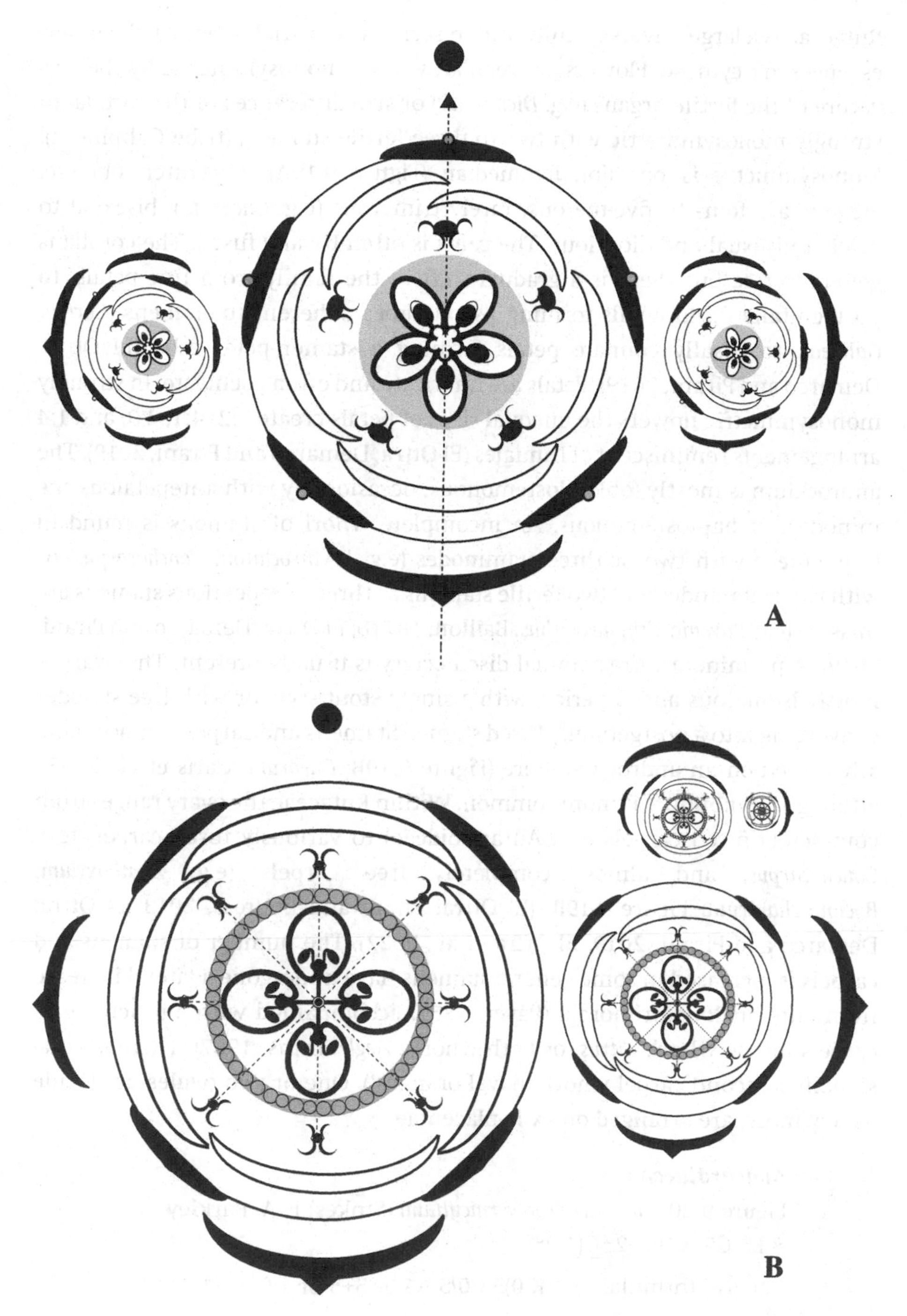

Figure 9.19 Rutaceae: A. *Galipea riedeliana*, partial inflorescence; *Boenninghausenia albiflora*, partial inflorescence; B. The broken line in B shows the attachment of stamens on the androgynophore.

Rutaceae is a large, diverse family. Inflorescences are variable but partial inflorescences are cymose. Flowers are regular, weakly monosymmetric by the curvature of the fertile organs (e.g. *Dictamnus*) or size differences of the corolla, to strongly monosymmetric with two to three fertile stamens (tribe Galipineae). Monosymmetry is occasionally median (Figure 9.19A), but often oblique. Flowers are four- to five-merous, rarely trimerous (e.g. *Cneorum*), bisexual to rarely unisexual and dioecious. The calyx is often basally fused. The corolla is generally free but there is a gradual shift in the family from free petals, to postgenitally fused petals forming pseudotubes adherent to stamens (*Correa*, *Galipea*), to basally connate petals forming a stamen-petal tube (El Ottra, Demarco and Pirani, 2019). Petals are imbricate and often cochleate. In strongly monosymmetric flowers the unequal sizes of petals create 3:2, 4:1, 2:3, and 1:4 arrangements reminiscent of Lamiales (El Ottra, Demarco and Pirani, 2019). The androecium is mostly (ob)diplostemonous, occasionally with antepetalous staminodes, or haplostemonous. An incomplete whorl of stamens is found in Galipeineae with two or three staminodes (e.g. *Erythrochiton*, *Conchocarpus*) or with five staminodes and two fertile stamens as three antepetalous stamens are missing (e.g. *Ravenia*, *Stigmatanthus*: Baillon, 1871b; El Ottra, Demarco and Pirani, 2019). A prominent intrastaminal disc nectary is usually present. The ovary is mostly isomerous and superior, with a single stout style or with free stylodes converging into a postgenitally fused stigma. Stamens and carpels are occasionally united on an androgynophore (Figure 9.19B; *Cneorum*: Caris et al., 2006), although a gynophore is more common. Within Rutaceae the ovary ranges from completely fused carpels (e.g. Aurantioideae) to variously fused carpels (e.g. *Conchocarpus*) and almost completely free carpels (e.g. *Zanthoxylum*, *Boenninghausenia*: Figure 9.19B) (El Ottra, Pirani and Endress, 2013; El Ottra, Demarco and Pirani, 2019; El Ottra et al., 2022). The number of stamens and carpels is increased in some genera: stamens have undergone a lateral increase from antesepalous primordia (Payer, 1857, accompanied with an increase of carpels in one whorl: *Citrus*, or both whorls: *Aegle*: Leins, 1967). There is occasionally a second carpel whorl ('navel oranges'). One or two ovules per locule (rarely more) are arranged on axile placentae.

Anacardiaceae

Figure 9.20 *Toxicodendron vernicifluum* (Stokes) F. A. Barkley

∗ K5 C5 A5+1–2° $\underline{G}$(1:2°)

General formula: ∗/↗ K(0)5 C0/5 A5 or 5+5 or 1:9° G (1)3–13

Inflorescences are paniculate with all flowers maturing simultaneously. Bracts and bracteoles are present but early caducous. In several genera, flowers are functionally unisexual by the late abortion of carpels or stamens (e.g. Wannan

Figure 9.20 *Toxicodendron vernicifluum* (Anacardiaceae). Particular flower with two staminodes (variable)

and Quinn, 1991; Gallant, Kemp and Lacroix, 1998). In staminate flowers of *Schinopsis* the gynoecium is completely aborted, while staminodes with reduced anthers are found in pistillate flowers (Gonzalez, 2016). The wind-pollinated *Amphipterygium* and *Pistacia* have reduced unisexual flowers, with a variable abortion in staminate and pistillate flowers and an unclear differentiation of bracts and sepals (Bachelier and Endress, 2007). Flowers are generally polysymmetric except for the gynoecium, which is pseudomonomerous. *Spondias* is isomerous in all whorls with carpels opposite the petals, or carpels are increased to ten in *S. pleiogyna* (Eichler, 1878); in *Pleiogynium* the carpel number ranges from five to thirteen (Wannan and Quinn, 1991). The androecium is usually diplostemonous (not obdiplostemonous as mentioned by Eichler, 1878). Haplostemony is found in a few genera, such as *Pistacia*, *Amphipterygium*, *Rhus* and *Toxicodendron*; in the absence of petals stamens alternate with sepals (*Amphipterygium*: Bachelier and Endress, 2007); in *Toxicodendron*, one to two antepetalous staminodes are occasionally present (Figure 9.20). In *Anacardium* and *Mangifera*, the androecium is reduced to a single large fertile stamen (opposite sepal one) and other stamens are reduced in a descending order starting from the fertile stamen, resulting in a monosymmetric flower. The number of stamens is rarely secondarily increased (*Sorindeia, Poupartia, Gluta*: Eichler, 1878; Wannan and Quinn, 1991). A well-developed intrastaminal disc nectary is present in most species, although nectariferous trichomes on the corolla are reported for *Anacardium occidentale* (Bernardello, 2007). A distinctive character of the family is that only one carpel is fertile (except for *Spondias*), with variable degrees of reduction of the other carpels. The fertile carpel is situated in a latero-anterior position opposite sepal one, leading to oblique monosymmetry. The sterile carpels are often completely reduced except for the styles (Figure 9.20; Ronse De Craene and Smets, 1998b; Gonzalez, 2016), or are completely absent (*Anacardium*). A single basal-axile ovule develops in the fertile carpel. Styles are either free or fused and opposite the locule.

Anacardiaceae are closely related to Burseraceae but differ in having one syntropous ovule, in contrast to the two epitropous ovules per locule in Burseraceae (Gonzalez, 2016; Bachelier and Endress, 2009).

Brassicales

The order contains seventeen families with a striking morphological diversity but a common chemical character (mustard oils, lost in *Koeberlinia* but also present in Putranjivaceae of Malpighiales: Rodman et al., 1998). No clear morphological synapomorphies exist for Brassicales. Several well-supported subclades can be recognized on a morphological and molecular basis (a tropaloid clade, a batoid clade, a moringoid clade and the 'core' Brassicales: Figure 9.9; Ronse De Craene and Haston, 2006; Soltis et al., 2018).

Tropaeolaceae is associated with *Bretschneidera* and *Akania* (Akaniaceae) as a basal clade of Brassicales. Both families share oblique monosymmetry (at least in early stages of the development of *Tropaeolum*), an octomerous androecium, as well as a hypanthium (Ronse De Craene and Smets, 2001b; Ronse De Craene et al., 2002). These characters are also found in Sapindaceae and represent a remarkable convergence (see Ronse De Craene and Haston, 2006). Moringaceae and Caricaceae are morphologically highly divergent, especially in their floral morphology, although they share a number of vegetative and floral characters (pentamerous flowers with a hypanthium, diplostemony, superior ovary with parietal placenta: Ronse De Craene and Smets, 1999a). Bataceae, Koeberliniaceae and Salvadoraceae represent a well-supported clade of small tetramerous, disymmetric flowers and a strong tendency for reduction (Ronse De Craene and Wanntorp, 2009). Most core Brassicales share tetramerous (disymmetric) flowers, a gynophore, curved embryos and campylotropous ovules, but there are several exceptions (Ronse De Craene and Haston, 2006).

The pentamery of Emblingiaceae and Pentadiplandraceae was interpreted as a reversal from tetramery (Soltis et al., 2005). A less parsimonious option is the repeated derivation of tetramery from pentamery in Setchellanthaceae, batoid clade (*Koeberlinia* and *Salvadora* are occasionally pentamerous) and core Brassicales. This makes more sense on a morphological ground and would represent a clear apomorphic tendency. Higher merisms evolved in Tovariaceae, Resedaceae and Gyrostemonaceae. *Pentadiplandra* functions as a morphological prototype for the deriviation of disymmetric Brassicaceae (see p. 228; Ronse De Craene, 2002). Brassicaceae *sensu lato* was recognized as a single family by APG II, although there is good morphological evidence to support a distinction between Capparaceae, Brassicaceae and Cleomaceae (Hall, Sytsma and Iltis, 2002; Ronse De Craene and Haston, 2006), and this approach was adopted by APG IV (2016).

When unisexual, flowers of Brassicales often show a high dimorphism (e.g. Caricaceae, Bataceae, Gyrostemonaceae).

Tropaeolaceae

Figure 9.21 *Tropaeolum majus* L.

↓[c] K5 C3:2 A5+3 $\underline{G}$(3)

Flowers of *Tropaeolum* arise singly in the axil of a leaf functioning as bract. There is a conspicuous hypanthium lifting calyx and corolla. A spur develops below the adaxial sepal within the hypanthial tissue, and is occasionally the most conspicuous part of the flower. The floral symmetry changes during development from oblique to median by the enlarging spur (Ronse De Craene and Smets, 2001b). Sepals have a quincuncial arrangement or become valvate at maturity. The genus shows much variation in the development of the petals, which are strongly clawed; the two upper (adaxial) petals are often strongly divergent from the three lower, which are smaller (e.g. *T. peregrinum*), or occasionally absent (e.g. *T. pentaphyllum*: Eichler, 1878). In *T. umbellatum*, the upper petals are reduced in size. The androecium consists of eight stamens with an unusual spiral initiation sequence. Ronse De Craene and Smets (2001b) discussed different interpretations for the origin of the octomerous androecium but concluded that two stamens opposite petals were lost linked with spatial constraints imposed by the gynoecium (asterisks in Figure 9.21). There is a striking analogy with the floral structure of *Koelreuteria*, at least in early stages of development (Ronse De Craene, Smets and Clinckemaillie, 2000). The ovary is always trimerous with a single large axile-apical ovule per carpel.

[c] ↘ in preanthetic flowers.

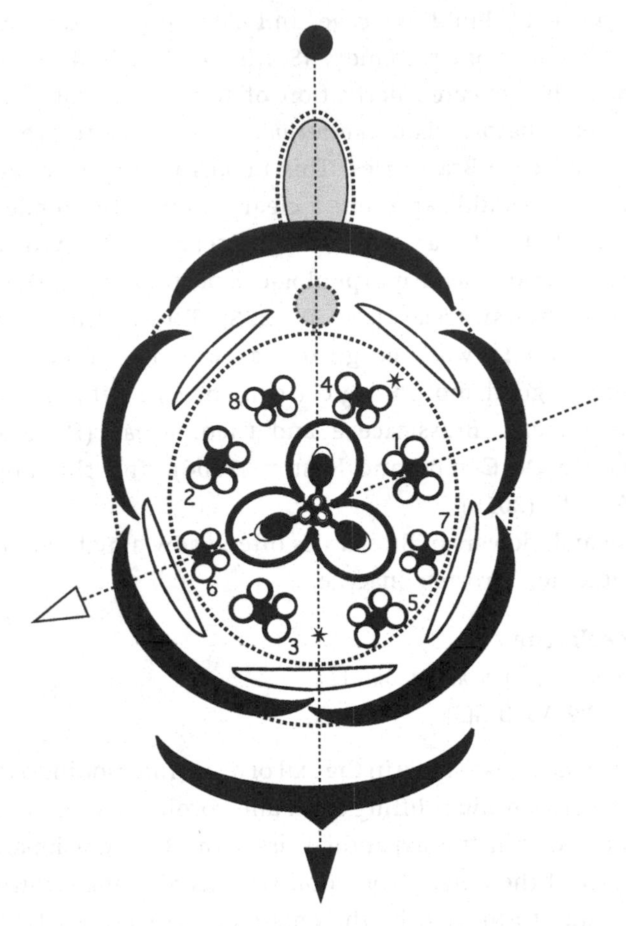

Figure 9.21 *Tropaeolum majus* (Tropaeolaceae). Numbers give order of stamen initation. White arrow, symmetry in young bud; black arrow, symmetry at maturity; asterisks: lost stamens

'Moringoid Clade'

Caricaceae

Figure 9.22A–B *Carica papaya* L., based on Ronse De Craene and Smets (1999a)

Staminate : ✶ K5 [C(5) A5+5] G0

Pistillate: ✶ K5 C(5) A0 $\underline{G}$(5)

Staminate and pistillate flowers are highly dimorphic and occur on the same or on different branches. Some cultivars are occasionally hermaphrodite (see Ronse De Craene et al., 2011). Staminate flowers are grouped in multiflowered dichasia with a strongly contorted corolla fused with the two stamen whorls by

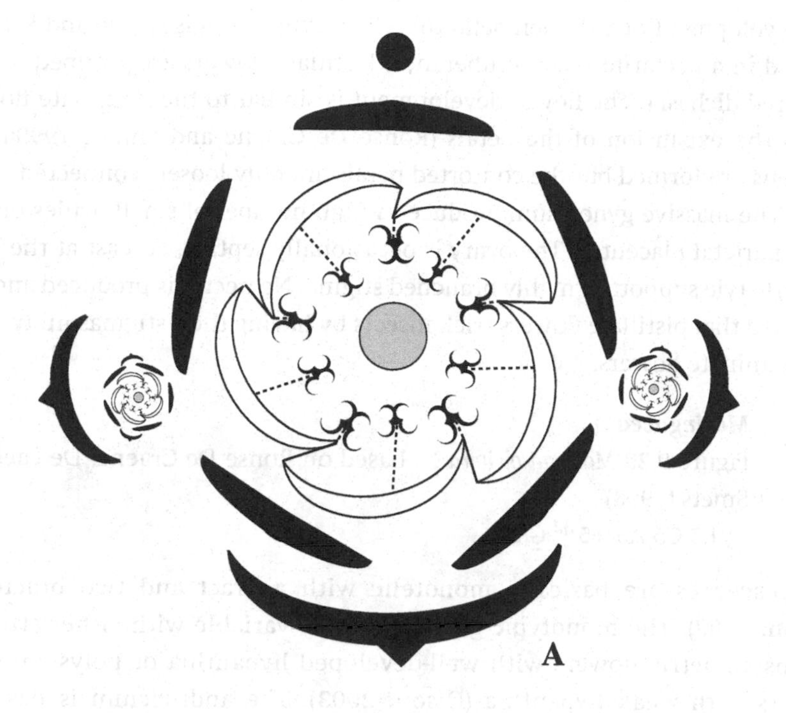

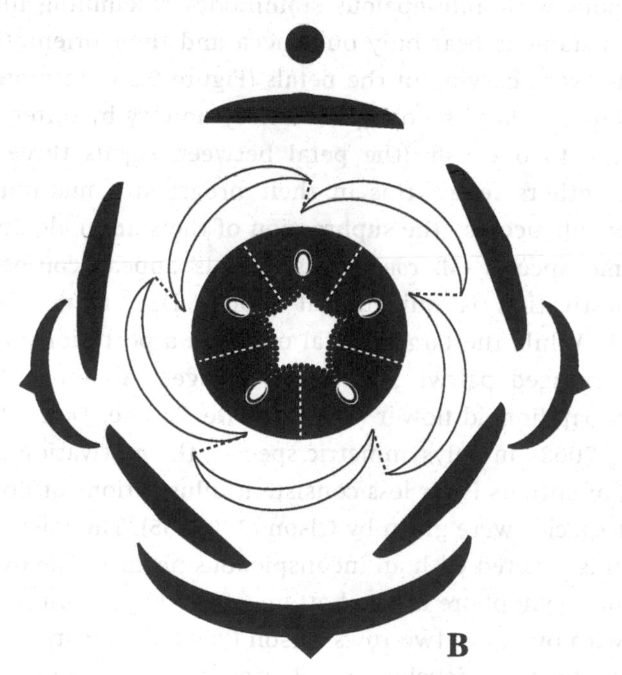

Figure 9.22 *Carica papaya* (Caricaceae): staminate partial inflorescence (A) and pistillate (B) flower. Note fusion of petals and stamen-petal tube.

the development of a stamen-petal tube. The gynoecium is sterile and is transformed in a nectariferous protuberance. Pistillate flowers are grouped in few-flowered dichasia. The flower development is similar to the staminate flowers up to the expansion of the petals (Ronse De Craene and Smets, 1999a). No stamens are formed but the contorted petals are only loosely connected at the base. The massive gynoecium produces a high number of small ovules on five large parietal placentae. The ovary is occasionally septate, at least at the base. A short style supports a highly branched stigma. No nectar is produced and it is assumed that pistillate flowers trick insects by having their stigmas mimicking the staminate flowers.

Moringaceae

Figure 9.23 *Moringa oleifera* L., based on Ronse De Craene, De Laet and Smets (1998)

↘ K5 C5 $A5^{\circ}+5^{1/2}$ $\underline{G}(3)$

Inflorescences are basically monotelic with a bract and two bracteoles (Olson, 2003). The monotypic genus is highly variable with either strongly monosymmetric flowers with well-developed hypanthia or polysymmetric flowers with weak hypanthia (Olson, 2003). The androecium is basically diplostemonous with antesepalous staminodes resembling filaments; the antepetalous stamens bear only one theca and their orientation is opposite the downward curving of the petals (Figure 9.23). In monosymmetric flowers, petals regulate the oblique monosymmetry by differences in size and curvature to one side (the petal between sepals three and five is larger) and anthers reflect this in their progressive maturation. This is occasionally enhanced by the suppression of the staminode opposite sepal four in some species (*M. concanensis*). Petals appear contorted in bud, although aestivation is quincuncial (Ronse De Craene, De Laet and Smets, 1998). While the largest petal occupies a posterior position, other petals are arranged pairwise as mirror images. As a result the flower resembles a papilionoid flower (cf. Ronse De Craene, De Laet and Smets, 1998; Olson, 2003). In polysymmetric species, the aestivation of petals and orientation of anthers is far less consistent. Illustrations of floral diagrams of different species were given by Olson (2003: 55). The lower slope of the hypanthium is covered with an inconspicuous nectary. The ovary is superior borne on a gynophore at the bottom of the hypanthium. Placentation is parietal with ovules in two rows. Olson (2003) demonstrated that monosymmetry is an early developmental process that is either accentuated later in development or repressed.

Figure 9.23 *Moringa oleifolia* (Moringaceae): partial inflorescence. Note the orientation of half-anthers.

'Batoid Clade'

The batoid clade consists of three families – Bataceae, Koeberliniaceae and Salvadoraceae – that are adapted to extreme halophytic and xerophytic habitats. All members share tetramerous flowers, diplostemonous in *Koeberlinia*, but haplostemonous in the other families, clawed petals and a dimerous ovary. Flowers of *Batis* are unisexual with a strong dimorphism between the male tetramerous flowers enclosed by a saccate calyx, and female flowers without perianth imbedded in an expanded receptacle (Ronse De Craene, 2005). Connecting characters are discussed in Ronse De Craene and Wanntorp (2009).

Salvadoraceae

Figure 9.24 *Salvadora persica* L., based on Ronse De Craene and Wanntorp (2009)
∗K(4) [C(4) A4] $\underline{G}$(1:1°)
General formula: ∗K(2)4(−5) C4(−5) A4(−5) G(2)

The small family with three genera shares haplostemonous, mainly tetramerous disymmetric flowers with reflexed petals. Flowers are arranged in

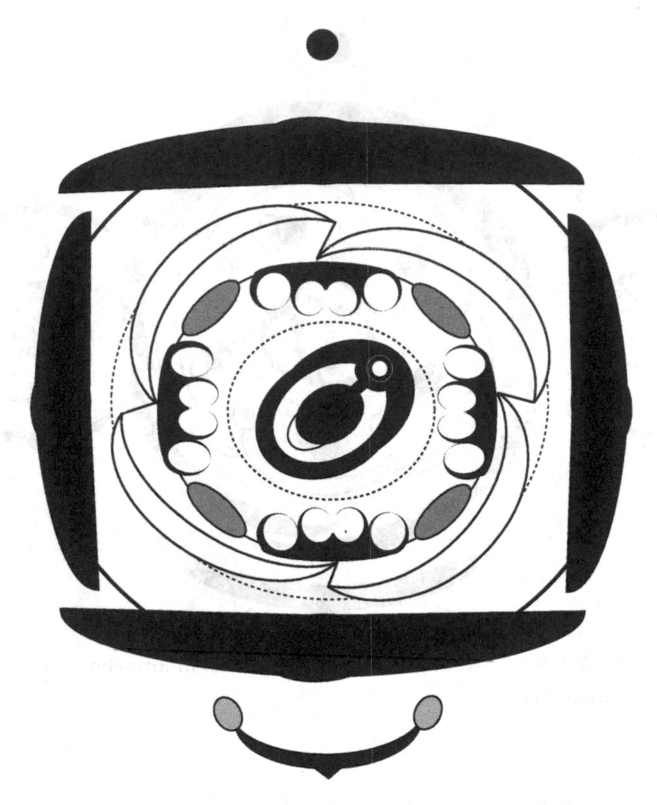

Figure 9.24 *Salvadora persica* (Salvadoraceae)

racemes and are subtended by a bract with glandular stipules. While stamens are free in *Azima*, stamens are fused in a tube in *Dobera* separate from the petals, while petal bases and stamens are connected by a hypanthial tube in *Salvadora* (Ronse De Craene and Wanntorp, 2009). Large nectaries alternate with the stamens (Figure 9.24) or are inserted outside the stamen tube in *Dobera*. Nectaries are not mentioned for *Azima* (Kubitzki, 2003). Nectaries have been reported to be staminodes, although this appears to be unlikely due to their variable position and very late appearance in the flower. The oblique ovary is inserted on a short gynophore; while two ovules are developed in *Azima* and *Dobera*, *Salvadora* has a single locule with a single basal ovule (Kshetrapal, 1970; Ronse De Craene and Wanntorp, 2009). The two ovules of *Dobera* and *Azima* are inserted on a parietal placenta and are separated by an apical false septum, giving the impression of a bilocular ovary. Floral developmental evidence indicates that the ovary is pseudomonomerous (Ronse De Craene and Wanntorp, 2009).

'Core Brassicales'

'Core' Brassicales comprise eight families, including members traditionally considered as a natural group based on distinct morphological characters (tetramery, disymmetry, mostly parietal placentation, a gynophore, campylotropous ovules: Ronse De Craene and Haston, 2006). However, these characters tend to be more widespread throughout the Brassicales.

Monosymmetry is widespread and always median (e.g. Emblingiaceae, Resedaceae, Cleomaceae, Capparaceae). As a result stamens tend to be often laterally increased on the abaxial side of the flower, while those on the adaxial side are not increased in number, are smaller, or are staminodial (e.g. Resedaceae: Sobick, 1983; Cleomaceae: Karrer, 1991; Patchell et al., 2011).

Capparaceae

Figure 9.25A *Capparis cynophallophora* L.
∗ K4 C4 A8+4+4+8+4+4 $\underline{G}(2)$

Figure 9.25B *Euadenia eminens* Hook. f.
↓ K4 C4 A5:6° $\underline{G}(2)$

General formula: ↓/⟷/∗ K(2)4(6) C0/2(4)6 A(2)4–∞ G2(–12)

Capparaceae with thirty-two genera have a highly diverse floral morphology (see Endress, 1992; Ronse De Craene and Smets, 1997a). Cleomoideae were placed in Capparaceae, but they represent a morphological intermediate between Capparaceae and Brassicaceae and should be recognized as a separate family (Hall, Sytsma and Iltis, 2002). Flowers are mostly tetramerous, (weakly) disymmetric to polysymmetric (*Capparis*) or strongly monosymmetric (e.g. *Euadenia*: Figure 9.25B; Karrer, 1991). The calyx is well developed and often bears nectary scales at the base (developing on a weak hypanthium). Petals are generally clawed and ephemerous with the stamens. Petals are occasionally absent (e.g. *Boscia*, *Ritchiea*). The androecium is variable, although the number of stamens tends to be constant within a genus, such as *Capparis* (Ronse De Craene and Smets, 1997a). A multistaminate androecium arises centrifugally on a ring primordium (a torus) with a clearly whorled sequence, in cases where numbers are moderately high (Figure 9.25A; Karrer, 1991; Ronse De Craene and Smets, 1997a). The gynoecium is mostly bicarpellate with parietal placentation and develops on a long gynophore. A style is absent or weakly developed. In some *Capparis* the number of carpels tends to be increased (e.g. *C. spinosa*: Ronse De Craene and Smets, 1997b).

Flowers of Cleomaceae are either polysymmetric or show strong monosymmetry with unidirectional flower initiation (Karrer, 1991; Erbar and Leins, 1997; Patchell et al., 2011). They share the same floral diagram as Brassicaceae,

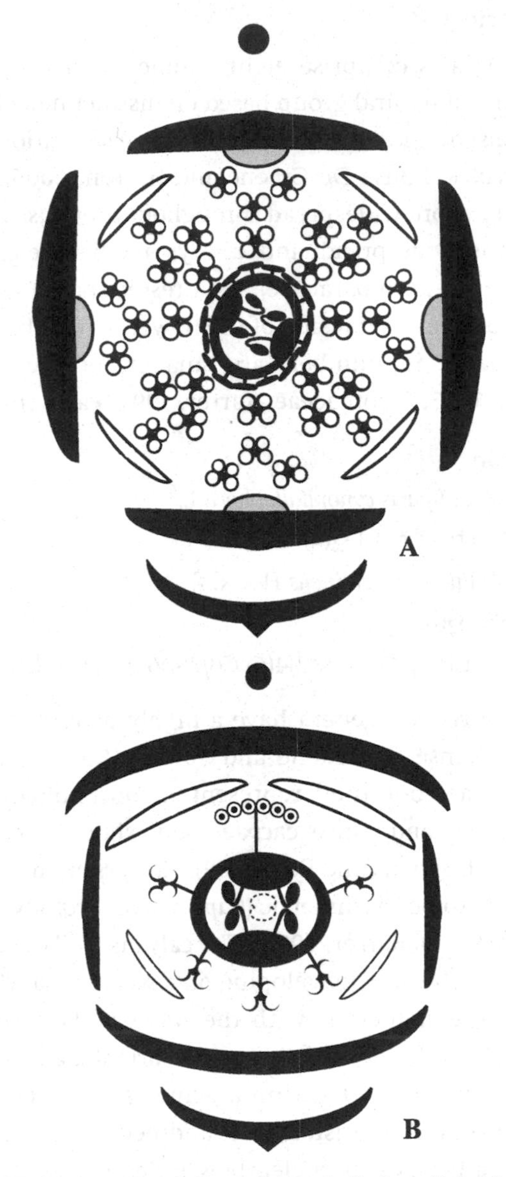

Figure 9.25 Capparaceae: A. *Capparis cynophallophora*; B. *Euadenia eminens*. In A ovary covered with peltate scales; position of stigma in B shown by broken contour

although stamen numbers can be laterally increased (e.g. *Polanisia*) or reduced (e.g. *Cleome*, *Dactylaena*). The adaxial side of the androecium is occasionally sterile or stamens are missing. In *Dactylaena* only one abaxial stamen is found with four adaxial staminodes (Karrer, 1991). Petals are strongly clawed and often unequal. A nectary is usually developed on the adaxial side, often as a crest on the androgynophore (e.g. *Cleome isomeris*: pers. obs.). Contrary to

Brassicaceae, Cleomaceae lack a false septum in fruit (Hall, Sytsma and Iltis, 2002). A number of floral diagrams of Capparaceae and Cleomaceae are provided by Karrer (1991).

Brassicaceae

Figure 9.26 *Pachyphragma macrophyllum* Busch

↔K4 C4 A2+4 G(2)

General formula: ↔(↓) K4 C(0)4 A2(0)+(2)4(−24) G2(4–6)

Figure 9.26 *Pachyphragma macrophyllum* (Brassicaceae). Note the absence of a subtending bract.

Inflorescences are generally racemose contrary to Capparaceae, which are more diverse. Bracts are generally missing. Brassicaceae (or Cruciferae) have one of the most readily recognizable floral diagrams shared by all members of the family (with very few exceptions enumerated by Endress, 1992). Flowers are tetramerous but appear to be dimerous and disymmetric. Because of a high similarity with the floral diagram of Papaveraceae, both families used to be associated into the Rhoeadales until fairly recently, although the difference in chemical composition was the main reason to separate them.

Monosymmetry is rare and is caused either by petals (e.g. *Iberis*) or stamens. In *Iberis amara* the abaxial petals are much longer than the adaxial petals, and this is regulated by late expression of *CYC* homologs (Busch and Zachgo, 2007). Monosymmetry corresponds with the arrangement of flowers in flattened racemes resembling umbels. Four sepals are arranged in a median and transversal position and are free (exceptionally fused). Four petals are clawed and arranged in diagonal position. They are rarely lacking (e.g. *Lepidium*). The androecium consists of six stamens, as in several Cleomaceae, with two outer smaller stamens opposite the lateral sepals and four inner stamens not exactly opposite the petals, but shifted towards the median line (tetradynamy). The genus *Lepidium* has the most variable stamen numbers: inner stamens are occasionally fused or missing, as are the outer stamens (Bowman and Smyth, 1998). As loss and fusion of stamens happen independently, the number of stamens fluctuates between two and six. Stamens are increased in number up to twenty-four in *Megacarpaea*, but their position is currently unknown. The gynoecium is almost always bicarpellate with parietal placentation. A false septum often divides the ovary in two parts and contributes with the persistent placental ridge (replum) to fruit formation and seed dispersal. Nectaries develop at the base of the stamens, often in variable position, wherever there is sufficient space. Bernardello (2007) distinguished four types, which can be grouped into an annular nectary, or a segmented nectary.

Endress (1992) showed that the Brassicaceae are highly uniform, compared to the highly diverse Capparaceae and considered the floral Bauplan of Brassicaceae to be ancestral for Brassicaceae and Capparaceae alike. At least some members of Cleomaceae, Capparaceae (e.g. *Steriphoma*) and Brassicaceae share the tetradynamous androecium of two and four stamens, and I believe this configuration to represent an apomorphic tendency. Hall, Sytsma and Iltis (2002) concluded on character reconstructions that an androecium of seven to fifteen stamens is plesiomorphic, but this is not helpful in understanding floral evolution, as they gave no indication of stamen position. There are two main interpretations for the origin of the peculiar flowers of Brassicaceae *sensu lato* (reviewed by Endress, 1992; Figure 9.27). The first interprets the flower as basically dimerous with six whorls (K2+2 C2^2 A2+2^2 G2: Figure 9.27A). This is based on the closer proximity of petals in a median plane and the occasional replacement of four inner stamens by two stamens (e.g. *Lepidium*, some *Cleome*). The double stamen and petal positions are interpreted as a result of dédoublement. The second hypothesis interprets the flower as basically tetramerous and five-whorled (K4 C4 A2+4 G2: Figure 9.27B). The first hypothesis is mainly discredited by the absence of clear dédoublement and the phylogenetic position of core Brassicales (embedded in a pentamerous, diplostemonous clade). The second hypothesis is endorsed by the enigmatic genus *Pentadiplandra* (Pentadiplandraceae), which can clarify the unusual floral diagram of

Brassicaceae. Flowers of *Pentadiplandra* are pentamerous and diplostemonous. However, they are pressed between bract and axis and are disymmetric in early development with the initiation of lateral sepals preceding the median sepals. Brassicaceae may have arisen through a similar process leading to the loss of two median antesepalous stamens (Figure 9.27B; Ronse De Craene, 2002). Androecial evolution of Capparaceae and Cleomaceae is more diverse, with a reduction of stamens up to one (e.g. *Dactylaena*), partial loss of stamens through monosymmetry (e.g. *Euadenia*, *Cleome*) and secondary increases of stamens (e.g. *Capparis*, *Polanisia*).

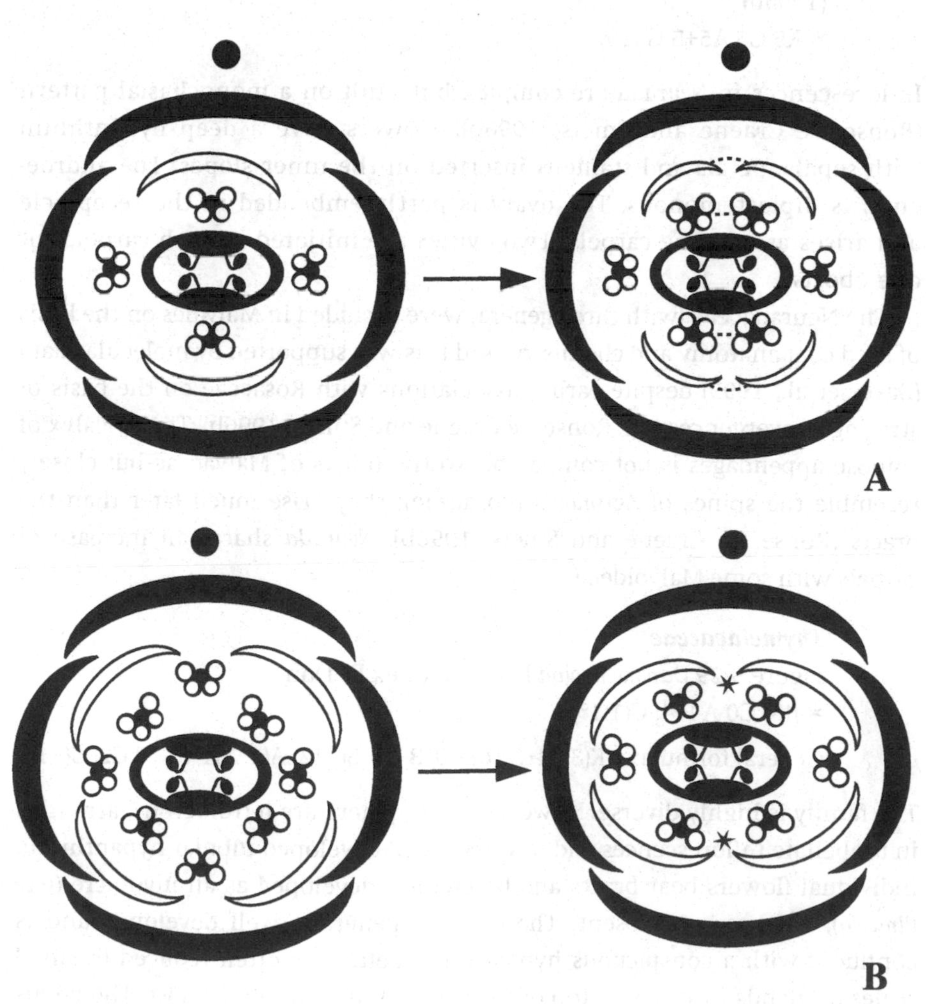

Figure 9.27 Diagrammatic representation of the derivation of the Brassicales floral Bauplan. A. dimerous prototype; B. tetramerous prototype

Malvales

The premolecular Malvales were restricted to five families (e.g. Cronquist, 1981), but were expanded to include ten families, with the exclusion of Elaeocarpaceae (e.g. Alverson et al., 1998). Von Balthazar et al. (2006) interpreted Malvales as basically diplostemonous (present in Neuradaceae, Thymelaeceae and Dipterocarpaceae), although several clades are characterized by a secondary stamen increase. However, indications for the existence of two stamen whorls were demonstrated for Cistaceae (Nandi, 1998) and Bixaceae (Ronse De Craene, 1989b).

Neuradaceae

Figure 9.28 *Neurada procumbens* L., based on Ronse De Craene and Smets (1996b)

∗ K5 C5 A5+5 Ğ (10)

Inflorescences in *Neurada* are complex but built on a monochasial pattern (Ronse De Craene and Smets, 1996b). Flowers have a deep hypanthium with sepals, petals and stamens inserted on the inner slopes. The androecium is diplostemonous. The ovary is partly embedded in the receptacle and arises as ten free carpels. Two ovules are initiated in each carpel, but one aborts.

The Neuradaceae (with three genera) were included in Malvales on the basis of seed coat anatomy and chemistry, and this was supported by molecular data (Bayer et al., 1999) despite earlier associations with Rosaceae on the basis of striking convergences (e.g. Ronse De Craene and Smets, 1996b). The epicalyx of spinose appendages is not comparable to the bracts of Malvaceae but closely resemble the spines of *Agrimonia* (Rosaceae); they arise much later than the bracts (Ronse De Craene and Smets, 1996b). *Neurada* shares an increase of carpels with some Malvoideae.

Thymelaeaceae

Figure 9.29 *Daphne bholua* Buch.-Ham ex D. Don

∗ [K4 C0 A4+4] G(1:1°)

General formula: ∗ K(3–)4–5(–6) C0/(3–)4–5(–12) A(1–2)4–8(–∞) G2–5(–12)

The family is highly diverse. However, most genera are tetramerous, arranged in umbellate inflorescences and possess a well-developed tubular hypanthium. Individual flowers bear bracts and bracteoles, developed as an involucre (e.g. *Pimelea*), or these are absent. The calyx is generally well developed and is confluent with a conspicuous hypanthium. Petals are often reduced to small scales or glands inserted on top of the tube, or are completely lost. The petals are occasionally bilobed or paired and have been variously interpreted as

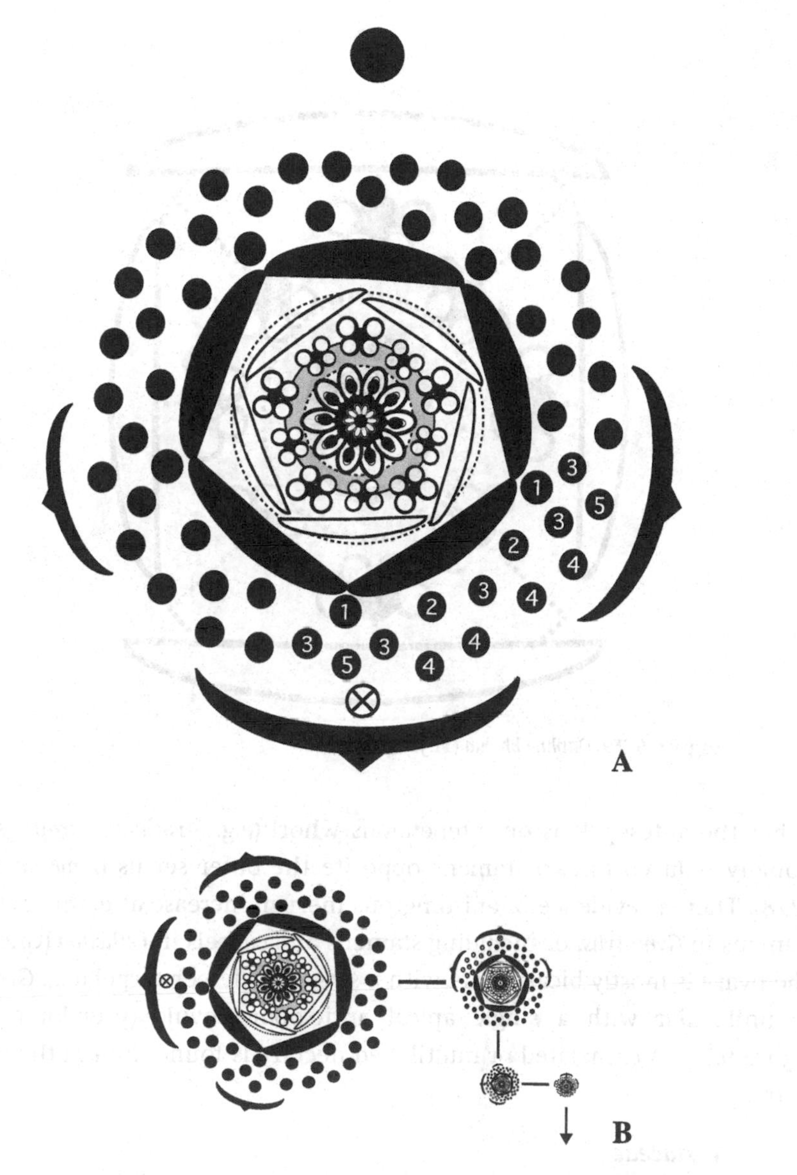

Figure 9.28 *Neurada procumbens* (Neuradaceae): A. flower; B. inflorescence. Black dots, outer spines with numbers referring to their sequence of initiation

glands, scales, receptacular outgrowths, stipules of the sepal lobes or staminodes (overview in Heinig, 1951), all of which are unlikely.

The androecium is diplostemonous and stamens are characteristically inserted on two levels on the hypanthium, those opposite the sepals being higher (*Gnidia*, *Daphne*). (Ob)haplostemony is found in some genera by loss of

Figure 9.29 *Daphne bholua* (Thymelaeaceae)

either the antesepalous or antepetalous whorl (e.g. *Struthiola*, *Drapetes*), occasionally reduced to two stamens opposite the outer sepals (*Pimelea*: Eichler, 1878). There is evidence of anisomerous merism increase affecting petals and stamens in *Gynostylis*, or including stamens and carpels in *Lethedon* (Gilg, 1894). The ovary is mostly bicarpellate with a single ovule per carpel (e.g. *Gyrinopsis*), or unilocular with a single apical antitropous ovule (pseudomonomery, Figure 9.29). A crenelated or multilobed nectary is found around the superior ovary.

Cistaceae

Figure 9.30 *Cistus salviifolius* L.

∗ K5 C5 A5$^{\infty}$+5 $\underline{G}$ (5)

General diagram: ∗ K3–5 C0/(3)5 A3–∞ G (2–)3–5(6–12)

Flowers are solitary or grouped in unipartite cymes with flowers arising in a zigzag pattern. Flowers are small to large, polysymmetric and subtended by a single bract. The calyx shows considerable variation between different species: it consists either of five subequal sepals with a clear 2/5 arrangement, or the two outer sepals are significantly smaller and occasionally absent. Petals are imbricate contorted and crumpled in bud. In species with only three sepals, two petals are opposite the sepals and the other three alternate with them (e.g. *C. ladanifer*). Eichler (1878) mentions a superposition of petals to sepals in some *Cistus* species. A similar arrangement is found in *C. creticus* and *C. salvifolius* where the outer sepals are highly different from the inner and usually do not enclose the bud (pers. obs.). The position of petals is influenced by size differences of sepals and the more or less high degree of contortion in the flower.

The androecium is usually polyandrous with centrifugal stamen initiation. Five antesepalous stamens arise earlier and more centrally than the other stamens. They are followed by five to thirteen antepetalous primordia and more primordia follow in a more or less regular arrangement on a ring primordium (Ronse De Craene and Smets, 1992a; Nandi, 1998). The androecium can be interpreted as complex haplostemonous or possibly diplostemonous with a secondary multiplication of stamens from antesepalous primordia, as is common in the Malvales. In large-flowered species, the number of primordia derived from the second whorl is usually much larger. The gynoecium consists of five (six to twelve) carpels in *Cistus*, but only of three carpels in other genera, with intruding parietal placentation. In *Cistus* the

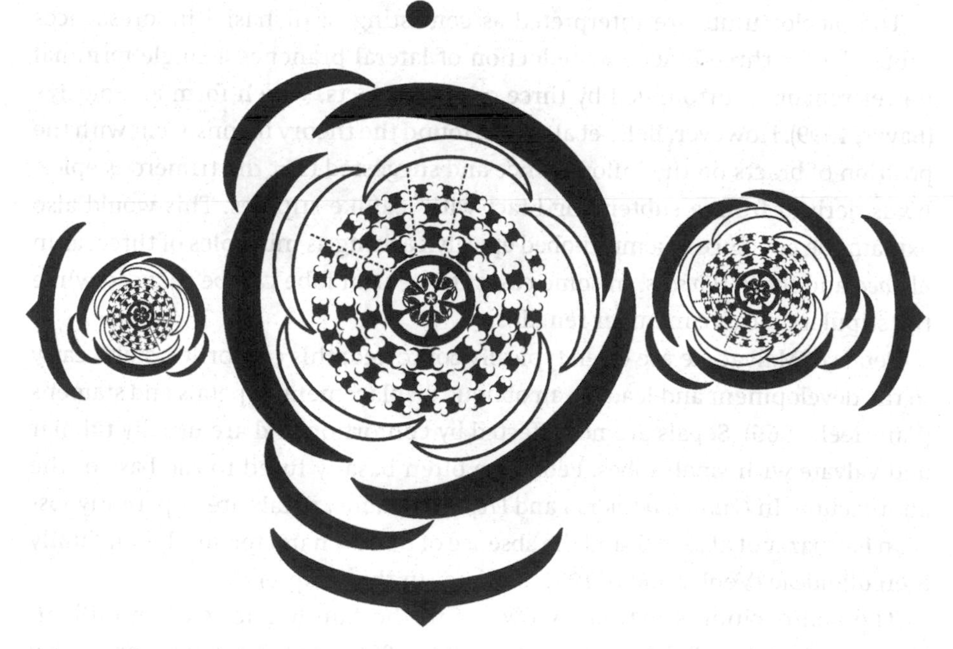

Figure 9.30 *Cistus salvifolius* (Cistaceae): partial inflorescence. Broken lines delimit groups of stamens.

placentae are deeply intruding and appear axile (Figure 9.30). Carpels alternate with petals and there is a single solid style. Ovules are inserted in pairs or in two rows on each placenta. No nectaries are found. Some genera have a tendency for the flowers to become trimerous (*Lechea*) or petals are occasionally lost (*Crocanthemum*), and this was interpreted as a reduced state linked with cleistogamy (Nandi, 1998).

Malvaceae

General formula: ✶ K5 C0/5 A5–∞ $\underline{G}$(2–3)5(–∞)

Molecular data have shown that there is no support for the traditional delimitation of four families: Malvaceae, Sterculiaceae, Bombacaceae and Tiliaceae. Sterculiaceae and Tiliaceae are largely polyphyletic, while Malvaceae *sensu stricto* is the only natural clade (Alverson et al., 1998, 1999; Bayer et al., 1999). Nine well-supported evolutionary lineages were identified and these are best treated as subfamilies, as a subdivision into separate families is clearly unpractical (Bayer et al., 1999; see comments on p. 70).

Core Malvales form a natural group that can be identified by a number of synapomorphies, including a typical dichasial inflorescence structure often including an epicalyx formed by three bracts ('bicolor units' *sensu* Bayer, 1999), trichomatic floral nectaries on the perianth (Vogel, 2000), a valvate calyx and basically obdiplostemonous androecium (Von Balthazar et al., 2004, 2006).

The bicolor units are interpreted as consisting of dichasial inflorescences subtended by three bracts. By reduction of lateral branches a single terminal flower remains, surrounded by three or more bracts, which form an epicalyx (Bayer, 1999). However, Bello et al. (2016) found the theory inconsistent with the position of bracts on the inflorescence and suggested that the trimerous epicalyx is derived from a subtending bract with leaflike stipules. This would also explain the presence of a multilobed epicalyx arising as multiples of three, as in *Althaea* and some *Hibiscus*. In some cases the central lobe can be reduced while the stipular epicalyx member remains.

Petals of Malvaceae are often typically contorted; this contortion starts early in the development and leads to an oblique displacement of petals and stamens (Van Heel, 1966). Sepals are not affected by contortion and are usually tubular and valvate with small lobes. Petals are often basally fused to the base of the androecium. In *Chiranthodendron* and *Fremontodendron* petals are apparently lost (Von Balthazar et al., 2006) and the absence of petals characterises the subfamily Sterculioideae (Venkata Rao, 1952; see later in this chapter).

The androecium is extremely diverse in the family and is often difficult to interpret. Von Balthazar et al. (2004, 2006) and Venkata Rao (1952) interpreted an obdiplostemonous arrangement as ancestral for Malvaceae.

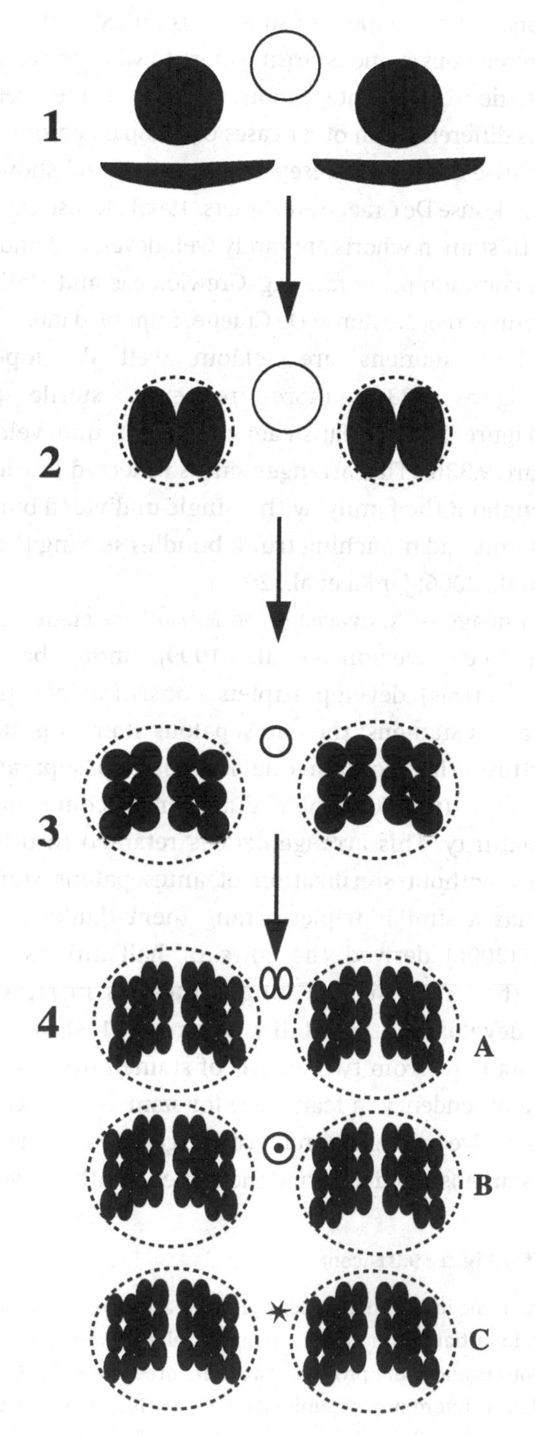

Figure 9.31 Model of sequence of androecial development in the Malvaceae, based on Janka et al. (2008), strongly modified. 1. Early stage with antesepalous stamen primordia (white dots) and antepetalous primordia (black dots); 2. Lateral division of antepetalous primordia; 3. Centrifugal differentation of secondary

The obdiplostemonous arrangement of stamens is linked with a progressive sterilization of the antesepalous stamens, arising more towards the centre of the flower following the initiation of the antepetalous stamens. I interpreted this kind of obdiplostemony as different from other cases of obdiplostemony where the antepetalous stamens arise after the antesepalous stamens and show a tendency for reduction (see p. 12; Ronse De Craene and Smets, 1995b; Ronse De Craene and Bull-Hereñu, 2016). Both stamen whorls are rarely well developed and produce few to many stamens on common primordia (e.g. Grewioideae and Matisieae: Van Heel, 1966; *Pachira* in Bombacoideae: Ronse De Craene, unpubl. data).

The antesepalous stamens are seldom well developed (e.g. some Sterculioideae, Figure 9.33A), more frequently sterile (Bombacoideae, Byttnerioideae, Figure 9.32A, Matisieae) or almost undeveloped or absent (Malvoideae, Figure 9.33B). This arrangement is reflected in the vasculature of the stamens throughout the family, with a single undivided bundle connecting the antesepalous unit and branching trunk bundles serving the other stamens (Von Balthazar et al., 2006; Janka et al., 2008).

In one major lineage of Malvaceae, the *Malvatheca* clade of Bombacoideae and Malvoideae (see Alverson et al., 1999), more basal genera (e.g. *Fremontodendron*, *Ochroma*) develop triplets consisting of one antesepalous and two antepetalous stamens; the antesepalous stamen is sterile and forms the tip of structures that contain one half of two separate antepetalous anthers, mimicking a true stamen! The anthers become long, twisted and fragmented at maturity. This arrangement is retained in other members of the clade with or without sterilization of antesepalous stamen primordia. *Ceiba pentandra* has a similar triplet arrangement (Janka et al., 2008). Von Balthazar et al. (2006) derived the rows of half-anthers found in many Malvaceae from the subdivision of long anthers as in *Fremontodendron* and the subsequent development of stalks. Figure 9.31 shows the centrifugal development of stamens from two whorls of stamen initials.

There is a general tendency in Malvaceae for stamens to become monothecal. However, this has evolved along different, convergent routes. In clades outside the *Malvatheca* clade, stamens proliferate and monothecal anthers evolved by splitting

---

Caption for Figure 9.31 (cont.)

primordia on the initial antepetalous primordia; 4. Lateral division of secondary primordia in tertiary primordia, each one developing into a half anther; A. the antesepalous primordium produces two half anthers: e.g. *Eriotheca, Adansonia*; B. the antesepalous primordium is staminodial: e.g. *Bombax, Pseudobombax, Goethea*; C. the antesepalous primordium is suppressed: most Malvoideae. Antesepalous primordium drawn to scale relative to antepetalous primordia.

of dithecal anthers; in the *Malvatheca* clade, monothecal anthers are the result of a compartimentalization of long anthers in rows (Alverson et al., 1998; Von Balthazar et al., 2006; Janka et al., 2008). The androecium of *Adansonia digitata* differs from other Malvaceae in the high number of small stamen primordia arising on a ring primordium (Janka et al., 2008). This obviously represents a secondary proliferation, leading to a much higher stamen number.

In Malvoideae, there is a strong tendency for complete loss of the antesepalous sector, which is occasionally present as small teeth in some genera (e.g. *Pavonia*, *Hibiscus*) and could represent the connective of the lost unit or sterile antesepalous stamen which is occasionally formed (Van Heel, 1966; Von Balthazar et al., 2004). In *Althaea officinalis* (Ronse De Craene, unpubl. data), stamen groups appear in antesepalous position, while in species without the antesepalous stamens, groups are distinctly antepetalous with each half developing distinct rows of stamens.

Almost all Malvoideae share a common early development of a primary ring primordium and the centrifugal initiation of two rows of secondary stamens opposite the petals (Figures 9.31, 9.33B). The ring primordium is broken up in five or rarely ten sectors on which secondary stamen primordia arise in two alternating rows. In many taxa, secondary stamens tend to be split radially or longitudinally in tertiary half-stamens, which can be sessile or connected on a common stalk. The centrifugal proliferation of stamens is made possible by upward growth of the initial ring primordium developing into a tube around the ovary. This development is unique in angiosperms (cf. Ronse De Craene, 1988) and leads to the formation of long staminal tubes from which the stylar lobes emerge centrally (e.g. *Hibiscus*).

The superior gynoecium is usually isomerous and develops five antepetalous carpels. Stamens and carpels are often united on an androgynophore or the ovary is stalked. There is the occasional proliferation of carpels in Malvoideae. In Malopeae (*Kitaibelia*, *Malope*), numerous small carpels arise as five groups in a waveline and are squeezed in the confinement of the stamen tube (Van Heel, 1995). Each carpel produces a single ovule and style. A second carpel whorl develops in Malvoideae-Ureneae, but the inner whorl develops no ovules (e.g. *Pavonia*, *Urena*) (Van Heel, 1978). The number of locules may be reduced, but this does not necessarily reflect a loss of carpels. A unilocular ovary is reported for *Waltheria* (Bayer and Kubitzki, 2003). Secondary apocarpy has evolved mainly in Sterculioideae and some other subfamilies (e.g. Endress, Jenny and Fallen, 1983; Jenny, 1988). Placentation is usually axile with ovules in two rows, often reduced to two lateral ascending ovules. Nectaries of Malvaceae are present on either sepal or petal bases, and consist of secretory trichomes (Vogel, 2000).

Byttnerioideae

Figure 9.32A *Theobroma cacao* L.

∗ K5 C5 A(5°+5^{2}) $\underline{G}$(5)

Byttnerioideae have retained the original obdiplostemonous arrangement. In *Theobroma*, antesepalous stamens develop into an erect staminode while antepetalous stamens divide in two lateral anthers facing away from each other. Stamens are rarely undivided, develop two monothecous anthers (*Byttneria*), or are more often in antepetalous triplets or higher numbers (Van Heel, 1966). Petals are usually cucullate and develop in a broad, cup-shaped basal part enclosing the anthers while the petal tip is spathulate and recurved backwards (Figure 9.32A; Bayer and Hoppe, 1990). Antepetalous stamens emerge before antesepalous staminodes, which are inserted more centrally, and the carpels are arranged opposite the petals, reinforcing the obdiplostemonous appearance of the androecium (Ronse De Craene and Bull-Hereñu, 2016).

Staminodes are occasionally lost and stamens can remain undivided leading to obhaplostemony (e.g. *Hermannia*). In *Theobroma* and allies, staminodes prevent selfing by mediating insect movement in the flower (Walker-Larsen and Harder, 2000). Flowers are rarely monosymmetric with unequal petals and stamens and gynoecium grouped in an androgynophore (*Kleinhovia*: Bayer and Kubitzki, 2003).

Tilioideae

Figure 9.32B *Tilia x europaea* L.

✶ K5 C5 A0+5$^{\infty}$ $\underline{G}$(5)

Subfamily Tilioideae is restricted to two genera, much fewer than in pre-molecular treatments. A subtending bract is typically fused to the inflorescence peduncle in a conspicuous wing, favoring dispersal. Sepals are free and bear glandular trichomes at the base. Petals are narrow and slightly cucullate with weakly imbricate aestivation. The androecium arises centrifugally in antepetalous groups forming two rows connected with one upper stamen, which is occasionally staminodial (*T. tomentosa*). In other Malvaceae, staminodes are usually in antesepalous position. Neighboring stamen fascicles are linked abaxially in a wave-line of stamens (cf. Van Heel 1966). The five carpels bear two ascending ovules per locule in *Tilia*.

Sterculioideae

Figure 9.33A *Sterculia coccinea* Jack

✶ K5 C0 [A(5+5) $\underline{G}$(3)][d]

Flowers are apetalous and functionally unisexual, and lack an epicalyx or staminodes (Alverson et al., 1999; Bayer et al., 1999). Stamens and carpels are connected by an androgynophore. In staminate flowers, stamens are compressed in bud and overtop the pistillode, resembling *Nepenthes* in bud (Ronse De Craene, unpubl. data). Venkata Rao (1952) described the androecium of *Sterculia foetida* as two-whorled with the alternisepalous stamens forming triplets. The androecium is variable in

[d] Shown as bisexual but with late abortion of one of the genders.

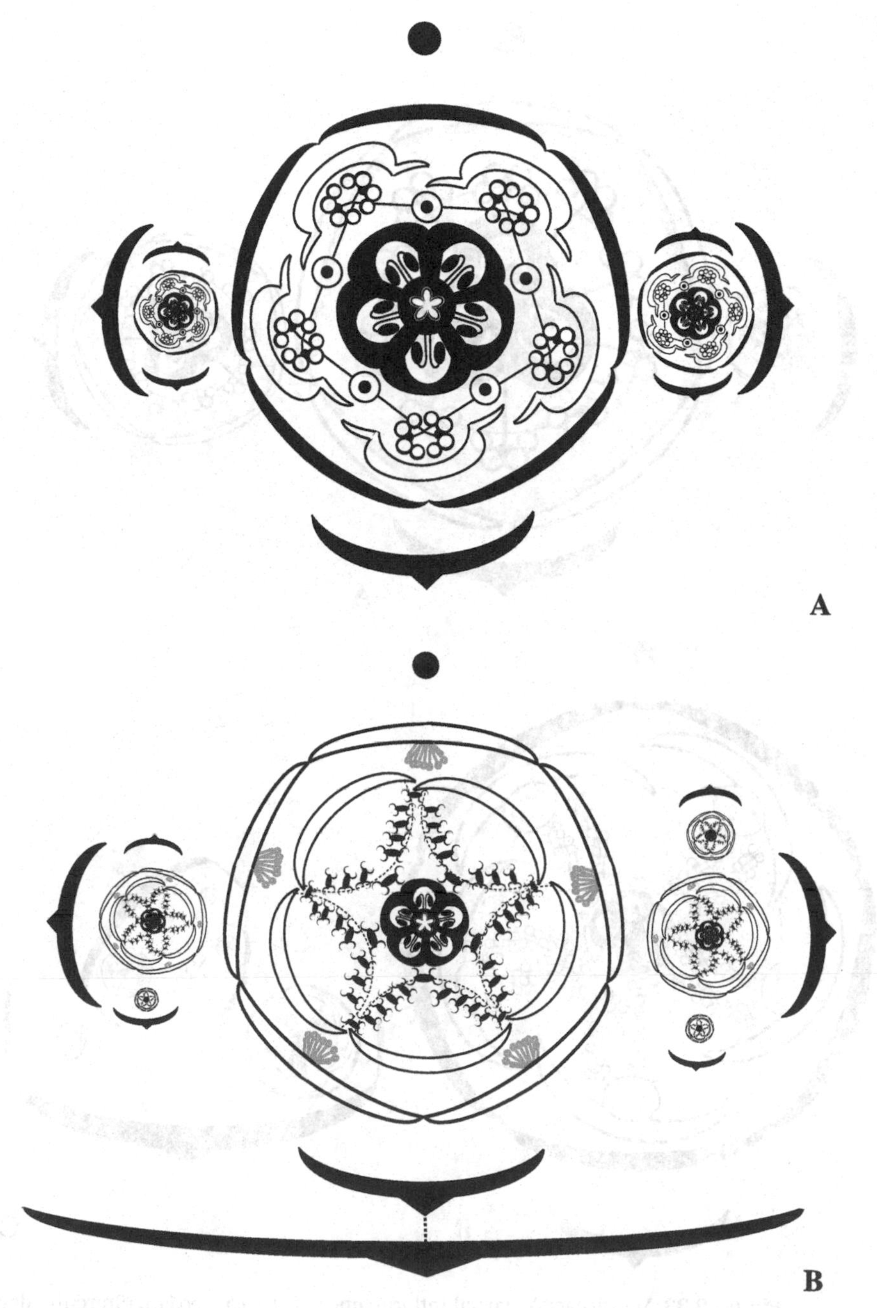

Figure 9.32 Malvaceae: A. partial inflorescence of *Theobroma cacao* (Byttnerioideae); B. *Tilia x europaea* (Tilioideae)

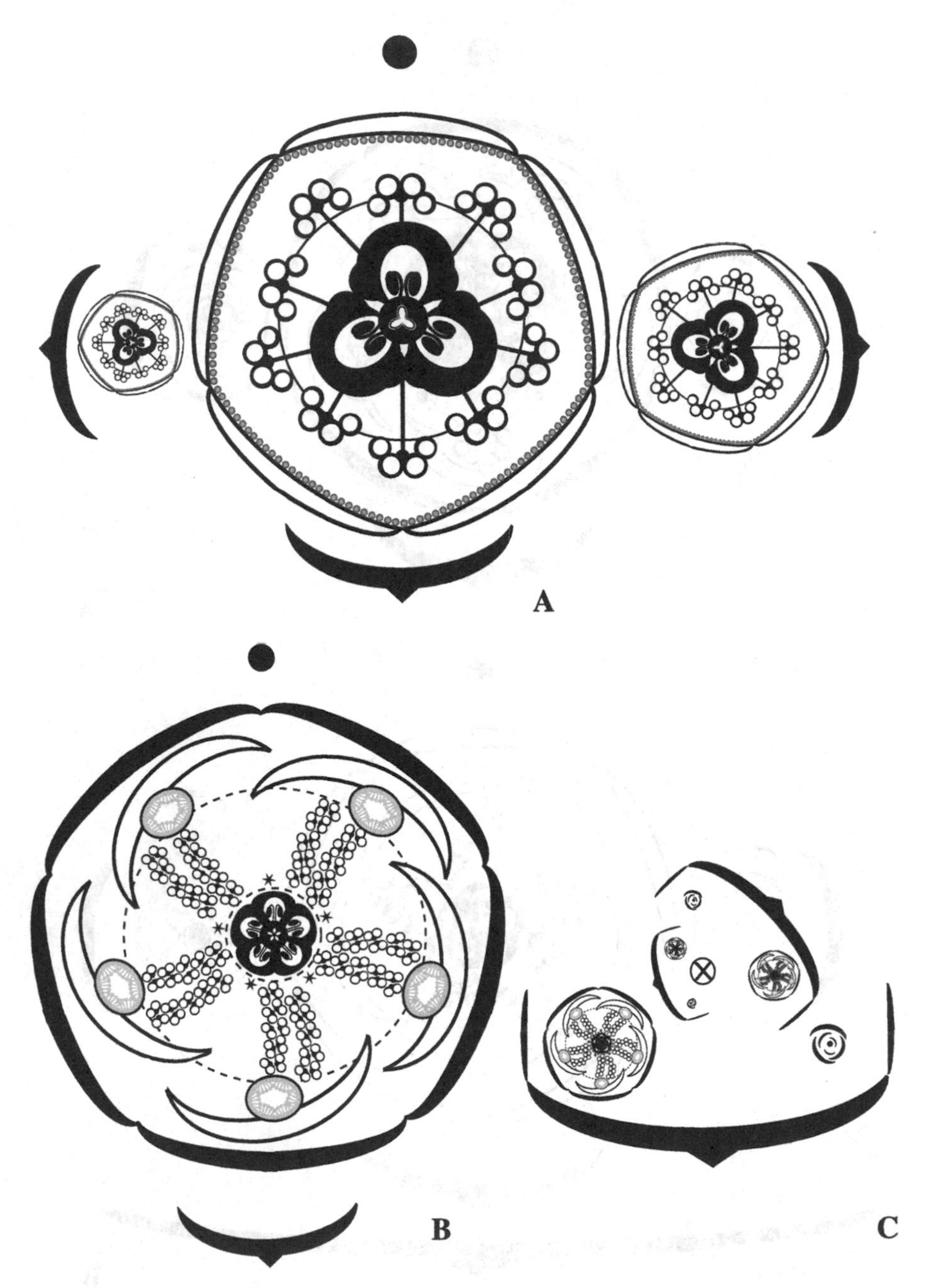

Figure 9.33 Malvaceae: A. partial inflorescence of *Sterculia coccinea* (Sterculioideae) and *Abutilon megapotamicum* (Malvoideae), flower (B) and partial inflorescence (C)

the subfamily with various increases of the antepetalous stamens to fifteen or more (e.g. *Brachychiton, Sterculia*) and loss of the antesepalous stamens (e.g. *Cola*, some *Sterculia*) (Van Heel, 1966). In *Cola*, stamens are paired with bisporangiate anthers arranged in a radial line (Van Heel, 1966; Bayer and Kubitzki, 2003). Carpels range from five to three in *Sterculia*. Secondary apocarpy evolved independently in the clade (Endress, Jenny and Fallen, 1983), and this is stressed at maturity by the development of individual carpophores. Trichomatic nectaries are variously spreading on the adaxial side of the sepal lobes and hypanthium, extending to the androgynophore.

Malvoideae

Figure 9.33B–C *Abutilon megapotamicum* (Spreng.) A. St. Hil. and Naudin

∗ K5 [C5 A(0+5$^{\infty}$)] $\underline{G}$(5)

Malvoideae correspond to the classical delimitation of Malvaceae and were placed with part of Bombacaceae in a *Malvatheca* clade (Alverson et al., 1999).

In *Abutilon*, an epicalyx is absent, while it is well developed in several other Malvoideae and consists of two, three or more lobes (Pluys, 2002; Bayer and Kubitzki, 2003; Bello et al., 2016). In *Goethea* the flower is enclosed by four large petaloid epicalyx lobes surrounding the red flowers (pers. obs.). The calyx is valvate. The corolla is basally adnate to the stamen column. The long staminal tube enclosing styles and carpels is a typical feature of Malvoideae (Ronse De Craene, 1988; Ronse De Craene and Smets, 1992a). The number of stamens tends to be correlated with the extent of development of the stamen tube, ranging from few (e.g. *Wissadula*) to many (e.g. *Abutilon, Hibiscus*). Stamens are disposed in long rows in antepetalous position. An antesepalous staminode is occasionally developed (e.g. *Goethea, Hibiscus*), more often lacking. Trichomatic nectaries are found in pockets on the calyx or opposite the petal lobes (e.g. *Abutilon*: Figure 9.33B). Floral development of numerous species of Malvoideae was studied by different authors (e.g. Payer, 1857; Van Heel, 1966, 1995; Von Balthazar et al., 2006).

9.3 Fabids

Fabids consist of three subclades: a nitrogen-fixing clade of Rosales, Cucurbitales, Fagales and Fabales; a 'COM-clade' of Celastrales, Oxalidales and Malpighiales; and a small basal clade of Zygophyllaceae and Krameriaceae not considered in this book (Figure 9.34; Schönenberger and Von Balthazar, 2006; Wang et al., 2009; Soltis et al., 2018). The common ancestor of the fabids had the ability for nitrogen fixation. This was retained in at least ten families (four in Rosales, three in Fagales, two in Cucurbitales, and Leguminosae) (Soltis et al., 2018). There is conflicting evidence about the position of the COM-clade, as phylogenetic analyses based on chloroplast genes place the clade with fabids, while

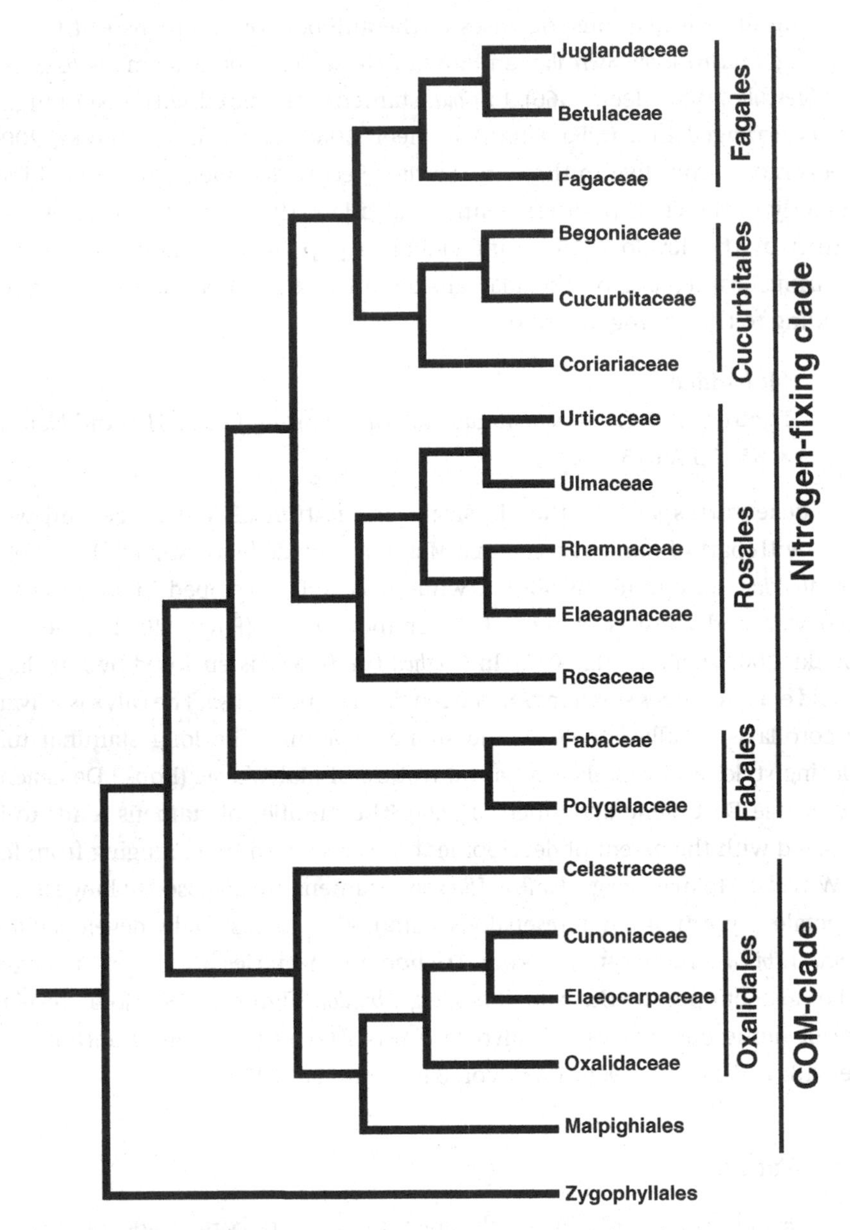

Figure 9.34 Phylogeny of fabids based on Wang et al. (2009)

nuclear and mitochondrial genes support a position with malvids (Figure 9.1). This could reflect an ancient introgressive hybridization between ancestors of fabids and malvids (Soltis et al., 2018). Morphological data tend to support a closer link with malvids, such as a tendency for a secondary stamen and carpel increase and contorted petals (Endress, Davis and Matthews, 2013).

9.3.1 COM-Clade: Celastrales-Oxalidales-Malpighiales

Celastrales

The order contains the two families Lepidobotryaceae and Celastraceae. Celastrales share several synapomorphies or apomorphic tendencies (summarized in Matthews and Endress, 2005). These include a tendency to produce proterandrous or functionally unisexual flowers, large petals in bud with often quincuncial aestivation forming the protective organs, mostly haplostemonous flowers (diplostemonous only in *Lepidobotrys*) or with antepetalous staminodes (e.g. *Parnassia*), tricarpellate gynoecia with commissural stigma and solid styles. Nectaries are often well-developed discs extending around the base of the gynoecium and protruding between the stamens. A short hypanthium is occasionally present and this is linked with the semi-inferior position of the ovary.

Celastraceae (incl. Hippocrateaceae, Brexiaceae, Stackhousiaceae, Plagiopteraceae, Canotiaceae, Parnassiaceae)

Figure 9.35A *Brexia madagascariensis* Thou. ex Ker Gawl

∗ K5 C5 A5+5° $\underline{G}$(5)

Figure 9.35B *Tonteleia sp.*

∗ K5 C5 [A3 $\underline{G}$(3)]

General formula: ∗ K(3–)4–5(–6) C(3–)4–5(–6) A(3–)4–5(–∞) G2–5

The circumscription of the Celastraceae has been controversial for a long time. The family is currently broadly circumscribed (see e.g. Simmons, 2004) and contains Hippocrateaceae, along with some smaller monotypic families. Matthews and Endress (2005) described floral morphology and anatomy of several representative species.

Flowers are generally polysymmetric and haplostemonous. Only *Apodostigma* has unequal petals and the androecium is polyandrous in *Plagiopteron* although its development is unknown (Simmons, 2004). A large receptacular disc nectary is present in several genera, and is sometimes extrastaminal (e.g. Hippocrateoideae). In Hippocrateoideae, stamen number is reduced to three, in alternation with three carpels (Figure 9.35B). The number of stamens is strongly linked to the merism of the gynoecium, which is enhanced by the position of the stamens inside the disc nectary (Matthews and Endress, 2005). My observations on the position of the stamens correspond to the description by Payer (1857) for *Salacia viridiflora* but differ from the diagram presented by Eichler (1878: 367). Payer accepts congenital fusion for four of initially five stamens; this can explain the position of two of the stamens opposite petals, comparable to the situation in *Mollugo*, Cucurbitaceae and *Hypericum* (see

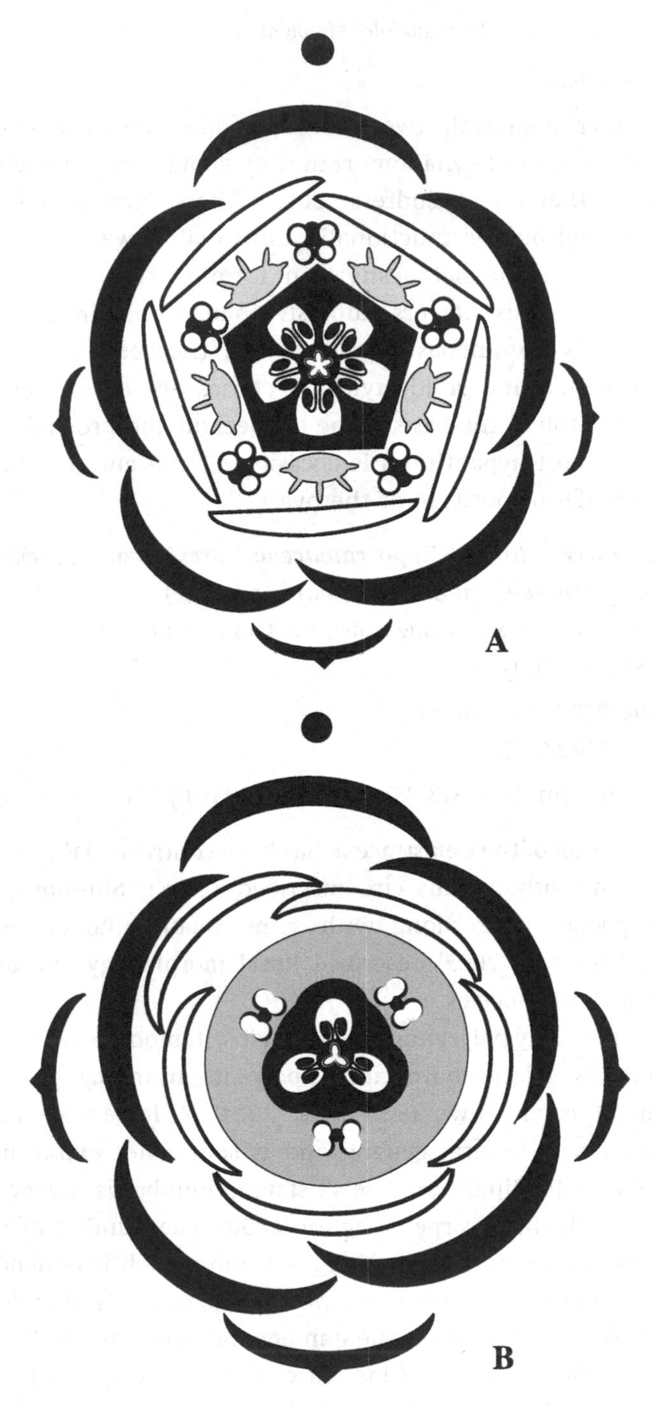

Figure 9.35 Celastraceae: A. *Brexia madagascariensis*; B. *Tonteleia sp.* Note the antepetalous staminodial nectaries in A.

Figures 9.35B, 9.41A, 10.19B). However, *Stackhousia* and *Tripterococcus* have three longer stamens and two shorter (the longer alternate with the three carpels). The condition in *Stackhousia* may be intermediate between the pentamerous and trimerous condition and trimery in the androecium could be the result of loss of stamens and a displacement of the remaining in alternation with the carpels (cf. Eichler, 1878). The extrorse stamens of *Tonteleia* are fused with the gynoecium into an androgynophore. *Brexia* and *Siphonodon* are reported to have antepetalous staminodes, although these are mostly interpreted as a receptacular disc nectary in *Brexia* (e.g. Simmons, 2004). The nectaries of *Brexia* (Figure 9.35A) resemble the staminodes of *Parnassia* and could be homologous with them, although they are initiated very late (Edgell, 2004). Both taxa share pointed extensions (more fingerlike in *Parnassia* with glistening tips) and a thick nectariferous base, in *Brexia* extending around the base of the ovary. Likewise, staminodes of *Parnassia* might be receptacular emergences with nectariferous tissue on the ventral side and glistening pseudostaminodes on the appendages (Endress and Matthews, 2006b). However, presence of a vascular bundle at the base of the structures supports a staminodial nature. Bensel and Palser (1975a) interpreted the fingerlike extensions of *Parnassia* and *Brexia* as individual staminodes and the whole structure as a fascicle and suggested that the androecium is derived from a fasciculate ancestor. Evidence for this are teratological cases with the replacement of the fingers by simple stamens (Wettstein, 1890, as cited in Matthews and Endress, 2005). However, this derivation is unlikely, as nearest clades are diplostemonous or haplostemonous.

The ovary is often trimerous (pentamerous in *Brexia*) with axile placentation and two ovules per carpel, reduced to one in *Stackhousia*. The ovary is rarely pseudomonomerous (*Pleurostylia*).

Malpighiales

The order is a huge, highly heterogeneous, and mainly tropical assembly of twenty-eight to circa thirty-seven families. Figure 9.36 is mainly based on Xi et al. (2012). Affinities of several families are partly unresolved as support for branches is weak. Finding common morphological characters is challenging (cf. Endress and Matthews, 2006a; Schönenberger and Von Balthazar, 2006; Endress, Davis and Matthews, 2013). Malpighiales are typical rosids with pentamerous flowers, free petals, diplostemony and a mostly superior three-to-five-carpellate ovary with free styles. However, the morphological diversity within the order is extraordinary, with several parallel evolutionary trends. When present, petals are contorted or imbricate. A hypanthium is common and nectaries are often distinct, rarely arranged as a disc. The ovary is either parietal or axile, with few

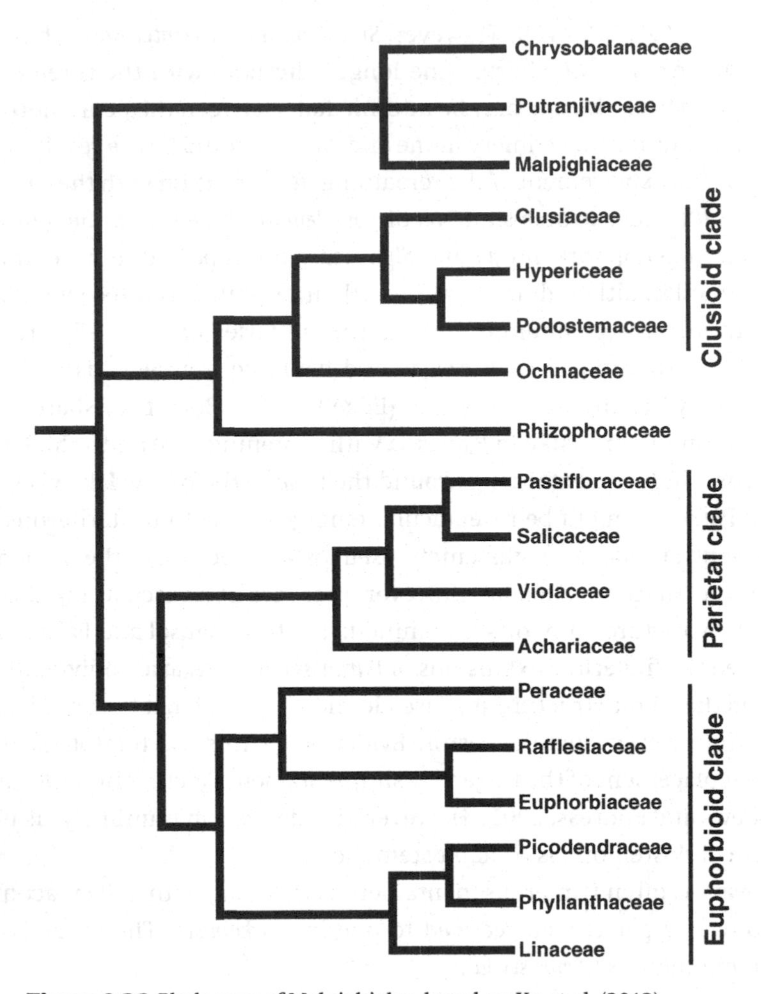

Figure 9.36 Phylogeny of Malpighiales, based on Xu et al. (2012)

anatropous ovules. Several Malpighiales are characterized by the presence of a corona, or various appendages on the petals. Oblique monosymmetry is found in Malpighiaceae and Chrysobalanaceae *sensu lato*.

There are some strongly supported associations of families sharing distinctive characters: (1) Clusioids, (2) Ochnoids, (3) Malpighioids, (4) a parietal clade and (5) Euphorbioids (Figure 9.36). Larger families have been split up, such as Euphorbiaceae in six entities, and Flacourtiaceae in Salicaceae and Achariaceae. It will be important to better circumscribe subordinal groups in the future, and attempts have been made to improve our knowledge of floral morphology in the order (e.g. Merino Sutter, Foster and Endress, 2006; Matthews and Endress, 2008, 2011).

Rhizophoraceae

Figure 9.37 *Bruguieria cylindrica* Blume

∗ K8 C8 A8+8 $\underline{G}$(2)

General formula: ∗ K4–5(–16) C4–5(–16) A8–∞ G2–20

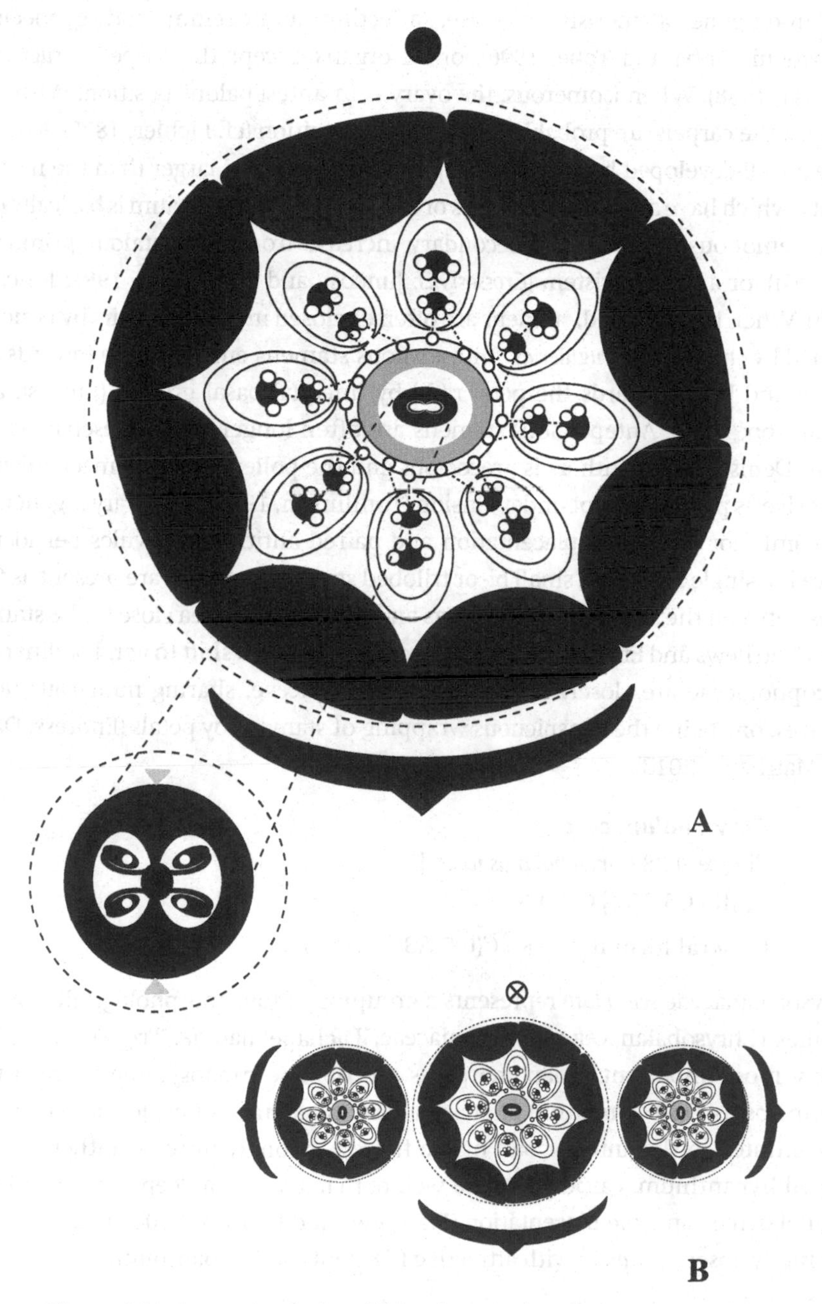

Figure 9.37 *Bruguieria cylindrica* (Rhizophoraceae): A. flower; B. partial inflorescence

Rhizophoraceae represents a small family of trees and shrubs specialized in a mangrove habitat. Inflorescences are built on a dichasial pattern consisting of one to a few flowers subtended by a bract and generally large bracteoles. Flowers are polysymmetric, but merism is variable. The most widespread genus, *Rhizophora*, is tetramerous, but *Bruguieria* fluctuates between hexa- and sixteen-merous (Tomlinson, 1986; Matthews and Endress, 2011). Especially the genus *Crossostylis* has undergone a merism increase, affecting androecium and gynoecium (Setoguchi, Ohba and Tobe, 1996), or all organs except the carpels (*Bruguieria*; Juncosa, 1988). When isomerous, the ovary is in antesepalous position. With two carpels, the carpels are probably in transversal position (cf. Eichler, 1878). Flowers have a well-developed hypanthium. Sepals are valvate and larger than the narrow petals, which have adaxial appendages or are hairy. The androecium is basically (ob) diplostemonous, or there is a secondary increase from antepetalous primordia (*Kandelia*), or a ring meristem (*Crossostylis*: Juncosa and Tomlinson, 1987; Juncosa, 1988). When two-whorled, stamens are often enclosed in unequal pairs by pouches formed by the petals (*Bruguieria*). Antesepalous stamens emerge more towards the centre and bend towards the petal radii by unequal basal growth (Juncosa and Tomlinson, 1987). Antepetalous stamens are often longer than antesepalous stamens. Dehiscence of anthers is precocious and the pollen dispersal mechanism is explosive as petals are kept under tension (Tomlinson, 1986). The ovary is generally (semi-)inferior with axile placentation and paired antitropous ovules per locule. There is a single style with small bi- or trilobed stigma. Nectaries are present as free lobes between the stamens (*Ceriops*) or as an intrastaminal area close to the stamen tube (Matthews and Endress, 2011). In *Rhizophora* there is a shift to wind pollination. Rhizophoraceae are closely related to Erythroxylaceae, sharing numerous floral features, one being the conspicuous wrapping of stamens by petals (Endress, Davis and Matthews, 2013).

Chrysobalanaceae

Figure 9.38 *Chrysobalanus icaco* L.

↘ [K5 C5 A5$^{\infty}$] $\underline{G}$(1:1°)

General formula: ↘ K5 C(0)5 A3–∞ G1–3

Chrysobalanaceae *sensu lato* represents a grouping of four morphologically diverse families (Chrysobalanaceae, Euphroniaceae, Dichapetalaceae, Trigoniaceae). The visibly most important synapomorphies are oblique monosymmetry (relatively rare in rosids) with a mixture of staminodes and stamens (fertile stamens being concentrated in the anterior part of the flower), strongly introrse anthers, a cup-shaped hypanthium, carpels that are well demarcated by a deep furrow and two parallel ovules on axile placentation (Matthews and Endress, 2008). Petals are free and rarely absent (*Lycania*) with attractive filaments and hypanthium.

Monosymmetry is associated with the retardation/sterilization of the posterior part of the androecium and gynoecium (Figure 9.38). In *Chrysobalanus* the gynoecium is pseudomonomerous with almost no trace of the second carpel at anthesis, although it is well visible in early stages, and the style is gynobasic through the extreme dorsal bulging of the fertile carpel. In other genera the gynoecium is tricarpellate and all carpels are fertile. Stigmatic lobes are carinal

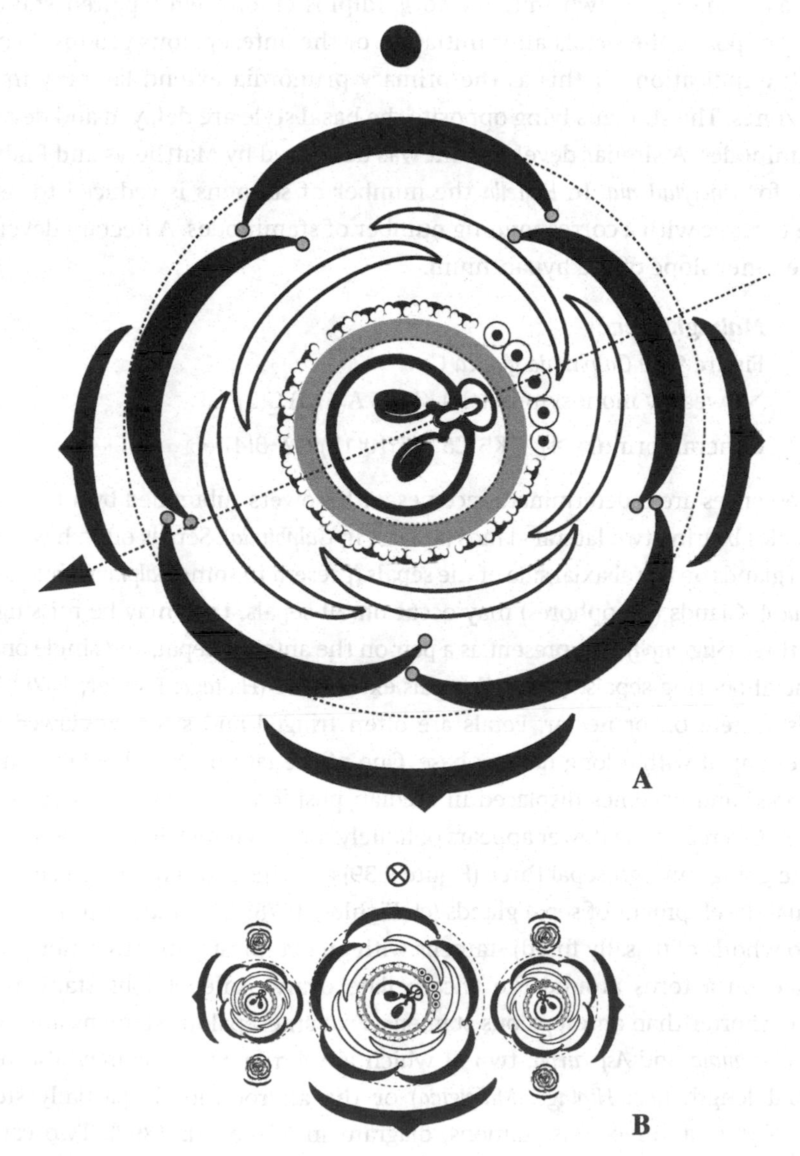

Figure 9.38 *Chrysobalanus icaco* (Chrysobalanaceae): A. flower; B. partial inflorescence

and two ovules are present in all taxa. A false septum develops occasionally between the two ovules (e.g. *Maranthes*).

Early stages of development of *Chrysobalanus* are regular (Matthews and Endress, 2008; pers. obs). A deep hypanthium places sepal lobes, petals and androecium as a rim around a cuplike depression. The numerous stamens in *Chrysobalanus* are arranged in a single ring, although they arise as five antesepalous groups (pers. obs.). Matthews and Endress (2008) interpreted the androecium as made up of two initial whorls (diplostemony with paired stamens arising opposite the petals after initiation of the antesepalous groups. I could see little indication for this as the primary primordia extend laterally in the petal zones. The stamens lying opposite the basal style are delayed and develop as staminodes. A similar development was described by Matthews and Endress (2008) for *Dactyladenia*. In *Hirtella* the number of stamens is reduced to seven down to three with a corresponding number of staminodes. A nectary develops on the inner slope of the hypanthium.

Malpighiaceae

Figure 9.39 *Galphimia glauca* Cav.

↘ (weakly monosymmetric) K5 C5 A(5+5) $\underline{G}$(3)

General formula: ∗/↘ K5 C5 A(5)10(15) G2–3(4)

Inflorescences are indeterminate racemes with flowers subtended by a bract and bracteoles bearing two lateral glands (at least in *Galphimia*). Sepals often have large paired glands on the abaxial side of the sepals (present in some *Galphimia* but not in *G. glauca*). Glands (elaiophores) may occur on all sepals, they may be missing on sepal three (*Stigmaphyllon*), present as a pair on the anterior sepal, and single on the two neighbouring sepals, or on all sepals except one (*Hiptage*: Eichler, 1878). The glands secrete oil or nectar. Petals are often fringed and strongly clawed, viz. paddle-shaped with a long narrow base. One of the latero-adaxial petals is more developed and becomes displaced in median position by a slight torsion of the pedicel. As a result the flower appears obliquely monosymmetric with the symmetry line going through sepal three (Figure 9.39) and this is often reinforced by the unequal development of sepal glands (cf. Eichler, 1878). The androecium consists of two whorls of (basally fused) stamens with an obdiplostemonous arrangement inserted on a torus around the tricarpellate ovary. Antepetalous stamens are slightly shorter than antesepalous stamens. Five antesepalous stamens are found in *Gaudichaudia* and *Aspicarpa*, two of which are fertile. Stamens may also be of unequal length (e.g. *Hiptage*, *Malpighia*) or the androecium is partially sterile (*Stigmaphyllon*: antesepalous stamens, diagram in Niedenzu, 1897). Two carpels are fertile in *Acridocarpus* (Stevens, 2001 onwards). Each carpel with individual style contains one pendent ovule. There is no nectary.

Figure 9.39 *Galphimia glauca* (Malpighiaceae)

'Clusioid Clade' (Including Clusiaceae, Hypericaceae, Podostemaceae, Calophyllaceae, Bonnetiaceae)

The circumscription of the large and heterogenous family Clusiaceae *sensu lato* has been problematic on a morphological basis, as it includes Podostemaceae, a highly derived family of plants adapted to aquatic habitats (Gustafsson, Bittrich and Stevens, 2002; Wurdack and Davis, 2009). Recognizing five families presents the least heterogeneous assemblage, but is not a perfect solution. Although morphologically heterogeneous, the clusioids share a number of prominent characters, which are less common in the rest of Malpighiales, such as superior ovaries with axile placentation and more than two ovules per locule, imbricate petals (often little differentiated from sepals), numerous stamens often arranged in antepetalous fascicles, in addition to shared vegetative characters, such as latex.

Clusiaceae

Figure 9.40A *Garcinia spicata* (Wight and Arn.) Hook

Staminate flower[e]: ∗ K2+2 C4 A4^3 G0

[e] No material of pistillate flowers was available.

Figure 9.40 *Clusia rosea* Jacq.; B. staminate flower; C. inflorescence; D. pistillate flower

Staminate: ∗K2+2 C2+2 $A4^{\infty}$ G0[f]

Pistillate: ∗ K2+2 C2+2 A°8 $\underline{G}(6)$[g]

Figure 9.40E *Symphonia globulifera* L.f.
∗ K5 C5 $(A5^{3})$ $\underline{G}(5)$

General formula: ∗ K2(3)4–5(–20) C(0–3)4–5(-8) A4–∞ G1–5(–20)

Flowers are extremely variable, represented mainly by the heterogeneous *Clusia* and *Garcinia* (Figure 9.40A–D; Gustafsson and Bittrich, 2002; Gustafsson, Bittrich and Stevens, 2002; Sweeney, 2008, 2010). Bracteoles are usually present as a transversal pair not differing from the sepals. The perianth and androecium are spiral throughout, often with decussate phyllotaxis and without clear boundaries between bracts, sepals and petals, an unusual condition in Pentapetalae (e.g. Gustafsson, 2000; *Clusia*, *Garcinia*). A feature common to all Clusiaceae are the thick, contorted or imbricate petals associated with antepetalous stamen fascicles (phalanges). The petals may become fused into a tubular corona in *C. gundlachii* (Gustafsson, 2000). Petals are often four in number (or three to ten) and stamens arise centrifugally in antepetalous fascicles or on a ring primordium. When arising on a ring primordium, four groups can still be recognized in early stages (Figure 9.40B; pers. obs.). In *Garcinia* the androecium is extremely variable but develops mostly from massive fascicles (e.g. Baillon, 1871a; Leins and Erbar, 1991; Sweeney, 2008, 2010). Androecia are rarely obhaplostemonous (e.g. *Havetiopsis*, some *Clusia*: Gustafsson and Bittrich, 2002).

Flowers are often unisexual (not in *Symphonia*) with a high divergence between staminate and pistillate morphs. While staminodes are usually present in pistillate flowers, pistillodes are either present or missing in staminate flowers (e.g. *Clusia*, *Garcinia*). Anthers may become highly modified in producing resin to attract small bees (Bittrich and Amaral, 1996). The number of carpels is variable in *Clusia*, ranging from five to eight. The superior gynoecium with axile placentation and paired ovules is more stable than the androecium. The style is very short to inexistent and the stigmatic lobes are strongly expanded. In staminate flowers of *Garcinia*, stamen fascicles alternate with a broad disc-like nectary (Figure 9.40A), which corresponds with antesepalous flaplike appendages alternating with staminodial phalanges in pistillate flowers (Sweeney, 2008). The nectaries are probably of receptacular origin, contrary to Hypericaceae and Bonnetiaceae, where they represent antesepalous staminodes (Sweeney, 2008, 2010).

[f] Arising as four weakly defined groups opposite the petals.

[g] There are four decussate whorls of bracts not differing much from the sepals.

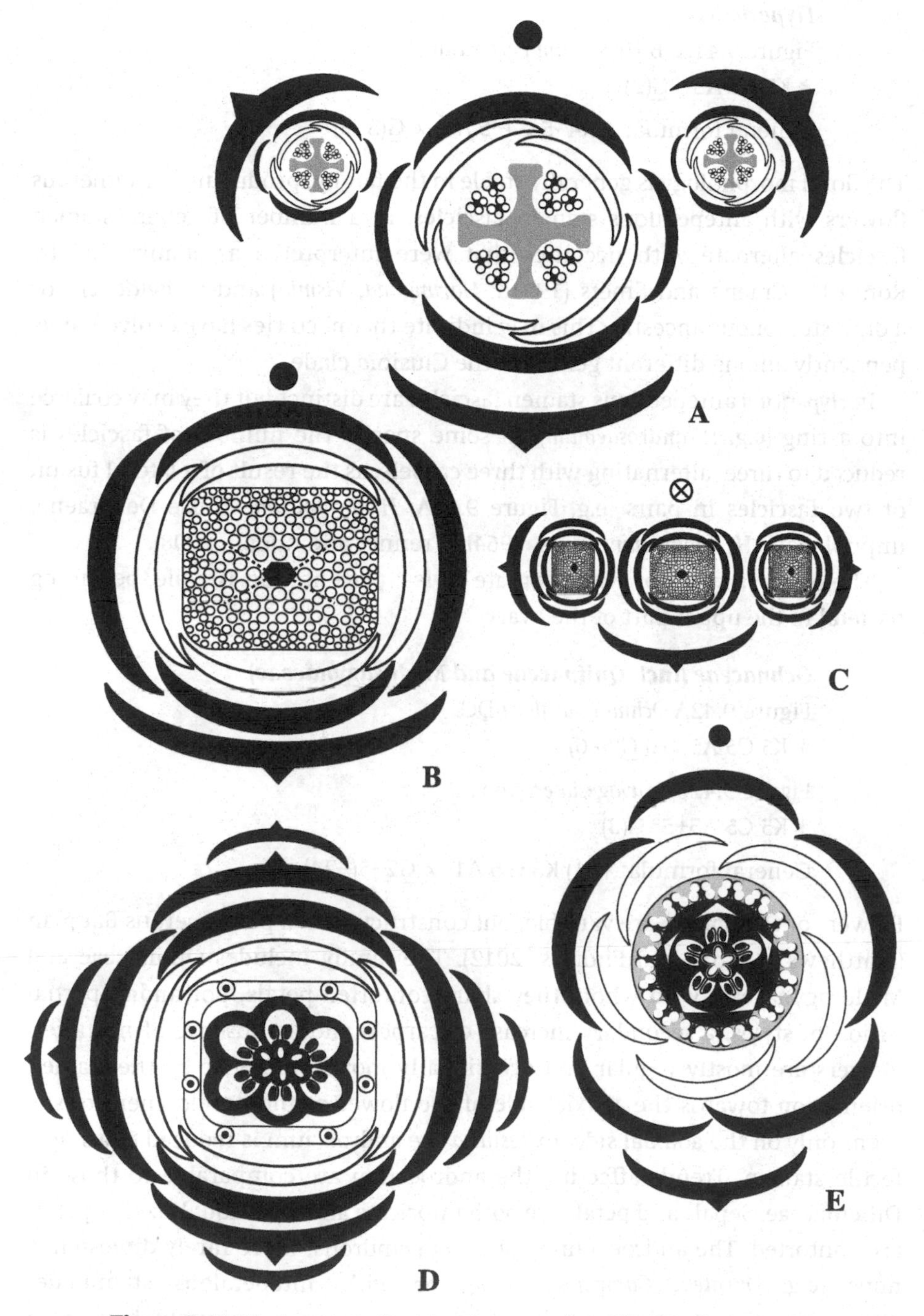

Figure 9.40 Clusiaceae: A. *Garcinia spicata*, staminate flower; *Clusia rosea*: B–C. staminate flower and partial inflorescence; D, pistillate flower; E. *Symphonia globulifera*. The broken line in B shows delimitation of four stamen groups.

Hypericaceae

Figure 9.41A, B *Hypericum perforatum* L.

$\ast$ K5 C5 A3^{∞} $\underline{G}$(3)

General formula: $\ast$ K4–5 C4–5 A3–∞ $\underline{G}$(3–5)

The floral morphology is generally stable in the family, producing pentamerous flowers with antepetalous stamen fascicles. In a number of genera stamen fascicles alternate with nectaries that were interpreted as staminodial by Ronse De Craene and Smets (1991d: *Harungana*, *Vismia*) and as evidence for a diplostemonous ancestry. This may indicate that nectaries have evolved independently among different genera of the Clusioid clade.

In *Hypericum* antepetalous stamen fascicles are distinct but they may coalesce into a ring (e.g. *H. androsaemum*). In some species the number of fascicles is reduced to three, alternating with three carpels, as the result of a lateral fusion of two fascicles in pairs (e.g. Figure 9.41A; *H. olympicum*: Ronse De Craene, unpubl. data; *H. aegypticum*: Leins, 1964b; Prenner and Rudall, 2008).

The ovary is superior with separate styles; placentation is axile, becoming parietal in the upper part of the ovary.

Ochnaceae (incl. Quiinaceae and Medusagynaceae)

Figure 9.42A *Ochna multiflora* DC.

$\ast$ K5 C5 A5^{∞}+0 $\underline{G}$(5–6)

Figure 9.42B *Sauvagesia erecta* L.

$\ast$ K5 C5 A5+5° $\underline{G}$(3)

General formula: $\ast$ (↓) K5 C 5 A1–∞ $\underline{G}$2–5(–24)

Flowers of Ochnaceae are variable, but constructed on a pentamerous Bauplan (Matthews, Amaral and Endress, 2012). The family includes Quiinaceae and Medusagynaceae, with whom they share contorted petals, polyandry, partial fusion of styles, a secondary increase of carpels and an absence of nectaries. Flowers are mostly regular and occasionally monosymmetric by the stamen orientation towards the adaxial side of the flower or the development of stamens only on the adaxial side. In *Testulea* the androecium is reduced to a single fertile stamen. Trends affecting the androecium are comparable to those in Dilleniaceae. Sepals and petals are both imbricate and of unequal size, or petals are contorted. The androecium is often polyandrous, more rarely diplostemonous (e.g. *Ouratea*, *Campylospermum*), or with antepetalous staminodes (*Sauvagesia*). The staminodes arise after the antesepalous stamens have been formed and shift outwards leading to an obdiplostemonous arrangement (Farrar and Ronse De Craene, 2013; Ronse De Craene and Bull-Hereñu, 2016).

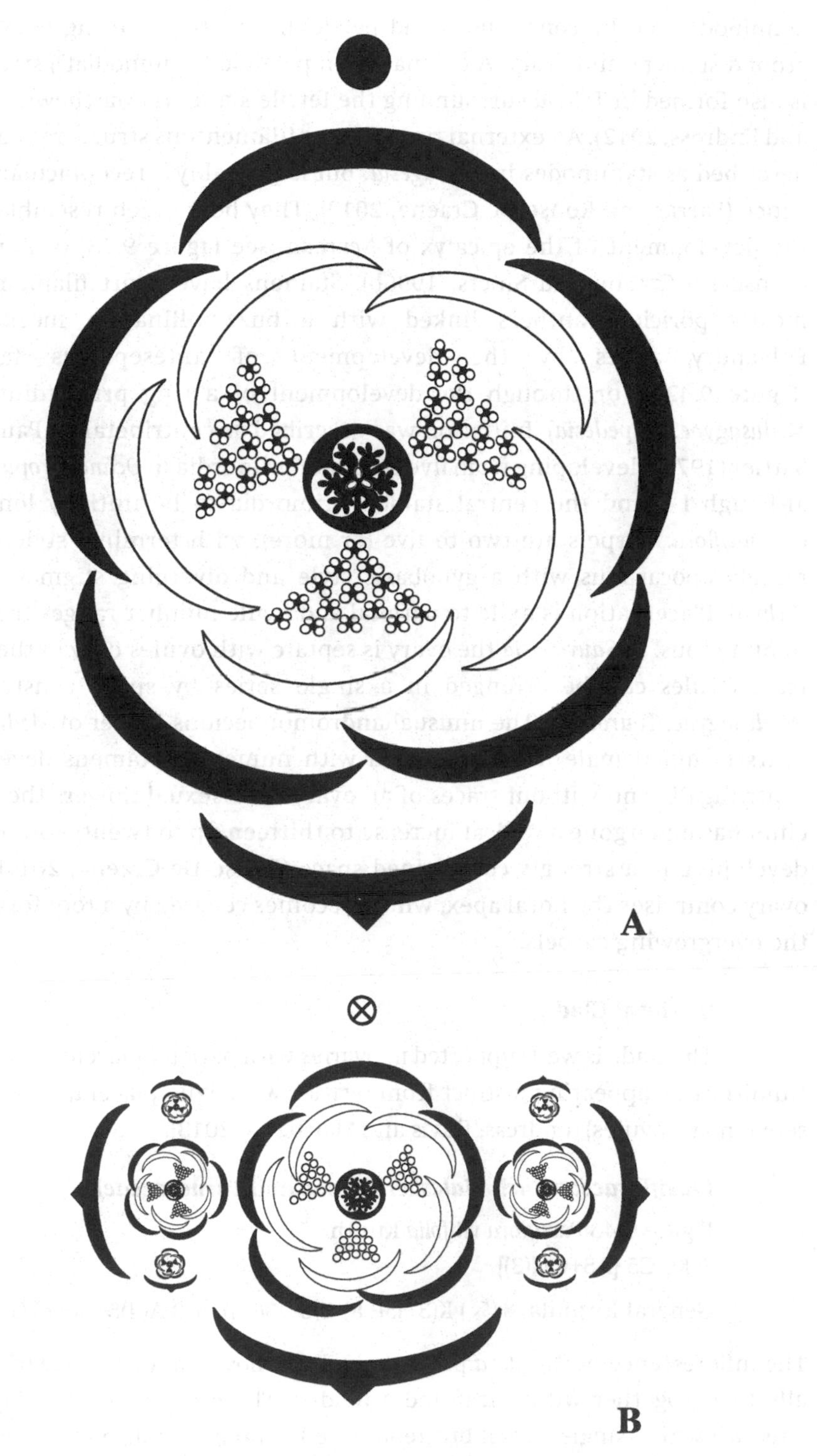

Figure 9.41 *Hypericum perforatum* (Hypericaceae): A. flower; B. partial inflorescence

Staminodes can be conspicuous and petaloid, sometimes fusing into a tube around stamens and ovary. A corona of ten petaloid (staminodial?) structures is also formed in *Tyleria* surrounding the fertile stamens (Matthews, Amaral and Endress, 2012). An external ring of outer filamentous structures has been described as staminodes in *Sauvagesia*, but is probably a receptacular emergence (Farrar and Ronse De Craene, 2013). They bear much resemblance to the development of the epicalyx of *Neurada* (see Figure 9.28) or *Agrimonia* (Ronse De Craene and Smets, 1996b). Stamens have short filaments and mostly poricidal anthers linked with a buzz-pollination mechanism. Polyandry arises by the development of antesepalous fascicles (Figure 9.42A), or through the development of a ring primordium (e.g. *Medusagyne*, *Cespedesia*). Initiation was described as centripetal by Pauzé and Sattler (1978), developing from five common primordia in *Ochna atropurpurea*, although I found the central stamen primordia to be initially longer in *O. multiflora*. Carpels are two to five (or more), with terminal style or seemingly apocarpous with a gynobasic style and diverging stigmatic lobes (*Ochna*). Placentation is axile to parietal and ovule number ranges from one to numerous. In *Sauvagesia* the ovary is septate with ovules only in the lower part. Ovules can be arranged in a single series by space constrictions (*Medusagyne*, *Touroulia*). The unusual andromonoecious flower of *Medusagyne* bears terminal male flowers covered with numerous stamens developing centrifugally and without traces of an ovary. In bisexual flowers the gynoecium has undergone a radical increase to thirteen up to twenty-four carpels developing in a strongly constrained space (Ronse De Craene, 2017b). The ovary comprises the floral apex, which becomes covered by a roof formed by the overgrowing carpels.

'Parietal Clade'

The clade is well supported in ovaries with parietal placentation (not in Humiriaceae) appearing distinct from ovaries with axile placentation (lack of septa, many ovules) (Endress, Davis and Matthews, 2013).

Passifloraceae (incl. Malesherbiaceae and Turneraceae)

Figure 9.43 *Passiflora vitifolia* Kunth

∗ K5 C5 [A5+0 $\underline{G}$(3)]

General formula: ∗(↖) K(3–)5(–8) C(3–)5(–8) or 0 A(4)5–8(–∞) G(2)3(–5)

The inflorescence of *Passiflora* produces a single flower at each node (occasionally two), together with a leaf and a tendril. The central axis develops into a tendril with a single lateral bracteole. The lateral bracteole encloses a flower

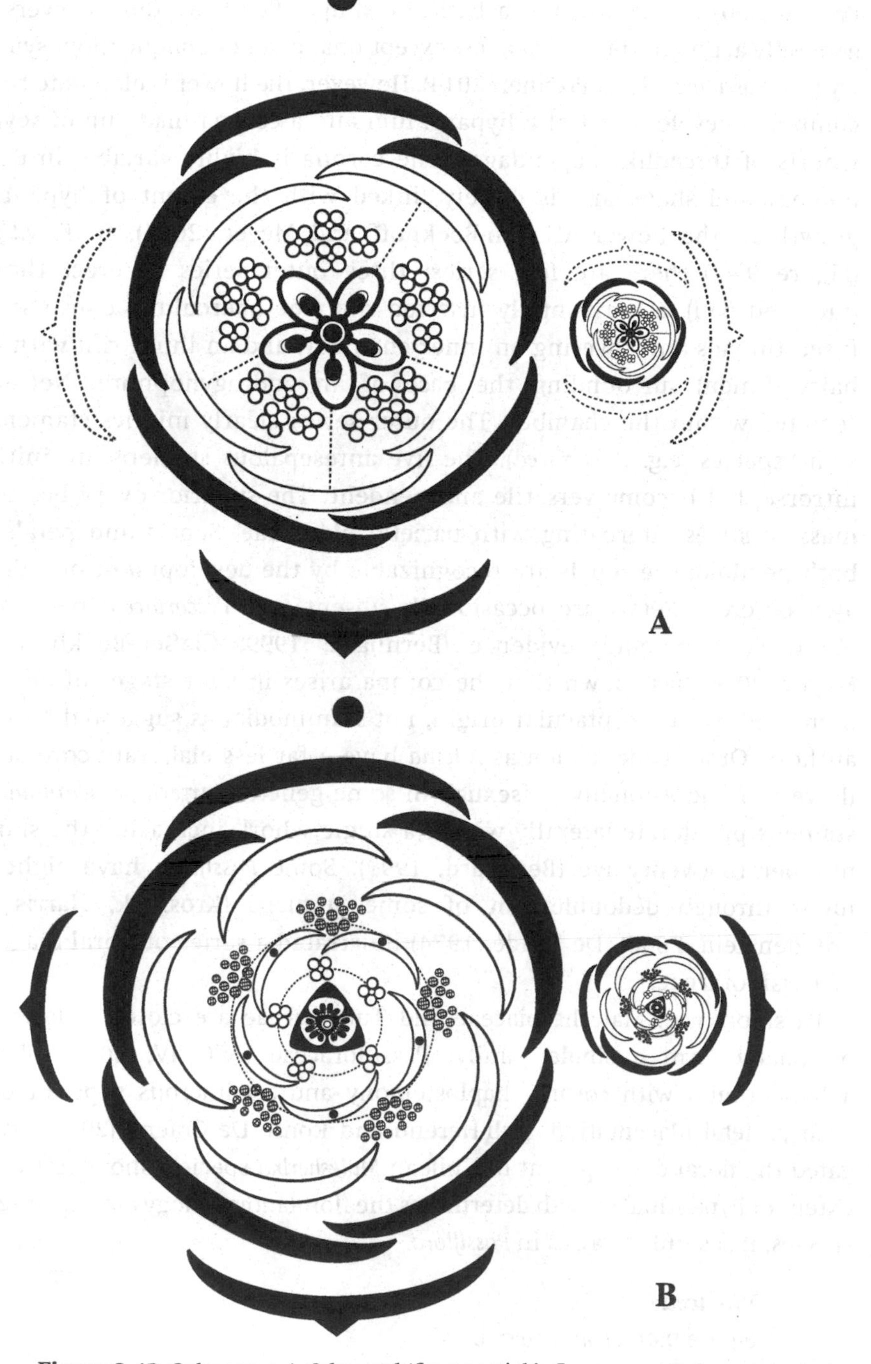

Figure 9.42 Ochnaceae: A. *Ochna multiflora*: partial inflorescence; B. *Sauvagesia erecta*: partial inflorescence. Separation of different fascicles shown by broken line in A.

together with two second-order bracteoles of the same size (cf. Eichler, 1878; Prenner, 2014). *Passiflora* has a basically simple floral diagram. Flowers are generally actinomorphic with a few exceptional cases of oblique monosymmetry (e.g. *Passiflora lobata*: Prenner, 2014). However, the flower is elaborate by the combined development of a hypanthium and a corona made up of several whorls of threadlike appendages. The corona is highly variable in development and shape and is closely linked with the extent of hypanthial growth in the flower (Claßen-Bockhoff and Meyer, 2016). In *P. vitifolia* (Figure 9.43) there are five series: three outer series of erect threads (radi and pali), a horizontally inserted rim (the operculum) consisting of fused threads and covering an inner chamber, and an inner rim with tiny hairs (limen) surrounding the base of an androgynophore. Nectar is secreted within the chamber. The outer series clearly mimics stamens in some species (e.g. *P. coriacea*). The five antesepalous stamens are initially introrse, but become versatile and pendent. The superior ovary has three massive styles alternating with parietal placentae. Sepals and petals are both petaloid; the sepals are recognizable by the development of a dorsal awn or crest. Petals are occasionally absent (e.g. *P. coriacea*: pers. obs.). Floral developmental evidence (Bernhard, 1999; Claßen-Bockhoff and Meyer, 2016) has shown that the corona arises in later stages of development and has a receptacular origin, not staminodial as suggested by some authors. Other genera such as *Adenia* have a far less elaborate corona and flowers are occasionally unisexual. In some genera (*Barteria*, *Smeathmannia*) stamens proliferate laterally within a single whorl, increasing the stamen number to twenty-five (Bernhard, 1999). Some *Passiflora* have eight stamens through dédoublement of some stamens (Krosnick, Harris and Freudenstein, 2006). De Wilde (1974) illustrated a series of floral diagrams of Passifloraceae.

Passifloraceae, Malesherbiaceae and Turneraceae are closely related and optionally form a single family, Passifloraceae (APG IV, 2016), sharing a hypanthium with corona, haplostemony and a trimerous superior ovary with parietal placentation. Bull-Hereñu and Ronse De Craene (2020) investigated the floral development of Chilean *Malesherbia* species, showing that the extent of hypanthial growth determines the floral morphology among different species, in a similar way as in *Passiflora*.

Violaceae

Figure 9.44 *Viola tricolor* L.

↓ K5 C1:2:2 A2:3 $\underline{G}(3)$

General formula: ∗/↓ K5 C5 A(3–)5 G(2–)3(–5)

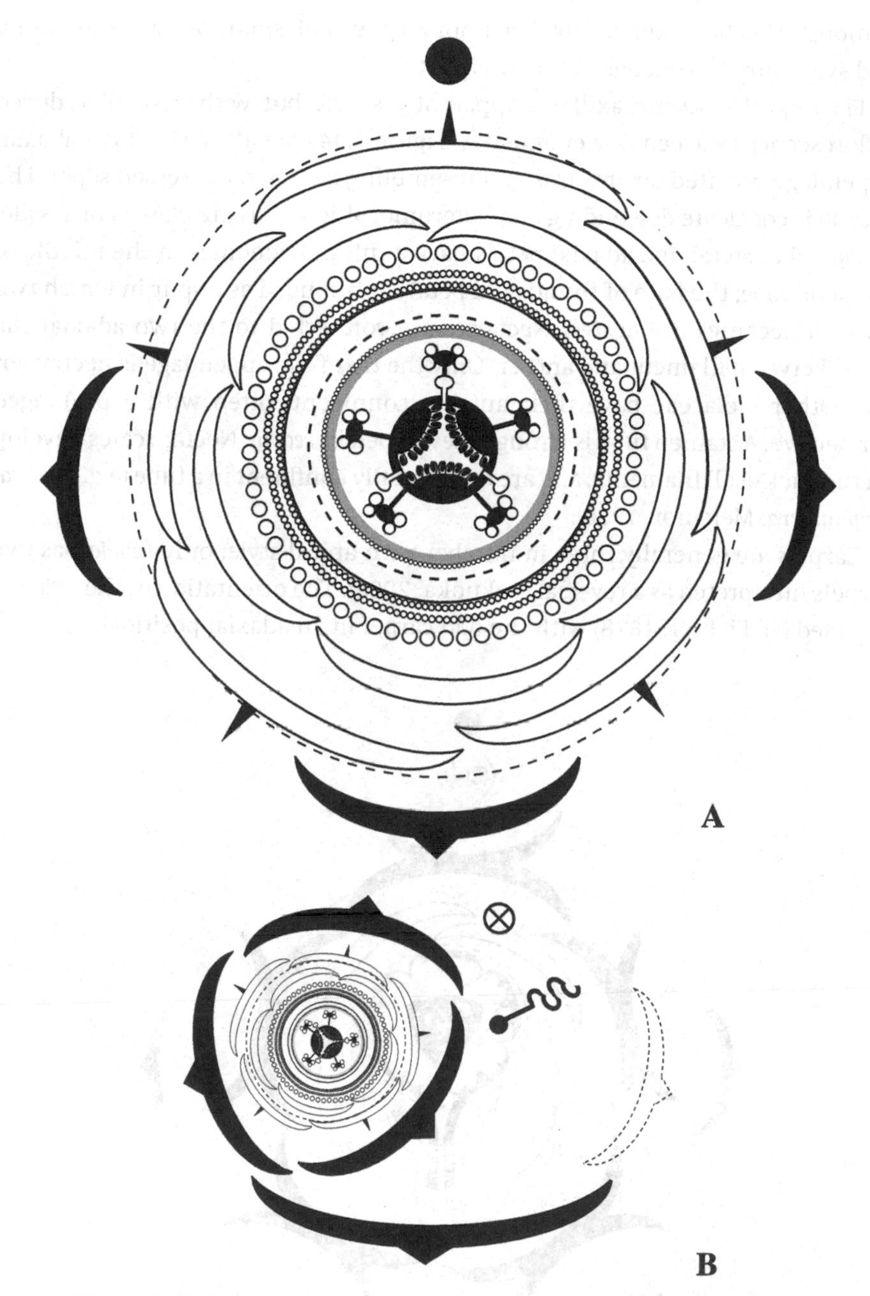

Figure 9.43 *Passiflora vitifolia* (Passifloraceae): A. flower; B. position of flower on a node with accompanying tendril

Violaceae represents a heterogenous family with poly- and monosymmetric flowers. Common characters are haplostemony, free or basally fused stamens with broad anthers and protruding connective, and superior gynoecium with parietal placentation. *Leonia triandra* has three stamens (Cronquist, 1981). As demonstrated by Tokuoka (2008) actinomorphy is plesiomorphic in the family and symmetry has changed at least eight times.

Flowers of *Viola* are axillary, apparently single but with a small reduced inflorescence between flower and axis (Figure 9.44). Sepals have a long abaxial appendage oriented downwards and resembling a second appressed sepal. The corolla is cochleate descending and heteromorphic. Petals are clawed or sessile. In *Viola* the lateral and adaxial petals have a tuft of trichomes in the middle of the petal lobe; the base of the adaxial petal is prolonged as a spur in which two filiform nectaries are nested. Nectaries are connected to the two adaxial stamens between filament and anther. Only the tip of the appendage is nectariferous. Other Violaceae have their anthers connivent, often with a prolonged connective. A stamen tube is strongly developed in *Leonia*. Nectar scales develop on the back of all filaments, and are occasionally confluent in a tube (e.g. *Rinorea*, *Hymanthera*: Melchior, 1925).

Carpels are generally three in number with apical style; only *Leonia* has five carpels interpreted as a reversal (Tokuoka, 2008). The orientation of the ovary is inversed (cf. Eichler, 1878) with the odd carpel in an adaxial position.

Figure 9.44 *Viola tricolor* (Violaceae): partial inflorescence

Salicaceae (incl. Scyphostegiaceae)

Figure 9.45 *Azara serrata* Ruíz and Pav. var. *serrata*

∗ K4–6 C0 A4$^{\infty}$–6$^{\infty}$ $\underline{G}$(4–6)

General formula: ∗ K0/(3)4–7 (8+) C0/3–8 A(2)4–∞ G2–9

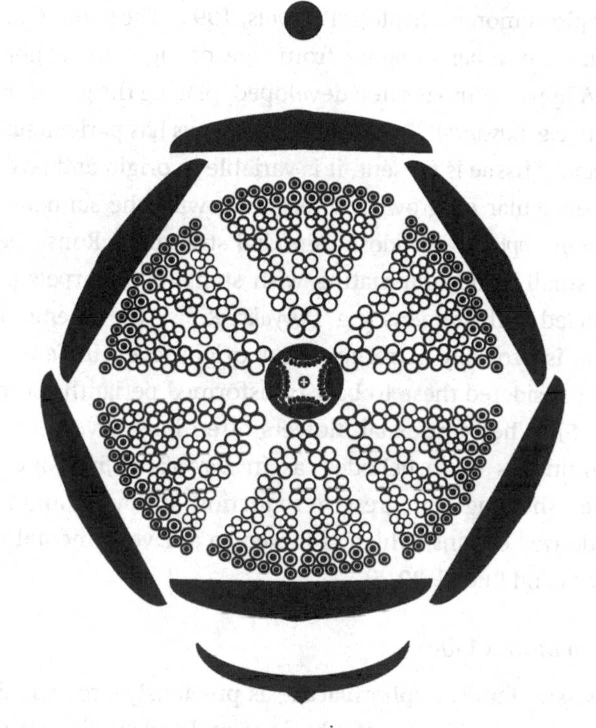

Figure 9.45 *Azara serrata* var. *serrata* (Salicaceae)

Molecular evidence has demonstrated that the pre-molecular Flacourtiaceae are not a natural entity that has to be split up in two clades belonging to two families with a previously much smaller circumscription, viz. Achariaceae and Salicaceae (Soltis et al., 2018). Traditional Salicaceae (*Salix* and *Populus*) were shown to represent extremes in floral reduction and belong close to members of former Flacourtiaceae. A morphological distinction between Achariaceae and Salicaceae was shown by Bernhard and Endress (1999) in that the former has centripetal polyandry, while in the latter stamens arise centrifugally.

The family is highly variable and difficult to define, because major defining characteristics (superior ovary with parietal placentation) are shared by other families. However, a series of traits is common in the family. Flowers are generally small, often unisexual with variable merism (occasionally hexa- to octomerous), with valvate calyx and rarely with corolla. When present, petals are small (e.g. *Homalium*) and they are frequently missing, as the attraction is taken over by stamens (e.g. *Azara*) or petaloid sepals (e.g. *Casearia*). An extreme reduction of the perianth occurs in *Salix* and *Populus*. Stamens are occasionally diplostemonous. When isomerous, stamens are obhaplostemonous alternating with large antesepalous nectaries (e.g. *Homalium*). When multistaminate, stamens arise centrifugally on a ring primordium or on common primordia (e.g. *Homalium abdessammadii* with obhaplostemonous triplets: Pauwels, 1993). The genus *Casearia* is highly variable in stamen number, ranging from five or eight to higher numbers in a single whorl. A hypanthium is often developed, placing the gynoecium in a semi-inferior position (e.g. *Casearia*). The gynoecium always has parietal placentation. In cases where nectary tissue is present, it is variable in origin and position, ranging from large, receptacular outgrowths alternating with the stamens (e.g. *Casearia bracteifera*, *Azara microphylla*: pseudostaminodial structures: Ronse De Craene and Smets, 2001a), small glands alternating with stamens or carpels (*Salix*) or large glands interspersed with stamens (e.g. *Dovyalis*: Ronse De Craene, unpubl. data). Glandular tissue is probably of receptacular origin, even in *Salix* where earlier interpretations considered these to be a transformed perianth (Cronk, Needham and Rudall, 2015). When unisexual, flowers often have divergent morphs with absence of staminodes or carpellodes, as in *Dovyalis caffra*. Nine genera form a 'salicoid clade', showing a progressive reduction of flowers and inflorescences leading to condensed catkins, which is linked to a developmental heterochrony (Cronk, Needham and Rudall, 2015).

'Euphorbioid Clade'

The massive family Euphorbiaceae as previously circumscribed was reorganized in at least six families on the basis of molecular evidence (Tokuoka and Tobe, 2006; Wurdack and Davis, 2009: Pandaceae, Euphorbiaceae s. str., Phyllanthaceae, Picodendraceae, Peraceae and Putranjivaceae). More recently, Rafflesiaceae with the largest angiosperm flowers have been added to the Euphorbioid clade together with Linaceae and Ixonanthaceae (Wurdack and Davis, 2009). Phyllanthaceae and Picodendaceae are sister families sharing biovulate carpels with Linaceae and Ixonanthaceae, while the other families have a single ovule per carpel (in Rafflesiaceae strongly derived). However, all Euphorbioid flowers share a combination of three characters not found elsewhere, which are antitropous hanging ovules, a nucellar beak and an obturator (Sutter and Endress, 1995; de Olivera Franca and Cavaleri De-Paula, 2017). Morphologically Euphorbioid flowers share many characters, including unisexual flowers with large separate nectar glands, and a superior capsular ovary with bifid styles (Endress, Davis and Matthews, 2013).

Euphorbiaceae

Figure 9.46A–C *Jatropha sp.* (Crotonoideae)

Staminate: ∗ K5 C5 A5+3 G0

Pistillate: ∗ K5 C5 A0 $\underline{G}$(3)

Figure 9.46 *Euphorbia mellifera* Aiton (Euphorbioideae). D. cyathium; E. partial cyme

General formula: ∗ K0/(3–)5–6 C0/5–6 A1–5–15–∞ G(1–)3(–20)

Euphorbiaceae *sensu stricto* contain four subfamilies that are not strongly supported.

Inflorescences are generally cymose or cymose-derived. However, floral diversity in the family is high, linked with unisexual flowers and a strong dimorphism between staminate and pistillate flowers. Merism can be highly variable, mostly tetra- or pentamerous, or more variable (tri- to octomerous). When trimerous, flowers are simple and apetalous (e.g. *Mercurialis*). Flowers generally have sepals with imbricate or valvate aestivation. Petals are present but rarely showy as in *Jatropha integerrima*, mostly equal to smaller than sepals (e.g. *Croton*) or frequently absent (e.g. *Manihot esculenta*). While petals are generally found in staminate *Croton*, they tend to become reduced or absent in corresponding pistillate flowers (Thaowetsuwan et al., 2020). Aestivation of petals is contorted to valvate. The androecium of staminate flowers is highly variable, either (ob)haplostemonous, (ob)diplostemonous (e.g. *Jatropha aconitifolia*, *Manihot*) or in multiple whorls (e.g. *Croton*) (Baillon, 1874; Michaelis, 1924; Thaowetsuwan, 2020; Thaowetsuwan et al., 2020). When two-whorled, the alternisepalous whorl is generally the outer and is shorter than the antesepalous whorl (e.g. *Jatropha*, *Cnidoscolus*). In some species the inner whorl consists of three stamens (Figure 9.46A; Nair and Abraham, 1962) or a single stamen (e.g. *Croton sp.*: Gandhi and Dale Thomas, 1983; Thaowetsuwan, 2020). Stamens are often monadelphous and fused into a column, and are often lifted on an androphore with clearly separate stamen whorls (e.g. *Caperonia*, *Ditaxis*: Michaelis, 1924). In *Croton alabamensis* centripetal stamen development is linked with the formation of a hypanthium (Thaowetsuwan et al., 2020). The gynoecium is generally strongly reduced with small carpellodes (e.g. *Cnidoscolus*) or without evidence or carpels (e.g. *Croton*). I suspect that stamens have replaced carpels in *Jatropha* or *Croton* through a process of homeosis, as one to three stamens occupy the expected position of the carpels (see also diagrams in Michaelis, 1924). When numerous, stamens develop in alternating whorls of five, although numbers can be unstable and fluctuating (e.g. *Codiaeum*, *Croton*: Baillon, 1874; Nair and Abraham, 1962; Gandhi and Dale Thomas, 1983;

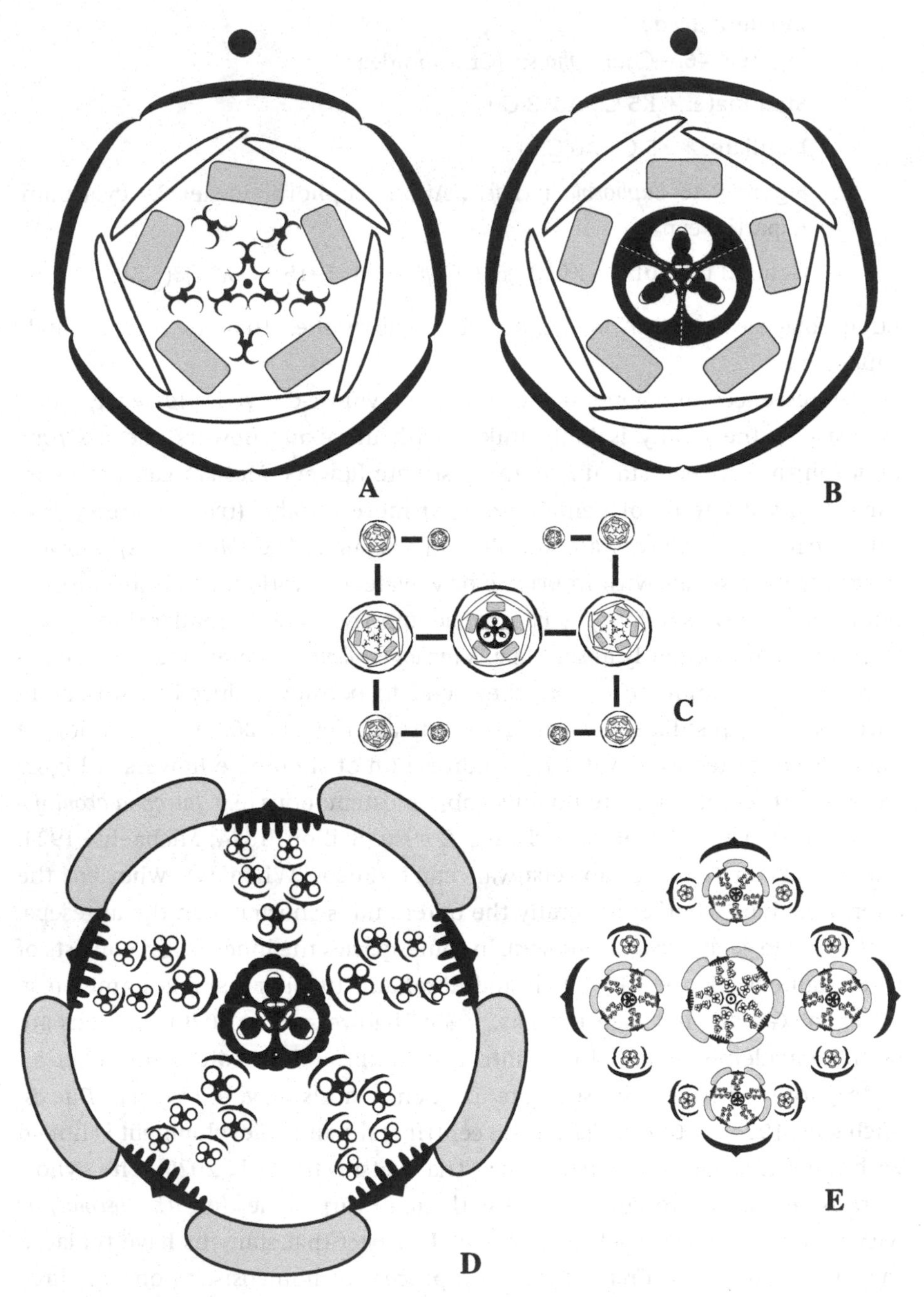

Figure 9.46 Euphorbiaceae: *Jatropha sp.*: staminate (A) and pistillate (B) flower; (C) partial inflorescence. *Euphorbia mellifera*: cyathium (D) and partial inflorescence with cyathia (E)

Thaowetsuwan, 2020). In some staminate flowers the stamen number can be very high and is chaotic (e.g. *Croton celtidifolius*: Thaowetsuwan, 2020). *Ricinus* is exceptional by the presence of highly branching stamen fascicles (Prenner and Rudall, 2008). Claßen-Bockhoff and Frankenhäuser (2020) demonstrated that these flowers represent in fact pseudanthia formed of fascicles of highly reduced staminate flowers. Floral nectaries are present, or in cases where flowers are reduced they are found on the inflorescence bracts. Nectar appendages are large, extrastaminal and situated in antesepalous or antepetalous position, occasionally in both (Michaelis, 1924; Thaowetsuwan et al., 2020). They are rounded or rectangular in shape or resemble staminodial structures and have been interpreted as such (as in Crotoneae: Cavaleri De-Paula et al., 2011), although Thaowetsuwan et al. (2020) demonstrated that they represent receptacular structures. Pistillate flowers usually resemble the staminate flowers up to the development of the perianth. In some taxa petals are present in staminate flowers but absent in pistillate flowers (e.g. *Croton sarcopetalus*: Freitas et al., 2001; Thaowetsuwan et al., 2020); pistillate flowers have an extra whorl of nectaries in alternisepalous position with a vascular connection that may be transformed petals indeed. Staminodes may be variously present or are more generally absent. The gynoecium is generally trimerous, rarely dimerous and has a generalized morphology in the Euphorbiaceae. A single pendent, antitropous ovule is located on axile placentation in each locule. The globular ovary is topped by a short style with bifid stigmatic arms opposite the locules. Exceptionally the ovary is reduced to a single carpel as in *Croton michauxii* (Thaowetsuwan, 2020).

Euphorbioideae are characterized by pseudanthia with highly reduced staminate and pistillate flowers. Prenner and Rudall (2007) demonstrated the progressive reduction of the perianth in staminate and pistillate flowers, culminating in *Euphorbia*. Staminate flowers are reduced to a single stamen (often with a constriction on the pedicel), while pistillate flowers are reduced to three carpels. The inflorescence of *Euphorbia* is called a cyathium and is a cymose inflorescence with terminal pistillate flower surrounded by male monochasia (Figure 9.46D). The cuplike inflorescence is formed by fusion of bracts and topped by large glands. Diagrams of inflorescences were shown in Stützel (2006). Similarly in *Dalechampsia* a cymose inflorescence gives rise to a complex pseudanthium enclosed by two petaloid bracts (Gagliardi, Cordeiro and Demarco, 2018).

Peraceae

Figure 9.47 *Clutia sp.* Staminate (A) and pistillate (B) flower

Staminate: ✶ K5 C5 A5 G0

Pistillate: ✶ K5 C5 A0 $\underline{G}$(3)

General formula: ✶ K5 C0/5 A5–20 $\underline{G}$(3)

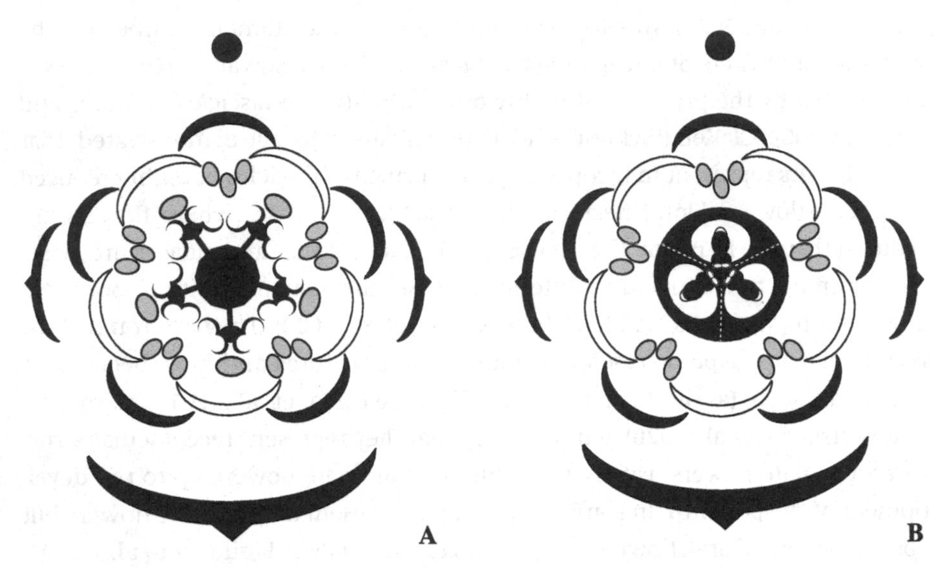

Figure 9.47 *Clutia sp.* (Peraceae): staminate (A) and pistillate flower (B)

Morphologically the family is comparable to Euphorbiaceae in several characteristics and trends. Flowers are grouped in male and female pseudanthia surrounded by showy bracts, when flowers are strongly reduced (e.g. *Pera*: Gagliardi, Cordeiro and Demarco, 2018). Flowers generally lack evidence of the other gender. In *Clutia* the five antepetalous stamens are fused into a column topped by a sterile hairy apex (Figure 9.47A). In *Pera*, a perianth is missing or the calyx is vestigial. The extrastaminal nectaries resemble staminodial structures and have been interpreted as such (Endress and Matthews, 2006b), although they probably represent receptacular outgrowths as in Euphorbiaceae.

Linaceae

Figure 9.48 *Linum monogynum* G. Forst.

∗ K5 [C5 A(5+5°)] $\underline{G}$(5)

General formula: ∗ K(4–)5 C(4–)5 A(4–)5–10 G2–5

In pre-molecular classifications (e.g. Cronquist, 1981) the family used to be broadly defined (including Humiriaceae, Ctenolophaceae and Ixonanthaceae) and was related to Geraniaceae on the basis of superficial similarities. Indeed, the flowers show many similarities to the aforementioned families. The current circumscription comprises Linoideae and Hugonioideae (Stevens, 2001 onwards).

Flowers are polysymmetric or weakly monosymmetric. Sepals are leaflike and equal in size with quincuncial aestivation. Petals are clawed and appear connected with the stamen tube by an erect basal elaboration situated against the tube. Flowers are (ob)diplostemonous, or haplostemonous with (e.g. Figure 9.48; *Reinwardtia*) or without staminodes (e.g. *Linum lewisii*: Narayana and Rao, 1976). The report of obhaplostemony in *Anisadenia* (e.g. Cronquist, 1981) is erroneous. Stamens and staminodes are basally fused; staminodes are small and toothlike in the species observed, but can be more developed. Staminodes are non-vascularized in *Anisadenia* and other Linaceae (Narayana and Rao, 1969, 1976). When in two whorls, stamens are of two lengths, the antepetalous being shorter, except for *Hugonia* where it is the opposite. Carpels are either antesepalous (in diplostemonous flowers e.g. *Hebepetalum*) or antepetalous (in obdiplostemonous flowers e.g. *Durandea*, *Linum*) (Matthews and Endress, 2011). The ovary fluctuates between three and five with six to ten pendent antitropous ovules, seemingly six- or ten-carpellate because locules are divided by false septa that do not reach the central axis in the upper part of the ovary (Figure 9.48; Narayana and Rao, 1969). The five styles are twisted in *Linum monogynum*.

Nectaries are extrastaminal. In *Linum*, five inconspicuous nectaries are situated at the base of the staminal tube, abaxially of the fertile stamens. In

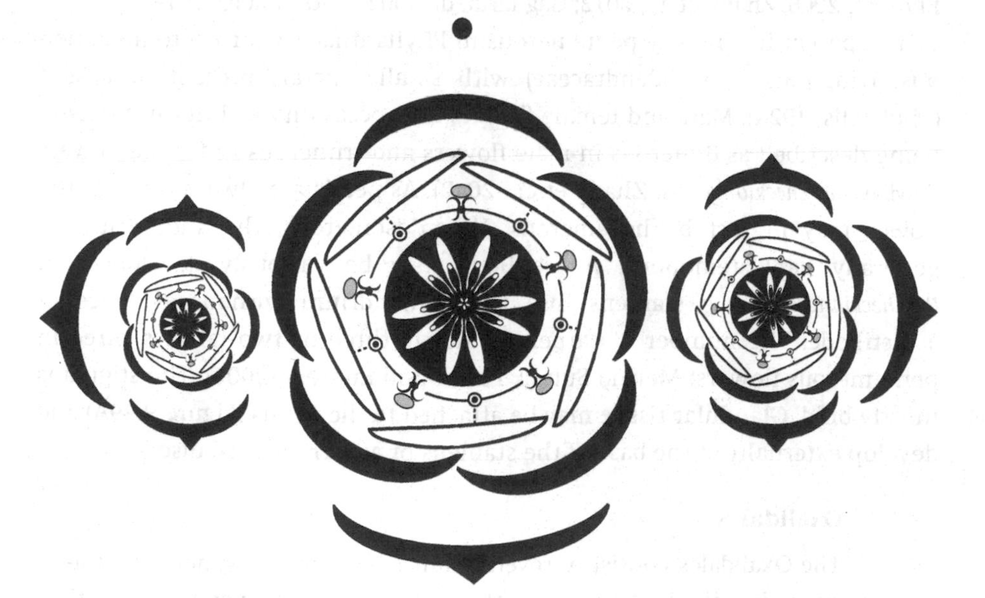

Figure 9.48 *Linum monogynum* (Linaceae): partial inflorescence. Petal bases are adherent to the stamen tube.

Anisadenia only one gland is developed below the adaxial antesepalous stamen (Narayana and Rao, 1969), while only two glands are present in *Linum perenne* (Narayana and Rao, 1976). Other Linaceae can have a prominent extrastaminal disc (e.g. *Philbornea*: Narayana and Rao, 1973).

Linaceae and Dichapetalaceae share a number of characters of ovules and placentation with Euphorbiaceae *sensu lato*. These include false septa in Linaceae and some species of Picodendraceae and Phyllanthaceae (Sutter and Endress, 1995; Merino Sutter, Foster and Endress, 2006).

Phyllanthaceae

Figure 9.49A–B *Leptopus chinensis* (Bunge) Pojark

Staminate: ∗ K5 C5 A5 $\underline{G}^{\circ}(3)$

Pistillate: ∗ K5 C5 A0 $\underline{G}(3)$

General formula: ∗ K2–8 C0/3–6(–9) A2–10(–19) G(1)2–5(–15)

Although molecular evidence places Phyllanthaceae in a clade different from Euphorbiaceae, flowers appear to be highly similar, except for ovule numbers, which are consistently two. Phyllanthaceae is sister to Picodendraceae and both families share a unique combination of characters with Euphorbiaceae including unisexual, apetalous flowers (often trimerous with bifid styles), large extrastaminal nectary glands, antitropous ovules with large obturator and nucellar beak, and explosive capsular fruits with similar anatomical details (Merino Sutter, Foster and Endress, 2006; Zhang et al., 2012; Gagliardi, de Souza and Albiero, 2014).

The perianth is mostly pentamerous in Phyllanthaceae (or tri- to nonamerous; trimerous in Picodendraceae) with smaller petals present or absent (Michaelis, 1924). Male and female flowers may occasionally differ in merism, being described as dimerous in male flowers and trimerous in female flowers (*Phyllanthus checkiangensis*: Zhang et al., 2012). As petals are always lacking, the flower may in fact be hexamerous. When isomerous, the androecium is generally haplostemonous. A pistillode may be variously developed. In *P. checkiangensis* two stamens fuse in a synandrium without evidence of a pistillode. The number of carpels is mostly three or two (always three in pentamerous flowers: Merino Sutter, Foster and Endress (2006). The stigma is mostly bifid. Glandular tissue may be attached to the petals (Figure 9.49A) and develop externally at the base of the stamens or as a crenelated disc.

Oxalidales

The Oxalidales consist of seven families. Flowers are generally tetra- or pentamerous in all whorls with the carpels in antepetalous position. Cephalotaceae is hexamerous throughout. Petals are either large in bud with

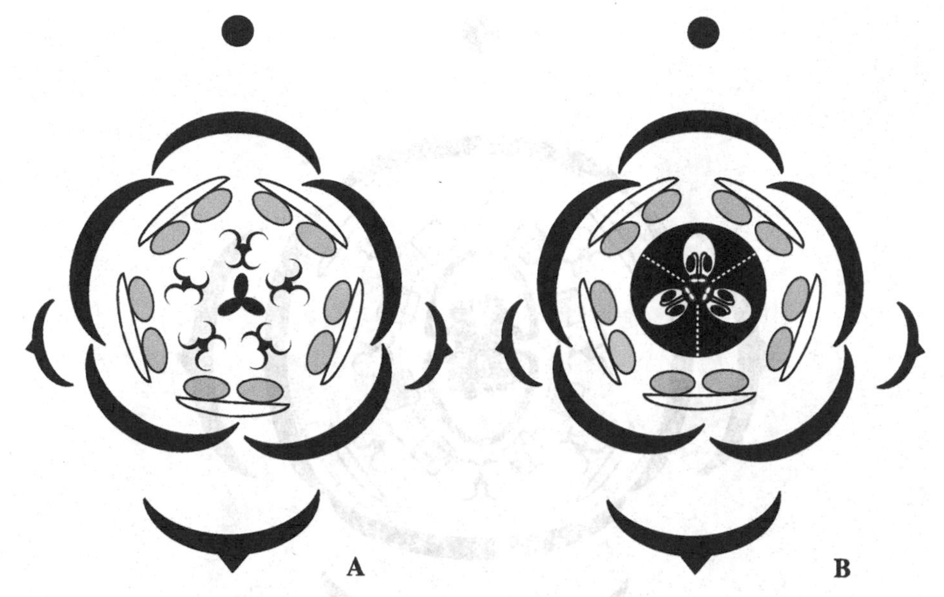

Figure 9.49 *Leptopus chinensis* (Phyllanthaceae): staminate (A) and pistillate flower (B). Note the glandular appendages on the petals.

imbricate (cochleate or contorted), rarely induplicate valvate aestivation (Cunoniaceae, Elaeocarpaceae), or are reduced or absent (e.g. Cunoniaceae). Two families, Cephalotaceae and Brunelliaceae, are apetalous. Petals are often elaborate and three-lobed (Cunoniaceae, Elaeocarpaceae). Stamens are often basally fused, while petals tend to become postgenitally fused at the base (Oxalidaceae and Connaraceae). The androecium is basically obdiplostemonous (present in all families) and occasionally haplostemonous. A secondary stamen increase occurs in some Cunoniaceae and Elaeocarpaceae, with antepetalous stamens in pairs. In Elaeocarpaceae, groups of stamens are wrapped by petals that have three vascular bundles (Matthews and Endress, 2002). Filaments tend to be longer than anthers and erect in bud. Carpels are mostly superior and, when isomerous, are always alternating with the sepals, but they are reduced to two or three in Cunoniaceae and Elaeocarpaceae. Styles are always carinal and each carpel contains either two ovules (rarely a single ovule), or two rows of parallel or superposed ovules on axile placentation. A hypanthium, when present, is weakly developed. Nectaries, when present, develop as separate glands or as an intrastaminal disc.

Cunoniaceae (including Davidsoniaceae, Baueraceae, and Eucryphiaceae)

Figure 9.50 *Caldcluvia paniculata* D.Don

∗ K4 C4 A4+4 $\underline{G}(2)$

General formula: ∗ (3)4–5(−10) C0/(3)4–5(−10) A(3)4−10/∞ G2−3−5(−14)

Figure 9.50 *Caldcluvia paniculata* (Cunoniaceae)

Floral morphology of the family is diverse with a mosaic of characters common to Oxalidales. Most Cunoniaceae have tetra- (penta-)merous flowers with a weak hypanthium and a valvate calyx. Petals are absent in more than half of the genera (e.g. *Davidsonia*, *Geissos*), while they are small in the other genera, and rarely larger (*Eucryphia*) (Dickison, 1975; Moody and Hufford, 2000a; Endress and Matthews, 2006b). Most taxa are obdiplostemonous. *Eucryphia* is tetramerous throughout with a secondary centrifugal stamen increase. Bull-Hereñu, Ronse De Craene and Pérez (2018) demonstrated that the number of stamens and carpels in *Eucryphia* is linked with the expansion of the floral receptacle. As the receptacle continues expanding after the development of the first stamens, additional stamens can develop in newly created spaces. This phenomenon of stamen increase was also observed in *Medusagyne* of Ochnaceae (Ronse De Craene, 2017b). Other Cunoniaceae occasionally have a lateral stamen increase from antesepalous primordia, while the antepetalous stamens are undivided (*Geissos*: Matthews and Endress, 2002).

Carpels are generally not isomerous, more frequently two (three) in median position, rarely more numerous as in *Eucryphia* (Dickison, 1978; Bull-Hereñu, Ronse De Craene and Pérez, 2018), variably in superior, half-inferior or inferior position. Fossil flowers of Cunoniaceae affinity have isomerous carpels in alternipetalous position (Schönenberger et al., 2001).

Nectaries tend to be separate inter- or intrastaminal structures (Figure 9.50; Dickison, 1975; Moody and Hufford, 2000a). In *Eucryphia* nectaries cover the receptacle between the stamens as small scales. Filaments are usually longer than petals at anthesis, resembling brush flowers (e.g. *Cunonia capensis*).

Elaeocarpaceae (incl. Tremandraceae)

Figure 9.51 *Crinodendron patagua* Molina

∗ K(2:3) C5 A5+5[2] $\underline{G}$(3)

General formula: ∗ K4–5 C0/4–5 A4–∞ G2–8(9)

The family used to be placed in Malvales because of superficial similarities in flower structure (Van Heel, 1966).

Sepals are valvate and basally united, breaking up in two unequal parts in *Crinodendron*. Petals are often folded and three-lobed and they enclose antepetalous stamens in bud in a comparable way to Rhizophoraceae (Figure 9.37; Endress and Matthews, 2006b). In *Crinodendron*, petals have a ventral ridge at the base forming two tubular access channels to the nectary (Matthews and Endress, 2002). Petals are absent in *Sloanea*. The androecium tends to be obdiplostemonous (e.g. *Platytheca*), with doubling of antepetalous stamens (*Aristotelia*, *Crinodendron*: Figure 9.51; Ronse De Craene and Smets, 1996a; Matthews and Endress, 2002). In *Tetratheca* the androecium consists of five pairs wrapped by the petals, but it is unclear whether the pairs are antesepalous or antepetalous stamens (Matthews and Endress, 2002). *Aristotelia fruticosa* has tetramerous haplostemonous flowers (Van Heel, 1966). A multistaminate androecium with centrifugal development is found in *Sloanea* and *Vallea* arising from antesepalous primary primordia (Van Heel, 1966; Matthews and Endress, 2002). Anthers are basifixed and mostly have poricidal dehiscence. A nectary is developed as a broad intrastaminal disc at the base of the ovary or surrounding the stamens, but is absent in *Sloanea* and Tremandraceae with specialized pollen flowers. The gynoecium consists of two or three carpels with a single style and ovules are apically inserted on axile placentae.

Oxalidaceae

Figure 9.52 *Averrhoa carambola* L.

∗ K5 C5 A(5+5°) $\underline{G}$(5)

General formula: ∗ K5 C5 A5+5/5° G(3–)5

Aestivation of petals is imbricate ascending or descending, or contorted. The androecium is obdiplostemonous and antepetalous stamens are occasionally

Figure 9.51 *Crinodendron patagua* (Elaeocarpaceae). Note the extrastaminal nectary and position of anthers relative to petals; the broken line refers to attachment of anthers to filament bases (white dots).

sterile (e.g. *Averrhoa*: Figure 9.52). Stamens are basally connate and nectaries occur as extrastaminal appendages of the antepetalous stamens or the staminodes are nectariferous at the base (pers. obs.).

Flowers are generally heterostylous. The superior gynoecium has free styles and a fused ovary with two or a single row of ovules on axile placentation. In other genera two rows of ovules are formed, but in *Averrhoa* the constricted locules only allow fewer ovules to develop on either row (cf. Matthews and Endress, 2002).

Oxalidaceae are closely related to Connaraceae, with whom they share several characteristics (see Matthews and Endress, 2002). Flowers of Oxalidaceae superficially resemble Geraniaceae and Linaceae; they share pentamerous, pentacyclic flowers with isomerous carpels, contorted petals, fused stamens, obdiplostemony with occasional reduction of antepetalous stamens and nectaries connected to the staminal tube. However, in Oxalidaceae the nectary is in antepetalous position, while in Geraniaceae it

Figure 9.52 *Averrhoa carambola* (Oxalidaceae): partial inflorescence. Note the antepetalous nectariferous staminodes.

is connected with antesepalous stamens (cf. Rama Devi, 1991a; Jeiter, Weigend and Hilger, 2017), and there is a strict separation between style and ovary in Geraniaceae.

9.3.2 *Remaining Fabids: Fabales, Rosales, Cucurbitales, Fagales*

Fabales

The order contains four families of unequal size with unresolved affinities: Surianaceae, Quillajaceae, Polygalaceae and Leguminosae (Fabaceae). Surianaceae and Quillajaceae have retained the basic formula of rosids (K5C5A5+5G̲5), while zygomorphy has variously affected the other families. A pentamerous ovary is probably the plesiomorphic condition for all Fabales, as it is present in both Surianaceae and Quillajaceae. Floral diagrams for the two families are presented in Bello, Hawkins and Rudall (2007). Gynoecia are (almost) apocarpous in Surianaceae and Quillajaceae, facilitating a reduction to the single carpel of Leguminosae.

Polygalaceae

Figure 9.53 *Polygala x dalmaisiana* Dazzler

↓ K3:2 C1:2:2 A(4+4) G̲(2)

General formula: K5 C(2–)3–5 A(2–)3–7(–10) G2–5(–8)

The Polygalaceae show a remarkable convergence with Leguminosae in developing monosymmetric pea-like keel flowers (Westerkamp and Weber, 1999). Prenner (2004b) mentioned developmental similarities between Polygalaceae and Leguminosae, such as racemose inflorescences, pressure of bracteoles on the flower development, suppression of organs and a comparable acquisition of monosymmetry. However, the Bauplan of the flower differs in several characteristics. In *Polygala* the inner lateral sepals are petaloid and contribute to floral display in a way comparable to the flag petal of Leguminosae. The keel is formed by a single abaxial petal, not two. The abaxial petal forms a protruding keel with a brushlike appendage, enclosing the stamens, while the four posterior petals are smaller, fused into lateral pairs and overlapping an entrance to the centre of the flower. Lateral petals are much smaller and are occasionally suppressed (Krüger and Robbertse, 1988; Prenner, 2004b). Petals are laterally fused with the stamen tube in an elaborate compound system creating a nectar chamber. The stamens are fused except for the adaxial side, which is fringed with hairs; they are curved within the keel together with the style. The abaxial antesepalous stamen and the adaxial antepetalous stamen are lost (Figure 9.53, asterisks) and stamens are arranged in two rows. There are two fused carpels on a gynophore and each bears an apical pendent ovule. The style is bilobed and often asymmetric with only the adaxial lobe developed and fertile. A nectary disc surrounds the ovary with access through an adaxial furrow in the stamen tube.

Other genera have more regular flowers or the floral display is provided by the petals as the calyx is inconspicuous (Westerkamp and Weber, 1999). Levyns (1949) presented a floral diagram for *Muraltia* with seven stamens; one stamen is situated abaxially opposite the keel and three outer antesepalous stamens are missing, indicating a different evolutionary pathway from *Polygala*. Baillon (1860) mentioned a different position for stamens in *Muraltia*: one stamen opposite sepal two and two stamens opposite the latero-anterior petals are missing. Most species present a secondary pollen presentation where the sterile abaxial lobe of the style functions as pollen presenter (Westerkamp and Weber, 1999).

Leguminosae

General floral formula: ↓/✶ K0–5(-6) C(0–3)–5(-6) A1–10 G1(2–16)

Leguminosae have traditionally been subdivided into three (sub)families: Papilionoideae, Mimosoideae and Caesalpinoideae. Molecular data have demonstrated the paraphyletic nature of Caesalpinioideae and the recent approach is to recognize the six subfamilies of Duparquetioideae, Cercidoideae, Dialioideae, Detarioideae, Faboideae and Caesalpinioideae, with

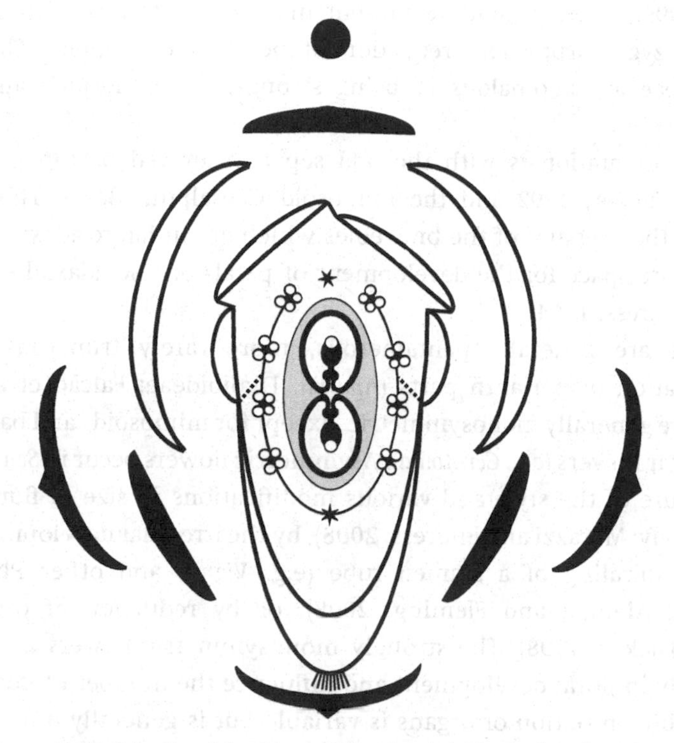

Figure 9.53 *Polygala x dalmaisiana* (Polygalaceae). The broken line shows the attachment of the staminal ring to petals.

Mimosoideae nested in the latter, despite clear morphological characterics (Legume Phylogeny Working Group, 2017). Most Leguminosae have twenty-one organs following the floral formula K5 C5 A5+5 G1 and share the absence of fused petals (except for mimosoids) and an androecium that is basically two-whorled (again with most variations in mimosoids and some basal Papilionoideae).

Inflorescences of Leguminosae are generally racemose. Bracts and bracteoles are usually well developed, the latter occasionally enclosing the floral bud and replacing the reduced calyx (Detarioideae). Bracteoles are more rarely absent (e.g. *Haematoxylum*, *Caesalpinia*, mimosoids). Prenner (2004a) reported several cases of bracteoles being initiated and aborting before maturity.

Leguminosae show a high level of variation in their flowers. The traditional divide of Leguminosae corresponds well with clearly defined flower shapes, which become more difficult to apply with the new classification. Compared to Papilionoideae with a typical papilionoid morphology (Westerkamp and

Weber, 1999), Caesalpinioideae present more open flowers with a variable extent of zygomorphy and reduction of petals and stamens. The former Mimosoideae are anomalous in being strongly actinomorphic and mostly polyandric.

Flower orientation is with the odd sepal in abaxial position (except in *Ceratonia*: Tucker, 1992 and the mimosoid Caesalpinioideae). This may be caused by the pressure of the bracteoles which are in latero-adaxial position, leaving more space for the development of petals on the adaxial side of the flowers (Endress, 1994).

Flowers are generally pentamerous, more rarely trimerous or with a proliferation of perianth parts (*Apuleia*, Dialioideae: Falcão et al., 2020). Flowers are generally monosymmetric, except for mimosoids and basal clades with regular flowers (e.g. *Ceratonia*). Asymmetric flowers occur in *Senna* mainly by curvature of the style and various modifications in size of floral organs (enantiostyly: Marazzi and Endress, 2008), by the irregular development of the keel and spiraling of a stamen tube (e.g. *Vignea* and other Phaseoleae: Etchevery, Alemán and Fleming, 2008), or by reduction of organs (e.g. *Labichea*: Tucker, 1998). The strongly monosymmetric flowers are differentiated early in floral development and influence the number of floral organs considerably. Initiation of organs is variable but is generally unidirectional. The calyx is well developed and imbricate, with the first sepal arising against the bract. Pressure caused by bract, bracteoles and floral axis is responsible for the variation in sepal initiation with increased pressure of bracteoles influencing the fate of calyx lobes and loss of organs (e.g. Detarioideae). Prenner and Klitgaard (2008) considered that the tetramerous calyx of *Duparquetia* is formed by loss of one latero-adaxial sepal due to pressure of bracteoles. In Dialioideae, a tetramerous calyx is often obtained by fusion of the two adaxial sepals (Falcão et al., 2020).

Petals are often large and clawed and generally free. Some Caesalpinioideae and Cercidoideae have more or less irregular flowers linked with a reduction of petals (*Gleditsia*) or have evolved a papilionaceous shape (*Cercis*, *Haematoxylon*). Contrary to Papilionoideae the standard petal is the inner one, and several features of Papilionoideae are absent. Petals are occasionally completely lost or formed and subsequently suppressed (e.g. *Dialium*: Tucker, 1998; *Ceratonia*: Tucker, 1992; *Apuleia*: Falcão et al., 2020). A hypanthium may be strongly developed, and various elaborations link androecium and petals in hiding access to the nectary (Endress, 1994).

The androecium is basically two-whorled and diplostemonous, but has been variously transformed, ranging from one to ten stamens, with variable number of staminodes or missing stamens (e.g. *Afzelia*, *Amherstia*, *Bauhinia*, *Cassia*,

Duparquetia, Petalostylis, Saraca, Senna: Tucker, 1988a, 1988b, 1996, 1998, 2000b, 2000c, 2002b; Prenner and Klitgaard, 2008; Falcão et al., 2020). Reduction of the androecium runs from the abaxial to the adaxial side of the flower and is often correlated with heterostaminody, as in *Senna* with two pollinating stamens, four fodder stamens and three staminodes (Tucker, 2003a). Reduction and loss of stamens affects the antepetalous stamens first (e.g. *Ceratonia, Tamarindus*: Tucker, 1992, 2000b). Stamens can be further reduced to four (exceptionally in adaxial position, *Duparquetia*: Prenner and Klitgaard, 2008), three (*Petalystylis*), two (*Labichea, Dialium*: Tucker, 1998; Falcão et al., 2020), or one (*Bauhinia divaricata*: Figure 9.55) with a variable number of staminodes.

A universal character for Leguminosae is the single superior carpel found in almost all taxa, with some exceptions in mimosoid Caesalpinioideae and Papilionoideae that may be caused by an increase in the floral apical meristem linked with polyandry (Paulino et al., 2014). The nectary is generally present as an intrastaminal disc around the ovary which usually bears a gynophore.

Caesalpinioideae

Typical Caesalpinoideae follow a unidirectional floral development as in *Cassia, Senna* or *Caesalpinia*, often with overlap in the initiation of whorls (e.g. Tucker, 1996, 2003a; Marazzi and Endress, 2008). Bracteoles are generally present, except in mimosoids, and small. Monosymmetric flowers generally have long filaments and exposed flowers, often with a variable number of staminodes.

Ceratonia represents an unusual genus in being unisexual, apetalous and haplostemonous and has been considered as a link to the polysymmetric mimosoids in sharing a same orientation of the median sepal on the adaxial side (Tucker, 1992).

Mimosoid Clade

Figure 9.54A *Calliandra haematocephala* Hassk.

✶ K(5) C(5) A∞:A°∞ $\underline{G}$1

General formula: ✶ K0–5 C0–5 A5–∞ G1(5)

Flowers are grouped in globular inflorescences and are subtended by a single bract. Flowers tend to arise synchronously (Tucker, 2003a). There is no pedicel. In *Calliandra* both calyx and corolla are fused for two-thirds. The attractiveness of the flower is caused by numerous stamens with long filaments in at least two series curled up in bud. The inner whorl is staminodial (or inner trichomes?) curved over the disclike nectary. Anthers are small and introrse. The single carpel is topped by a long style and dish-shaped stigma.

The relationship between mimosoids and its sister lineages are fuzzy (Legume Phylogeny Working Group, 2017). The clade differs from the other Leguminosae by a number of structural characters: absence of bracteoles, orientation of the flower, which has the median sepal in adaxial position as in most angiosperms, the radial symmetry of flowers, which is usually associated with a secondary stamen increase and globular inflorescences, and the development of sepal and petal tubes (see Tucker, 2003a). The basal genera *Pentaclethra* and *Dimorphandra* share several characters with other Caesalpinioideae but also characters of mimosoids, such as lost bracteoles, an imbricate calyx with the median sepal adaxially, shorter sepals in bud and dimorphic androecium (De Barros et al., 2017). There is much variation in the initiation sequence of sepals (simultaneous, unidirectional, helical), which are imbricate, while petals arise generally simultaneously and have a valvate aestivation (Ramírez-Domenech and Tucker, 1990). Calyx development is generally followed by a process of equalization leading to a valvate calyx (De Barros et al., 2017). Petals are generally much shorter, and this is correlated with long showy filaments of the polyandrous flowers, analogous with some Myrtaceae (brush flowers: Endress, 1994; see p. 207).

There is much variation in the number of stamens because of a secondary stamen increase. Stamen arrangement is generally diplostemonous or haplostemonous (e.g. *Mimosa*, *Neptunia*, *Calliandra*: Gemmeke, 1982; Prenner, 2004c). In *Dimorphandra* and *Mora* the antesepalous stamen whorl consists of staminodes, while in *Pentaclethra* the antepetalous whorl is sterile (De Barros et al., 2017). A secondary stamen increase is linked with alternisepalous primary stamen primordia (e.g. *Lysoloma*, *Accacia*) or the development of a ring meristem (e.g. *Albizia*, *Accacia*, *Calliandra sp.*) (Gemmeke, 1982; Derstine and Tucker, 1991). The stamens are increased in centripetal direction but also partly centrifugally (Derstine and Tucker, 1991).

Detarioideae

Figure 9.54B *Afzelia quanzensis* Welw., based on Tucker (2002b)
↓ K5 C4:1 A5+2 $\underline{G}1$

The subfamily shows the highest variation in floral diversity among Leguminosae and this is linked with elaborate zygomorphy and open-access flowers with various adaptations for pollination. Flowers show some resemblance with Capparaceae in the development of spreading petals and long filaments.

The subfamily is characterized by large petaloid bracteoles enclosing the bud. This is linked with a progressive reduction of petals and stamens. In some taxa, adaxial sepals become fused during development or the fifth sepal becomes absorbed in a larger adaxial sepal (e.g. *Saraca*, *Schotia*,

Afzelia, Sindora: Tucker, 2000c, 2001b, 2002b, 2003c); this may be a preliminary step for a transition to a tetramerous calyx. Petals are generally differentiated with the adaxial petal largest and the others variously reduced. In some cases all but one petal are initiated but suppressed (e.g. *Aphanocalyx, Monopetalanthus*: Tucker, 2000a), or petals are initiated but they are arrested in mid-development (e.g. *Crudia*: Tucker, 2001a). Petal rudiments can be variously present (e.g. Figure 9.54B; *Amherstia, Tamarindus*: Tucker, 2000b). *Saraca* represents an unusual homeotic transformation of petals into stamens and this is accompanied by variable loss of antesepalous stamens (Tucker, 2000c).

The androecium ranges from ten to two fertile stamens (by a progressive reduction starting with the adaxial and antepetalous stamens (Tucker, 2002b). In *Sindora* the two fertile stamens are in latero-adaxial position with all other stamens staminodial (Tucker, 2003c).

Cercidoideae

Figure 9.54C *Bauhinia divaricata* L.: staminate flower

↓K(5) C2 A(1:4°+5°) $\underline{G}^{\circ}1$

Flowers are generally bisexual, occasionally dioecious, as in *Bauhinia malabarica* and *B. divaricata* (Figure 9.54C; Tucker, 1988a). *Bauhinia* shows a strong variation in petal and stamen number, the latter ranging from ten to one. In *Bauhinia divaricata* petal number ranges from two to three narrow appendages (Figure 9.54C), although five petals are initiated (Tucker, 1988a); there is only a single functional stamen and nine staminodes (Tucker, 2003a). Sepals are fused in a single calyptrate structure that dehisces abaxially under pressure of the single stamen (Figure 9.54C). In *Bauhinia galpinii* stamen reduction is correlated with lateral dédoublement of staminodes (Endress, 1994, 2008b). *Cercis* strongly resembles papilionoid Papilionoideae with the difference that petal aestivation is ascending as opposed to descending with the adaxial petal innermost (Eichler, 1878).

Papilionoideae

Figure 9.55A *Swartzia aureosericea* R. S. Cowan, based on Tucker (2003b)

↓ K(5) C1 A∞ $\underline{G}1$

Figure 9.55B *Strongylodon macrobotrys* A. Gray

↓ K(5) C2:2:1 A(5+4):1 $\underline{G}1$

General formula: ↓ K(3–)5 C(1–)5 A(9)10–∞ G1

Inflorescences are mostly racemose (plesiomorphic condition), but cymose forms are occasionally found as a derived condition (Tucker, 1999a). Flowers are nearly always subtended by a bract and two bracteoles.

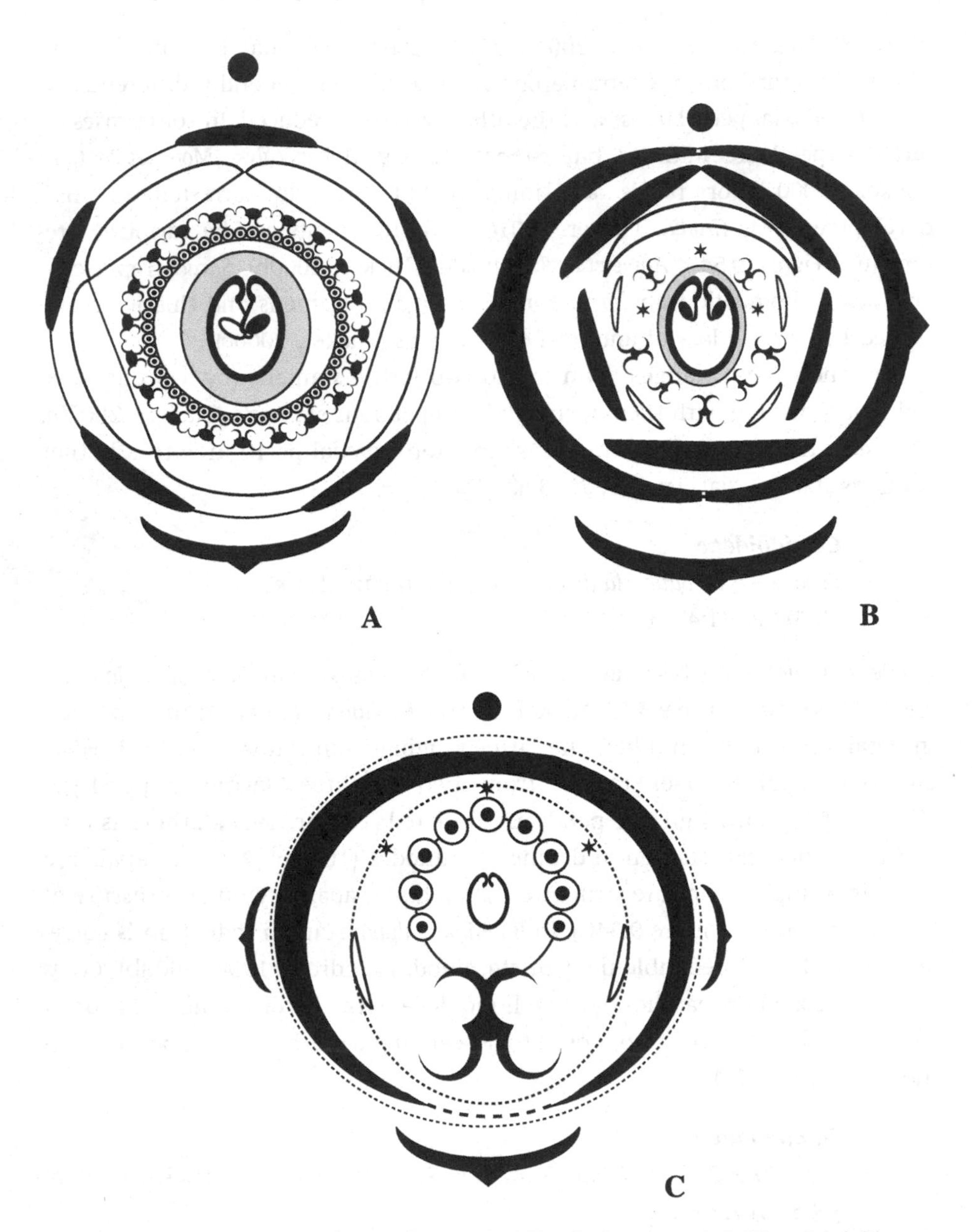

Figure 9.54 Leguminosae flowers. A. *Calliandra haematocephala* (Caesalpinioideae – Mimosoids). Note the inner ring of staminodes; B. *Afzelia quanzensis* (Detarioideae); asterisks refer to lost stamens; C. *Bauhinia divaricata* (Cercidoideae): staminate flower. The dehiscence line of the calyx tube is shown by a broken line; asterisks refer to lost petals.

Most Papilionoideae are median monosymmetric with a surprisingly uniform ontogeny and rarely with loss or increase of floral organs (compared with other subfamilies). In Phaseoleae flowers are often asymmetric by the twisting of the stamen tube. The papilionate floral shape is almost generalized in the subfamily and epitomizes a highly successful bee pollination syndrome (Westerkamp and Weber, 1999). The adaxial petal is mostly the largest (standard, flag or vexillum) and petals are arranged in cochleate descending aestivation. Anterior petals are usually distally connivent in the characteristic keel enclosing androecium and gynoecium. The mechanism of pollination is explosive and stegnotribic. Flowers of *Strongylodon* are inverted as an adaptation to hanging inflorescences. This adaptation is frequently observed in Papilionoideae with a climbing habit, leading to a nototribic deposition of pollen compared to keel flowers (Amara-Neto, Westerkamp and Melo, 2015). Main variation is found in fusion of members of the androecium, either all ten (in nectarless flowers), or as nine + one, where one adaxial stamen is separated from the tube by two slits allowing access to nectar at the inner side of the tube (e.g. Figure 9.55B, *Vicia*). All organs have a strong unidirectional development starting early in ontogeny, but there are several exceptions to this rule (e.g. Tucker, 1984), although the flag petal becomes the largest before anthesis. An obvious exception is *Cadia* with radially symmetrical flowers. As several Papilionoideae have a regular symmetry at mid-development, Tucker (2002a) interpreted the condition in *Cadia* as a neoteny (a retention of the juvenile state), contrary to Citerne, Pennington and Cronk (2006) who saw this not as an evolutionary reversal, but as a homeotic transformation in which all petals have acquired the genetic identity of the abaxial petal (see p. 59).

Flowers are mostly isomerous for perianth and androecium with two stamen whorls; reductions of organs are limited and are restricted to tribes Sophoreae and Swartzieae (Tucker, 1988b). The genus *Swartzia* is basal in the subfamily. It shares the single well-developed banner with members of the amorphoid clade, but the remaining petals are variously developed and abort in early stages of development (MacMahon and Hufford, 2005). Flowers of *Swartzia* are extremely heterogenous without, with one, or with two adaxial petals and a high number of stamens arising on a complete or partial ring meristem. In partial meristems a variable number of much larger abaxial stamens arise before the ring meristem (Figure 9.57A; Tucker, 2003b). This phenomenon is best interpreted as a partial increase of the stamens, affecting the abaxial stamens to a variable extent. The androecium has thus partially complex polyandry linked with free adaxial stamens. Other early diverging papilionoid legumes appear to be variable in their development (e.g. *Petaladenium*), preceding any canalization in the generalized papilionoid flower morphology (Prenner et al., 2015).

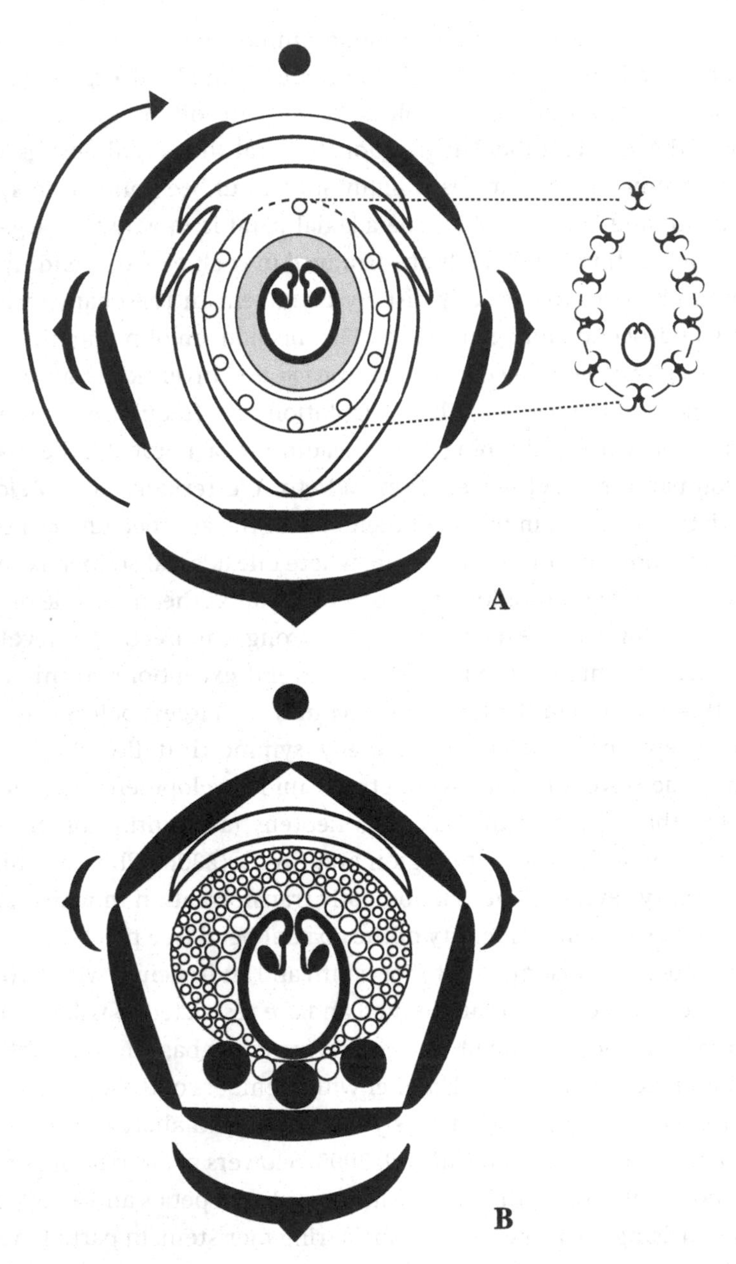

Figure 9.55 Leguminosae – Papilionoideae: A. *Strongylodon macrobotrys*. The broken line represents the gap separating the odd adaxial stamen. B. *Swartzia aureosericea*. Black dots, outer larger stamens; white dots, inner smaller stamens

Rosales

The order contains nine families and is strongly supported. The circumscription of Rosales has changed dramatically since Cronquist (1981). Soltis et al. (2005) enumerated a hypanthium and a reduction of endosperm as main synapomorphies. Stevens (2001 onwards) included the hypanthial nectary, petals that are clawed when present, a valvate calyx and single apotropous ovule per carpel. The order is believed to be basically apetalous and wind-pollinated (cf. Ronse De Craene, 2003). This is reflected in families formerly placed in Urticales, which share a simple floral formula of K5 C0 A5 $\underline{G}(2)$. Within Urticales there is a progressive reduction of flowers linked with unisexuality, loss of a hypanthium, reduction of the perianth and pseudomonomery (e.g. Urticaceae, Moraceae: Bechtel, 1921; Leite, Mansano and Teixera, 2018). Loss of a hypanthium occurred in the clade comprised of Cannabaceae, Moraceae and Urticaceae, while it is retained in Ulmaceae. Flowers are reduced to five or fewer stamens and bicarpellate unilocular ovaries with a single basal or apical ovule (Judd and Olmstead, 2004).

Rosaceae occupies a basal position in the order and this is linked with a unique kind of polyandry including petals of staminodial origin. I suspect that petals were secondarily derived from stamens on two occasions, once in Rosaceae and a second time in the Rhamnaceae and Dirachmaceae, corroborating earlier theories (e.g. Bennek, 1958). It would be interesting to study the genetic background for petal expression in Rosales.

Diplostemonous flowers are rare in the order (some Rosaceae, *Shepherdia* in Elaeagnaceae), although it is probably plesiomorphic. Obhaplostemonous flowers have mainly evolved in Rosaceae, Rhamnaceae and Elaeagnaceae, while 'Urticales' are haplostemonous (Ronse De Craene, 2003).

Rosaceae

Figure 9.56A *Potentilla fruticosa* L.
∗ K5 C5 A10+10+5+5 $\underline{G}\infty$

Figure 9.56B *Spiraea salicifolia* L.
∗ K5 C5 A10+5+5 $\underline{G}(5)$

Figure 9.56C *Sanguisorba tenuifolia* Fisch. Ex Link
∗ K4 C0 A4 $\underline{G}1$

General formula: ∗ K(3–)5(–10) C0 or (3–)5(–10) A(1–)5–10–∞ G1–2–5–∞

Flowers are mostly arranged in cymose units or in racemose inflorescences. In racemes terminal flowers precede lower flowers (botryoids: e.g. *Sanguisorba*, *Gillenia*, *Photinia*: Evans and Dickinson, 2005; Wang et al., 2020).

Flowers are generally pentamerous, more rarely tetramerous (Sanguisorbinae, trimerous in *Cliffortia*: Eichler, 1878). Aestivation of sepals is generally valvate. An epicalyx is developed in several genera and alternates with sepal lobes. The epicalyx resembles the calyx closely (e.g. *Potentilla*: Pluys, 2002), is reduced to a laciniate appendage (e.g. *Geum*), or develops as spines with a centrifugal development (e.g. *Agrimonia*: Ronse De Craene and Smets, 1996b). The epicalyx has been interpreted as stipular in origin (e.g. Trimbacher, 1989), although homologies are not always clear (see p. 31).

Rosaceae have either apetalous flowers or petaliferous flowers with a well-developed hypanthium (floral cup) reminiscent of Myrtaceae. Petals bear a short claw and have a variably imbricate aestivation. When petals are present, they are closely connected to the androecium and are interpreted as having a staminodial origin in the family and to represent the upper stamens of complex alternisepalous primordia (Ronse De Craene, 2003). In *Cecrocarpus* and *Neviusia* stamens occupy the position of petals in other genera (Kania, 1973; Lindenhofer and Weber, 2000). A plausible explanation is that petals have been replaced with stamens through a process of homeosis. In the *Pygeum* and *Maddenia* groups of *Prunus*, mature petals are little differentiated from sepals, but they initially resemble stamens. The stronger link with sepals can be a reduced B-gene expression in the petals at later stages of development (Wang et al., 2021). These shifts in organ identity resemble similar shifts in Papaveraceae (p. 152). The androecium is extremely variable in the family by the presence of reductive trends and secondary elaborations. The androecium of the Rosaceae appears to be unique in the angiosperms in the development of alternating decamerous or pentamerous whorls linked with the growth of the hypanthium (described as cyclic polyandry by Ronse De Craene and Smets, 1987). In most Rosaceae the androecium consists of several whorls (often four or five with twenty to twenty-five stamens arranged as 10+5+5+5) that develop thanks to the growth of the hypanthium. The upper (outer) stamen whorl is always decamerous and closely linked with the petal (parapetalous stamens). Two more whorls of five stamens are generally formed in antesepalous and antepetalous position (Figure 9.56B). However, the number of stamens formed can fluctuate strongly, even within a same species and depends on reductions of the length of the hypanthium or an earlier onset of the development of the gynoecium (*Crataegus*: Evans and Dickinson, 1996). Kania (1973) interpreted the increase of stamens as secondary and derived from a diplostemonous ancestry. However, when the androecium is limited to ten stamens (e.g. *Crataegus*), these consist of the upper pair only and do not have a diplostemonous arrangement. Clearly diplostemonous androecia occur in *Agrimonia*, *Stephanandra* and *Plagiospermum*, although the number of stamens is variable and can be higher

(Kania, 1973; Ronse De Craene, unpubl. data). *Filipendula* is also unique in the development of ten radial rows of stamens (Ronse De Craene, unpubl. data). In *Potentilla*, petals occupy the upper section of fascicles (ridges: Innes, Remphrey and Lens, 1989), extending centripetally on the slopes of the hypanthium. The number of stamens can be increased dramatically, as in *Rubus* or *Rosa*, through reduction of size of primordia and development of a ring primordium. Ronse De Craene and Smets (1992a) and Ronse De Craene (2003) argued that the cyclic development of stamen whorls is not different from a secondary multiplication of stamens on common primordia. The common primordia are rarely well differentiated from the hypanthium (e.g. in *Potentilla*, *Crataegus*, *Fragaria*: Sattler, 1973; Innes, Remphrey and Lens, 1989; Evans and Dickinson, 1996; Ronse De Craene, 2003). Fluctuation in stamen number and development of whorls is linked to the size of stamen primordia and the extent of growth of the hypanthium in a very similar pattern as found in Myrtaceae (Ronse De Craene and Smets, 1991a). An alternative interpretation was presented by Lindenhofer and Weber (1999a, 1999b, 2000), that polyandry in Rosaceae is derived from a spiral androecium. They considered a spiraeoid pattern of 10+5+5 stamens as original and interpreted diplostemony as derived by reduction. They especially rejected the notion of dédoublement that was used to explain the paired position of stamens. Indeed, a process of splitting is never observed and stamen numbers can occupy paired or unpaired positions (e.g. *Stephanandra*, *Aruncus*, *Agrimonia*: Lindenhofer and Weber, 1999a; Ronse De Craene, unpubl. data). The interpretation of common primordia, including the petals (Ronse De Craene, 2003), makes a discussion of a derivation from basal diplostemony unnecessary, although diplostemony is probably plesiomorphic in Rosales. There is no support for an ancestral spiral polyandry in Rosaceae by comparison with the sister groups of the family.

Flowers can also be strongly reduced, often in relation to wind pollination (subtribe Sanguisorbinae e.g. *Acaena*, *Sanguisorba*, *Cliffortia*). Flowers are generally tetramerous, petals are absent and the gynoecium is unicarpellate. Stamens replace petals through homeosis (e.g. *Alchemilla*: Ronse De Craene, 2003) or stamens are opposite the sepals (e.g. *Sanguisorba*, *Sibbaldia*: Figure 9.56C). The tetramerous calyx tends to have a strong influence on the sequence of initiation of antesepalous stamens, which emerge in a sequence of two and are often reduced to two stamens by the loss of one tier (e.g. *Acaena*: Wang et al., 2020). *Aphanes arvensis* has a single stamen in median position (K4 A1 G1: Kania, 1973).

Much of the earlier classifications of the Rosaceae are based on chromosome numbers and fruit types. The gynoecium is basically isomerous with five antesepalous carpels and two (rows of) collateral ovules per carpel, as is a generalized condition in the Maloideae (Evans and Dickinson, 2005). Ovules

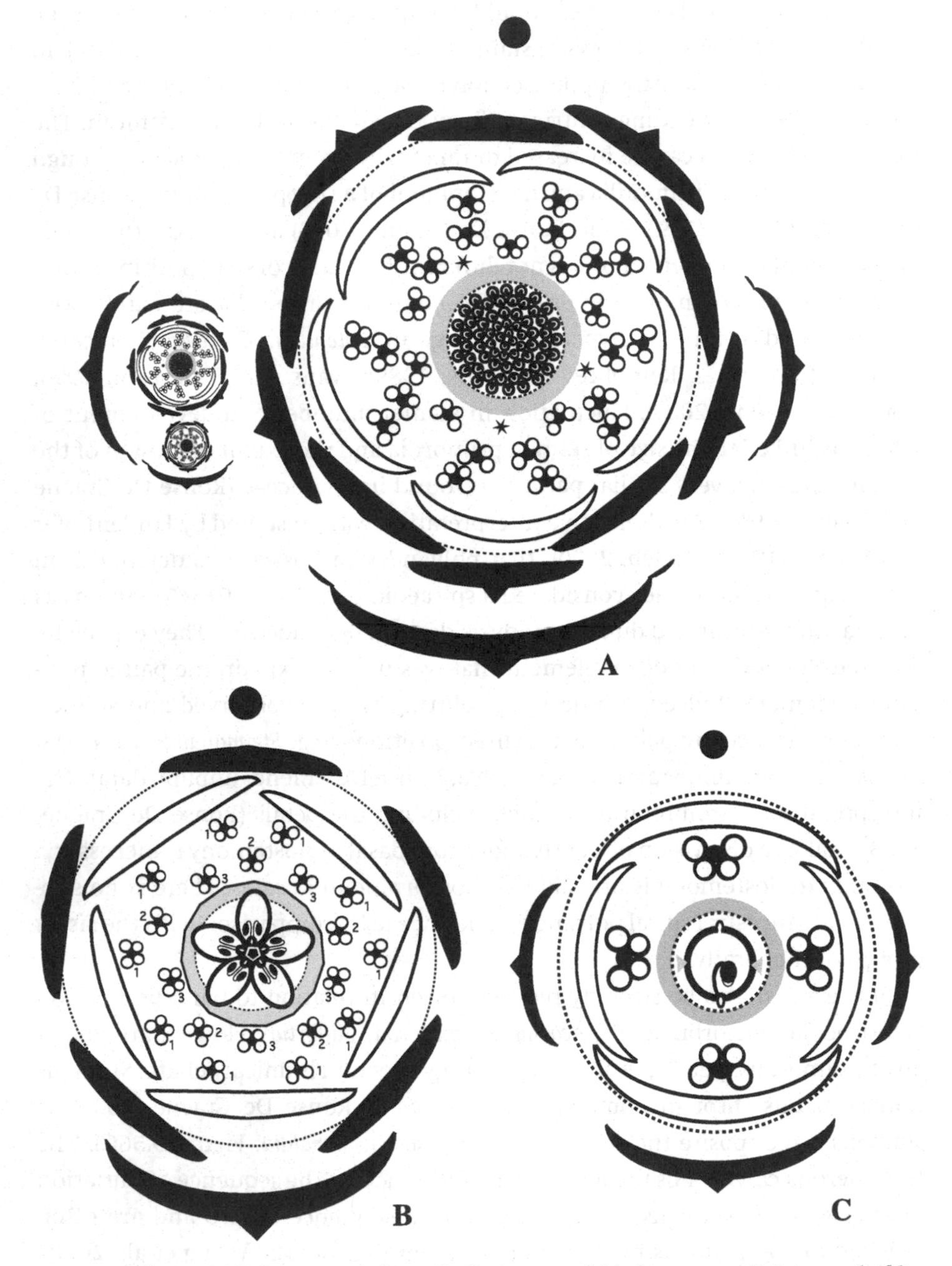

Figure 9.56 Rosaceae: A. *Potentilla fruticosa*: partial inflorescence; B. *Spiraea salicifolia*; C. *Sanguisorba tenuifolia*. Asterisks in A refer to empty stamen positions.

become superposed in *Crataegus* and *Mespilus*. A false septum is occasionally formed (e.g. *Amelanchier*: Steeves, Steeves and Randall Olson, 1991; Rohrer, Robertson and Phipps, 1994). Gynoecia are often apocarpous leading to follicular fruits (e.g. *Spiraea*: Figure 9.56B), become reduced to a single carpel (e.g. *Prunus, Sanguisorba*) or increase in number (e.g. *Rubus, Fragaria*). The ovary is occasionally syncarpous, as in most Maloideae (Rohrer, Robertson and Phipps, 1994). The ovary is superior but can become half-inferior or inferior by invagination and coalescence with the margins of the hypanthium (Steeves, Steeves and Randall Olson, 1991; Evans and Dickinson, 2005). A reduction to a single carpel is found in Prunoideae and in taxa with smaller reduced flowers. In several Rosoideae (e.g. *Rubus, Potentilla*), carpels are reduced in size with a single ovule, and this correlates with a secondary increase of several whorls of apocarpous carpels (Kania, 1973). Styles are single and situated opposite the locule. In *Alchemilla* the style is gynobasic.

An intrastaminal nectary develops on the inner slope of the hypanthium, encircling the ovary. In *Spiraea* the nectary develops as a crenelated rim that was interpreted as staminodial (e.g. Eichler, 1878) although there is no evidence for this.

Rhamnaceae

Figure 9.57 *Ceanothus dentatus* Torr. and A. Gray
∗ K5 C5 A5 -G(3)-
General formula: ∗ K4–5 C0/4–5 A4–5 G2–3(5)

Flowers are small and tetra- to penta- (hexa-) merous, grouped in racemes with small abortive bracts and bracteoles. A hypanthium is present and the ovary can be variously embedded in the receptacle. Sepals are valvate and usually pigmented. Petals arise as tiny appendages of the stamens (often from common primordia: e.g. Bennek, 1958; Sattler, 1973; Medan and Hilger, 1992; Ronse De Craene and Miller, 2004) and remain inconspicuous in the flower. In *Ceanothus* petals are hoodlike. Some genera lack petals altogether and have a petaloid calyx and hypanthium (e.g. *Colletia, Pomaderris*). The androecium is always obhaplostemonous. Most Rhamnaceae have little pigmentation and have greenish to yellowish flowers. The gynoecium contains two to three (rarely five) carpels with axile placentation. The ovary is superior or more frequently (half-) inferior with three carpels and a single, often deeply cleft style. Each locule has one-two basal ovules per placenta. In *Ceanothus* one ovule per locule is developed (Figure 9.57), while the number of fertile placentae can be variable in the family. For example, in *Colubrina* with a trimerous ovary, one septum bears two ovules, another one ovule, and a third none (Medan

Figure 9.57 *Ceanothus dentatus* (Rhamnaceae)

and Hilger, 1992). The central part of the flower is mainly occupied by a large disc nectary, which is especially well developed in cases where the ovary is inferior.

Closest relatives of Rhamnaceae are Dirachmaceae and Barbeyaceae (Richardson et al., 2000). Dirachmaceae shares several characters with Rhamnaceae, including obhaplostemony, although several unique features separate the family (Ronse De Craene and Miller, 2004).

Elaeagnaceae

Figure 9.58 *Elaeagnus umbellata* Thunb.

✶ K4–5 C0 A4–5 $\underline{G}$ 1

General formula: ✶ K4–5(-6) C0 A4–5–8 $\underline{G}$ 1

The small family of three genera is characterized by the absence of a corolla, although this was questioned by Rao (1974) on anatomical grounds.

Flowers are tetra-, penta- or hexamerous and are grouped in clusters in the axil of leaves. Flowers are bisexual only in *Elaeagnus*. Sepal aestivation is always valvate. In *Hippophae*, lateral sepals are much smaller than median sepals and were described as bracteoles by Eichler (1878) who interpreted the flowers as

Figure 9.58 *Elaeagnus umbellata* (Elaeagnaceae): A. flower; B. inflorescence

dimerous. Stamens alternate with sepal lobes (*Elaeagnus*, *Hippophae*), or the androecium is diplostemonous (*Shepherdia*). The filament is very short. Stamens and sepals are connected by a long hypanthium that runs down below the ovary. In *Elaeagnus*, the ovary appears to be inferior but the ovary wall is not connected with the external wall. The appearance of an inferior ovary is stressed by the presence of a nectary on the hypanthial slope, surrounding the style. This arrangement resembles some unicarpellate Rosaceae to a great extent (e.g. *Sanguisorba*: Wang et al., 2020). In *Shepherdia* well-developed nectary lobes alternate with the stamens or surround the carpel at the level of attachment of the sepal lobes (Baillon, 1870). The ovary is unicarpellate with single, basal, anatropous ovule. Sepals, hypanthium and ovary are covered with peltate hairs. Elaeagnaceae share many similarities with reduced flowers of Rosaceae, such as apetaly, valvate tetramerous calyx, alternate stamens, and single carpel.

Ulmaceae

Figure 9.59 *Zelkova serrata* (Thunb.) Makino, based on Okamoto, Kosuge and Fukuoka (1992)

✶ K5–6 C0 A5–6 $\underline{G}$(1:1°)

General formula: ✶ K(2–)5(–9) C0 A(2–)5(–16) G2(–3)

Figure 9.59 *Zelkova serrata* (Ulmaceae): partial inflorescence

Compared to other members of 'Urticales', Ulmaceae possess a relatively well-developed flower. Flowers are bisexual to unisexual and arranged in cymose inflorescences or as solitary flowers (pistillate inflorescences). Pistillodes and staminodes are variously developed or rudimentary. Merism can be variable with fluctuating numbers of sepals and petals in the same inflorescence (Figure 9.59). Petals are always absent, although vascular bundles may persist in the receptacle (Bechtel, 1921). Stamens are of the same number (rarely more or double in *Holoptelea* and *Ampelocera*) and opposite sepals (Todzia, 1993). Flowers of *Ampelocera* are unusual in being tetramerous and having an additional alternisepalous stamen whorl (Leme et al., 2018). This is reflective of an ancestral diplostemony in Rosales. Filaments are straight or inflexed in bud. The bicarpellate ovary is pseudomonomerous with one apical ovule. Okamoto, Kosuge and Fukuoka (1992) studied the floral development of pistillate *Zelkova serrata* and demonstrated that the single ovule arises laterally on a parietal placenta. This is not visible in mature stages by fusion of tissue. The sterile carpel has an empty locule. The two styles are generally well developed and carinal.

Cucurbitales

The order contains seven families. Matthews and Endress (2004) provided several floral characters in support for the order, except for Anisophylleaceae, which has more morphological affinities with Cunoniaceae and might be misplaced (see Endress and Matthews, 2006a, 2006b). While Corynocarpaceae and Coriariaceae appear to be more basal with more generalized rosid characters, several characters link Cucurbitaceae, Begoniaceae, Tetramelaceae and Datiscaceae (Soltis et al., 2018). These include unisexual flowers, often with a strong dimorphism between staminate and pistillate flowers, an inferior ovary with intruded parietal placentation and a high number of ovules (cf. Judd et al., 2002).

Cucurbitaceae

Figure 9.60A–B *Cucurbita palmata* S. Watson

Staminate ✶ K(5) C(5) A$1^{1/2}$: ($2^{1/2}$)($2^{1/2}$)G0

Pistillate ✶ K(5) C(5) A5° Ğ(3)

General formula: ✶ K(3)5(6) C(3)5(6) A(1–)3–5 G2–3(–5)

Flowers are grouped in various inflorescences or are solitary. Flowers are unisexual with variable reductions of the other gender. In *Cucurbita palmata*, flowers are grouped in compound axillary racemes, consisting of a basal branch with a single pistillate flower only, and another branch with several staminate flowers. In *Cucurbita*, small staminodes are present in pistillate flowers, while no trace of the gynoecium is found in staminate flowers (Figure 9.60A). Petals are free or fused, which is a rare phenomenon in rosids, although this is probably due to hypanthial growth as there is a connection with the sepals (at least in *C. palmata*). The androecium rarely consists of five free stamens opposite the sepals (haplostemony), which represents the basic condition in the family. Some genera have five monothecal stamens. Chakravarty (1958) demonstrated the existence of transitions between dithecal and unithecal stamens. In *Xerosicyos pubescens* four antesepalous stamens have either dithecal or monothecal anthers in variable proportions (Matthews and Endress, 2004). More often there are only three stamens, two are dithecal and inserted between two sepals, and the third is unithecal and opposite a sepal (Figure 9.60A). Floral anatomical and developmental evidence (e.g. Payer, 1857; Chakravarty, 1958; Leins and Galle, 1971) have shown that the peculiar androecium is derived by the pairwise fusion of four monothecal stamens, while a single stamen remains free. Pollen sacs are characteristically twisted and convoluted and anthers appear asymmetrical. Stamens can be either free or variously fused at the base or in a synandrium (not in *C. palmata*). An extreme case is *Cyclanthera* with the anthers forming a continuous horizontal dehiscence line, which is derived by fusion of three stamens (Chakravarty, 1958). As a result flowers resemble staminate flowers of some Menispermaceae (p. 155).

The ovary is inferior, (two-) three (-five) carpellate and has three parietal placentae with two rows of ovules. Ovaries are unilocular or a false septum runs to the centre of the ovary and the locules are filled with parenchymatous tissue. The style is often branched and twisted as in Begoniaceae.

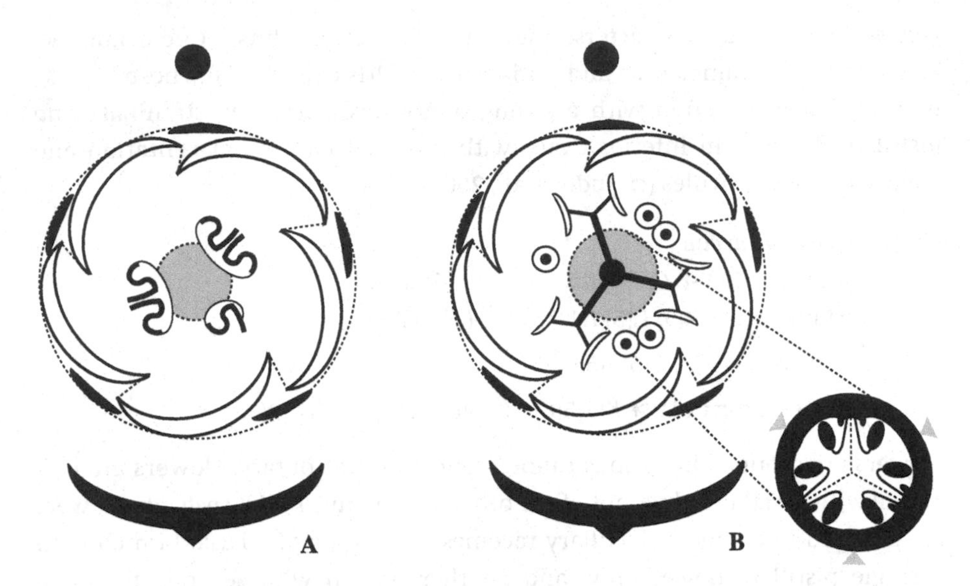

Figure 9.60 *Cucurbita palmata* (Cucurbitaceae): staminate (A) and (B) pistillate flower

Two types of nectaries are found in the family: mesenchymatous nectaries as an intrastaminal hypanthial disc, and trichomatous extrastaminal glands (Vogel, 1997); the latter were interpreted as substitutes for the ancestral disc nectaries. In *Cucurbita*, the nectary is found on the inner slope of the hypanthium, in the male flower accessible between the filament bases and in the female flower around the base of the style (Nepi and Pacini, 1993).

Begoniaceae

Figure 9.61A–B *Begonia albo-picta* Bull

Staminate: ↓/⟷[h] K2+2 A12 G0

Pistillate ✶ K5/2:3 A0 Ğ(3)

General formula: staminate ↓/⟷ K2–4(5–10) C0(5) A4–∞ G0; pistillate ✶/⟷ K2–6 C0 G2–3(–5)

Flowers are arranged in cymes with staminate and pistillate flowers on different inflorescences of the same plants (rarely dioecious), or are mixed with staminate flowers developing first. Bracteoles drop off early, leaving a scar. Merism is variable in the genus, even between staminate and pistillate flowers on a same inflorescence, ranging from dimerous to pentamerous or hexamerous. Petals are absent and the calyx is petaloid. Staminate flowers are mostly tetramerous (or apparently dimerous because of the difference between inner and outer calyx whorls (cf. Ronse De Craene

[h] Weakly monosymmetric to disymmetric.

and Smets, 1990b). The calyx consists of an outer whorl of large petaloid parts, valvately arranged in bud and compressing the flower. The inner transversal sepals are much smaller and have a tendency to become lost in several species (e.g. *B. heracleifolia*). In *Begonia pleiopetala* sepals are increased to eight by an addition of whorls. The calyx is rarely fused and tubular (*B. argenteomarginata*: pers. obs.). Stamens are numerous, or much lower in number (down to eight or four) and arranged on an inflated receptacle. Initiation is centripetal in regular to irregular whorls on an elliptical receptacle (Ronse De Craene and Smets, 1990b). The androecium is regular with extrorse to latrorse stamens, or monosymmetric with all stamens facing one of the median sepals. Staminate *Begonia* flowers offer pollen as reward and are without nectary, and stamens appear flattened with a broad connective (Matthews and Endress, 2004). Pistillate flowers offer no reward and deceive visitors by mimicking stamens with their contorted stigmas. Only a few bird-pollinated Begonias produce nectar in pistillate flowers (e.g. *B. ferruginea*: Vogel, 1998c). Merism in pistillate flowers is mostly five with a quincuncial arrangement, hexamerous with an imbricate arrangement of an outer and inner whorl (e.g. *B. galinata*, *B. bogneri*), or tetramerous (e.g. *B. mannii*) to dimerous (e.g. *B. ampla*). Arrangement of sepals in bud is similar to staminate flowers, with the outer sepals pressing the inner sepals in one plane. This arrangement is found in dimerous pistillate flowers (e.g. *B. incana*: Matthews and Endress, 2004). The ovary is inferior, mostly of three, rarely two or four carpels with parietal placentae. Ovaries have one to three wings as an extension of the locular space. The development of the wing is independent of the placental development. The placentation is frequently described as axile in the literature, which is erroneous. In *B. albo-picta* one of the placentae is more strongly developed than the two others, creating the impression of an axile placentation (Figure 9.61B). A high number of small ovules is produced. Styles are free opposite the carpels, and often have dichotomously branched stigmas that can be extensively twisted and contorted. No stamens are found in pistillate flowers.

The basal monotypic *Hillebrandia sandwicensis* differs from *Begonia* in having pentamerous staminate and pistillate flowers with reduced petals. The gynoecium is isomerous opposite the sepals (Matthews and Endress, 2004). Gauthier and Arros (1963) interpreted the petals as sterile stamens because of their strong similarity.

Begoniaceae share the centripetal stamen development with Datiscaceae (Ronse De Craene and Smets, 1990b). Absence of a gynoecium in staminate flowers may be linked with the apparently unordered development, which is rare in rosids.

Fagales

The order consists of seven families of which Fagaceae, Betulaceae and Juglandaceae are the largest. Flowers are mostly highly reduced and well

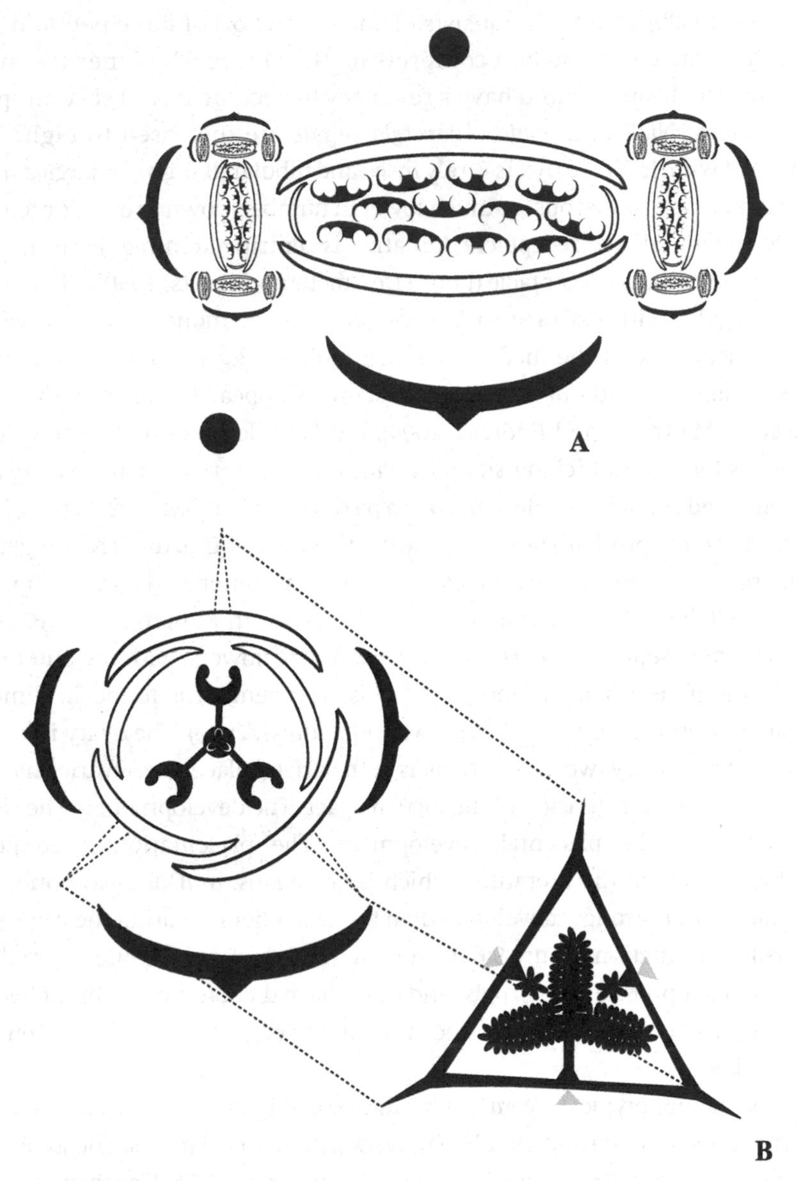

Figure 9.61 *Begonia albo-picta* (Begoniaceae), A. staminate partial inflorescence; B. pistillate flower. Note the presence of abaxial crests on the inferior ovary and apparently axile placentation by reduction of two placentae.

adapted to wind pollination, but in some groups there is a reversal to insect pollination. The character syndrome includes hexamerous unisexual flowers with staminate and pistillate flowers on separate branches or on separate plants, reduction of the perianth and generalized absence of petals, reduction of the number of carpels and ovules (one-two per locule) and increase of size of styles. The order is characterized by intricate dispersal mechanisms involving fusions of bracts with flowers (e.g. Juglandaceae, Betulaceae) and associations of several bracts and flowers into cupules (e.g. Nothofagaceae, Fagaceae) (see e.g. Abbe, 1935, 1938, 1974; Oh and Manos 2008). Endress (1967, 1977) pointed out several similarities between Hamamelidaceae and Fagales but these appear to be adaptive rather than a reflection of affinitiy.

Flowers of Fagales are well represented in the fossil record with a mix of adaptations to insect and wind pollination (e.g. Friis, Pedersen and Schönenberger, 2006; Crepet, 2008). Fagales form the core of the former order Amentiferae, which were thought to be ancestral among flowering plants (e.g. Eichler, 1875). However, it appears that the reduced flowers are adaptive and derived.

Betulaceae

Figure 9.62A–B. *Carpinus betulus* L., based on Endress (2008b) and Abbe (1935)

Staminate: ⟷ K0 A4–6 G0

Pistillate: ⟷ K4–8 C0 $\underline{G}$(2)

General formula: ⟷/✱ K0/1–6 C0 A(1–)4(–6) G2(–3)

Pistillate flowers are arranged in pairs in reduced dichasia without central flower. A central flower is present only in *Betula* (Lin, Zheng and Chen, 2010). Each flower is subtended by a bract and two bracteoles, which are postgenitally fused into a single unit. The calyx is reduced in size and has a variable number of parts; four larger sepals are found in median and transversal position, but with a variable number of smaller parts in between (Endress, 2008b). Sepal lobes are colleter tipped and have strong bundles. Endress (2008b) argued that the variability of the sepals is linked with their reduction and change of function; the increase in parts may be the result of a subdivision of the four major sepals. The two carpels bear a single ovule each.

There is diversity in the number and arrangement of bracts in the family. Only the subtending bracts are formed in pistillate *Betula*, while the flower only has two sepals that become arrested in development. Sepals are absent in *Alnus* (Lin, Zheng and Chen, 2010).

The staminate catkin is built on the grouping of partial dichasial systems of three flowers (Abbe, 1935). There is much variation in the number of stamens of individual flowers and in the extent of development of the perianth. Staminate flowers lack a perianth in *Carpinus* as well as *Corylus*, while two sepals are found in *Betula*, and four in *Alnus*. In *Carpinus* the filament is branched and bears a pair of thecae topped with hairs. In *Corylus* anthers appear monothecal by the strong division of the filament.

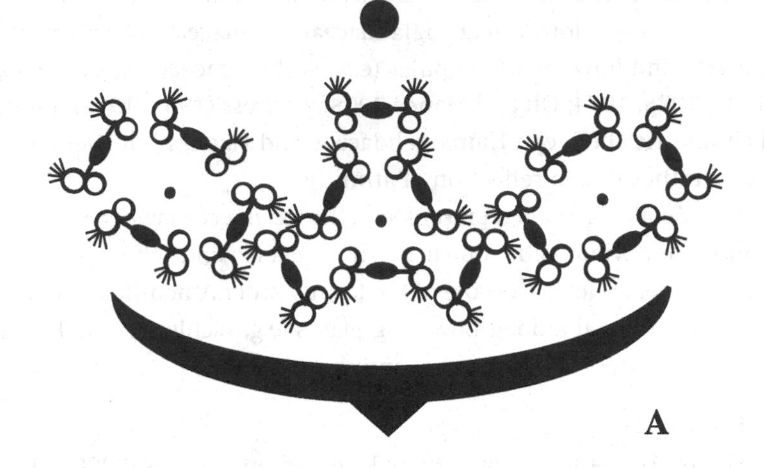

Figure 9.62 *Carpinus betulus* (Betulaceae): staminate (A) and pistillate (B) partial inflorescence

Abbe (1935, 1938) interpreted the ancestral merism of Betulaceae as hexamerous. This relates the merism of Betulaceae to the condition of Fagaceae with a stronger floral reduction in the former.

Fagaceae

Figure 9.63A–B *Castanea sativa* Mill.

Staminate ✶ K3+3 A3^3+3 G(3°)

Pistillate ✶ K3+3 A°3^3+3 Ğ(6)

General formula: ✶ K 6(-9) C0 A(3)6–12(−90) G2–6(−9)

Flowers in Fagaceae are often pentamerous to hexamerous with an arrangement of parts in two whorls. There has been discussion whether flowers of Fagaceae are hexamerous or trimerous (see e.g. Endress, 1977; Okamoto, 1983), but it is reasonable to consider them as hexamerous because the well-nested position of the family within rosids (analogous to the condition in apetalous Polygonaceae). Staminate and pistillate flowers in *Castanea* are formed on separate catkin-like inflorescences or occasionally on the same with intermediate partial inflorescences. Staminate flowers are clustered in dichasia of up to seven to eight flowers, while pistillate inflorescences usually consist of three flowers surrounded by several bracts. Fey and Endress (1983) demonstrated that the four-valved cupules of *Castanea* are derived from complex highly contracted dichasia, consisting of suppressed lateral axes with their subtending bracts. Third-order bracts are fused with a pair of fourth-order bracts to form one of the valves (numbered in Figure 9.63B). Staminate flowers in five genera of Fagaceae, including *Castanea*, usually have twelve stamens and the same number of staminodes in pistillate flowers (cf. Okamoto, 1983). The stamens opposite the outer sepals are arranged as triplets, although Baillon (1876a) argued that the position of the extra stamens is variable. Most other genera have six stamens only, opposite the sepals. In *Quercus* flowers are four- to five-merous with sequential initiation (Sattler, 1973).

There is high diversity in cupule morphology in Fagaceae with considerable discussion on the origin and evolution of cupules as derived from complex dichasial systems (reviewed by Oh and Manos, 2008). There are two major types, the dichasium-cupule, as in *Castanea* or *Fagus*, enclosed by triangular valves, and the flower-cupule, as in *Quercus or Lithocarpus*, which is a single flower surrounded by a valveless cupule. The phylogeny demonstrates at least three independent derivations of valveless cupules from dichasial predecessors. In *Fagus*, the central flower of the dichasium is lost, leaving two lateral flowers enclosed by two bracts each, forming the valves (Fey and Endress, 1983). In *Quercus*, the dichasium is reduced to one terminal flower surrounded by several bracts of sterile flowers, forming scales on the acorn.

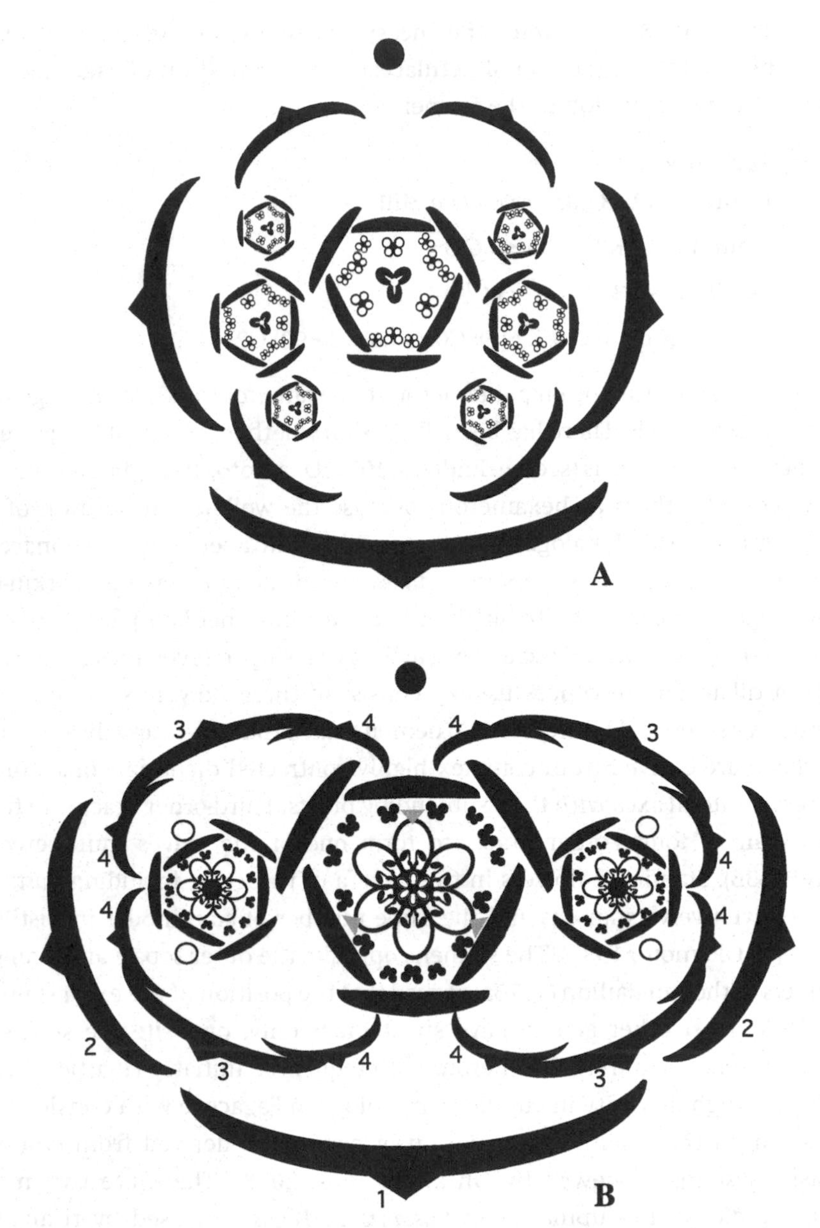

Figure 9.63 *Castanea sativa* (Fagaceae): staminate (A) and pistillate (B) partial inflorescence. Numbers indicate order of initiation of bracts; white dots represent aborted flower buds.

Nothofagaceae (southern beech) have often been included in Fagaceae. They occupy a basal position in Fagales and share a comparable cupule system with *Fagus*.

Juglandaceae (incl. Rhoipteleaceae)

Figure 9.64A–B *Juglans regia* L., based on Lin et al. (2016) and pers. obs.

A. Staminate: ✶ K 4(−5) A12+6+(1−)2 G0

B. Pistillate: ✶ K 4(−5) A0 Ğ(2)

General formula: ✶ K 0–4(−5) C0 A(3−)6–25 G 2(3–4)

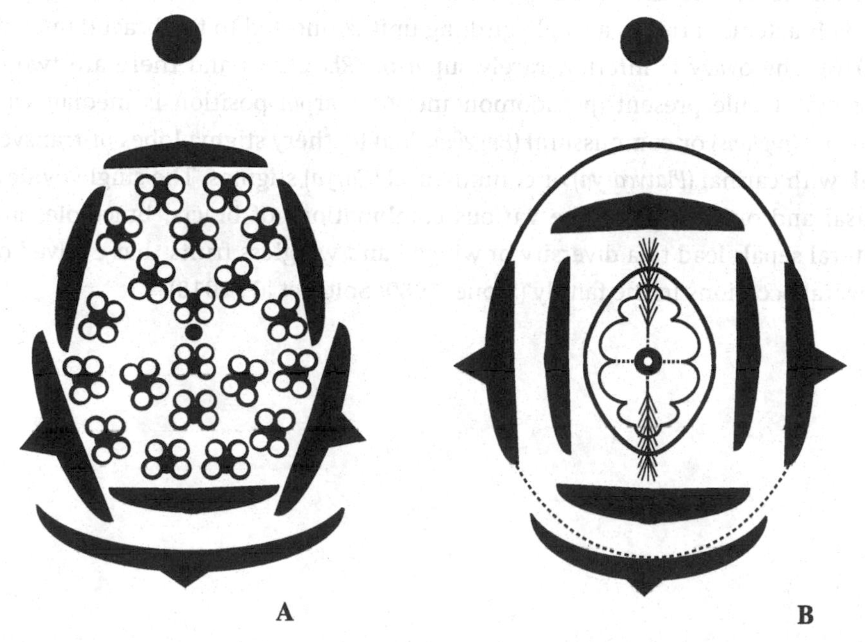

Figure 9.64 *Juglans regia* (Juglandaceae). A. staminate flower; B. pistillate flower

Flowers are unisexual, with male catkins containing numerous flowers separate from female flowers (solitary, in triplets or spikes). The perianth is reduced to small lobes or is missing. The interpretation of the perianth has been controversial as flowers are subtended by a conspicuous bract and bracteoles that appear to be integrated in the perianth. Lin et al. (2016) interpreted the staminate flower as having two trimerous whorls of tepals and a single bract. However, the perianth more likely consists of four (or fewer) sepals and two bracteoles that are closely associated with the flower, supporting earlier interpretations of an association of bracteoles and sepals (e.g. Eichler, 1878; Manning, 1948; Stone, 1989). The association of bracteoles in the flower leads to an elongated receptacle in staminate flowers (Manning, 1948). In *Carya*, *Platycarya* and *Pterocarya* sepals are missing, while bracteoles are also missing in *Platycarya*. The stamen number fluctuates between three and four (*Carya*) or eight to sixteen or more (up to one hundred), arranged in one to three whorls. In

Juglans an outer whorl of twelve stamens is arranged as pairs opposite the sepals, followed by six stamens opposite sepals and bracteoles, and one (two) central stamens (Lin et al., 2016). The flower is often interpreted as primitively polyandrous (e.g. Manning, 1948; Stone, 1989), although this is unlikely as an increased stamen number is a response to wind pollination.

Pistillate flowers consist of two carpels surrounded by a perianth similar to staminate flowers and variously combined with bract and bracteoles. In *Juglans regia* bracteoles arise as a single girdling unit connected to the bract (Lin et al., 2016). The ovary is inferior, rarely superior (*Rhoiptelea*) and there are two or a single locule present (pseudomonomery?). Carpel position is median with carinal (*Juglans*) or commissural (*Engelhardtia*) feathery stigma lobes or transversal, with carinal (*Platycarya*) or commissural (*Carya*) stigmas. The single ovule is basal and orthotropous. The various combinations of bracts, bracteoles and lateral sepals lead to a diversity of winged and wingless fruits that evolved on several occasions in the family (Stone, 1989; Soltis et al., 2018).

10

Caryophyllids

How to Reinvent Lost Petals

The caryophyllid clade or Caryophyllales *sensu lato* contains forty families of diverse size grouped in two major clades (Figure 10.1; APG, 2016; Yao et al., 2019). A natural Caryophyllales ('Centrospermae') has been recognized for a long time, mainly on the basis of embryological and phytochemical characteristics (e.g. Mabry, 1977). Inclusion of molecular characteristics has increased the size of the order dramatically by adding carnivorous plant families (including Droseraceae and Nepenthaceae), knotweeds (Polygonaceae) and some halophytic groups (Frankeniaceae, Tamaricaceae and Plumbaginaceae). Except for carnivorous families, these taxa were often associated with Caryophyllales in the past. Whether these taxa should be grouped with core Caryophyllales as one order, Caryophyllales, or be separated as Polygonales is a matter of preference. Throughout this book I use the second option.

(Core) Caryophyllales have a number of floral features in common that clearly separate the clade from other Pentapetalae. Flowers are basically pentamerous and mostly apetalous with a derived insect pollination syndrome. The ovary regularly has a free-central or basal placentation by break-up of septa. Styles are typically separate as stylodes or there are well-developed style branches on a common trunk. Ovules and seeds are characteristically curved (campylotrous) and the reserve tissue of seeds often contains perisperm. Nectaries are typically situated on the inner side of the stamens or stamen tube. The circumscription of certain families has strongly changed based on the molecular phylogeny (Molluginaceae, Phytolaccaceae, Portulacaceae), resulting in the creation of several new families.

However, the addition of families previously thought to be unrelated alters the context in which to approach the characteristics of the group. Members of Polygonales contain families where petals are present (e.g. Tamaricaceae, Droseraceae) and others without (e.g. Polygonaceae, Nepenthaceae). Another

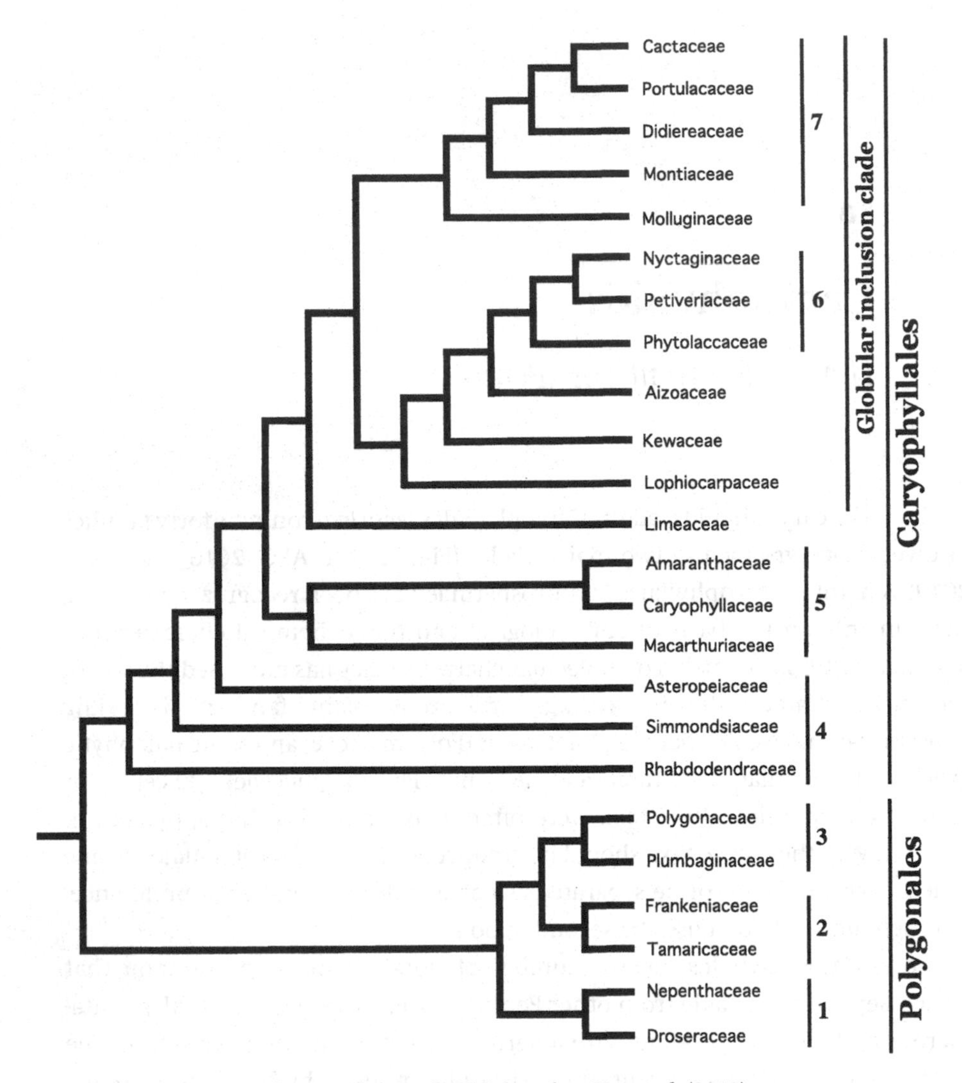

Figure 10.1 Phylogenetic tree of caryophyllids based on Yao et al. (2019). 1. carnivorous clade; 2. tamaricoid clade; 3. polygonoid clade; 4. basal grade; 5. caryophylloid clade; 6. phytolaccoid clade; 7. Portulacineae clade

characteristic, which appears as an apomorphic tendency, is the absence of bracteoles and the lateral position of the outer sepals (Ronse De Craene, 2008). Placentation is mostly parietal, spreading on the ovary floor, or basal.

Caryophyllids emerge from the core eudicot polytomy in a weak sister group relationship with Dilleniales or Berberidopsidales as subsequent sisters to asterids (Soltis et al., 2018). Floral and vegetative morphologies indicate that caryophyllids have become widely adapted to extreme environments with unique floral mechanisms.

10.1 Polygonales

'Carnivorous Clade'

Molecular data have demonstrated that five carnivorous families (Droseraceae, Nepenthaceae, Drosophyllaceae, Dioncophyllaceae and Ancistrocladaceae) are closely related (see e.g. Williams, Albert and Chase, 1994). They share a syndrome of glandular hairs and leaf apices becoming specialized in trapping insects. However, floral morphology is highly variable, with loss of petals in Nepenthaceae. Ancistrocladaceae and Dioncophyllaceae are basically diplostemonous.

Nepenthaceae

Figure 10.2A–B *Nepenthes alata* Blanco (staminate) and *N. hirsuta* Hook. f. (pistillate)

Staminate: ∗ K4 A4+4+1[a] G0

Pistillate: ∗ K4 A0 G̲(4)

General formula (combined): ∗ K (3)4–(5–6) A4–24 G(3)4

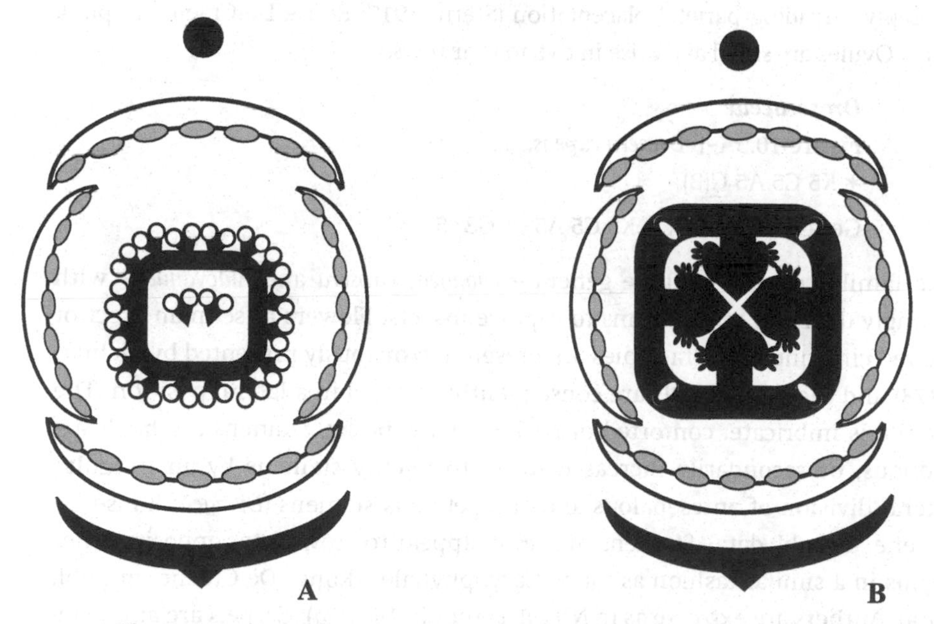

Figure 10.2 Nepenthaceae. A. *Nepenthes alata*: staminate flower; B. *N. hirsuta*: pistillate flower

[a] Apical stamen not present in all flowers.

Inflorescences appear racemose, although flowers can be paired and all flowers mature synchronously, suggesting a derivation from a cymose inflorescence (cf. Macfarlane, 1908). Flowers are unisexual on separate plants, subtended by a single bract, and are mostly tetramerous. The perianth is similar in pistillate and staminate flowers and is interpreted as a calyx with an initiation of two alternating pairs; there are rarely two whorls of three sepals (Eichler, 1878). Sepals bear large glands on the adaxial side (Macfarlane, 1908; Endress, 1994). Stamens range from four to twenty-four, often with a terminal anther and no trace of the gynoecium. Stern (1917) described the development of the stamens as alternating whorls of four. The extrorse stamens are grouped on a common synandrium with an expanding androphore. In cases where only four stamens are present, they alternate with the sepals. With eight stamens a second whorl is inserted more towards the centre of the flower (Ronse De Craene, unpubl. data). A pair or a single stamen may be apically present or a stub is visible in young stages. In older buds all thecae (with two pollen sacs) are arranged equidistantly as anther tissue is embedded in the common column. This arrangement closely resembles staminate flowers of Myristicaceae. Pistillate flowers have four carpels opposite the sepals and broad mostly sessile stigmatic lobes. The placentation is generally described as axile, but is in reality a deeply intruding parietal placentation (Stern, 1917; Ronse De Craene, unpubl. data). Ovules are small and arise in two to four rows.

Droseraceae

Figure 10.3A–B *Drosera capensis* L.

∗ K5 C5 A5 $\underline{G}(3)$

General formula: ∗ K5 C5 A5–∞ G3–5

The family consists of three genera - *Dionaea*, *Drosera* and *Aldovanda* - with strongly different mechanisms to capture insects. Flowers arise in an erect or curved cincinnus. No bracteoles are present (erroneously presented by Eichler, 1878) and the outer sepals are consequently inserted in a lateral position. The corolla is imbricate, contorted or cochleate ascending. Stamens are haplostemonous, or secondarily increased to up to twenty stamens by an irregular lateral division of antesepalous and antepetalous stamens (*Dionaea*: Ronse De Craene, unpubl. data). Stamens in *Drosera* appear to grow faster opposite inner sepals in a similar fashion as many Caryophyllales (Ronse De Craene, unpubl. data). Anthers are extrorse as in *Nepenthes* (not in *Dionaea*). Carpels are antepetalous when isomerous, or most often three with parietal-basal placentation. In *Dionaea*, ovules spread over the ovary floor in five indistinct groups. Styles are short, carinal and variously branched, comparable to Polygonaceae. Nectaries appear to be absent in the family.

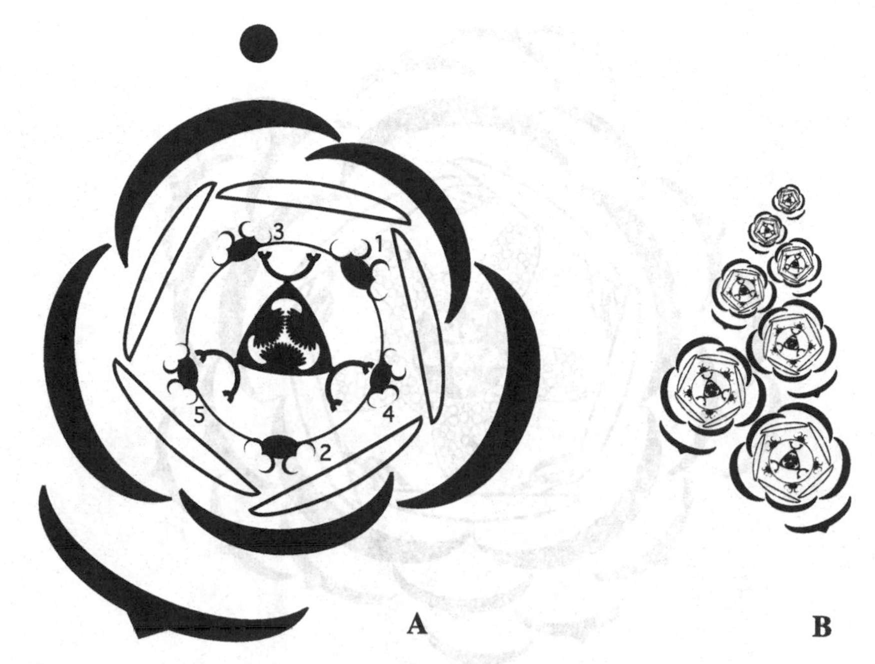

Figure 10.3 *Drosera capensis* (Droseraceae): A. flower, numbers refer to initiation sequence of stamens; B. inflorescence

'Tamaricoid Clade'

Tamaricaceae and Frankeniaceae share several floral characters besides their halophytic adaptations, including fused sepals (long tubes in *Frankenia*), free contorted petals (often with basal adaxial appendages overlapping each other in a direction opposite to the petal margins), a basically diplostemonous androecium with stamen tube, and a superior ovary with parietal placenta.

Tamaricaceae

Figure 10.4 *Reaumuria vermiculata* L., based on Ronse De Craene (1990)

$\ast$ K(5) C5 A(5+5$^{\infty}$) $\underline{\text{G}}$(5)

General formula: $\ast$ K4–5C4–5A(4–)5–10(–∞) G3–5

Inflorescences are racemose or flowers are single and terminal surrounded by several bracts (*Reaumuria*). The latter was interpreted as a derivation from a cymose inflorescence by loss of lateral flowers (Ronse De Craene, 1990). Bracteoles are typically absent and the outer sepals are in lateral position (Eichler, 1878). Petals have a variable aestivation (contorted to quincuncial) and broad lateral appendages are found in *Reaumuria* and *Hololachna*, but not in the other genera (Ronse De Craene, 1990). Stamens are clearly of different

Figure 10.4 *Reaumuria vermiculata* (Tamaricaceae)

length with an obdiplostemonous arrangement (*Myricaria*) or the androecium is often haplostemonous in *Tamarix*. The androecium of *Reaumuria* is polyandrous with a diplostemonous ground plan and a centrifugal proliferation of antepetalous sectors as broad wings covered with stamens (Ronse De Craene, 1990). Stamens are extrorse to introrse and basally fused. The inside of the stamen tube of *Tamarix* develops into a nectary that can develop into various shapes (Zohary and Baum, 1965). Carpels are antepetalous when isomerous; with three carpels two are oriented in an adaxial position (Eichler, 1878). The ovary contains two to several ovules grouped at the base on parietal placentae, which are occasionally intruding but never connected in the middle. The styles are carinal (*Tamarix*, *Reaumuria*) or commissural (*Myricaria*). Tamaricaceae superficially resemble some Malvales (Cistaceae) in the similar gynoecium and contorted aestivation.

Frankeniaceae

Figure 10.5A–B *Frankenia laevis* L.

∗ K(5) C5 A5+1 $\underline{G}$(3)

General formula: ∗K4–5 C4–5 A(3)–6(–24) G3

Inflorescences are terminal dichasia with lateral flowers enclosed by transversal bracteoles. A second set of bracteoles arises within each set and the four

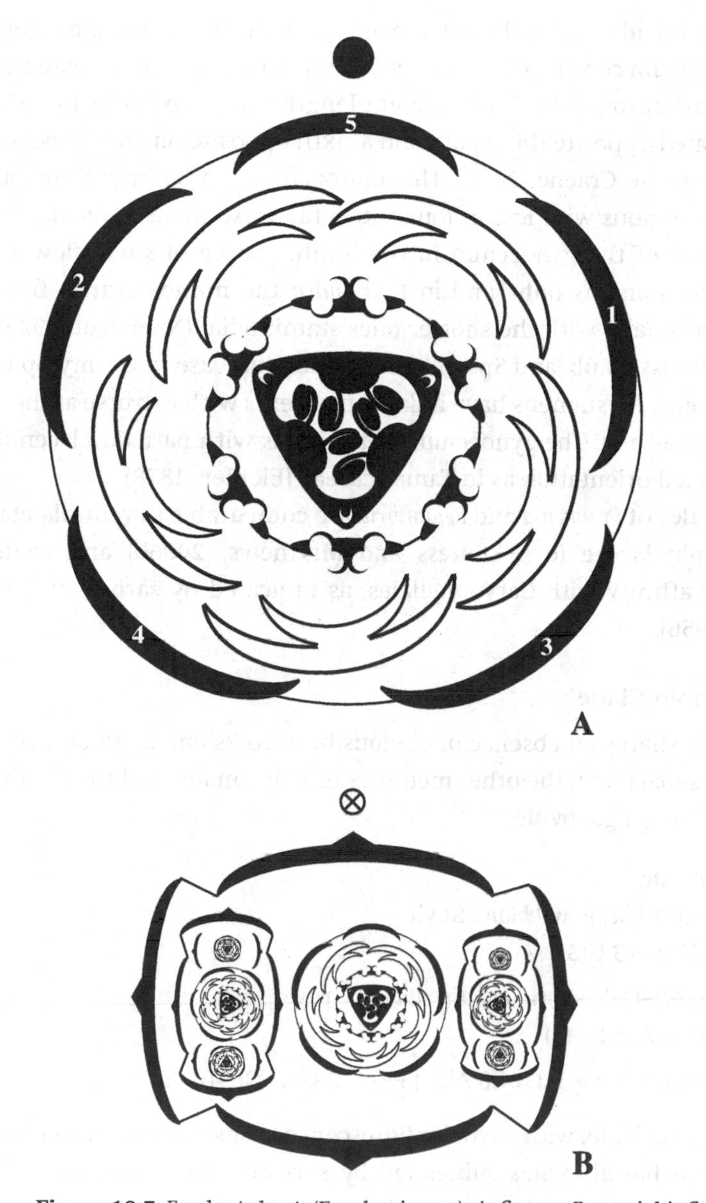

Figure 10.5 *Frankenia laevis* (Frankeniaceae): A. flower; B. partial inflorescence

bracteoles are united at the base. The calyx is tubular and valvate with an unusual initiation of two pairs followed by a single sepal between the first pair (Ronse De Craene, unpubl. data), contrary to the 2/5 sequence reported by Payer (1857) for *F. pulverulenta*. This corresponds with a transition of a decussate arrangement of bracts to a pentamerous calyx. Petals have a contorted aestivation and a broad basal scale-like appendage interlocking neighbouring petals. The androecium consists of six stamens (rarely three or more) apparently arranged in two trimerous whorls of different length. However, in reality five stamens are situated opposite the sepals, and a sixth opposite one of the petals (Payer, 1857; Ronse De Craene, 2018). The androecium can be interpreted as basically diplostemonous with loss of four antepetalous stamens. This may be caused by pressure of the gynoecium in the limited space of small flowers. *F. boissieri* has five stamens only, and in *F. triandra* the number varies from three to six, occasionally with the shorter ones staminodial (Niedenzu, 1925). *Frankenia persica* (Boiss.) Jaub. and Spach has a stamen increase to twenty up to twenty-four stamens. All stamens have inflated filaments with extrorse anthers and are fused into a tube. The gynoecium is trimerous with parietal placentation and an inversed orientation as in Tamaricaceae (Eichler, 1878).

The ventral scales of *Frankenia* and *Reaumuria* are comparable to ventral petal scales in Caryophyllaceae (e.g. Endress and Matthews, 2006b) and could indicate a closer affinity with Caryophyllales, as indicated by earlier authors (e.g. Friedrich, 1956).

'Polygonoid Clade'

The clade shares an absence of obvious bracteoles and a lateral insertion of the outer sepals with the other members of Polygonales. It differs in the basal position of the single ovule.

Polygonaceae

Figure 10.6A *Rheum webbiana* Royle
✶ K3+3 C0 A6+3 $\underline{G}$(3)

Figure 10.6B–C *Persicaria lapathifolia* (L.) Gray
✶/⟷ K5 C0 A5+1 $\underline{G}$(2)

General diagram: ✶ K2–6C0 A(2–)6–9(–∞) G(2–)3(–4)

Inflorescences are variable, with partial inflorescences consisting of cluster-like dichasial or monochasial cymes subtended by saccate bracts (Figure 10.6C, Brandbyge, 1993). The bracteoles are considered to be fused into an ocreola and shield one flower or a group of flowers; free bracteoles are present in *Triplaris* and *Coccoloba* (Eichler, 1878). Larger sheathing bracts may surround groups of flowers, which appear axillary at the nodes. Flowers are basically apetalous with a mostly petaloid calyx. Some clades are wind pollinated with inconspicuous greenish perianth (e.g. *Rumex*, *Emex*), while insect pollination has developed in all major subfamilies. Flowers are hermaphrodite, and occasionally unisexual with aborted anthers and ovules (e.g. *Fallopia japonica*, *Rumex*). Although the perianth of Polygonaceae is always spirally initiated, there is

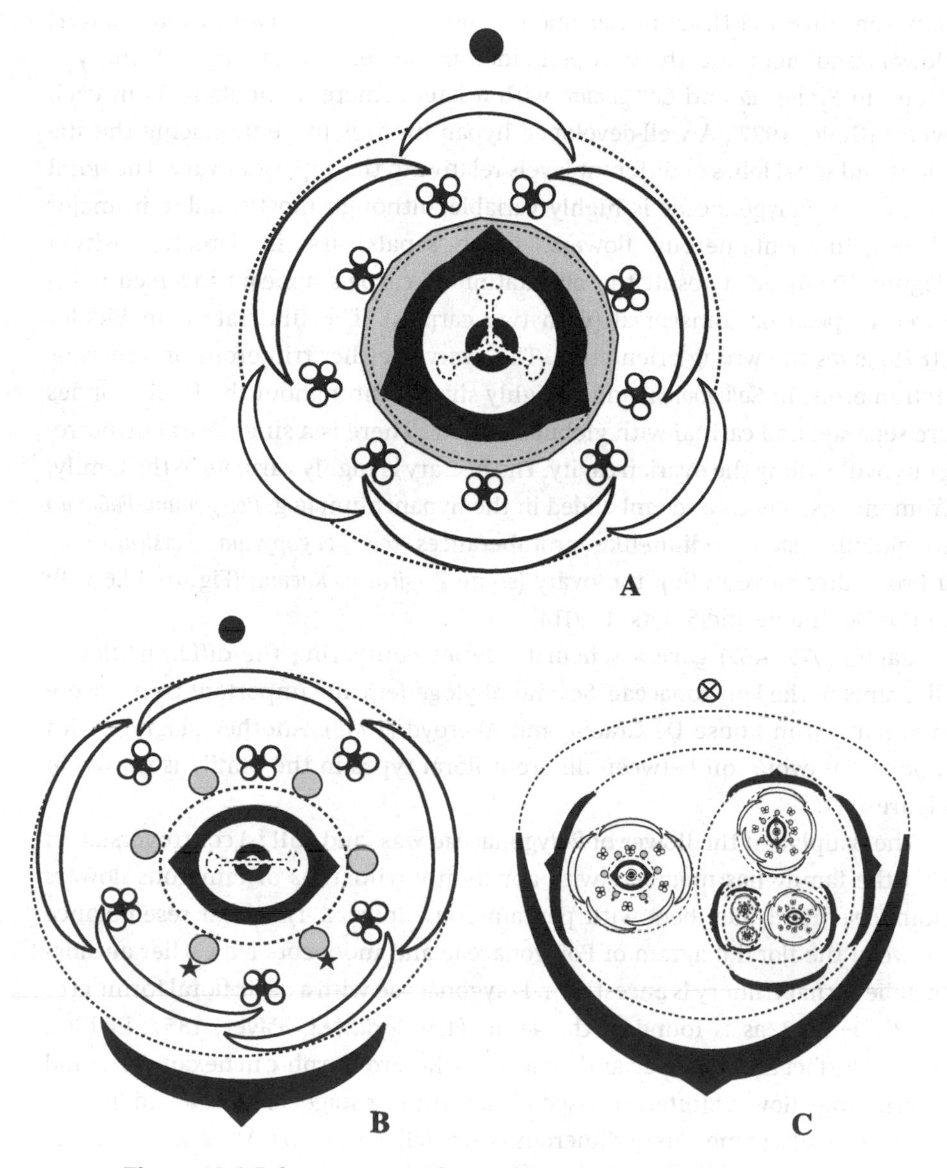

Figure 10.6 Polygonaceae: A. *Rheum webbiana*; B–C. *Persicaria lapathifolia*, flower and partial inflorescence

a tendency for a cyclic arrangement. The outer perianth parts are often well differentiated from the inner, especially in hexamerous and tetramerous flowers at fruiting stage (e.g. *Rumex*, *Triplaris*), or the differentiation is gradual along a quincuncial aestivation with a difference in petaloidy (e.g. *Polygonum*, *Persicaria*: Ronse De Craene and Akeroyd, 1988). Stamen number ranges between nine and three in hexamerous flowers, six and two in tetramerous flowers and eight and three in pentamerous flowers. Secondary polyandry is found in *Symmeria* and *Calligonum* with a lateral increase of stamens in each whorl (Galle, 1977). A well-developed hypanthium is present, placing the stamens and sepal lobes at different levels relative to the superior ovary. The floral diagram of Polygonaceae is highly variable, although mostly stable in major clades. In pentamerous flowers, outer sepals are in lateral position (Figure 10.6B). As a result the orientation of carpels appears inversed (with three carpels) or transversal (with two carpels). The illustration in Eichler (1878) gives the wrong orientation. The ovary is either trimerous or dimerous (tetramerous in *Calligonum*) and is highly similar throughout the family. Styles are separate and carinal with globular stigma. There is a single basal orthotropous ovule filling the ovarian cavity. The nectary is highly variable in the family, from inconspicuous and embedded in the hypanthium (e.g. *Polygonum*, *Fallopia*) to globular staminodium-like protuberances (e.g. *Fagopyrum*, *Persicaria*) or a broad disc surrounding the ovary (some *Persicaria*, *Rheum*) (Figure 10.6A, B; Ronse De Craene and Smets, 1991b).

Galle (1977: 467) gave a schematic series comparing the different flower diagrams of the Polygonaceae. Several phylogenetically important trends were summarized in Ronse De Craene and Akeroyd (1988). Another diagram with a potential evolution between different floral types in the family is shown in Figure 10.7.

The Bauplan of the flower of Polygonaceae was (and still is) controversial, in that the family has members with apparently trimerous or dimerous flowers (Rumiceae) and members with pentamerous flowers. The high resemblance between the floral diagram of Polygonaceae and monocots led earlier authors to believe that trimery is ancestral in Polygonaceae with a basic floral formula of P3+3 A3+3 G3, as is found in the genus *Pterostegia* (e.g. Payer, 1857; Eichler, 1878). The fact that both perianth whorls are heteromorphic in hexamerous and tetramerous flowers (often stressed at the fruiting stage) creates an additional impression of a trimerous or dimerous flower (cf. Figure 1.7). Although ancestral trimery was questioned by some authors (e.g. Bauer, 1922), it was generally accepted by others (e.g. Geitler, 1929; Laubengayer, 1937; Ronse De Craene and Akeroyd, 1988) because of seemingly convincing evidence in the perianth and androecium for a transition of trimerous flowers to pentamerous flowers. In pentamerous flowers the third perianth part is half outer–half inner and often bears combined morphological characters of an outer and inner whorl, even in the anatomy (Ronse De Craene and Akeroyd, 1988). The presence of outer stamen pairs was another argument in favour of basic trimery, linking Polygonaceae with outer stamen pairs found in basal angiosperms. However,

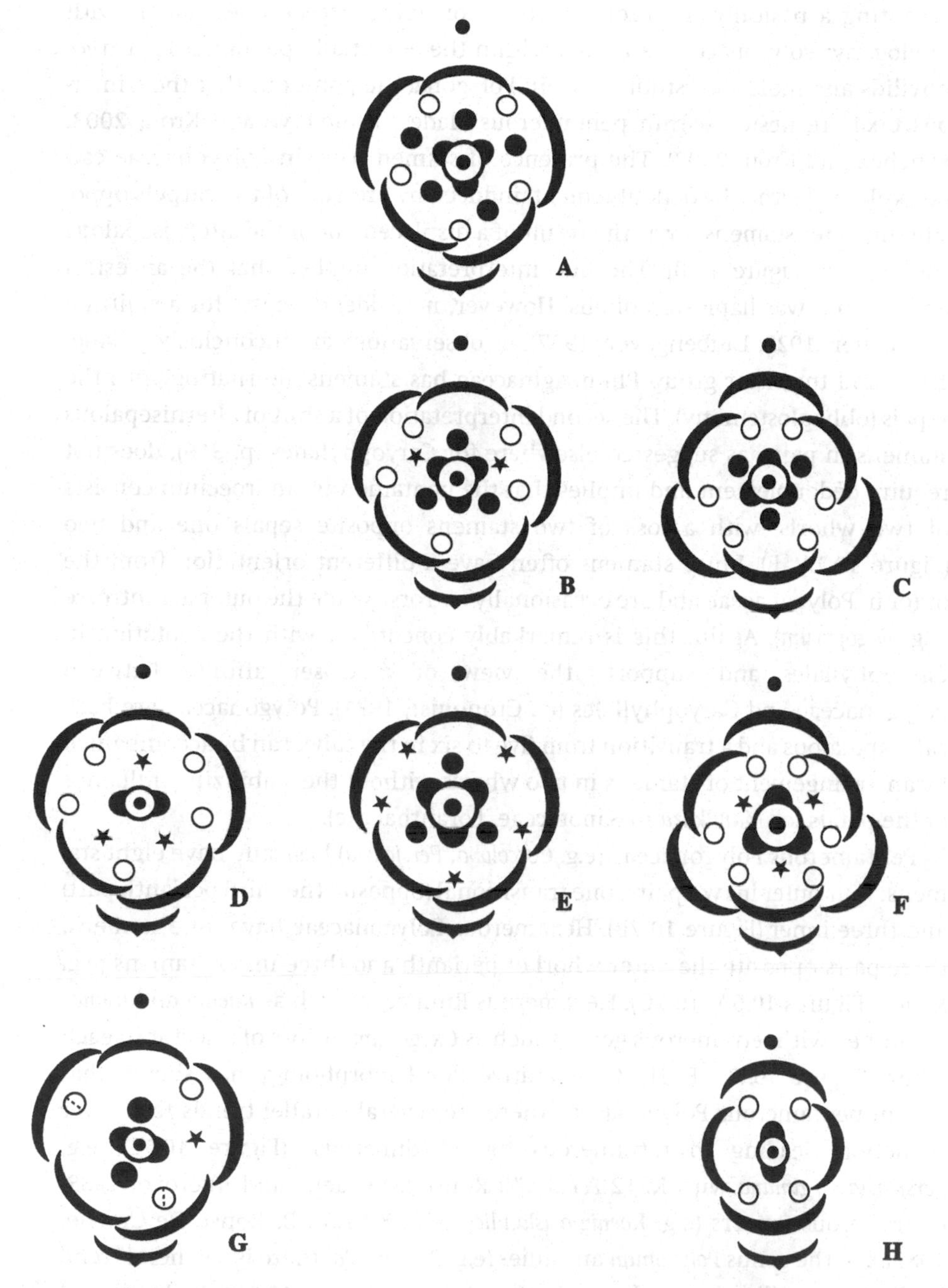

Figure 10.7 Evolution of the flower in Polygonaceae. A. ancestral pentamerous flower of Polygonaceae; B. pentamerous flower with eight stamens; C. hexamerous flower of *Rheum* with nine stamens; D. *Persicaria sp.* with five stamens and two carpels; E. *Polygonum sp.* with three stamens; F. *Rumex sp.* with six stamens; G. *Polygonum molliaeforme* with three stamens – two stamens derived by a pairwise fusion of two pairs; H. tetramerous flower of *Oxyria digyna*. White dots: outer alternisepalous stamens; black dots: inner antesepalous stamens; asterisks: lost stamens

accepting a basically trimerous Bauplan for Polygonaceae does not fit with phylogeny: Polygonaceae is nested within the essentially pentamerous caryophyllids and molecular studies within Polygonaceae point out that the trimerous taxa are nested within pentamerous clades (Lamb-Frye and Kron, 2003; Sanchez and Kron, 2008). The presence of stamen pairs in Polygonaceae can be explained either by dédoublement, induced by pressure of the carpels opposite the outer stamens, or as the result of a displacement of the alternisepalous stamens (cf. Figure 1.3B). The first interpretation implies that the ancestral androecium was haplostemonous. However, no evidence exists for a splitting (e.g. Bauer, 1922; Laubengayer, 1937), or observations are inconclusive (Galle, 1977) and the sister group Plumbaginaceae has stamens alternating with the sepals (obhaplostemony). The second interpretation of a shift of alternisepalous stamens in pairs, as suggested elsewhere for Caryophyllales (p. 316), does not require dédoublement and implies that the pentamerous androecium consists of two whorls, with a loss of two stamens opposite sepals one and two (Figure 10.7A, B). Inner stamens often have a different orientation from the outer in Polygonaceae and are occasionally extrorse while the outer are introrse (e.g. *Fagopyrum*). Again, this is remarkably concordant with the evolution in Caryophyllales and supports the view of a closer affinity between Polygonaceae and Caryophyllales (cf. Cronquist, 1981). Polygonaceae are basically apetalous and a transition from five to six in the calyx can be accompanied by an arrangement of stamens in two whorls without the stabilizing influence of the petals (cf. *Manilkara* in Sapotaceae, Loranthaceae).

Pentamerous Polygonaceae (e.g. *Coccoloba*, *Persicaria*) basically have eight stamens: four outer in two pairs, one transitional (opposite the third perianth part) and three inner (Figure 10.7B). Hexamerous Polygonaceae have nine stamens, three pairs opposite the outer whorl of perianth and three inner stamens (e.g. *Rheum*: Figures 10.6A, 10.7C). Hexamerous Rumiceae, such as *Rheum* and *Rumex* are linked with tetramerous genera such as *Oxygonum* by loss of a sector in each whorl (Figure 10.7C, F, H). Comparative floral morphology has shown that within pentamerous Polygonaceae there are several parallel trends for flower reduction, leading to tetramerous (pseudo-dimerous) (Figure 10.7D; e.g. *Persicaria virginiana* with K2+2 A4+1 G2: Ronse De Craene and Akeroyd, 1988) or trimerous flowers (e.g. *Koenigia islandica* with K3 A3 G3: Ronse De Craene, 1989a). In the genus *Polygonum* and allies (e.g. *Polygonella*) there is a general trend leading to sterilization or loss of outer stamens (Figure 10.7E). Reduction of stamens is often correlated with a reduction of the ovary to two carpels. (Figure 10.7D; cf. Bauer, 1922; Ronse De Craene and Akeroyd, 1988; Ronse De Craene, Hong and Smets, 2004). The outer alternisepalous stamens may be replaced by single stamens, as in *Polygonum molliaeforme* (Figure 10.7G; Ronse De Craene, Hong and Smets, 2004).

In summary, changes in merism are confusing in the Polygonaceae because of the small, reduced flowers with easy transitions to an increase (e.g. *Calligonum*) or a reduction (*Oxyria, Koenigia*), and the confusion of trimery with hexamery and of dimery with tetramery.

Plumbaginaceae

Figure 10.8 *Ceratostigma minus* Stapf ex Prain

∗ K(5) [C(5) A(5)] G̲(5)

Figure 10.8 *Ceratostigma minus* (Plumbaginaceae): partial inflorescence

Partial inflorescences develop as monochasial units. Two bracteoles are usually present, except in *Armeria*.

De Laet et al. (1995) studied the floral development of a selection of species. They found that five common stamen-petal primordia alternate with the sepals in all species. Common zonal growth leads to a stamen-petal tube, except in *Plumbago*, where petals and stamens are separate. The ovary bears a single basal anatropous ovule and five stigmatic lobes facing the sepals. At anthesis the ovule becomes inversely orientated towards the top of the ovary by coiling of the funiculus and becomes connected with a stylar obturator. Nectary tissue is closely associated with the androecium and is confined to the inner area of the stamen tube.

Plumbaginaceae share more morphological characters with some Caryophyllales than with its sister group Polygonaceae, including the single basal anatropous ovule. Friedrich (1956) points to the similarity between *Aegilitis* and *Stegnosperma* (Stegnospermaceae) in the stamen tube and attachment of petal lobes on the abaxial side of it.

10.2 Caryophyllales

Caryophyllales consist of a basal grade and two sister clades, a 'caryophylloid clade' and the 'globular inclusion clade' (Figure 10.1; Brockington et al., 2009, 2015; Yao et al., 2019). Floral diagrams of the Caryophyllales have the following aspects in common: they are basically pentamerous (rarely tetramerous) with a well-developed imbricate calyx, often without petals and with superior ovary, and nectaries are associated with the stamen bases. In many families of the order the perianth is basically derived from sepals and different degrees of petaloidy have evolved independently in different clades. The five sepals arise in a 2/5 sequence and have an imbricate aestivation (except for Nyctaginaceae). Several Caryophyllales have a petaloid perianth with pronounced sepal characteristics very similar to the condition in Polygonaceae (viz. sepals hooded or with a dorsal appendage, green on the outside and petaloid on the inside, divergence between margins and central vein: Ronse De Craene and Akeroyd, 1988; Brockington et al., 2013; Ronse De Craene, 2013). Especially the outer sepals are characteristically hooded or bear a dorsal crest, with petaloidy developing on the margins of the sepals, leaving the midrib green. Mixed-green sepals characterize basal clades such as Macarthuriaceae, Caryophyllaceae and Amaranthaceae; in more derived clades sepals are fully petaloid (Portulacineae clade, phytolaccoid clade), or petaloid appendages develop from staminodes on several occasions (e.g. Aizoaceae, Lophiocarpaceae p.p.; for a discussion see Ronse De Craene, Smets and Vanvinckenroye, 1998; Ronse De Craene, 2013).

Petaloids[b] have arisen several times in the Caryophyllales (Ronse De Craene, 2008, 2013; Brockington et al., 2009). There has been much uncertainty about the homology of petals in Caryophyllales, which are hard to define and are 'true petals', transformed 'sepals' or derived from stamens (See Ronse De Craene, 2013). I initially interpreted Caryophyllales as having lost their petals after the divergence of Rhabdodendraceae (Ronse De Craene, 2013), as several basal families (e.g. Simmondsiaceae, Microteaceae, Physenaceae) are apetalous

[b] I prefer not to use the term 'petals' in (core) Caryophyllales because of their variable homology.

(cf. Brockington et al., 2009). Based on floral morphological and developmental evidence in Caryophyllaceae (discussed in Ronse De Craene, Vanvinckenroye and Smets, 1998), the general assumption is that petaloids have been reinvented from a staminodial whorl. This interpretation was based on the strong association of petaloids with the androecium (arising from the same primordium, often as part of the same complex primordium, or fusion with a staminal tube: Ronse De Craene, Vanvinckenroye and Smets, 1998).

However, more recent studies have convinced me that there is another explanation for the strong resemblance of petaloids with staminodes in Caryophyllaceae, and put the early loss of petals in the Carophyllales in doubt. In Caryophyllaceae petaloids are delayed in their initiation and increasingly undergo the influence of the stamens, as to become absorbed in staminal tissue from early stages. This shift represents a case of heterochrony, as the more precocious initiation of stamens strongly affects the petaloids which show a tendency for reduction (see Ronse De Craene and Wei, 2019; Wei and Ronse De Craene, 2019, 2020). Petaloids are variously present or absent throughout the basal grade of Caryophyllales, but only become completely lost at the separation of the 'globular inclusion clade'. It is clear that within Portulacineae and the raphide clade, petaloids have been repeatedly reinvented, either as coloured sepals (e.g. phytolaccoid clade, Portulacineae) or as staminodes as part of complex androecia (e.g. Aizoaceae, Lophiocarpaceae).

The androecium of Caryophyllales represents the greatest diversity among angiosperms next to Malvaceae, ranging from a single stamen to very high numbers arising in a centrifugal sequence (Ronse De Craene, 2013). Trying to understand the ancestral androecium of Caryophyllales and its evolution has been challenging because of the apparent morphological difference from other Pentapetalae (see Hofmann, 1993; Ronse De Craene and Smets, 1993, 1994; Ronse De Craene, Vanvinckenroye and Smets, 1997; Ronse De Craene, Volgin and Smets, 1999; Ronse De Craene, Smets and Vanvinckenroye, 1998; Ronse De Craene, 2013).

The basic androecial configuration as widespread in Pentapetalae is two whorled with stamens in the antesepalous and antepetalous sectors. Earlier authors interpreted the variation of the androecium as a derivation of a classical diplostemonous stamen arrangement, and floral diagrams were misused in abundance to support this view (e.g. Walter, 1906; Lüders, 1907; Franz, 1908; Friedrich, 1956). However, the pattern of stamen initiation throughout the Caryophyllales is different from diplostemony by the more precocious initiation of either the antesepalous or alternisepalous stamen whorl, the inversed spiral development of the antesepalous stamens and the centrifugal initiation of the second whorl (see Ronse De Craene, 2013; Wei and Ronse De Craene, 2019, 2020). This led Ronse De Craene, Smets and Vanvinckenroye (1998) to describe the androecium of Caryophyllaceae as a different type called *pseudodiplostemony*.

The androecial configuration in Caryophyllales appears unique by a combination of the following factors (see Ronse De Craene, 2013):

(1) Alternisepalous stamens become frequently shifted in pairs opposite sepals one, two and three, and more rarely opposite sepals four and five, similar to Polygonaceae (e.g. Phytolaccaceae). Ronse De Craene, Vanvinckenroye and Smets (1997) and Ronse De Craene, Smets and Vanvinckenroye (1998) interpreted these pairs as an ancestral condition originating with the derivation of pentamerous flowers from trimerous flowers. However, in the context of the relationships of Caryophyllales this seems unlikely. Alternatively, a more likely suggestion for the occurrence of pairs is a shift of the alternisepalous stamens towards the middle of the sepal (Figure 1.3B; cf. Ronse De Craene, 2013).

(2) The androecium has an unusual development with a very rapid initiation linked with an inversion in the direction of development of the antesepalous stamens. This pattern of development is widespread in Caryophyllales, but also in several Polygonales (Ronse De Craene, unpubl. data). The antesepalous stamens grow fastest opposite sepals four, five, and three, followed by those opposite sepals two and one. This sequence is inversely correlated with reductions and losses of stamens, as the more strongly developing stamens are the last to be lost when the androecium becomes reduced (Caryophyllaceae, Macarthuriaceae: Ronse De Craene, Smets and Vanvinckenroye, 1998; Ronse De Craene, 2013; Ronse De Craene and Wei, 2019).

(3) This unusual development is correlated with pressures of the calyx and gynoecium, affecting the fate of stamens in a confined space (Ronse De Craene and Wei, 2019; Wei and Ronse De Craene, 2020). The arrangement and number of carpels highly influence the number and position of antesepalous stamens, which tend to alternate with the carpels. The more common stamen losses in the antesepalous whorl are always those opposite the outer sepals, facing the carpels, and often leading to eight stamens (e.g. Caryophyllaceae, *Limeum*, *Macarthuria*: Hofmann, 1993; Brockington et al., 2013; Ronse De Craene, 2013; Ronse De Craene and Wei, 2019).

(4) Either the antesepalous or the alternisepalous whorl emerges first, and additional stamens always arise centrifugally, occasionally by a secondary increase linked with the expansion of the receptacle.

In all cases with more than ten stamens, more stamens develop centrifugally and the outer stamens may be sterile or petaloid. Some taxa have a unique combination of three whorls of stamens arising in a centrifugal sequence (5+5+5). This arrangement is found in *Kewa salsoloides* (Figure 10.14) and some Petiveriaceae (Figure 10.18A), but also in Caryophyllaceae where the petals often develop in a similar way.

Figure 10.9 shows the main evolutionary trends in the androecium of Caryophyllales (cf. Ronse De Craene, 2013), starting with a pentacyclic isomerous ancestral flower (Figure 10.9A; cf. Wei and Ronse De Craene, 2020). Two main shifts have affected the relative position of the stamen whorls:

(1) Antesepalous stamens have been reinforced in the basal grade and Caryophylloid clade, leading to a reduction and loss of alternisepalous stamens (and petals) (Figure 10.9C; Ronse De Craene, 2013).
(2) The alternisepalous stamens have been reinforced in the 'globular inclusion clade', arising before and more centrally to antesepalous stamens. This has led to a predominance of obhaplostemony in Aizoaceae and Nyctaginaceae (Figure 10.9G). A more pronounced development of the alternisepalous sectors leads to an external shift of antesepalous stamens (Figure 10.9D) and their eventual loss (Figure 10.9F, G). A secondary stamen increase can arise at any level and affect one or two whorls (Figure 10.9E). The reduction of carpels has an obvious influence on the stamen number as stamens are preferentially lost in the radius of carpels (Figure 10.9B, G).

Flowers with eight stamens are often found in Phytolaccaceae, Nyctaginaceae (e.g. *Bougainvillea*), Polygonaceae (p. 310), Molluginaceae, Limeaceae and Macarthuriaceae, and occasionally in Caryophyllaceae.

The ancestral gynoecium consists of five carpels, generally in an antesepalous position, although occasionally shifting to an antepetalous position and this has been demonstrated to be linked with subtle shifts during the development of sepals and stamens (see Wei and Ronse De Craene, 2020). The number is often reduced to four (in tetramerous flowers), three, two or a single carpel (see Ronse De Craene, 2021).

The ovary is basically septate with axile or basal placentation, but the septa become reduced or dissolve in several clades, leaving a free-central placentation (e.g. Caryophyllaceae, Montiaceae) or a basal placentation (e.g. Amaranthaceae). By expansion of the receptacle, a parietal placentation is formed (e.g.

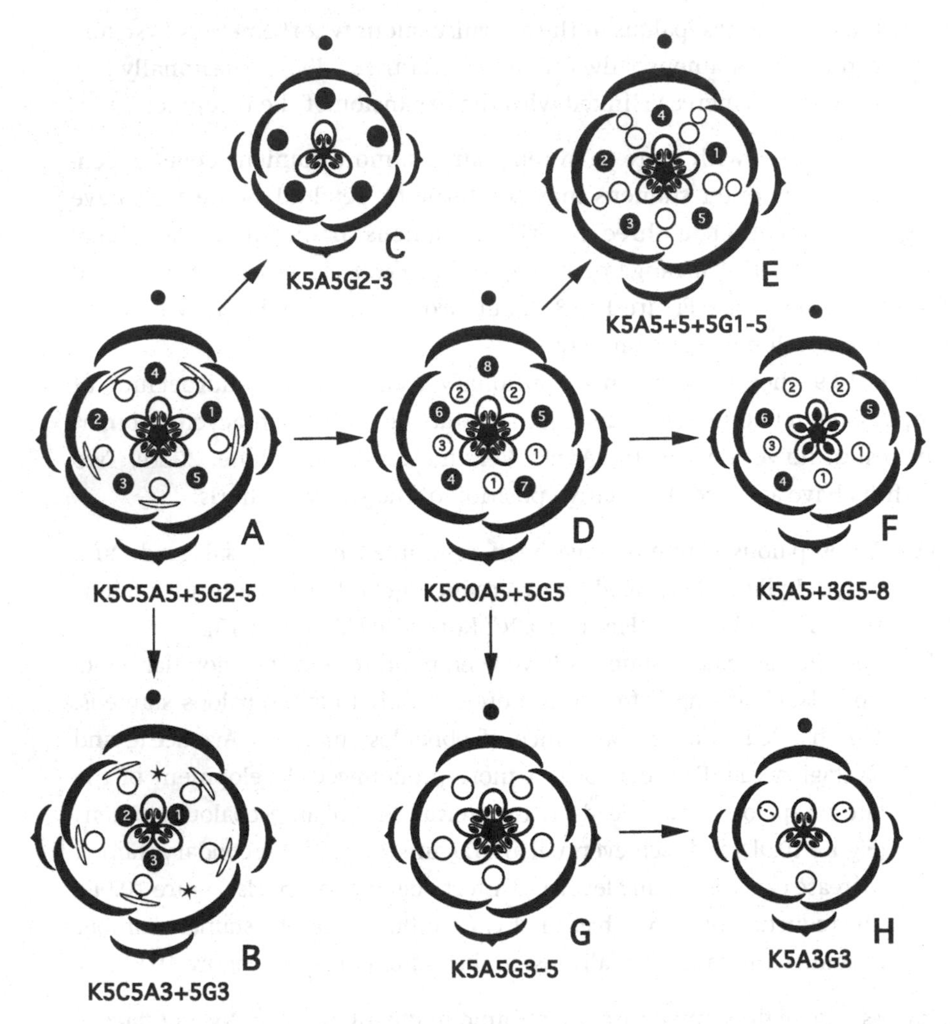

Figure 10.9 Schematic presentation of the main steps of the floral evolution of (core) Caryophyllales (fluctuation of carpel numbers shown). A. ancestral condition with petal whorl closely associated with the androecium; B. reduction of carpel number and loss of two stamens; C. reduction of alternisepalous sector leading to haplostemony; D. inward shift of alternisepalous stamens and delayed initiation of antesepalous stamens; E. Radial division of upper alternisepalous stamens; F. pairwise shift of alternisepalous stamens opposite sepals and loss of two antesepalous stamens; G. loss of antesepalous stamens leading to an obhaplostemonous flower; H. pairwise shift of two stamen pairs in alternation with carpels

Portulacaceae, Cactaceae). As reviewed by Ronse De Craene (2021), the diversity of gynoecia is linked with differential growth rates between the carpel wall and floral apex. A secondary carpel increase occurred in some *Phytolacca* (Phytolaccaceae), *Pleuropetalum* (Amaranthaceae) and Cactaceae, with several carpels arising in a single whorl. This higher number is associated with an increase in the diameter of the flower and the number of stamens. A reduction to a single carpel is characteristic of Petiveriaceae and Nyctaginaceae.

Basal Grade

The basal grade consists of six monotypic families (Yao et al., 2019), representing a mixture of reduced apetalous wind-pollinated (Simmondsiaceae, Microteaceae, Physenaceae) and insect-pollinated petal-bearing flowers (Rhabdodendraceae, Asteropeiaceae, Stegnospermaceae).

Simmondsiaceae

Figure 10.10A–B *Simmondsia chinensis* C. K. Schneid.

Staminate: ✶ K5 C0 A5+3[c] G0

Pistillate: ✶ K5 C0 A0 G̲(3)

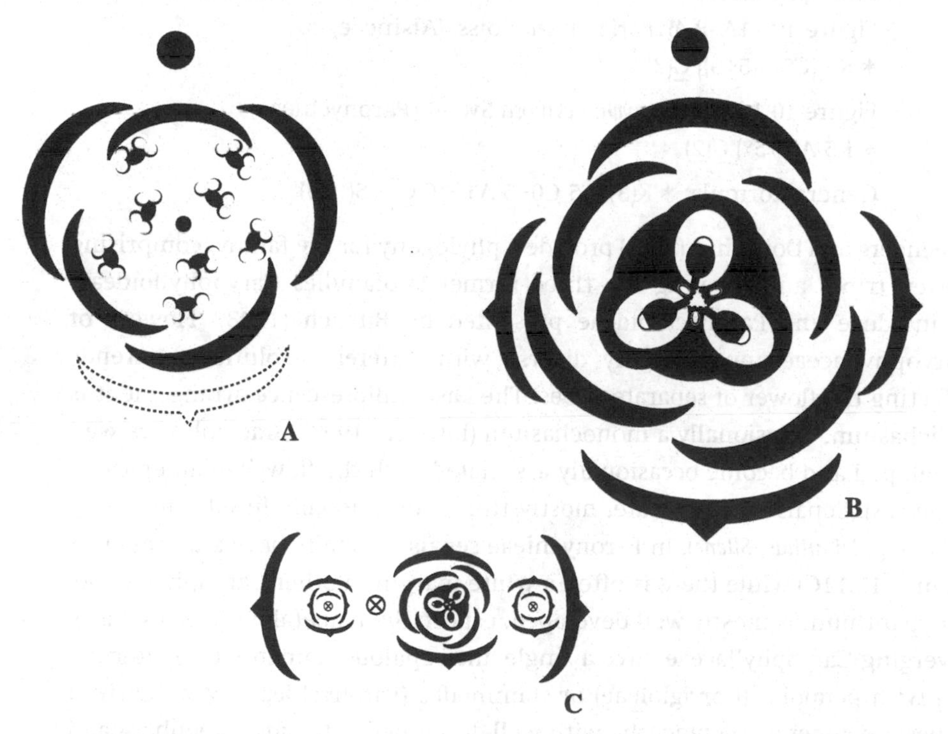

Figure 10.10 *Simmondsia chinensis* (Simmondsiaceae): staminate (A) and pistillate (B) flower; C. arrangement of pistillate flower on the main stem

[c] The number of inner stamens is variable, ranging from one to five. Köhler (2003) gives a range of eight to sixteen.

Simmondsiaceae occupy a basal position in Caryophyllales, immediately behind Rhabdodendraceae with which they share wind-pollinated flowers (Figure 10.1; Brockington et al., 2009). Flowers are (tetra-) penta- to (hexa-)merous. *Simmondsia* is dioecious with staminate inflorescences grouped in axillary clusters. Staminate flowers lack bracteoles and outer sepals are lateral. The extrorse stamens are arranged in two distinct girdles; the outer is complete and alternisepalous while the inner is generally reduced in number (Ronse De Craene, unpubl. data). Pistillate flowers are solitary and pendent. Two bracteoles are closely connected with the strongly imbricate calyx and outer sepals are arranged in median position. The three carpels bear an apical-axile placenta with one (two) ovules and three reflexed styles (Köhler, 2003). Either one or two ovules are formed on a placenta filling the locular cavity (Ronse De Craene, 2021). A nectary is absent.

Caryophylloid Clade

Caryophyllaceae

Figure 10.11A–B *Arenaria tmolea* Boiss. (Alsineae)
✱ K5 [C5 A(5+5)] $\underline{G}$(3)

Figure 10.11C *Chaetonychia cymosa* Sweet (Paronychieae)
✱ K5 A(5+5°) $\underline{G}$(2)

General formula: ✱ K(3)4–5 C0–5 A1–10 $\underline{G}$2–5(–10)

Greenberg and Donoghue (2011) provide a phylogeny for the family, comprising eleven tribes and covering the three former subfamilies Caryophylloideae, Alsinoideae and Paronychioideae presented by Bittrich (1993). Flowers of Caryophyllaceae appear highly diverse with different evolutionary trends affecting the flower of separate tribes. The basic inflorescence arrangement is a dichasium, occasionally a monochasium (Bittrich, 1993). Bracteoles are well developed and become occasionally associated with the flower in an epicalyx (*Dianthus*). Sepals are imbricate, mostly free or occasionally fused into a long tube (e.g. *Dianthus*, *Silene*). In Paronychieae sepals generally have a dorsal crest (Figure 10.11C) while there is often a white margin in other Caryophyllaceae. A hypanthium is mostly well developed, especially in apetalous species. Early diverging Caryophyllaceae have a single antesepalous stamen whorl alternating with petaloids (Corrigioleae) or staminodes (Paronychieae). More derived tribes are generally pentacyclic with well-developed petaloids. Greenberg and Donoghue (2011) suggested that stamens were increased in number up to ten because these are only found in derived groups, although this is highly unlikely and structurally impossible. In Caryophylleae (e.g. *Dianthus*) petals, stamens and carpels are lifted on an anthophore separate from the calyx (Figure 1.5D).

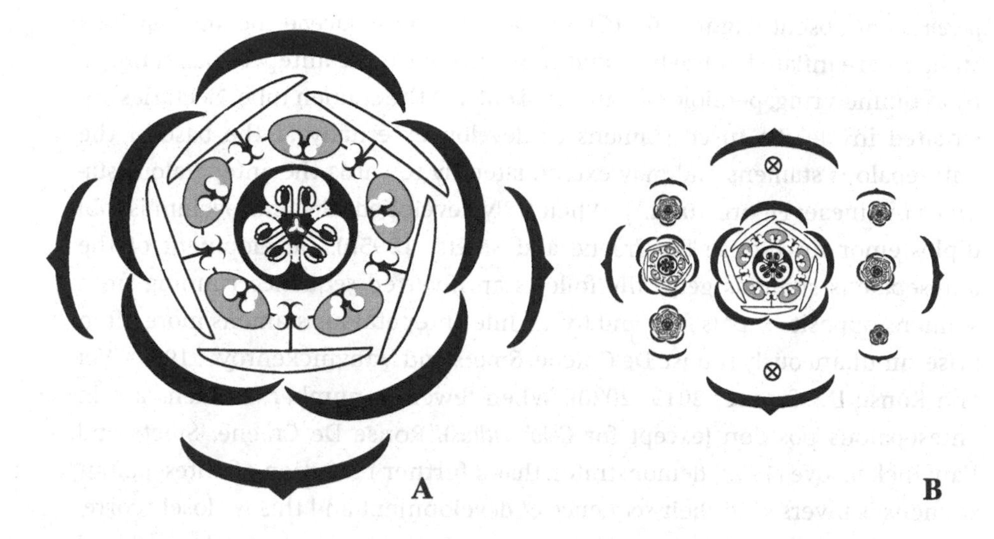

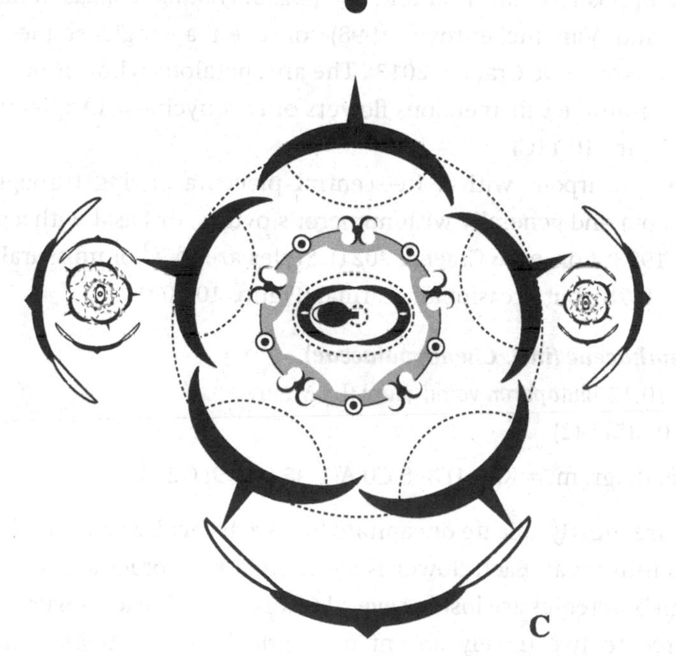

Figure 10.11 Caryophyllaceae: *Arenaria tmolea*, A. flower; B. partial inflorescence; C. *Chaetonychia cymosa*: partial inflorescence. Note the large stipules on bracts and depressions opposite the sepals.

Petaloids vary much in shape; they are generally clawed with a bifid apex, and there is often an appendage at the top of the claw (Wei and Ronse De Craene, 2019). Petaloids are occasionally reduced, replaced by stamens (*Scleranthus perennis*) or absent (Figure 10.11C). In most Caryophyllaceae the antesepalous stamens are inflated at the base and connected with the antepetalous stamens by a common ring; petaloids are inserted outside the stamen tube. Nectaries are situated inside the fused stamens or develop externally at the base of the antesepalous stamens and may extend laterally towards the antepetalous stamens (Alsineae: Figure 10.11A). When fully developed, the androecium is (ob) diplostemonous (Ronse De Craene and Smets, 1995b). Development of the antesepalous stamens generally follows an inverted sequence running from stamens opposite sepals four and five, while antepetalous stamens more often arise simultaneously (Ronse De Craene, Smets and Vanvinckenroye, 1998; Wei and Ronse De Craene, 2019, 2020). When fewer in number, stamens are in antesepalous position (except for *Colobanthus*). Ronse De Craene, Smets and Vanvinckenroye (1998) demonstrated that a further reduction of antesepalous stamens is inversed to their sequence of development and this is closely correlated with the carpel numbers. The number of stamens can be further reduced to two, situated opposite sepals four and five (e.g. *Scleranthus annuus*: Ronse De Craene, Smets and Vanvinckenroye, 1998), or even a single stamen (e.g. *Scleranthus brockei*: Ronse De Craene, 2013). The antepetalous whorl is occasionally present as staminodes in apetalous flowers of Paronychieae (Appleton and Schenk, 2021; Figure 10.11C).

The ovary is syncarpous with a free-central placenta arising through the breakdown of septa and generally with numerous ovules, or basal with a single ovule (Bittrich, 1993; Ronse De Craene, 2021). Styles are in a commissural position (cf. Eichler, 1878), but occasionally carinal (Figure 10.10A).

Amaranthaceae (incl. Chenopodiaceae)

Figure 10.12 *Blutaparon vermiculare* (L.) Mears

∗ K5 C0 A(5) $\underline{G}$(2)

General diagram: ∗ K(0–1) 3–5 C0 A(1–)5 (–8–9) G2–3 (–6)

Inflorescences are mostly spicate or capitate but are basically cymose. Flowers are bisexual to unisexual. Each flower is subtended by a bract and two bracteoles, although bracteoles are lost in several Chenopodioideae. Sepals (tepals) are mostly three to five (rarely absent or single in some *Amaranthus* and *Salicornia*), with imbricate aestivation, free or basally connate, with the two inner occasionally shorter or absent. Sepals can be petaloid or are greenish and are often persistent and hardened in fruit. In female flowers of *Spinacia* and *Atriplex* two sepal lobes are accrescent and resemble bracteoles (Flores-Olvera et al., 2011). Stamens are equal in number as the sepals and opposite to them (rarely fewer). *Pleuropetalum* is exceptional in having a higher number of

Figure 10.12 *Blutaparon vermiculare* (Amaranthaceae)

stamens and carpels by the expansion of the floral meristem (Ronse De Craene, Volgin and Smets, 1999; Ronse De Craene, 2021). Stamens are free (in Chenopodioideae) or more commonly fused into a tube. Interstaminal appendages (pseudostaminodes: Eliasson, 1988) are frequently present on the tube and have various forms and at least two independent *de novo* origins (Sánchez-del Pino et al., 2019). The presence of these appendages is inversely correlated with broad filaments (Eliasson, 1988). The ovary consists of two or three carpels and is unilocular with basal placentation, highly similar to some Caryophyllaceae or Plumbaginaceae. A single ascending ovule is generally present, although the number is higher in *Celosia* and *Pleuropetalum*. In *Pleuropetalum* several ovules develop centrifugally on a gynoecial dome and are interpreted as the result of a secondary multiplication (Ronse De Craene, Volgin and Smets, 1999). The development of the ovules is decoupled from the ovary wall, which develops stylar tissue, although evidence of a basal connection is found in several species (Vrijdaghs, Flores-Olvera and Smets, 2014; Sánchez-del Pino et al., 2019; Ronse De Craene, 2021). Floral nectaries are formed as a ring at the inner base of the filaments or extending between the filaments. Nectaries are absent in several unisexual flowers and there is a tendency for reduction linked with wind pollination (*Chenopodium*, *Salicornia*, *Atriplex*).

'Raphide Clade'

Lophiocarpaceae

Figure 10.13 *Corbichonia decumbens* (Forsk.) Exell

$\ast$ K5 C0 A(5°+5°+10°+10+5+5+5)[d] $\underline{G}$(5)

General formula: $\ast$/$\longleftrightarrow$ K4–5 C0 A4–∞/∞° $\underline{G}$2 or 5

Figure 10.13 *Corbichonia decumbens* (Lophiocarpaceae), partial inflorescence.

The family groups two morphologically highly different genera, *Corbichonia* and *Lophiocarpus*. In pre-molecular classifications *Corbichonia* used to be associated with Molluginaceae, while *Lophiocarpus* was a member of Phytolaccaceae. *Corbichonia* resembles Aizoaceae to a great extent by the development of numerous series of stamens and petaloid staminodes (Brockington et al., 2013). While the five imbricate sepaloids are green, the petaloid staminodes arise in three series forming a total of twenty petaloids. Stamens and staminodes arise centrifugally on complex primordia. The ovary is globular with axile placentation and five stylodes (Ronse De Craene, 2021). Flowers of *Lophiocarpus* are reduced to five imbricate sepals with four stamens, two carpels

[d] The stamen whorls read in inversed order as a result of a centrifugal development with central stamens and peripheral staminodes.

and a single basal ovule. A floral diagram is provided by Eckardt (1974). The ovary is arranged obliquely and stamens are arranged symmetrically in relation to the bifid styles with one stamen opposite sepal one and the other alternating with the remaining sepals.

Kewaceae

Figure 10.14 *Kewa salsoloides* (Burch.) Christenh.
✶ K5 C0 A5+5+5 G(5)

Figure 10.14 *Kewa salsoloides* (Kewaceae). The numbers refer to the centrifugal initiation of stamens.

With the disintegration of the former Molluginaceae, a new genus *Kewa* has been segregated from the former genus *Hypertelis* and placed in its own family with eight species (Christenhusz et al., 2014). Flowers are isomerous in all whorls.

The perianth of *Kewa* is distinctive in the differentiation of the green outer sepals and inner petaloid sepals. Petaloidy is linked with an expansion of the sepal margins, and as sepals are in a quincuncial arrangement, the third sepal is half-petaloid (Figure 10.14).

Kewa is unusual in Pentapetalae in possessing three stamen whorls. Brockington et al. (2013) demonstrated that the androecium develops centrifugally as in Caryophyllaceae, with the outer stamen whorl arising from common primordia. However, contrary to petals of Caryophyllaceae, which arise on stamen-petal primordia as a result of a delay in initiation (p. 315; Wei and Ronse De Craene, 2019), I believe the extra stamen whorl is the result of an increase linked to the expansion of the receptacle, similar to the related Phytolaccaceae and Petiveriaceae (cf. Ronse De Craene, 2018). However, in other species, such as *K. bowkeriana*, there are only five alternisepalous stamens, corresponding to the first-formed stamens of *K. salsoloides*. The gynoecium bears five stylodes and has two rows of ovules on an axile placentation. Little is known about the floral morphology of other species of *Kewa*.

Aizoaceae

Figure 10.15 *Mesembryanthemum nodiflorum* L.

∗ K 5 C0 A5°+10°+10°+ . . . + 5°+10°+10°+10+5[e] –G– (5)

General floral diagram: ∗ K (3–)5(–8) C0 A(1–)5– 5^{∞} G(1–)2–5

Figure 10.15 *Mesembryanthemum nodiflorum* (Aizoaceae). Note the false septa in the locules; the broken line delimits five common stamen primordia.

[e] The stamen whorls read in inversed order as a result of a centrifugal development with central stamens and peripheral staminodes.

Aizoaceae is a large family with about four to five subfamilies. Flowers are often pentamerous with a 2/5 arrangement or tetramerous with sepals in a decussate arrangement (rarely with six or up to eight sepals). In most subfamilies, outer sepals continue the sequence of the leaves and are different from the inner sepals, being thicker and hoodlike. The complex androecium of Aizoaceae has been known for a long time (e.g. Payer, 1857; Ihlenfeldt, 1960). The androecium consists basically of five alternisepalous (complex) stamens, which remain simple (e.g. *Plinthus*), divide in pairs (*Galenia*), centrifugally in triplets (e.g. Aizooideae), or in large groups of stamens (Mesembryanthemoideae, Ruschoideae: Ihlenfeldt, 1960; Hofmann, 1973, 1993). Exceptionally there are two whorls of five stamens, or stamens are laterally increased in some species of *Sesuvium*. Taxa with a higher number of stamens tend to develop a ring primordium (e.g. *Mesembryanthemum, Aptenia*). Outer primordia may develop in coloured staminodes, effectively replacing a corolla (Figure 10.15). The number of stamens and staminodes developing centrifugally can be variable or is consistent. Ihlenfeldt (1960) and Haas (1976) listed various possibilities for differentiation into fertile inner stamens and outer staminodes with a progressive or abrupt transition. This is strongly taxon bound, but is highly similar to the development in *Corbichonia*. Species without staminodes usually have an adaxially coloured calyx (e.g. *Sesuvium, Tetragonia*), while it is green in taxa with staminodes. The ovary is generally isomerous with axile placentation, and a sterile central apical part can sometimes be strongly developed (columella: Haas, 1976). The ovary is often half-inferior or inferior by the development of an extensive hypanthium lifting androecium and sepals and stretching the carpellary tissue in a horizontal plane. As a result the original axile placentation can become parietal with the addition of false septa (Ronse De Craene, 2021). A nectary develops on the inner side of the stamen-petal tube.

Phytolaccoid Clade

Phytolaccaceae

Figure 10.16A *Phytolacca dodecandra* L'hérit., based on Ronse De Craene, Vanvinckenroye and Smets (1997)

∗ K5 C0 A5+5 $\underline{G}$(5)

Figure 10.16B *Ercilla volubilis* A. Juss.

∗ K5 C0 A5+3–4 $\underline{G}$(5)

General formula: ∗ K(4–)5 C0 A (5–)8(–30) $\underline{G}$3–5(–17)

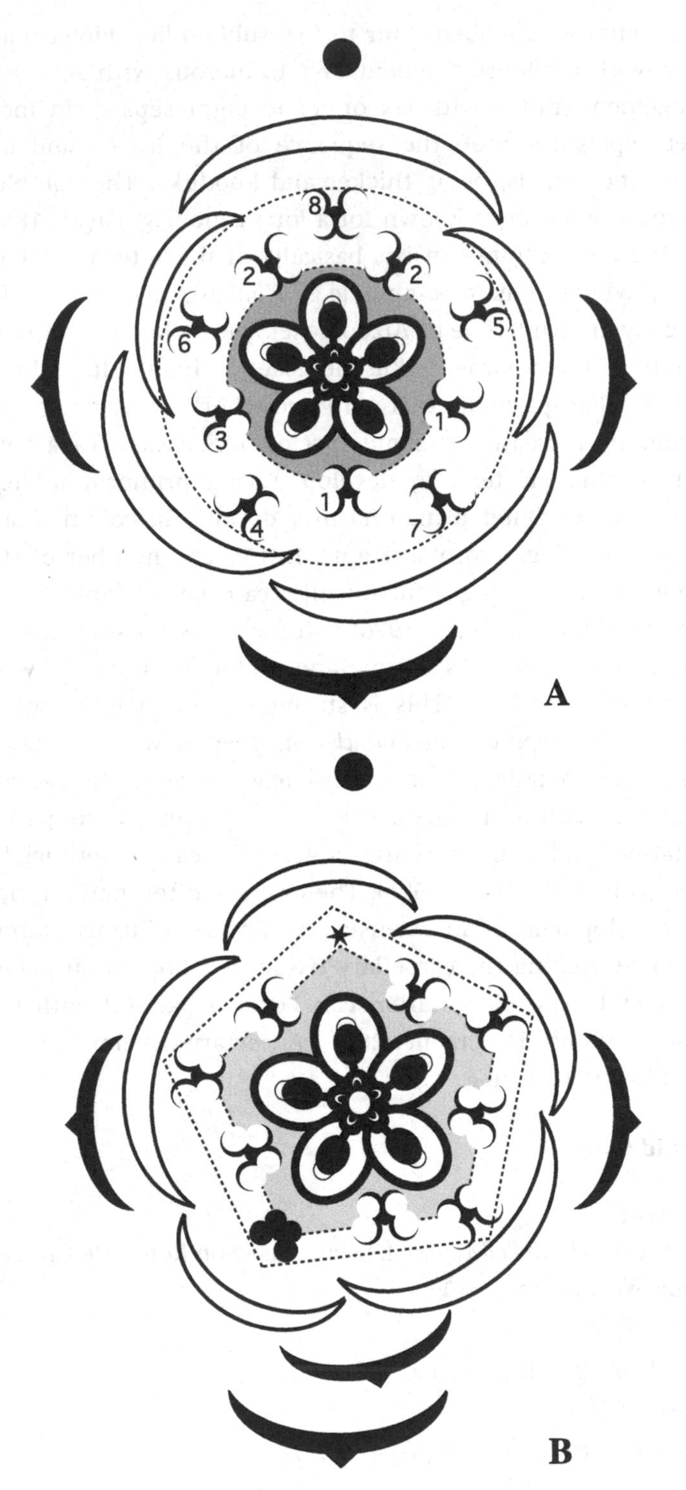

Figure 10.16 Phytolaccaceae: A. *Phytolacca dodecandra*; B. *Ercilla volubilis*. The numbers refer to order of initiation of stamens.

Phytolaccaceae as traditionally defined are paraphyletic, as subfamily Rivinoideae is more closely related to Nyctaginaceae (Brockington et al., 2009) and recognized as family Petiveriaceae (Yao et al., 2019), with which it shares unicarpellate flowers. The family is associated with two other families – Sarcobataceae and Agdestidaceae – with the former having much reduced unisexual flowers. *Microtea* has been removed as a monotypic family of the basal grade.

Flowers are pentamerous (or more rarely tetramerous) and arranged in racemose inflorescences (Rohwer, 1993b). Sepals are inconspicuous or petaloid, mostly with quincuncial aestivation. *Phytolacca dodecandra* with ten to twelve stamens occupies a pivotal position in the family (Figure 10.16A; Ronse De Craene, Vanvinckenroye and Smets, 1997): upper stamens arise simultaneously in two pairs opposite the outer sepals; a single stamen arises sideways of sepal three, before three other stamens initiate in sequence opposite sepals four, three and five. Two more stamens may arise opposite sepals one and two. *Ercilla* and some *Phytolacca* species (e.g. *P. esculenta*) with eight stamens have no stamens opposite sepals one and two (except for alternisepalous stamens converging in pairs) or one is staminodial. It is best to interpret the androecium of *P. dodecandra* and all Phytolaccaceae as derived from two initial whorls, with an upper alternisepalous whorl with stamens converging in pairs opposite sepals one and two (contrary to the interpretation of Ronse De Craene, Vanvinckenroye and Smets, 1997), and an antesepalous whorl arising in an inversed sequence and centrifugally relative to the first whorl (Ronse De Craene, 2013).

Carpel numbers are closely linked with the number of upper stamens and tend to alternate with them. In some species of *Phytolacca* the number of carpels has been secondarily increased by the inflation of the floral apex and this affects the number and arrangement of inner stamens, which are occasionally paired as to alternate with the carpels (*P. americana*: ten stamens and ten carpels; *P. esculenta*: eight stamens and eight carpels). A secondary centrifugal stamen increase can also be superimposed on this basic arrangement (e.g. *P. dioica*, *P. acinosa*: Hofmann, 1993; Ronse De Craene, Vanvinckenroye and Smets, 1997; Ronse De Craene, 2013).

The ovary is basically syncarpous, with a single basal-axile ovule per carpel, but can become secondarily apocarpous (Rohweder, 1965) by the expansion of the floral apex. The floral apex functions as an extragynoecial compitum in this case (see Endress, 2019; Ronse De Craene, 2021). A short gynophore may be present. The placenta is deeply embedded in the apical meristem, appearing to arise separately from the carpel wall (Ronse De Craene, Vanvinckenroye and Smets, 1997). A nectary is usually situated as a rim between the base of the gynoecium and the inner stamens.

Nyctaginaceae

Figure 10.17 *Bougainvillea spectabilis* Willd.

∗ K5 C0 A(5+3) G̲1

General formula: ∗ K(3–)4–5(–7) C0 A(1–)5(–40) G1

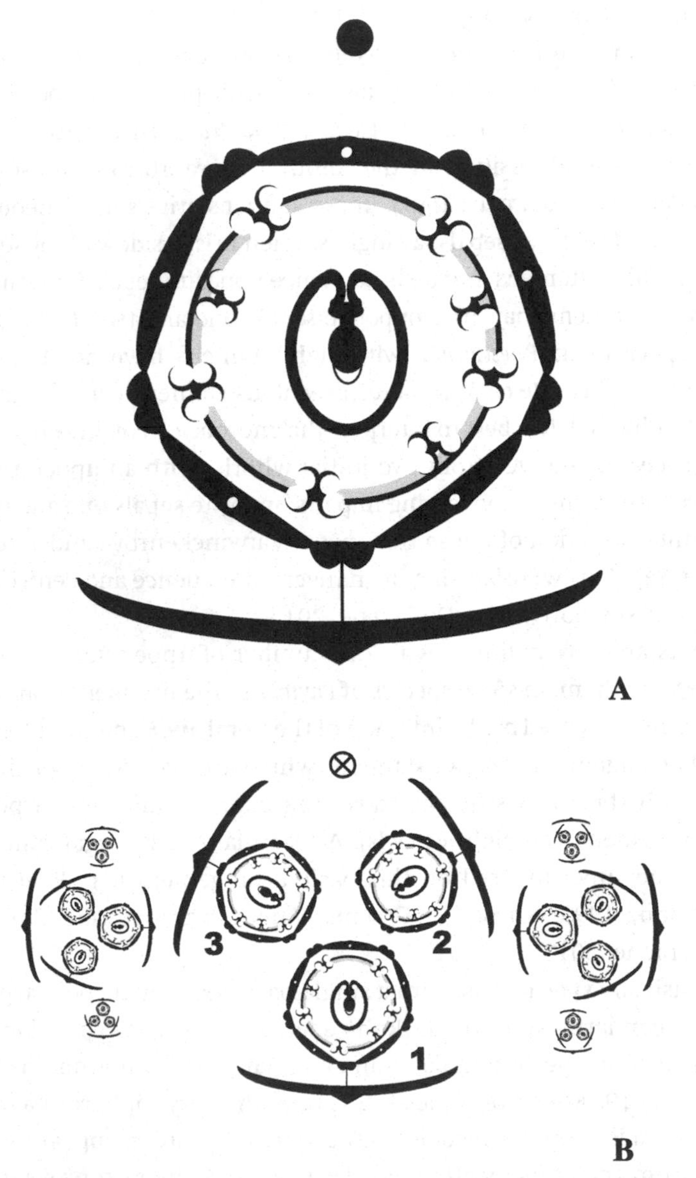

Figure 10.17 *Bougainvillea glabra* (Nyctaginaceae): A. flower; B. inflorescence. The numbers refer to the sequence of maturation of flowers in a triplet.

The family is distinctive by its fused perianth tube consisting of an upper petaloid section that shrivels at fruiting and a lower persistent section enclosing the ovary (anthocarp).

Flowers of *Bougainvillea* are typically arranged in three-flowered dichasia with the terminal flower not clearly delimited from the other two. Flowers are enclosed by large, often petaloid bracts (involucre). In some cases bracts may be confused with sepals, as in *Mirabilis* where five fused bracts enclose the flower in the manner of sepals. This has occasionally been regarded as an incipient acquisition of a bipartite perianth, as bracts cannot be distinguished from sepals (Brockington et al., 2009). The single flower in *Mirabilis* is the central remainder of an originally five-membered inflorescence surrounded by an involucre (Sattler and Perlin, 1982). The perianth is pentamerous, rarely tetramerous with valvate or plicate aestivation (appearing contorted). The anthocarp was variously interpreted as a perianth tube (e.g. Rohweder and Huber, 1974) or a hypanthium (Buxbaum, 1961: 'Achsenbecher'). The fact that stamens are variously fused to the tube and the dual nature of the perianth point to a nature of a hypanthium.

The number of stamens in Nyctaginaceae varies from (one) two or three (*Boerhaavia*, *Oxybaphus*) to up to forty (some *Pisonia*). Eight stamens occur in some *Pisonia* and in *Bougainvillea* (Figure 10.17; Vanvinckenroye et al., 1993). Stamens in *Bougainvillea* are initiated in a 3/8 sequence, following the 2/5 initiation of the perianth (Sattler and Perlin, 1982; Vanvinckenroye et al., 1993). Five stamens alternate more or less with the sepals while three are positioned more or less opposite the inner sepals. At maturity stamens have different lengths. Most other Nyctaginaceae have five stamens alternating with the sepals (e.g. *Mirabilis*). When lower than five, stamen numbers can be variably in alternation or opposite sepals (*Boerhaavia*, *Oxybaphus*). Stamens are free or monadelphous. A nectary is situated on the adaxial side of stamens or below the gynoecium (Zandonella, 1977). The ovary is always superior and unicarpellate with single basal ovule.

Fiedler (1910) provided several diagrams of the flower of Nyctaginaceae, concentrating on the androecium that he interpreted as basically diplostemonous; higher numbers as in *Pisonia* were seen as the result of dédoublement. However, his diagrams have little value as he analysed flowers with a preconceived approach that does not correspond to the real development and structural arrangement of the flower. It is possible that higher numbers of stamens result from a centrifugal increase, as in Phytolaccaceae.

Petiveriaceae

Figure 10.18A *Trichostigma peruvianum* (Moq.) H. Walt.
∗ K2+2 C0 A4+2+2+4 $\underline{G}$1

Figure 10.18B *Petiveria alliacea* L.
↓K4 C0 A6 $\underline{G}$1

General formula: K4–5 C0 A4–∞ $\underline{G}$1

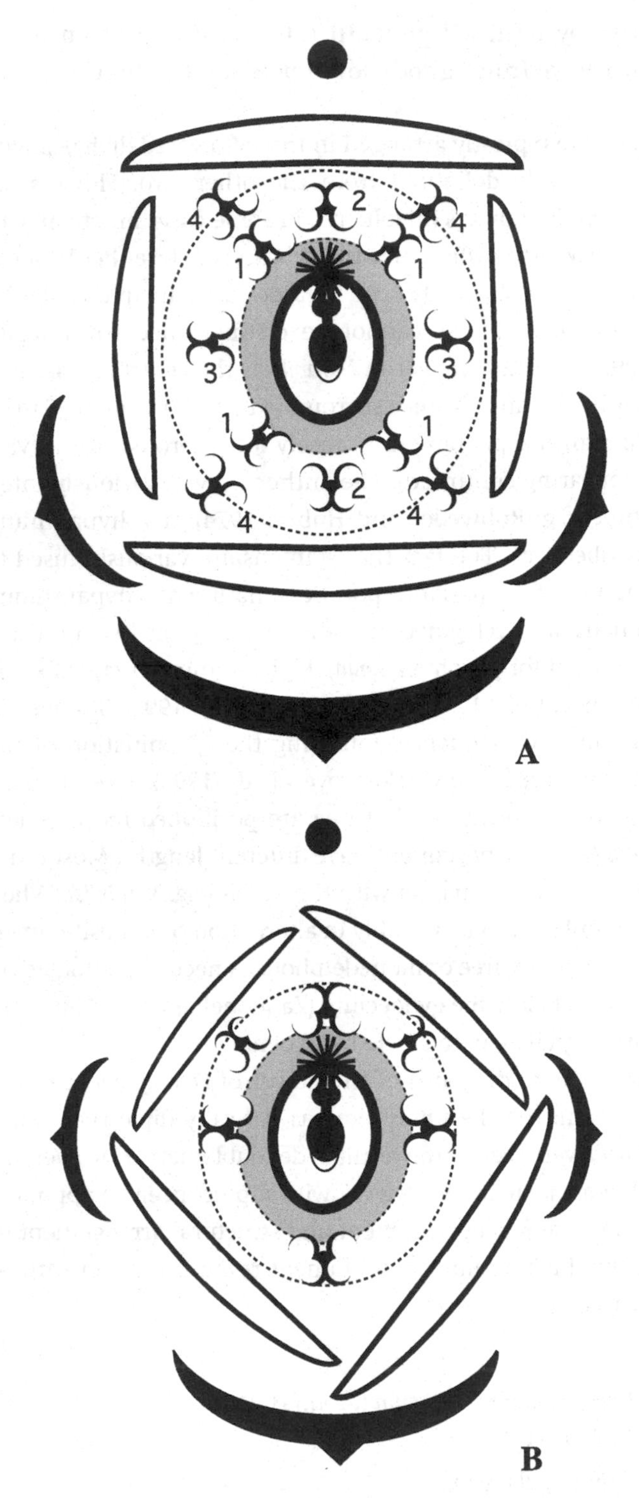

Fig 10.18 Petiveriaceae: A. *Trichostigma peruvianum*; B. *Petiveria alliacea*. The numbers refer to the order of initiation of stamens.

Flowers are arranged in racemes with bract and bracteoles that can be shifted close to the flower. Flowers are either pentamerous (e.g. *Seguieria*) or tetramerous (apparently dimerous; e.g. *Trichostigma, Rivina*). The androecium of *Trichostigma* and related genera is comparable to the three-whorled pentamerous arrangement of *Kewa* (Brockington et al., 2013). Initiation is centrifugal and this was interpreted as a tendency for reduction (Ronse De Craene and Smets, 1991c), as some Petiveriaceae have only four alternisepalous stamens (e.g. *Hilleria, Rivina*). Many stamens are found in *Seguieria*. The flower of *Petiveria* is monosymmetric with four diagonally inserted sepals, six stamens and an oblique ovary position. Ronse De Craene and Smets (1991c) interpreted the distorted position as a result of a reduction of the size of the bracteoles, influencing the position of four sepals and the rest of the flower. *Monococcus echinophorus*, an Australian dioecious endemic, shares the same floral arrangement as *Petiveria*; staminate flowers have a multistaminate androecium arising in four alternisepalous groups and without trace of a carpel, while pistillate flowers lack stamens (Vanvinckenroye, Ronse De Craene and Smets, 1997). The gynoecium is generally unicarpellate with a flattened crest-like stigma and without style. No nectaries are present in Petiveriaceae (Rohwer, 1993b).

Molluginaceae

Figure 10.19A *Glinus lotoides* L.
✱ K5 C0 A(3°:1+5) $\underline{G}$(3)

Figure 10.19B *Mollugo verticillata* L.
✱ K(4)5 A3–5 $\underline{G}$(3)

General formula: ✱ K(4)5 A°0–5 A(3–)4–5(–∞) G(1)2–5

Recent molecular phylogenies have shown Molluginaceae to be paraphyletic, containing at least five clades (*Macarthuria, Limeum, Corbichonia* in Lophiocarpaceae, *Kewa* (*Hypertelis* p.p.) and Molluginaceae s. str., including *Glinus* (Cunéoud et al., 2002; Brockington et al., 2009). The five sepals have quincuncial aestivation and are often hooded and petaloid at the margins. Petals are absent. In most Molluginaceae stamens are in one or two whorls, and the antesepalous whorl ranges from two or three to five, which can be partially sterile or absent (Figure 10.19A; Hofmann, 1993; Brockington et al., 2013). The androecium of *Glinus lotoides* is extremely variable, occasionally comparable to some *Corbichonia* (Hofmann, 1973). In my material only the alternisepalous whorl was complete, with a variable number of staminodes and odd stamen opposite sepals (Figure 10.19A). The alternisepalous whorl is longer and inserted more centrally. In *Mollugo* the number of stamens ranges from five (*M. cerviana*) to three (*M. nudicaulis, M. verticillata*:

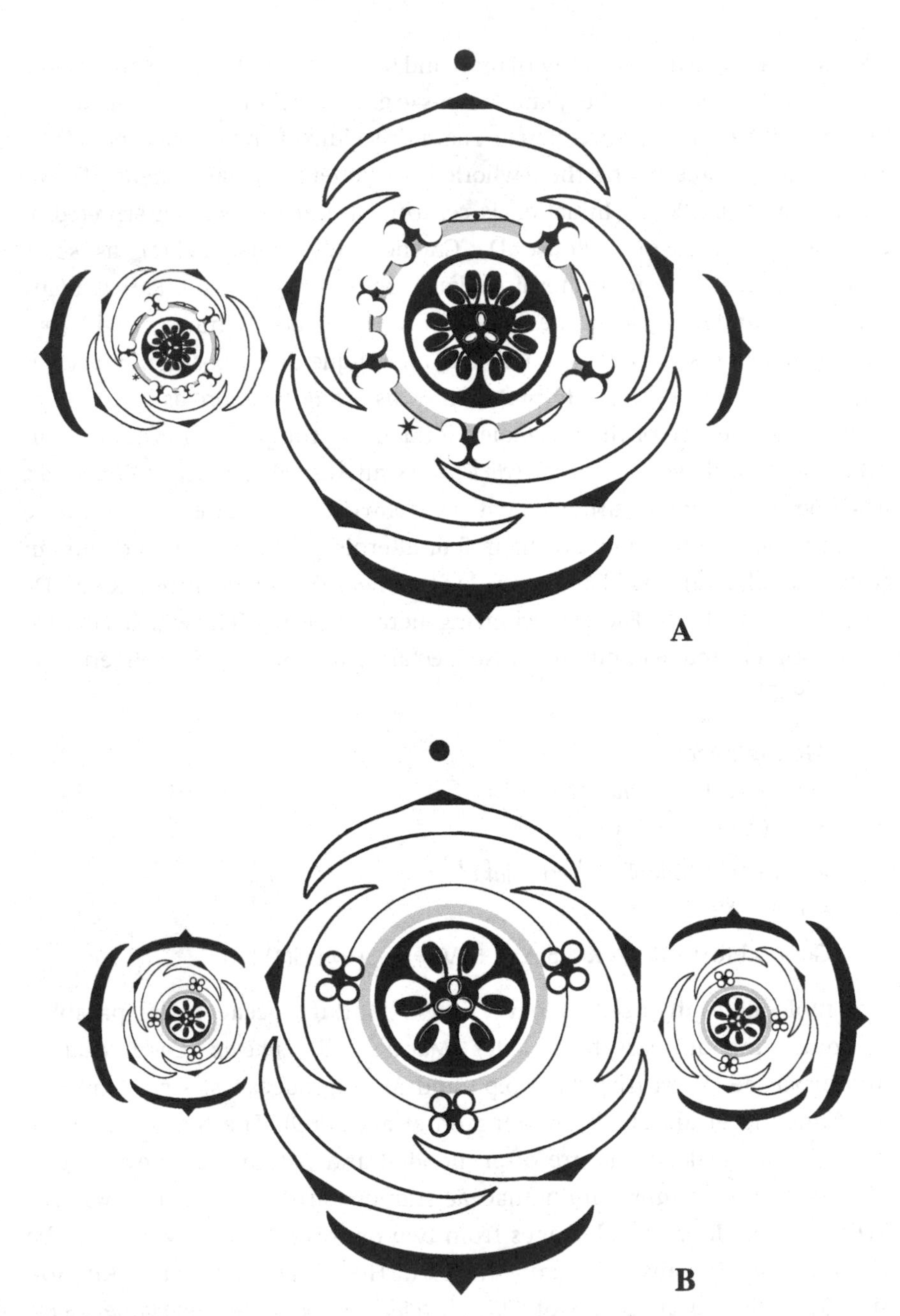

Figure 10.19 Molluginaceae: A. *Glinus lotoides*, partial inflorescence. Note the incomplete antesepalous stamen whorl. B. *Mollugo verticillata*, partial inflorescence

Figure 10.19B), derived from a pairwise amalgamation of four stamens (Batenburg and Moeliono, 1982; cf. Cucurbitaceae). The three remaining stamens alternate with the trimerous gynoecium.

The ovary is isomerous with antesepalous carpels or is reduced to three (two, rarely one: *Adenogramma*). Placentation is axile with narrow partitions and styles are carinal, very short to absent.

Portulacineae Clade

The Portulacineae clade (suborder Portulacineae *sensu* Nyffeler and Oggli, 2010) consists of eight families, as the traditional Portulacaceae are recognized as being paraphyletic. Inflorescences are mostly cymose with dichasia deviating into monochasia (Carolin, 1993). All members of the clade except Cactaceae are pentamerous or tetramerous and apetalous with sepals enclosed by two (three) large median bracts forming an involucre. The persistent involucre has often been interpreted as a calyx (sepaloids) and the sepals as a corolla (petaloids; e.g. Eichler, 1878; Franz, 1908; Milby, 1980; Carolin, 1993), although morphological evidence suggests that the petaloids are homologous to sepals as found in other Caryophyllales (Ronse De Craene, 2013). The petaloids, when five in number, always arise in a 2/5 sequence with the outer in a lateral position. With four petaloids they arise in a 1/2 sequence with similar arrangement. The androecium of the Portulacineae is highly variable, ranging from one to many stamens. With multistaminate androecia stamens arise centrifugally on a meristematic ring primordium. In other Portulacineae the initiation of the ten upper primordia is less clear by the development of a ring primordium (e.g. *Calandrinia*, *Anacampseros*, *Portulaca*: Hofmann, 1993; Vanvinckenroye and Smets, 1996, 1999). This increase is superimposed on the initial whorl of ten to thirteen stamens. As in other Caryophyllales the inner side of the stamen tube develops as a nectary. An apparent synapomorphy for the clade is the single umbrella-like style with long stylar lobes corresponding to the number of carpels.

Montiaceae (incl. Hectorellaceae)

Figure 10.20A–B *Lewisia columbiana* (J. T. Howell) Rob.

$\ast$ K7[f] C0 A(5) $\underline{G}$(3)

Figure 10.20C *Claytonia perfoliata* (Donn. ex Willd.) Howell

$\ast$ K5 C0 A(5) $\underline{G}$(3)

General formula: $\ast$ K(4–)5(–17) A(1–)5–∞ $\underline{G}$2–5(–8)

The inflorescence of *Lewisia* is a complex dichasium, while it is monochasial (cincinnus) in *Claytonia*. Most Montiaceae possess a calyx of four or five petaloids. *Lewisia* has a higher and variable number of tepals, caused by an expansion of the receptacle by the pressure of the involucral bracts. This leads to an increase of inner tepals (up to seventeen) and stamens (Dos Santos and Ronse

[f] K up to seventeen and G up to eight in some *Lewisia*.

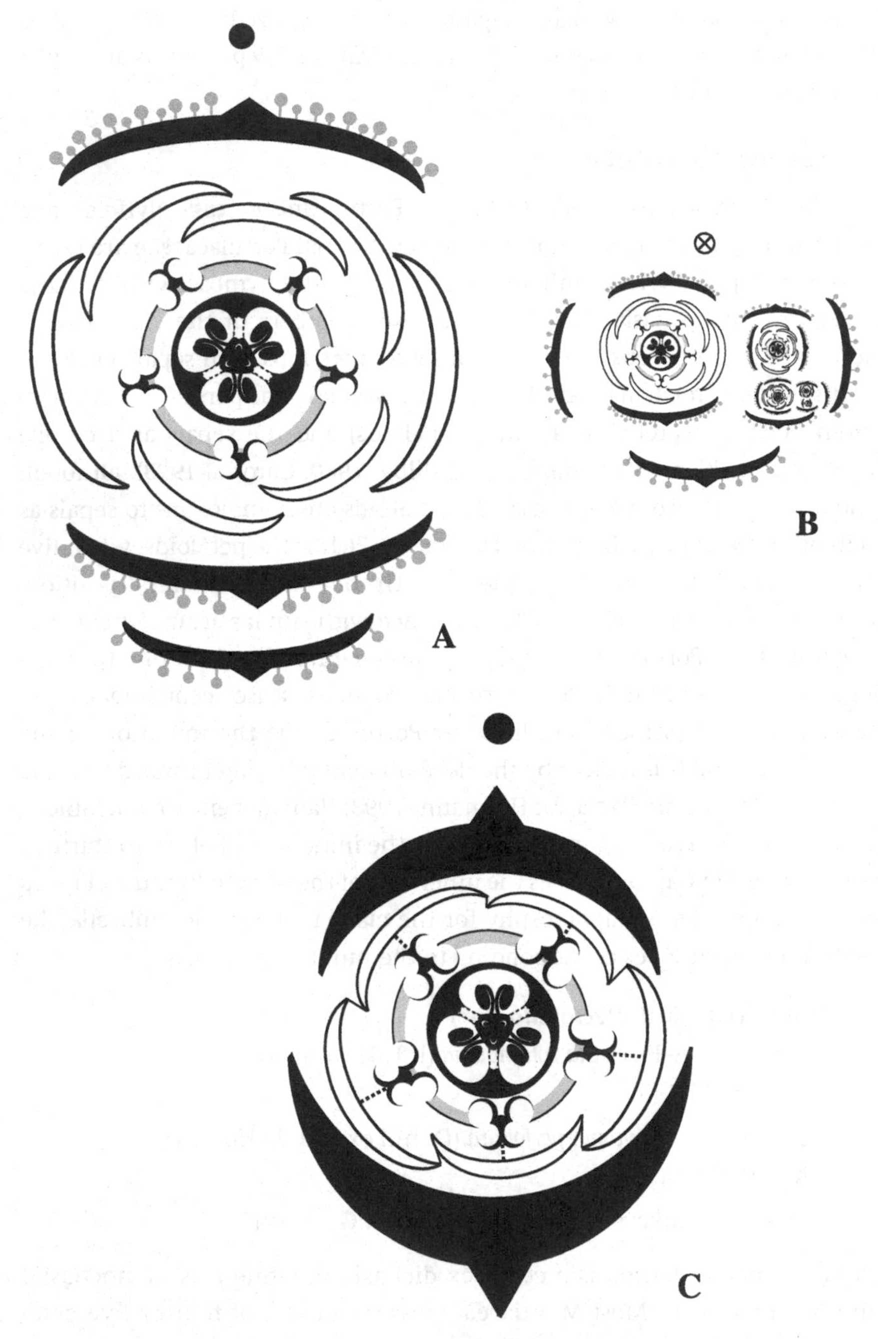

Figure 10.20 Montiaceae: *Lewisia columbiana*, A. flower; B. partial inflorescence; C. *Claytonia perfoliata*

De Craene, 2016). The number of involucral bracts can also be much higher in the genus (e.g. *L. rediviva*: Franz, 1908). In *Claytonia* (*Montia*) the five stamens are antesepalous but can be reduced to three opposite sepals three, four and five (*M. fontana*: Payer, 1857). *Lyallia* has four sepals and three stamens. Stamen numbers are centrifugally increased in *Lewisiopsis* and *Cistanthe*.

Sepals of *Claytonia* arise basipetally from common primordia with the stamens (Milby, 1980; Hofmann, 1993). This is caused by a shift of the protective function towards the involucre, linked with a retardation of the sepal development and their absorption by the staminal tissue. The gynoecium as in the majority of Portulacineae is often trimerous (rarely dimerous or pentamerous) and develops septa. The ovules are attached on a central column by the later disintegration of the septa (Ronse De Craene, 2021).

Portulacaceae

Figure 10.21 *Portulaca grandiflora* Hook., partly based on Soetiarto and Ball (1969); A. flower; B. relationship of flower to the main stem

∗ K(4–)5(–8) C0 A(4–)8–∞ -$\underline{\text{G}}$-(4–)5(–8)

The circumscription of Portulacaceae has considerably changed with the molecular phylogeny, resulting in the splitting of the family in Montiaceae, Anacampserotaceae, Talinaceae and Portulacaceae, leaving a single genus: *Portulaca*.

The involucre surrounds large bright petaloids. In cultivated *P. grandiflora* the number of petaloids can be higher as they are arranged in several series (Soetiarto and Ball, 1969). Stamens are inserted in a ring and develop centrifugally. In a few species stamen numbers are lower (four to eight: *Portulaca kuriensis*, Ronse De Craene, 2021) but arising on a ring primordium. The ovary consists of many ovules on axile placentation or parietal placentation that develops by the breakdown of the septa in the middle (Ronse De Craene, 2021). Perianth and stamens persist around a pyxidium.

Cactaceae

Figure 10.22 *Echinopsis formosa* (Pfeiff.) Jacobi

∗ P∞ A∞ Ğ (7)

General formula: ∗/ ↓ P∞ A∞ Ğ (2–∞)

Cactaceae is the sister clade to Anacampserotaceae and Portulacaceae. Cactaceae differ from other Portulacineae in that they are not pentamerous and do not share a bipartite involucre, but have a spiral, indefinite perianth.

Figure 10.21 *Portulaca grandiflora* (Portulacaceae). A. flower; B. relationship of flower to main stem; rounded structures represent upper leaves. In A the broken line represents the outline of the stigmatic lobes.

Flowers are formed solitary on short shoots, rarely in racemes (e.g. *Leuenbergia*, *Pereskia*: Rosas-Reinhold et al., 2021). Flowers are generally polysymmetric, more rarely monosymmetric by the bending of the hypanthium and unequal distribution of stamens (e.g. *Schlumbergera*, *Selenicereus*). Contrary to other Portulacineae the perianth is indefinite, with several bracts progressively shifting into petaloid tepals, sepaloids and petaloids. However, in *Leuenbergia* and *Pereskia* there is an abrupt transition between bracts and petaloids. The numerous spirally inserted tepals of Cactaceae could be derived by a division of

Figure 10.22 *Echinopsis formosa* (Cactaceae)

initially fewer perianth parts (as in *Lewisia*) or by addition of bracts in the confines of the flower (Ronse De Craene, 2008, 2013). Rosas-Reinhold et al. (2021) discussed different hypotheses for genetic shifts leading to the unusual perianth of Cactaceae. These do not explain the structural difference of the flower, only the petaloidy of organs involved. Most Cactaceae have the inferior ovary sunken within vegetative axial tissue that can expand as a rim around the flower (pericarpel: see Rosas-Reinhold et al., 2021). The pericarpel differs from a hypanthium as it represents vegetative tissue, although the distinction between flower and stem becomes blurred. The androecium is multistaminate and develops centrifugally on a ring primordium (see also Boke, 1963, 1966; Ross, 1982; Leins and Schwitalla, 1985). Through the development of a deep hypanthium the gynoecium occupies a (half-) inferior position and the stamens are placed in several tiers on the slopes of the hypanthium. The number of carpels ranges from (two-) four to five, up to twenty, centred around a central residual floral apex (Boke, 1963). Placentation is basal-laminar because septa fail to extend beyond small outgrowths (Leins and Erbar, 2010). The ovules arise on the carpel wall at the base of the septa and, depending on the genus, extend beyond the septa in a wave line, or become restricted to one or two ovules (*Pereskia aculeata*). Ovary development is similar to certain Aizoaceae by the horizontal extension of the floral receptacle (Ronse De Craene, 2021). Style and stigma formation is similar to other Portulacineae in being umbrella-like with many stylodial branches. The inner side of the androecium produces a narrow nectary.

Leuenbergeria and *Pereskia* are the earliest diverging clades of Cactaceae and look less cactus-like than the other genera, differing in inflorescence, abrupt transition between bracts and tepals, presence of a superior or (half-)inferior ovary, low ovule number (*Pereskia*), axile placentation (*Leuenbergia*) and vegetative characteristics (Rosas-Reinhold et al., 2021). In *Pereskia* five to eight androecial areas are demarcated on the ring primordium alternating with the inner tepals and (half-)superior carpels (Boke, 1963; Ross, 1982; Leins and Schwitalla, 1985). Although this could represent the basic condition in Cactaceae, it does not correspond with the position of stamens in the nearest sister groups where stamens have an antesepalous position.

11

Asterids

Tubes and Pseudanthia

The asterids *sensu* APG IV (2016) contain about one third of all angiosperms (Soltis et al., 2018). Asterids are subdivided into two main groupings: basal asterids (a grade) consisting of Cornales and Ericales, and euasterids (with lamiids and campanulids) (Figures 11.1, 11.2; e.g. Judd and Olmstead, 2004). Spichiger et al. (2002) distinguished between archaic asterids, superior hypogynous asterids, and superior epigynous asterids.

Characters common in a majority of taxa of asterids are sympetaly with adnate stamens (stamen-petal tube), unitegmic, tenuinucellate ovules, cellular endosperm formation, terminal endosperm haustoria, pollen that is released at the trinucleate stage and the presence of iridoids (e.g. Judd and Olmstead, 2004; Erbar and Leins, 2011). Apart from sympetaly, the other characters cannot be used in floral diagrams.

Most taxa share the development of a ring primordium in early stages of development (Erbar, 1991), indicating a generalized syndrome of petal development (see p. 42).

Core asterids (euasterids) share sympetaly, a bicarpellate gynoecium and haplostemony, while the basal orders Cornales and Ericales are much more variable with a basically diplostemonous androecium and the occasional development of sympetaly. However, they show several apomorphic tendencies pointing towards euasterids.

11.1 Basal Asterids: Cornales and Ericales

Cornales

The basalmost order of asterids (Figure 11.2) contains seven families. All share small sepals, a half-inferior ovary, an epigynous disc

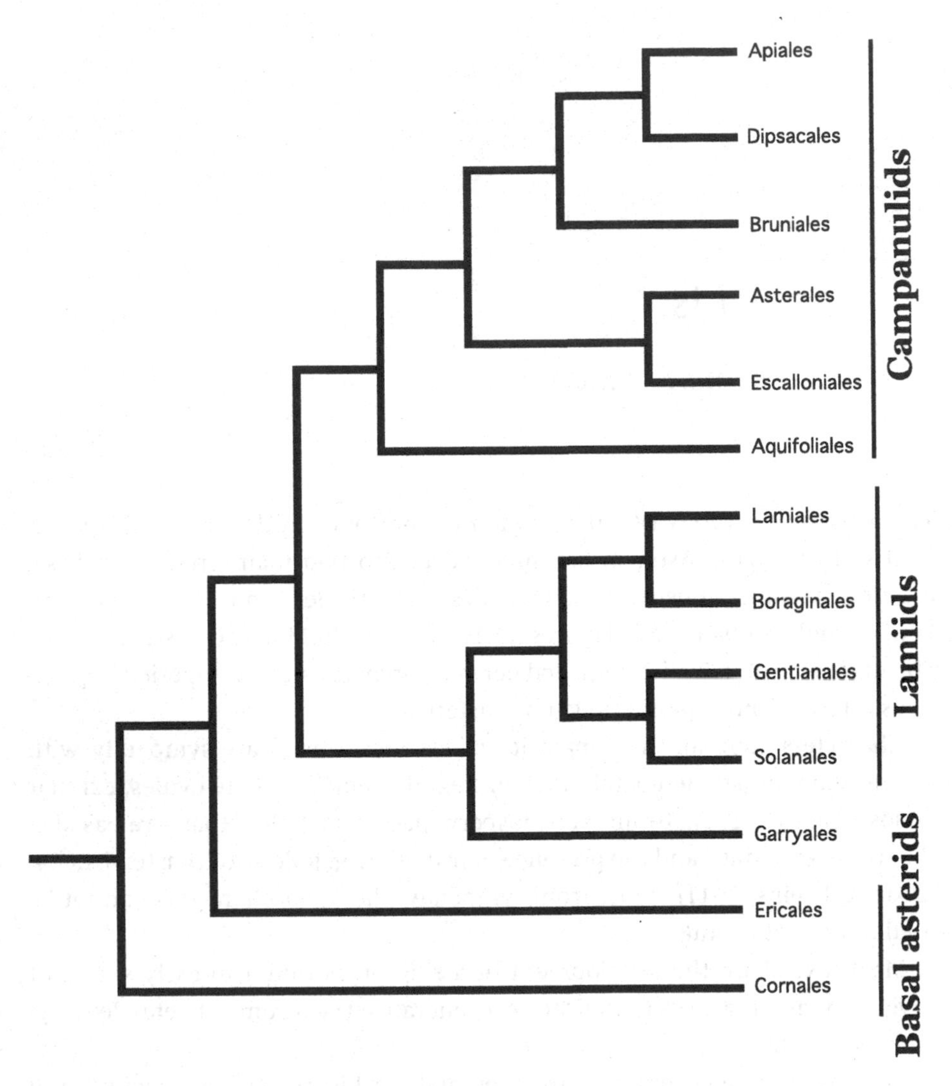

Figure 11.1 Phylogenetic tree of asterids, based on Soltis et al. (2018). The position of Boraginales is debatable.

nectary and often drupaceous fruits (Soltis et al., 2018). Soltis et al. (2005) argued that the half-inferior ovary of Hydrangeaceae evolved from epigyny common in the clade. Most Cornales have free petals, while there are indications of an incipient early sympetaly *sensu* Erbar (1991), which is mostly expressed in some Loasaceae. Hydrangeaceae are strongly linked with Loasaceae in floral characters (see Roels, Ronse De Craene and Smets, 1997; Hufford, 1989a, 1989b, 1998, 2001), with the same pattern of polystemonous androecia, although the androecium is more complex in

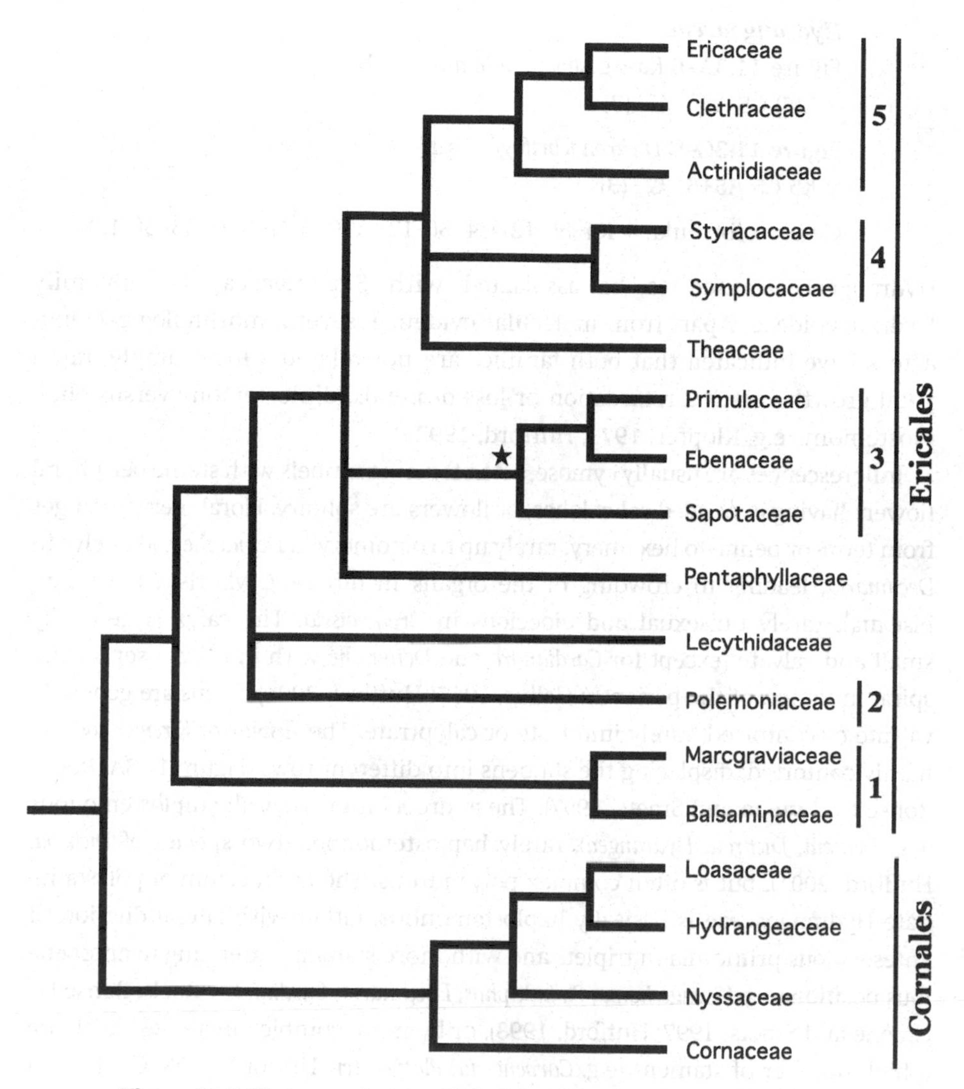

Figure 11.2 Phylogenetic tree of Ericales and Cornales, based on Soltis et al. (2018). 1. balsaminoid clade; 2. polemonioid clade; 3. primuloid clade; 4. styracoid clade; 5. ericoid clade. Asterisk: development of obhaplostemony

Loasaceae subfamily Loasoideae. Both families share a stamen increase with a development that shifts between a centrifugal and centripetal direction (e.g. Hufford, 1990, 1998; Ge, Lu and Gong, 2007). Gynoecium and hypanthium development are highly similar with the development of a 'corolla ring primordium' (Erbar, 1991; Erbar and Leins, 2011), also present in other Cornales. Development of a cuplike hypanthium in early stages of development in all Cornales links the clade with euasterids.

Hydrangeaceae

Figure 11.3A–B *Kirengeshoma palmata* Yatabe
∗ K(5) C5 $A5^3$ –G– (3)

Figure 11.3C–D *Dichroa febrifuga* Lour.
∗ K5 C5 A5+5 –G– (3)

General formula: ∗ K4–5(–12) C4–5(–12) A10–∞ Ğ/-G- (2–)3–5(–12)

Hydrangeaceae used to be associated with Saxifragaceae in subfamily Hydrangeoideae. Apart from molecular evidence, several morphological characters have indicated that both families are not related - for example, rapid petal growth versus a retardation or loss of petals, diplostemony versus obdiplostemony (e.g. Klopfer, 1973; Hufford, 1992a).

Inflorescences are usually cymose, often forming umbels with sterile peripheral flowers having expanded calyx lobes, or flowers are solitary. Floral merism ranges from tetra- or penta- to hexamery, rarely up to octomery in *Deinanthe* and twelve in *Decumaria*, leading to crowding of the organs in different whorls. Flowers are bisexual, rarely unisexual and dioecious in *Broussaissia*. The calyx is generally small and valvate (except for *Cardiandra* and *Deinanthe* with imbricate sepals and spiral initiation of the perianth: Gelius, 1967; Hufford, 2001). Petals are generally valvate or contorted, rarely imbricate or calyptrate. The flower of *Kirengeshoma* is highly contorted, displacing the stamens into different rows (Figure 11.3A; Roels, Ronse De Craene and Smets, 1997). The androecium is basically diplostemonous (e.g. *Deutzia*, *Dichroa*, *Hydrangea*), rarely haplostemonous (two species of *Dichroa*: Hufford, 2001), but is often complex polyandrous. The androecium of polystaminate Hydrangeaceae is basically haplostemonous, either with lateral division of antesepalous primordia in triplets and with more stamens extending in antepetalous position (e.g. *Kirengeshoma*, *Philadelphus*, *Decumaria*, *Cardiandra*: Roels, Ronse De Craene and Smets, 1997; Hufford, 1998), or by more complex increases involving a high number of stamens (e.g. *Carpenteria*, *Platycrater*: Hufford, 1998; Ge, Lu and Gong, 2007). This indicates that increase of antesepalous stamen primordia probably prevents the development of the antepetalous stamens. In *Carpenteria*, numerous stamens arise on complex U-shaped primordia extending in a wave-line in antepetalous sectors of the flower (Hufford, 1998), resembling the development in Malvaceae. In *Deinanthe* and *Platycrater*, numerous stamens arise on a ring primordium, linked with identifiable primordia in antesepalous position, and covering the hypanthial slope (Hufford, 1998; Ge, Lu and Gong, 2007). Placentation is intruding parietal to axile, usually U-shaped with many ovules confined to the upper part of the ovary. A variable hypanthial growth is responsible for the position of the ovary, which is halfway emergent with usually free stylodes or a single style with long stylar lobes (Hufford, 2001). A disc nectary is intrastaminal to epigynous.

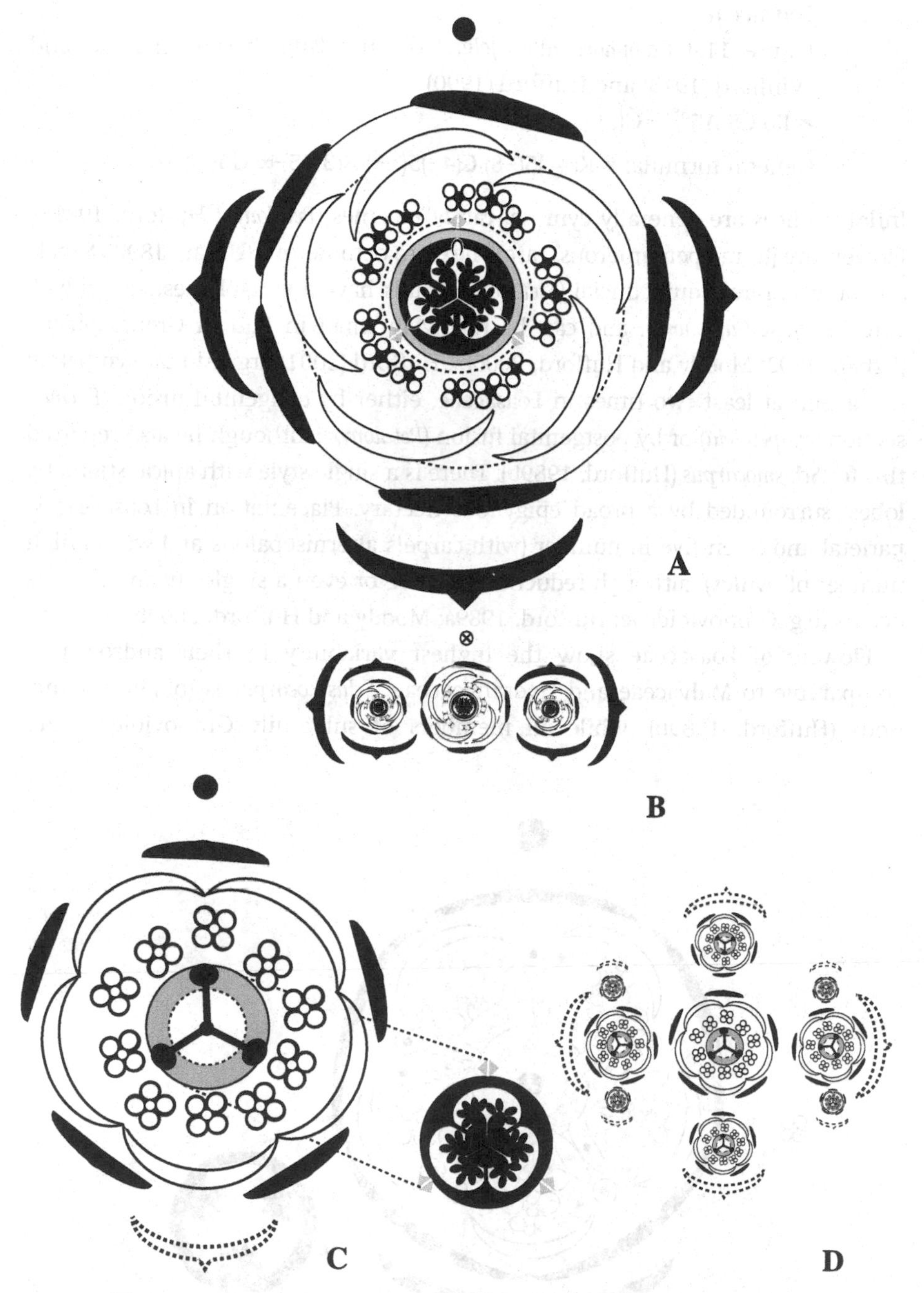

Figure 11.3 Hydrangeaceae: *Kirengeshoma palmata*, A. flower; B. partial inflorescence, *Dichroa febrifuga*; C. flower; D. partial inflorescence

Loasaceae

Figure 11.4 *Cajophora hibiscifolia* Urb. and Gilg, based on Leins and Winhard (1973) and Hufford (1990)

$\ast$ K5 C5 A5$^{5°:\infty}$ $\breve{G}$(3)

General formula: $\ast$ K(4–)5(–8) C(4–)5(–8) A(3–)5–∞ G3–5(–7)

Inflorescences are generally cymose, rarely racemes (*Petalonyx*: Hufford, 1989a). Flowers are (tetra-) pentamerous, rarely hexa- to octomerous (Urban, 1892). Sepals are valvate, rarely quincuncial (*Mentzelia*). Petals have a variable aestivation (valvate in most *Cajophora*), and can be relatively small in bud in Gronovioideae (Urban, 1892; Moody and Hufford, 2000b). Hufford (2001) argued that sympetaly has arisen at least two times in Loasaceae, either by congenital fusion (*Eucnide* section *Sympetaleia*) or by postgenital fusion (*Petalonyx*), although he also reported this for *Schismocarpus* (Hufford, 1989b). There is a single style with apical stigmatic lobes, surrounded by a broad epigynous nectary. Placentation in Loasaceae is parietal and often five in number (with carpels alternisepalous and with a high number of ovules), although reduction to three or even a single subapical ovule occurs (e.g. Gronovioideae: Hufford, 1989a; Moody and Hufford, 2000b).

Flowers of Loasaceae show the highest variability in their androecium, comparable to Malvaceae and Hydrangeaceae. *Schismocarpus* is (ob)diplostemonous (Hufford, 1989b), while all members of subfamily Gronovioideae are

Figure 11.4 *Cajophora hibiscifolia* (Loasaceae). Antesepalous complex stamen primordia differentiate in staminodial scales and spread out opposite the petals. Broken lines: limits of stamen group

haplostemonous, occasionally with some staminodial stamens (Urban, 1892; Moody and Hufford, 2000b), and most other taxa are polyandrous. The development of polyandrous flowers is highly complex, with continuity between antesepalous and antepetalous stamen groups, but the development can be linked to complex antesepalous (or more rarely antepetalous) groups of stamens (Hufford, 1990). A single stamen primordium followed by two lateral stamen primordia always precedes the initiation of more stamens. In *Mentzelia* and *Eucnide* development of the androecium runs centripetally from antesepalous forerunners, and stamens cover the slope of a deep hypanthium. In *Mentzelia* a variable number of first-formed stamens is staminodial, but this pattern is accentuated in other genera of subfamily Loasoideae (e.g. *Cajophora, Loasa*), with differentiation of antepetalous fertile stamens and conversion of the five initial stamen primordia into antesepalous staminodes (Figure 11.4). Antepetalous stamens arise as an overflow of the antesepalous sectors and initiation of stamens is a mixture of centripetal and centrifugal development, comparable to some polyandrous Hydrangeaceae (e.g. *Platycrater*). Stamens in the antepetalous stamen fascicle consist of stamens of two adjacent mounds (e.g. Brown and Kaul, 1981; Hufford, 1990; contrary to the interpretation of Leins and Winhard, 1973), and this is comparable to the stamen development in Malvaceae. In *Blumenbachia*, the pattern of initiation appears to be reversed with antepetalous common primordia arising before the antesepalous staminodes. One could possibly interpret this as an inversion of centres of initiation linked with the retardation of staminodial primordia.

The three outer staminodes (corresponding to the first formed stamens of the antesepalous common primordia) are connected into a single cucullate structure, while the two inner staminodes become elongated and cover the stigma (e.g. *Loasa, Cajophora*, Hufford, 1990). In other genera the two sets of staminodes are more variable in number (five to seven) and the shape of the staminodes can be variously elaborate (e.g. *Huidobria, Nasa*: Hufford, 2003). As the nectary is epigynous, staminodes may function as nectar containers (e.g. Brown and Kaul, 1981; Smets, 1988). Flowers of *Schismocarpus* are buzz pollinated and lack nectaries (Hufford, 1989b). Distally forked filaments are found in some *Mentzelia* and are reminiscent of comparable structures in Hydrangeaceae (e.g. *Deutzia*: Hufford, 2003).

Ericales

Ericales contain twenty-two families and were considerably extended compared to the pre-molecular classifications, incorporating several smaller orders (Theales, Ebenales, Sarraceniales, Primulales, Polemoniales, Ericales *sensu stricto*: Figure 11.2; Anderberg, Rydin and Källersjö, 2002; Schönenberger et al., 2005). Delimitation of Ericales, although well supported on a molecular basis, tends to be far less clear on a morphological basis. Apart from the recognition of a number of well-supported clades (balsaminoid, polemonioid, primuloid, styracoid, ericoid), the inter-familial relationships of Ericales are not well supported (Figure 11.2; Schönenberger, Anderberg and Sytsma, 2005). Soltis et al. (2005) recognized two synapomorphies: gynoecia with protruding diffuse placentae and theoid leaf teeth.

The order can be considered as a transitional group (together with Cornales) between asterids and rosids. Compared to euasterids, sympetaly is not universal and generally limited, and the basic floral diagram is similar to rosids: K5 C5 A5+5 G5. Several fossil flowers have been identified as belonging to Ericales (e.g. Keller, Herendeen and Crane, 1996; Schönenberger and Friis, 2001; Martínez-Millán, Crepet and Nixon, 2009; Schönenberger et al., 2012; Crepet, Nixon and Weeks, 2018). Within Ericales this diagram tends to be retained in the Ericales *sensu stricto* (Ericaceae and satellite families). Some families (Polemoniaceae, Balsaminaceae, Tetrameristaceae) are haplostemonous. Sterile stamens occur in an antesepalous position (Sapotaceae, primuloid clade), or more rarely opposite the petals (e.g. Diapensiaceae). Obhaplostemony (with or without sterile antesepalous stamens), while rare for the angiosperms, is common in the primuloid clade and Theaceae. Except for Symplocaceae, double stamen positions often occur in antepetalous position (e.g. Fouquieriaceae, Ebenaceae, Styracaceae, Pentaphyllaceae: Saunders, 1939; Dickison, 1993; Schönenberger and Grenhagen, 2005; Zhang and Schönenberger, 2014). Some families such as Actinidiaceae, Lecythidaceae and Theaceae have undergone a far-reaching specialization with secondary stamen multiplications on a ring primordium and a reversal to a spiral flower (Van Heel, 1987; Tsou, 1998; Tsou and Mori, 2007). A lateral increase of stamens in a single girdle is a common feature in several families and represents an apomorphic tendency (Actinidiaceae, Sarraceniaceae, Marcgraviaceae, Pentaphyllaceae, *Omphalocarpum* in Sapotaceae). A recurrent character is the fact that petals are often large and sepal-like and arranged in an imbricate aestivation, and in many families of Ericales, sepals, petals and androecium arise in a spiral sequence (Ronse De Craene, 2008).

Schönenberger, Anderberg and Sytsma (2005) reconstructed morphological characters on their phylogeny and concluded that sympetaly has arisen on more than one occasion. They also assumed that diplostemony is derived and that haplostemomy is plesiomorphic because it is present in basal groups. This assumption is erroneous as diplostemony is present in several subclades and there is no morphological evidence to support the *de novo* development of a second whorl, even if the most parsimonious reconstructions suggest this.

Monosymmetry is rare (some Ericaceae, Polemoniaceae, Balsaminaceae) and is expressed at a late stage of development. There is a strong tendency for the formation of petal tubes, including petals with or without stamens (e.g. Ericaceae, Lecythidaceae, Ebenaceae, Sapotaceae, Primulaceae, Theaceae). The development of a stamen-petal tube tends to be inconsistent and the attachment of stamens to the petal tube may fluctuate within a family (e.g. Polemoniaceae: Schönenberger, 2009) or the tube may be very short (e.g. Theaceae). The corolla is rarely reduced or lost and this may be linked to the ubiquitous animal pollination in the clade. The gynoecium is generally syncarpous, often isomerous with five carpels and axile placentation, or reduced to three (-two). Comparison of gynoecium development in several Ericales shows an identical invagination of septa, which may or may not connect in the middle of the ovary (pers. obs.). In the primuloid clade, placentation is free-central, but this is probably derived by a reduction of the septa. A disc nectary is frequently present and closely linked with the base of the ovary (e.g. Ericaceae) or nectaries are variously developed (e.g. Primulaceae: gynoecial or trichome nectary) or are absent (e.g. Actinidiaceae, Ebenaceae) (Bernardello, 2007).

'Balsaminoid Clade'

Balsaminaceae

Figure 11.5 *Impatiens platypetala* Lindsey

↑K3C5A5 G̲(5)

Balsaminaceae is part of a basal clade with Marcgraviaceae and Tetrameristaceae. Von Balthazar and Schönenberger (2013) enumerate a number of floral morphological and embryological characters shared between the families.

The family contains two genera *Hydrocera* and *Impatiens*. The floral diagram of *Impatiens* (Figure 11.5) has a number of interesting features. The presence of bracteoles is controversial as they are either considered absent and lateral adaxial organs are designated as sepals (Eichler, 1878; Caris et al., 2006; Janssens, Smets and Vrijdaghs, 2012), or interpreted as bracteoles that have become incorporated in the flower (Von Balthazar and Schönenberger, 2013). Evidence for the presence of bracteoles is their presence in the other families of the balsaminoid clade, their earlier initiation in the flower, the presence of empty spaces where lateral abaxial sepals would be expected and their smaller size at maturity. As a result, one can conclude that the abaxial lateral sepals are suppressed, resulting in a trimerous calyx (Von Balthazar and Schönenberger, 2013). Additional suppression of the two latero-abaxial sepals in several *Impatiens* species (Caris et al. 2006), occasionally with the retention of the

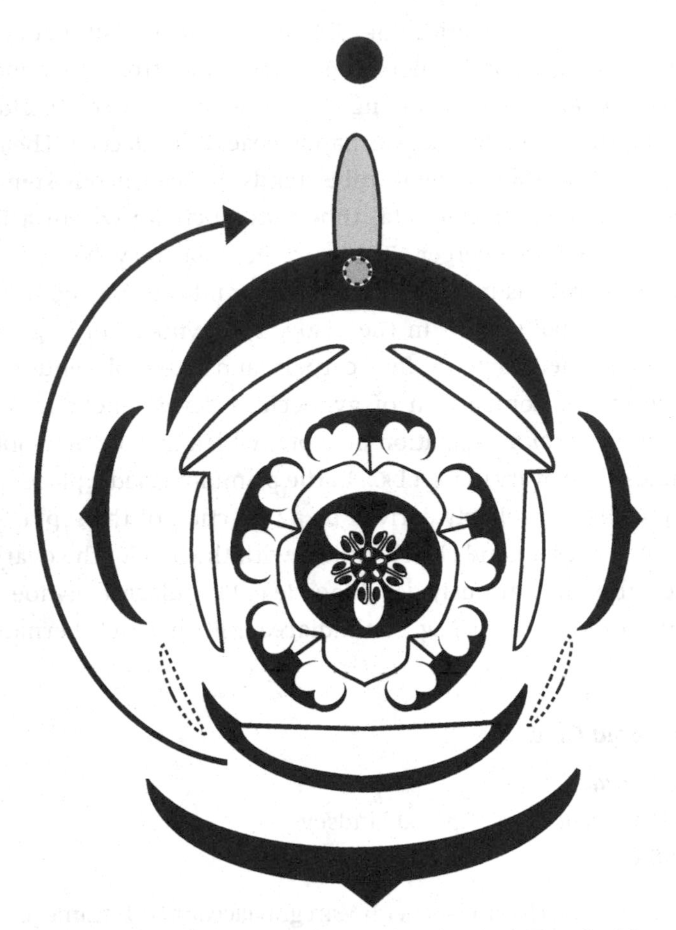

Figure 11.5 *Impatiens platypetala* (Balsaminaceae). The adaxial petal is half sepaloid.

vasculature (Grey-Wilson, 1980), would lead to a single adaxial sepal (see discussion in Von Balthazar and Schönenberger, 2013). The flower is resupinate at maturity, adding confusion to the interpretation of the nature of organs. The posterior sepal has a basal nectar spur arising late in development. Five petals are present; the laterals are basally fused in pairs in *Impatiens*, sometimes forming a single unit, while they are free in *Hydrocera* (Janssens, Smets and Vrijdaghs, 2012). The anterior petal is larger and appears to have a dual morphological nature of calyx and corolla. Rama Devi (1991b) and Caris et al. (2006) thought the petal to have arisen by the congenital fusion of the antero-lateral sepals and the anterior petal, resulting in a hybrid organ. The flowers used for the present floral diagram had a dual nature in the abaxial petal being half petaloid with a dorsal crest. It is possible that increased monosymmetry has broken down any clear delimitation of whorls with gene activity of sepals invading the anterior petal. In some species, the anterior petal continues the spiral sequence of the sepals and behaves as a sepal (Caris et al., 2006). Interestingly, the bracteoles, petaloid sepals and all petals have a similar *SEP3*-like gene expression, which is normally only present in the three inner whorls of eudicot flowers (heterotopic petaloidy: Geuten et al., 2006). A similar expression was also reported for *Marcgravia* (Marcgraviaceae), the putative sister group of Balsaminaceae.

The five antesepalous stamens are free but develop ventral appendages on their filaments that become connivent around the ovary. Monosymmetry is also expressed in the androecium, as the posterior stamen is smaller than the lateral stamens (possibly sterile?). Anthers open latrorsely together with abundant secretion on the inside of the filaments while pollen is loaded between the flaps and anthers (proterandry with secondary pollen presentation?). However, pollination mechanisms may be variable between different species depending on the strength of the postgenital fusion of anthers (Von Balthazar and Schönenberger, 2013). Anthers remain attached around the developing fruit while the filaments break off. There are always five carpels with two rows of ovules, with a tendency for reduction to a single row or a single ovule per carpel (Caris et al., 2006; Janssens, Smets and Vrijdaghs, 2012). There is no distinct or very short style and stigmatic tissue forms commissural lobes.

The floral structure of *Hydrocera* appears less strongly monosymmetric and corresponds to the plesiomorphic condition of Balsaminaceae: five sepals are present with the antero-lateral sepals of similar size as the laterals, and there is no fusion between the petals (Eichler, 1878; Janssens, Smets and Vrijdaghs, 2012). The ancestral flower of Balsaminaceae was probably regular and haplostemonous.

'Polemonioid Clade'

Polemoniaceae

Figure 11.6 *Polemonium acutiflorum* Willd. (pers. obs. and Caris, 2013 for P. *coeruleum*)

∗ K(5) [C(5) A5] G(3)

General formula: ∗/↓ K(4–)5(–6) C(4–)5(–6) A(3–)5(–6) $\underline{G}$(2)–3(–5)

Polemoniaceae form a clade with Fouquieriaceae. Inflorescences are determinate and bracts are variably developed without bracteoles. Flower shape is highly diverse with different lengths of stamen-petal tubes as an adaptation to different pollinators. Flowers are mostly actinomorphic, with a few median monosymmetric genera (e.g. *Cobaea*), pentamerous or tetra- to hexamerous in

Figure 11.6 *Polemonium acutifolium* (Polemoniaceae): partial inflorescence. Note the hairy bases of petals and filaments.

Linanthus (Byng, 2014). Flowers are always haplostemonous with stamens of slightly unequal length; in *Phlox* these are arranged at different levels similar to some Solanaceae. In *Polemonium* a short stamen-petal tube is formed and in its upper part the hairy base of the filaments form protruding ridges compartimentalizing the floral tube into five sections (Schönenberger, 2009). Flowers are characterized by late sympetaly (Erbar, 1991). Petals are very slow to develop in *Polemonium coeruleum*, giving the impression of arising after the stamens, but are contort at maturity (Caris, 2013). This is in contrast to Fouquieriaceae with petals continuing the spiral sequence of the calyx with quincuncial aestivation (Schönenberger and Grenhagen, 2005). The ovary is always tricarpellate with strongly developed septa and ovules in two irregular rows within each locule. Dawson (1936) considered the nectary staminodial and the androecium originally two-whorled, as abortive staminal traces are found opposite the petals, occasionally supplying a five-lobed nectar disc. The presence of diplostemony in Fouquieriaceae (Schönenberger and Grenhagen, 2005) suggests that ancestors of the clade had two whorls of stamens.

Lecythidaceae (including Scytopetalaceae)

Figure 11.7 *Napoleonaea vogelii* Hook. and Planch.

✶ K(5) [C(5)A80°+40°+5$^{2°}$+5^{2}][a] Ğ(5)

General formula: polysymmetric flowers: ✶ K(3–)4–5 C4–5(–18) A∞G4–5(–8); monosymmetric flowers ↑K6 C6 A∞ G(3)4–6

This large tropical family of mainly trees is highly diverse in floral structure. Flowers are usually large and merism tends to be variable with tetra- or hexamery being the commonest number. Complexity is expressed mainly in the androecium, which is polyandric as a rule. Very often, the flower is strongly monosymmetric by the development of an abaxial staminodial hood covering the other stamens. The gynoecium is semi-inferior to inferior with axile placentation. Flowers of Lecythidaceae are adapted to

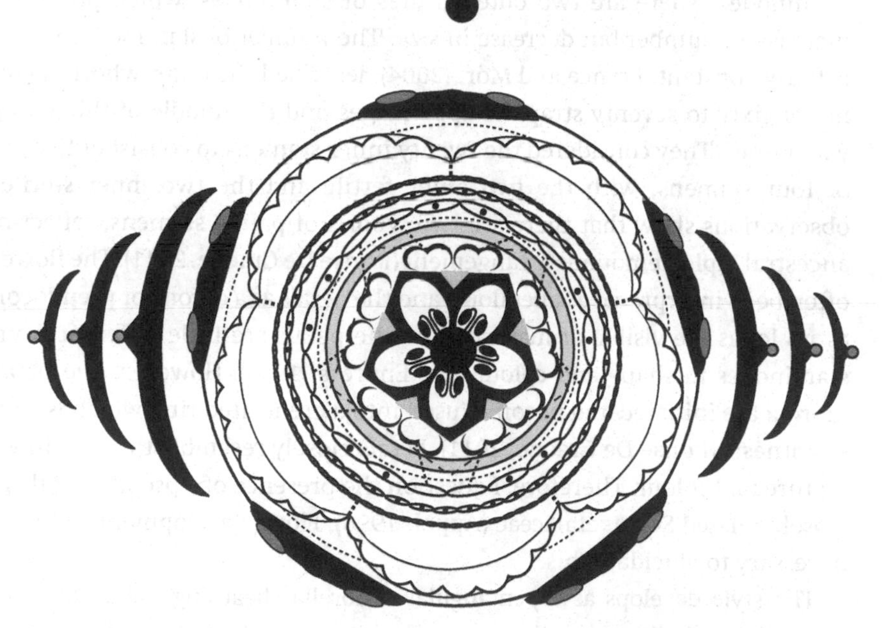

Figure 11.7 *Napolaeonaea vogelii* (Lecythidaceae). Curving of stamens over the nectary reflected by broken line connecting anthers to their insertion base. The pentagonal broken line delimits the stigma covering the anthers.

[a] Ten fertile half stamens in pairs and ten staminodes in between.

a wide range of pollinators with specialized mechanisms mainly linked to the androecium (Tsou and Mori, 2007). Nectar is produced by an intrastaminal disc (e.g. *Napoleonaea, Barringtonia*) or occasionally from a modified part of the androecium, which acts as fodder or provides nectar for pollinators (several Lecythidoideae).

The androecium arises as a ring primordium with centrifugal initiation of a very high number of stamens. In polysymmetric genera such as *Gustavia* or *Barringtonia*, stamens cover the ring primordium in an unordered mass (Endress, 1994; Tsou and Mori, 2007). Monosymmetry is variously expressed in the family, including perianth, androecium and gynoecium to variable degrees. A flap of sterile tissue is detached from the anterior side of the ring primordium and grows over the fertile stamens. A diplostemonous androecium is found in *Cariniana micrantha*.

Napoleonaea, a West African genus at the base of the Lecythidaceae, is included as a representative of polysymmetric flowers in the family (Figure 11.7; Ronse De Craene, 2011). The androecium is highly complex with several series of threadlike or flap-like staminodes, surrounding ten fertile monothecal stamens inserted in five pairs and alternating with five pairs of staminodes. There are two outer girdles of staminodes, which progressively increase in number but decrease in size. The number of staminodes per whorl is fairly constant. Prance and Mori (2004) described the outer whorl as consisting of sixty to seventy straplike appendages and the middle of thirty to forty wider ones. They considered the twenty inner stamens to consist of five groups of four stamens, with the two outer fertile and the two inner sterile. My observations show that there are two whorls of paired stamens, reflecting an ancestral diplostemonous arrangement (Ronse De Craene, 2011). The flower has often been interpreted as apetalous, and the petals as a corona or pseudocorolla, as no lobes are visible at maturity and the petals are little different from the staminodes in shape and colour (e.g. Endress, 1994). However, five petal primordia are initiated and become fused into a crenelated rim, which is reflexed at anthesis (Ronse De Craene, 2011). Petals closely resemble the staminodes in texture and colour. Therefore, I question the presence of a pseudocorolla in the closely related Scytopetalaceae (Appel, 1996). Floral developmental studies are necessary to elucidate this.

The style develops as a pentangular umbrella sheathing the anthers, which are bent in a hollow space between nectary and ovary. As the flower is hanging downwards, *Napoleonaea* could share the same pollination mechanism with Sarraceniaceae, another family in Ericales (Ronse De Craene, unpubl. data). However, pollinators of *Napoleonaea* are remarkably small thrips that find shelter in the inner chamber between style and curved filaments (Frame and Durou, 2001).

Pentaphyllaceae (incl. Ternstroemiaceae)

Figure 11.8 *Cleyera japonica* Thunb.

∗ K5 C5 A5^5 $\underline{G}$(2–3)

General formula: ∗ K5 C5(–10) A(5)10–∞ G(1)2–5

Figure 11.8 *Cleyera japonica* (Pentaphyllaceae). Stamens develop laterally from common antesepalous primordia; bracteoles with terminal gland; numbers give sequence of perianth initiation.

The family shares several morphological similarities with Theaceae, although they differ in embryological and certain floral characters (Soltis et al., 2018). The immediate sister family is Sladeniaceae, but it is poorly known. Flowers are single in the axil of leaflike bracts. Both sepals and petals arise in a spiral sequence, in continuation of two bracteoles, and are imbricate in bud. In some genera, petals and sepals are superposed, as petals continue the 2/5 sequence of the sepals without interruption (Ternstroemieae). The number of corolla lobes may vary with up to ten petals. In *Cleyera*, the first petal arises opposite sepal one, but becomes shifted in alternate position due to variable growth. The androecium is mostly multistaminate, except for *Pentaphylax*, which is haplostemonous. Stamens are uniseriate in *Cleyera* but they arise laterally from five antesepalous primordia (Ronse De Craene, 2018). In *Ternstroemia* a ring primordium is formed with stamens preferentially developing laterally from the antesepalous initials; in addition more stamens arise centripetally (Zhang and Schönenberger, 2014). In *Eurya* and *Visnea* stamens develop apparently in two whorls with paired antepetalous stamens (Payer, 1857; Zhang and Schönenberger, 2014). It is possible that these represent antesepalous triplets rather than two whorls, as in *Kirengeshoma* (Figure 11.3A). In some genera, flowers are unisexual with a non-functional or absent ovary in staminate flowers, and a reduced androecium in pistillate flowers (e.g. *Eurya*). An intrastaminal disc is present in *Cleyera*, but is absent in other genera such as *Ternstroemia*. The ovary is superior and carpels develop sequentially, often with one carpel conspicuously smaller. Placentation is axile and there is a single style, topped with carinal stigmatic lobes.

'Primuloid Clade'

Sapotaceae

Figure 11.9.A *Payena leerii* (Teijsm. and Binn.) Kurz
✱ K2+2 C(4+4) A8+8 G(8)

Figure 11.9.B *Manilkara zapota* L.
✱ K3+3 C(6) A6°+6 $\underline{G}$(6–9)

Figure 11.9.C *Palaquium amboinense* Burck
✱ K3+3 C(6) A6+6 $\underline{G}$(6)

Figure 11.9.D *Sideroxylon inerme* L.
✱ K5 C(5) A5°+5 $\underline{G}$(5)

Diagrams based on Caris (2013) and Kümpers et al. (2016)
General formula: ✱ K4–5(–12) C4–6(9–18) A4–6(12–∞) G1–6(15–30)

Flowers are generally bisexual or unisexual and arranged in fascicles, occasionally solitary (*Manilkara*). Bracts and bracteoles are present but generally overlooked at maturity. Although pentamery represents the ancestral condition, a reduction to tetramery or increases to a higher merism (mainly hexamery or octomery) have occurred repeatedly through different evolutionary pathways (Kümpers et al., 2016). Increased merism may affect all organs equally (e.g. *Palaquium*, Figure 11.9C), or may affect different organs (stamens, petals, ovary: *Madhuca*, *Payena*, Figure 11.9A), or only carpels. The calyx is mostly imbricate and in a single whorl but can be biseriate with two whorls with the outer whorl valvate and the inner one imbricate in bud (*Palaquium*, Figure 11.9C). With merism increase the sepal whorl (and occasionally the petals also) becomes effectively rearranged in two whorls of two, three or four sepals (*Payena*, *Manilkara*).

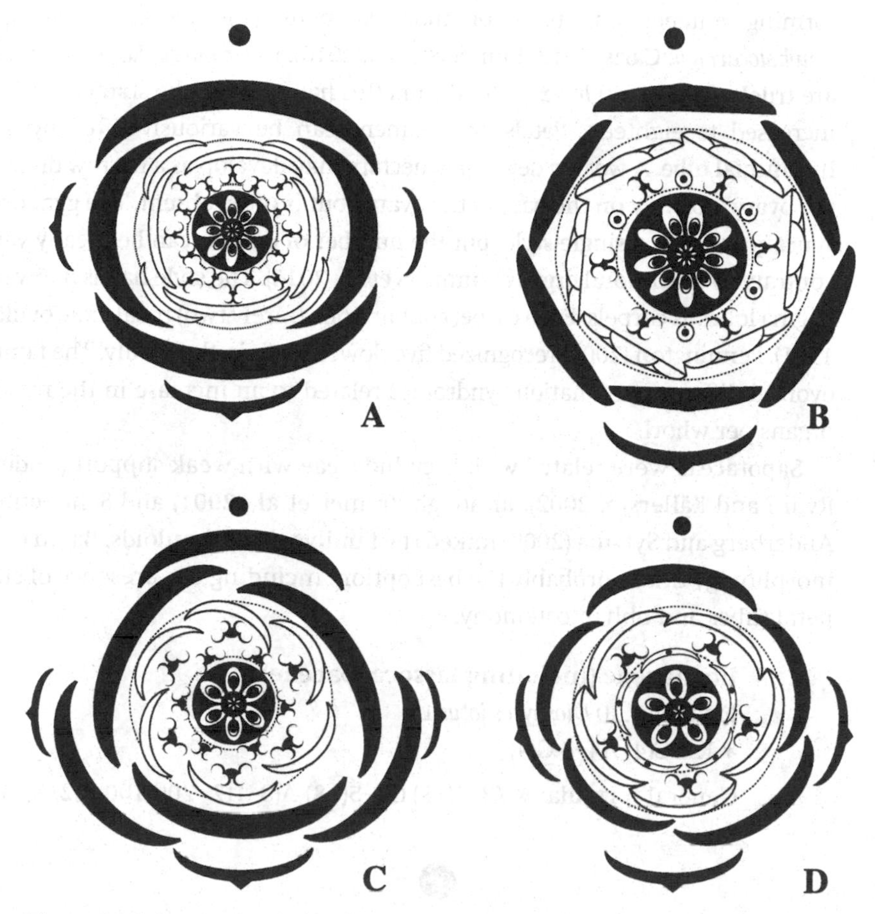

Figure 11.9 Sapotaceae. A. *Payena leerii*; B. *Manilkara zapota*, note trilobed petals; C. *Palaquium amboinense*; D, *Sideroxylon inerme*

The corolla is always fused, arising as the result of late sympetaly, and the number of lobes can be either the same as sepals, or two to three times that of the sepals. Petal lobes are often divided in a median and two lateral segments (*Manilkara*, *Mimusops*). The lateral lobes spread horizontally while the median lobe remains erect and clasps the opposite extrorse anther (Pennington, 2004). Lobes have been confused with staminodes (e.g. Endress and Matthews, 2006b) as they develop in a way comparable to a lateral dédoublement of stamens. Diplostemony is found only in tribe Isonandreae (e.g. *Payena*, *Palaquium*: Figure 11.9A, C). In other tribes the antesepalous stamens are sterile or absent and fertile stamens are superposed to the corolla lobes (obhaplostemony: *Sideroxylon*, Figure 11.9D; *Pouteria*). Staminodes can take all sizes and are occasionally fertile in *Manilkara*. Flowers become obdiplostemonous by an inward shift of antesepalous stamens/ staminodes. In some tribes, the number of stamens is increased laterally by forming antepetalous pairs or more stamens (e.g. *Madhuca*, *Magodendron*, *Omphalocarpum*: Caris, 2013; Kümpers et al., 2016). In *Labourdonnaisia* the six petals are trilobed (eighteen lobes in total), and this has induced the stamens to become increased to eighteen. Petals and stamens can be variously lifted by a short hypanthial tube. A weakly developed nectary may develop as a narrow disc around the ovary (possibly on the base of the ovary), but is often absent. The gynoecium is syncarpous with a single style, but the number of carpels can be greatly variable, contrary to other floral organs (Kümpers et al., 2016). The style has as many carinal stigma lobes as carpels. Each carpel contains one (rarely two) basal-axile ovules (Ng, 1991). Pennington (2004) recognized five flower types in the family. The family has evolved different pollination syndromes related to an increase in the number of organs per whorl.

Sapotaceae were related with Lecythidaceae with weak support (Anderberg, Rydin and Källersjö, 2002), although Bremer et al. (2001) and Schönenberger, Anderberg and Sytsma (2005) linked the family with primuloids. Based on floral morphology this is probably the best option, including the presence of stamen-petal tubes and obhaplostemony.

Ebenaceae (including Lissocarpaceae)

Figure 11.10 *Diospyros lotus* L.

∗ K 4 C(4) A4+4 $\underline{G}$(4)

General formula: ∗ K3–5(–8) C3–5(–8) A(3–)12–20(–100) G2–8(–16)

Figure 11.10 *Diospyros lotus* (Ebenaceae), inflorescence. Note the partially developed false septa.

The family includes four genera, of which the large genus *Diospyros* is highly variable. Inflorescences are cymose and axillary. *D. lotus* has occasionally bisexual flowers, but flowers are mostly unisexual and heteromorphic, with a reduced pistillode in staminate flowers and a different merism with fewer (antesepalous) staminodes in pistillate flowers. Flowers are often tetramerous (octomerous in *Lissocarpa*). The calyx is basally fused and accrescent; the corolla is sympetalous with reflexed lobes, and stamens are inserted in the lower part of the tube. Stamens are mostly diplostemonous, often with paired stamens in antepetalous position, or stamens are in fascicles. When reduced to a single whorl, the position of stamens is opposite the sepals. In staminate *D. lotus* stamens are increased centripetally by a radial subdivision of the eight initial stamen primordia (Caris, 2013). The carpels have two ovules on apical-axile placentae, generally with a false septum (Ng, 1991; Caris, 2013). There is a single short style or styles are free. A disc nectary is rarely developed (Wallnöfer, 2004).

Primulaceae sensu lato *(including Maesaceae, Theophrastaceae, Myrsinaceae, Samolaceae)*

General floral formula: ∗K4–5(–9) C(0)4–5(–9) A4–5(–9) G3–5

The circumscription of Primulaceae and allied families (previously Primulales) was revisited by Källersjö, Bergqvist and Anderberg (2000) and Anderberg, Rydin and Källersjö (2002) among others, who found that the traditional circumscription of Primulaceae was artificial, as Myrsinaceae appears to be nested within the family. The different clades represent a natural entity with strong morphological support and obvious synapomorphies. Main connecting characters are obhaplostemony (occasionally with a whorl of antesepalous staminodes variously developed as petaloid appendages or as reduced stubs, sympetaly, and an ovary with free-central placentation (Anderberg and Stahl, 1995). The development of a stamen-petal tube is different from other asterids, developing after the formation of common stamen-petal primordia (cf. Erbar and Leins, 2011). The placentation is conspicuous as a broad central column with ovules in a single or several rows. The tip of the placenta is usually extended as a stalk penetrating the base of the single style. The stalk can be short and blunt (e.g. Myrsinoideae) or very long and filamentous (e.g. *Soldanella*: Caris, 2013). The number of carpels making up the ovary is often difficult to verify, as no septa are present and the ovary wall develops as a single ring primordium. However, the number of stylar lobes, fruit dehiscence patterns, ridges between the ovules and vascular anatomy give indications of the number of carpels (Sattler, 1962; Ronse De Craene, Smets and Clinckemaillie, 1995). Obhaplostemony is shared with

Sapotaceae within the primuloid clade. Filaments can be variously fused in a staminal tube or fused to the corolla, resulting from common basal growth.

The division in subfamilies is reflected in the nectaries. Most Theophrastoideae have trichome nectaries (also found in *Anagallis*, *Aegiceras*, *Glaux* and *Lysimachia* of Myrsinoideae), which were interpreted as basal in primuloid taxa by Vogel (1997). Gynoecial nectaries are found in *Maesa*, *Samolus*, and Primuloideae (Vogel, 1997). Presence of stamen-petal primordia was interpreted as a derived condition by Sattler (1962) and Ronse De Craene, Clinckemaillie and Smets (1993) as a delay in the inception of either petals or stamens. Floral evolution has proceeded independently in the different subfamilies with several transitional genera (Källersjö, Bergqvist and Anderberg, 2000). The basalmost Maesoideae have bracteoles (absent in other primuloid families) but lack staminodes and have a semi-inferior ovary (Caris et al., 2000). Some Theophrastoideae (e.g. *Clavija*) and Myrsinoideae are dioecious, while heterostyly occurs mainly in Primuloideae.

Theophrastoideae (incl. Samolus)

Figure 11.11 *Bonellia macrocarpa* (Cav.) B. Ståhl and Kallersjö

$\ast$ K5 [C(5)A5°]+(5) $\underline{G}$(3)

Flowers are generally tetra- to pentamerous, arranged in cymose to racemose inflorescences. Sepals are lifted by a hypanthium and have a cochleate aestivation (Caris and Smets, 2004). All Theophrastoideae possess staminodes that arise with the petals on the margins of a ring primordium and become incorporated in the petal tube. Contrary to other Primuloid subfamilies, stamens and petals do not arise from common primordia (except *Samolus*). Anthers are commonly extrorse (except *Samolus*: Källersjö, Bergqvist and Anderberg, 2000). Caris and Smets (2004) did not discuss carpel numbers as the gynoecium arises as a ring primordium. However, stylar lobes or fruit valves can be an indication of carpel number as two lobes were seen in *Deherainia*, three in *Jacquinia* and *Clavija*, and five valves in *Samolus*. A large number of ovules develop on the central placentation, except for *Clavija* with fewer ovules. Ovules are not embedded in placental tissue and the tip of the column is weakly developed.

Samolus differs in several characteristics from other Theophrastoideae (half-inferior ovary, synsepaly with mostly quincuncial aestivation, obvious stamen-petal primordia, gynoecial nectary, capsular fruits versus berries, free cochleate sepals, no obvious stamen-petal primordia, trichome nectaries). However, they share other characteristics, such as the external staminodes (occasionally absent in *Samolus*), the fusion of bract with the pedicel and a similar placenta with many ovules (Caris, 2013).

Figure 11.11 *Bonellia macrocarpa* (Primulaceae-Theophrastoideae)

Primuloideae

Figure 11.12 *Soldanella villosa* Darracq

∗ K5 [C(5) A5°+5] G(5)

Flowers are grouped in cymose inflorescences, generally with a fused calyx and corolla. In *Soldanella*, sepal lobes are free and the corolla is divided in a variable but higher number of segments (up to twenty). Petals are usually imbricate, although this is not visible in mature flowers of *Soldanella*. Obhaplostemony is the common androecial configuration, although staminodial antesepalous scales are present in some genera, including *Soldanella*. The staminodes of *Soldanella* become embedded in the corolla tube (Caris, 2013). The scales stand opposite a separate vascular bundle in the corolla that was interpreted as remnant of the vascular supply of staminodes (e.g. Saunders, 1939). Scales on the corolla tube are present in *Androsace* and *Douglasia* but lack vascular tissue (Anderberg and Ståhl, 1995). The ovary has a basal nectary consisting of stomata that occasionally extends to the base

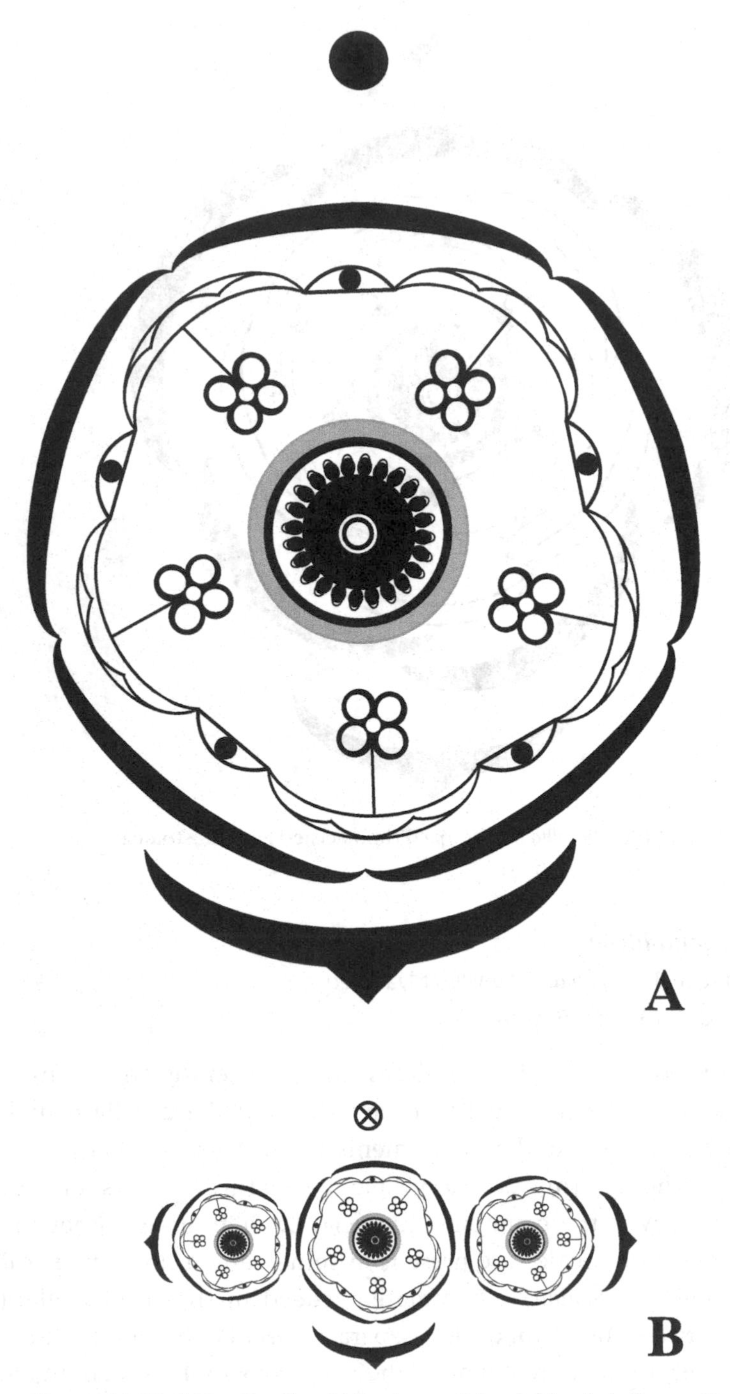

Figure 11.12 *Soldanella villosa* (Primulaceae-Primuloideae): A. flower; B. partial inflorescence. Note the formation of lobes on the fused corolla.

of the style (*Dionysia*: Cano, 2012). The central placentation bears a high number of ovules that are not embedded in placental tissue.

Reflexed corolla lobes are present in *Dodecatheon*, resembling *Cyclamen* (Myrsinoideae) as an adaptation to a similar buzz pollination mechanism (Källersjö, Bergqvist and Anderberg, 2000).

Clinckemaillie and Smets (1992) pointed to the high numbers of similarities between the floral diagrams of Plumbaginaceae and Primulaceae, which appears to be the result of homoplasy. The main differences are the basal placentation against the free central placentation and location of nectary tissue, besides a different floral development. The superficial similarity is comparable to the recognition of Rhoeadales linking Brassicaceae with Papaveraceae.

Myrsinoideae (incl. Lysimachiaceae, Aegicerataceae, Coridaceae)

Figure 11.13 *Anagallis arvensis* L.

∗ K5 [C(5)A5] $\underline{G}$(5)

General formula: ∗ (↓) K(3)4–5(−7) C(3)4–5(−7)/0 A(3)4–5(−7)+(4°–5°) G5?

Flowers appear solitary in the axil of a single bract in *Anagallis*, although other Myrsinoideae have a racemose inflorescence. Flowers are mostly pentamerous; in *Trientalis* (*Lysimachia*) *europaea*, there is an unstable heptamery. Flowers usually have a well-developed corolla (fused at the base), but it is absent in *Glaux*. Calyx and corolla are contorted or imbricate. In *Embelia* petals are exceptionally free (Anderberg and Ståhl, 1995). The filaments are hairy in *Anagallis* (trichome nectaries). In some *Lysimachia* glandular tissue consists of oil glands present on the corolla and stamens. Other members of Myrsinoideae have hanging flowers with poricidal anthers that are mainly buzz pollinated. The ovules are numerous in several series or in a single series of (3) 4–6 and are generally embedded in the placental tissue without or with a short placental tip (e.g. *Lysimachia*, *Hymenandra*, *Myrsine*, *Parathesis*, *Embelia*: Ma and Saunders, 2003; Caris, 2013). Staminodes are generally absent (present in *Myrsine*). Stamens and petals developed on common primordia in all Myrsinoideae that were investigated (Ronse De Craene, Smets and Clinckemaillie, 1995; Ma and Saunders, 2003; Wanntorp et al., 2012; Caris, 2013). The two basal genera *Stimpsonia* and *Ardisiandra* appear transitional between Primuloideae and Myrsinoideae in lacking embedded ovules and in having a similar gynoecial nectary in *Ardisiandra* (Wanntorp et al., 2012). *Coris* is unusual in having an epicalyx consisting of

Figure 11.13 *Anagallis arvensis* (Primulaceae-Myrsinoideae)

five spines and a monosymmetric corolla with imbricate aestivation (Ronse De Craene, Smets and Clinckemaillie, 1995).

Theaceae

Figure 11.14 *Stewartia pseudocamellia* Maxim., based on pers. obs., Erbar (1986) and Tsou (1998)

$\ast$ K5 C5 A5$^{\infty}$ $\underline{G}$(5)

General formula: $\ast$ K(4)5(−14) C(4)5–7 A5$^{\infty}$–∞ G(2)3–5(−10)

Theaceae was considered as a major family of Theales (Cronquist, 1981), an amalgamation of families now placed in various orders. Four subfamilies used to be recognized, but Theaceae is now restricted to three tribes, while Bonnetiaceae is placed in the clusioid clade of Malpighiales, and

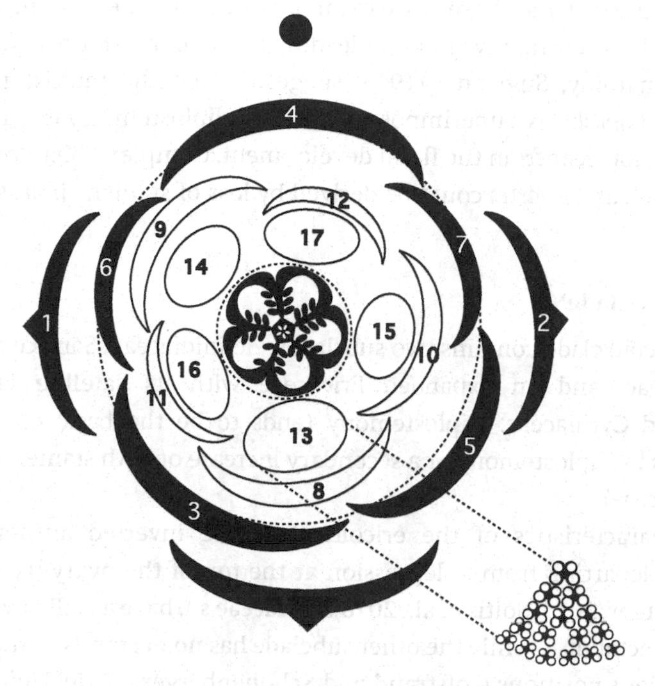

Figure 11.14 *Stewartia pseudocamellia* (Theaceae). Numbers give the order of initiation from bracteoles onwards; common stamen primordia divide centrifugally in many stamens (shown for stamen group labeled 13).

Tetrameristaceae and Ternstroemioideae are placed in other clades of Ericales, and Asteropeiaceae in Caryophyllales (Stevens, 2001 onwards).

Flowers are solitary or grouped in axillary cymes. The distinction between bracteoles, sepals and petals is unclear with all organs arising in a spiral sequence. Tsou (1998) recognized two major groups on the basis of perianth phyllotaxis and differentiation. The first group, including *Camellia*, has a variable number of large sepals (ten to fourteen) and petals (five to seven) arising in a spiral sequence, with bracts, sepals and petals intergrading into each other (Sugiyama, 1991; Tsou, 1998). The multistaminate androecium arises centrifugally on a ring primordium and stamens are often basally connected into a tube, as are the petals (Payer, 1857; Tsou, 1998). The second group, where *Stewartia* belongs, is pentamerous with a better distinction between sepals and petals (still with spiral initiation) and stamens are grouped in antepetalous fascicles (Figure 11.14; Erbar, 1986; Tsou, 1998). Stamens develop centrifugally on the five stamen fascicles.

The gynoecium is tri- to five carpellate with basal-axile placentation. The invaginating septa are weakly fused in the centre, at least for part of the genera.

Styles are often distinct and stigmas are carinal. The development of the flower is accompanied by a central invagination leading to a half inferior ovary. Based on the floral anatomy, Sugiyama (1995) suggested that the multistaminate androecium of *Camellia* is superimposed on an obdiplostemonous Bauplan, although this is not clear from the floral development. Complex obhaplostemonous flowers such as *Stewartia* could be derived by loss of stamens in antesepalous sectors.

'Ericoid Clade'

The ericoid clade contains two subclades, Actinidiaceae, Sarraceniaceae and Roridulaceae, and an expanded Ericaceae with its satellite families Clethraceae and Cyrillaceae. Diplostemony tends to be the basic condition, with reductions to haplostemony or a secondary increase of both stamen whorls (e.g. Sarraceniaceae).

Frequent characteristics of the ericoid clade are inverted anthers and a hollow style departing from a depression at the top of the ovary (Löfstrand and Schönenberger, 2015; Soltis et al., 2018). Ericaceae s.l. have a well developed intrastaminal nectar disc, while the other subclade has no nectaries or these are rare and in various positions (Löfstrand and Schönenberger, 2015). Unique for Ericales is the petal tube in Ericaceae which is (mostly) separate from the stamens, while other Ericales generally have a stamen-petal tube when fusion occurs.

Actinidiaceae

Figure 11.15 *Saurauia serrata* DC

✶ K5 [C(5)A(∞)] G(5)

General formula: ✶ K(3–)5(–9) C(3–)5(–8) A10–∞ G3–5(–∞)

The family contains the three genera *Clematoclethra*, *Saurauia* and *Actinidia*.

Flowers are arranged in cymes or are solitary. All genera share a short stamen-petal tube. Both sepals and petals are imbricate, with petals continuing the 2/5 initiation of the sepals (Figure 11.15; Caris, 2013).

Clematoclethra is diplostemonous, while other genera are multistaminate (Dickison, 1972; Löfstrand and Schönenberger, 2015). *Saurauia* shows a secondary lateral stamen increase (up to fifty or more) with all stamens fused at the base. The first stamens are paired in antesepalous position and more stamens develop laterally and centrifugally forming up to three tiers (Ronse De Craene, unpubl. data). Van Heel (1987) describes a lateral stamen increase in *Actinidia melandra* as developing from a single whorl in alternation with the petals. In a similar way to *Diospyros* additional stamen tiers may become

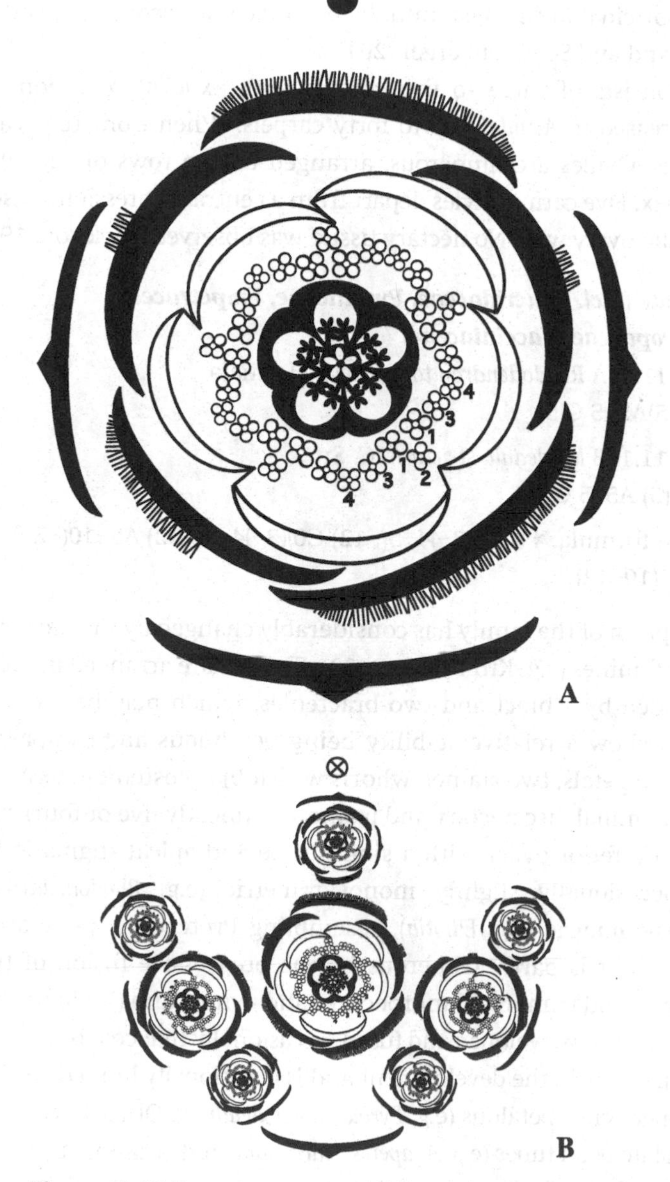

Figure 11.15 *Saurauia serrata* (Actinidiaceae): A. flower; B. partial inflorescence. Numbers give the order of initiation of the stamens. Note the hairs partially covering the abaxial side of the sepals.

detached centripetally by a radial subdivision, leading to four or five girdles of stamens as described in *A. deliciosa* (Caris, 2013). The pattern of development in Actinidiaceae corresponds to the common Ericales pattern of a lateral stamen increase as found in other families such as Pentaphyllaceae, Marcgraviaceae and Sarraceniaceae and represents an apomorphic tendency. Anthers are U-shaped and poricidal in *Saurauia*, initially extrorse but becoming inverted at anthesis (Löfstrand and Schönenberger, 2015).

The ovary consists of three to five carpels with axile placentation, or is secondarily increased in *Actinidia* up to forty carpels. When isomerous carpels are antesepalous. Ovules are numerous, arranged in two rows on an inflated central floral apex. Five carinal styles depart from a central depression caused by the bulging of the ovary wall. No nectary tissue was observed (Dickison, 1972).

Ericaceae (incl. Epacridaceae, Pyrolaceae, Empetraceae, Monotropaceae, Vacciniaceae)

Figure 11.16A *Rhododendron tolmachevii* Harmaja
↓ K5 C(5)A5+5 $\underline{G}$(5)

Figure 11.16B *Macleania stricta* A. C. Sm.
∗ K5 C(5) A5+5 Ğ(5)

General formula: ∗ (↓) K(3–)4–5(–12) C0/(3–)4–5(–12) A5–10(–24) G(3)4–5(10–12)

The circumscription of the family has considerably changed by the inclusion of several smaller families (e.g. Kron et al., 2002). Flowers are arranged in racemes and are subtended by a bract and two bracteoles, which may be reduced or absent. Flowers show a relative stability being pendulous and campanulate, with mostly fused petals, two stamen whorls with (ob)diplostemonous arrangement, an intrastaminal disc nectary and isomerous (mostly five or four) carpels in a superior to inferior ovary with a single style and apical stigmatic lobes. Flowers are occasionally slightly monosymmetric (e.g. *Rhododendron*), or strongly monosymmetric (e.g. *Elliotia*), resembling Proteaceae (pers. obs.). In *Tripetaleia* the flower is partly trimerous by the progressive fusion of two of the petals in pairs and trimery of androecium and gynoecium (Nishino, 1988).

The calyx is variously developed and fused, occasionally reduced to small lobes. Sympetaly occurs early in the development and is occasionally limited, leading to flowers that appear choripetalous (e.g. *Pyrola*, *Rhododendron*). Other Ericaceae produce long urceolate petal tubes (e.g. *Agapetes*, *Vaccinium*) and occasionally calyptras (*Richea*). Petals are absent in *Empetrum* and *Ceratiola*.

The obdiplostemonous arrangement of stamens is caused by a displacement during development (e.g. Leins, 1964a). Loss of antepetalous stamens leads to haplostemony in some *Rhododendron*. *R. semibarbatum* has a dimorphic androecium with three fertile stamens and two staminodes playing a role in limiting access to the nectary (Ono, Dohzono and Sugawara, 2008). The androecium is reduced to two (with two staminodes) in *Oligarrhena* (Byng, 2014). As common for the clade anthers are often inversed with the morphological lower part on

Figure 11.16 Ericaceae: A. *Rhododendron tolmachevii*; B. *Macleania stricta*. Note the ribs protruding from the receptacle.

top. There are often appendages or awns, which are associated with a poricidal dehiscence common to the family.

The gynoecium is mostly isomerous and antepetalous (exceptionally antesepalous) and develops congenitally fused carpels with intruding septal lobes (Caris, 2013). These lobes remain visible at the top of the style as commissural stigmatic appendages. As septa develop as intruding lobes, septation is only complete in the lower region of the ovary where centrally protruding placentae are formed. Ovules range from a single per locule to a high number of small ovules. In *Gaylussacia* the carpel number is increased to ten by the development of false septa (Eichler, 1875). A nectar disc is usually present as a multilobed ring around the base of the ovary or on top in the case of an inferior ovary (Figure 11.16B). In Styphelioideae the nectary consists of antepetalous scales, which are occasionally confluent. Cronquist (1981) and Caris (2013) interpret the scales as staminodial structures.

Rhododendron and *Enkianthus* have an unusual floral diagram in that the odd petal is not abaxial but adaxial (cf. Eichler, 1875). The combination of a reduction of bracteoles and smaller sepals places the lateral petals outside the other petals in bud; flowers are slightly to strongly monosymmetric with a difference in size between adaxial and abaxial petals and stamens.

11.2 Lamiids: Gentianales, Solanales, Boraginales and Lamiales

Figure 11.17 shows the phylogenetic tree of lamiids (Euasterids I) based on Soltis et al. (2018). Apart from Lamiales and Boraginales, relationships of families are fairly well resolved (Stevens, 2001 onwards). Bremer et al. (2001) list the following synapomorphies for the lamiids: hypogynous flowers, late sympetaly (with a few transitions to early sympetaly: Leins and Erbar, 2011), fusion of stamen filaments with the corolla tube, capsular fruits, opposite leaves and entire leaf margin. These characters are recurrent within the order. In addition, the calyx is well developed, often accrescent, stamens range from five to two, the ovary is mostly superior with axile placentation, surrounded by a broad nectar disc, and ovules are generally unitegmic. Lamiales have undergone an evolution to median monosymmetry, while the other orders are generally polysymmetric. When monosymmetric, the symmetry is oblique in Gentianales, Solanales and Boraginales. Flowers are haplostemonous (exceptionally with higher numbers: *Dialypetalanthus* and *Theligonum* in Rubiaceae) and an isomerous or bicarpellate gynoecium.

The basal clades of lamiids (Icacinales, Metteniusales, Garryales) have features that are rare or absent elsewhere. The frequent occurrence of pseudomonomery may indicate that the bicarpellate ovary of lamiids may have evolved from five-carpellate pseudomonomerous ancestors (González and Rudall, 2010; Endress and Rapini, 2014). Interestingly, similar flowers are found in some Cardiopteridales of the campanulids (Stemonuraceae, Cardiopteridaceae: Tobe, 2012; Kong et al., 2018), suggesting a common apomorphic tendency.

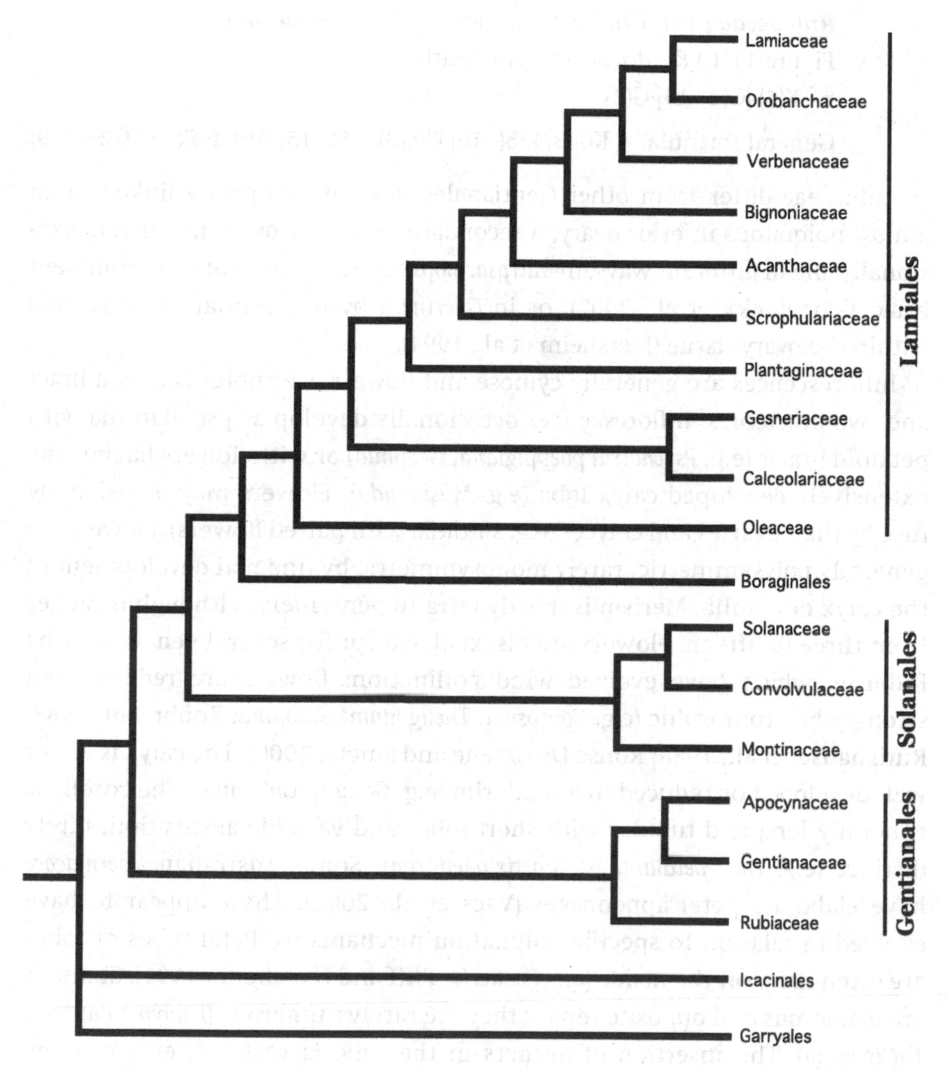

Figure 11.17 Phylogenetic tree of lamiids (euasterids I), based on Soltis et al. (2018)

Gentianales

The order Gentianales is a well-supported clade which consists of five families: Rubiaceae, Gentianaceae, Loganicaceae, Gelsemiaceae and Apocynaceae (Backlund, Oxelman and Bremer, 2000). Characteristics in common include interpetiolar stipules (with colleters), sympetaly with usually contorted corollas (rarely imbricate or valvate), a scarcity of (slight) zygomorphy and unisexual flowers, near absence of wind pollination and plasticity in the gynoecium (Nicholas and Baijnath, 1994; Backlund, Oxelman and Bremer, 2000).

Rubiaceae (incl. Dialypetalanthaceae, Theligonaceae)

Figure 11.18 *Rondeletia laniflora* Benth.

∗[b] K(4) [C(4) A4] Ğ(2)

General formula: ∗ K0/(3)4–5(–15) C0/(3)4–5(–15) A(3)4–5(–∞) G2–5(–9)

Rubiaceae differ from other Gentianales in early sympetaly linked to an almost ubiquitous inferior ovary. A secondarily superior ovary has arisen occasionally and in different ways, in *Mitrasacmopsis* by development of a prominent beak (Groeninckx et al., 2007), or in *Gaertnera* by differential zonal growth within the ovary tissue (Igersheim et al., 1994).

Inflorescences are generally cymose and flowers are subtended by a bract and two bracteoles. Inflorescences occasionally develop as pseudanthia with petaloid bracts (e.g. *Psychotria poeppigiana*, *Geophila*) or with flowers having one extensively developed calyx lobe (e.g. *Mussaenda*). Flowers may occasionally fuse by their ovaries and calyces (e.g. *Mitchella* with paired flowers). Flowers are generally polysymmetric, rarely monosymmetric by unequal development of the calyx or corolla. Merism is mostly tetra to pentamery, although it ranges from three to fifteen. Flowers are bisexual, except for several genera of tribe Rubieae, which have evolved wind pollination: flowers are reduced and strongly heteromorphic (e.g. *Coprosma*, *Theligonum*, *Galopina*: Robbrecht, 1988; Rutishauser et al., 1998; Ronse De Craene and Smets, 2000). The calyx is either well developed or reduced to a weak rim (e.g. *Galium*, *Galopina*). The corolla is generally long and tubular with short lobes and variable aestivation, rarely distinct (e.g. *Dialypetalanthus*, *Mastixiodendron*). Some Australian *Spermacoce* have elaborate petal appendages (Vaes et al., 2006). These appear to have evolved in relation to specific pollination mechanisms. Petal tubes or lobes are often hairy on the inside (e.g. *Paederia*: Puff and Igersheim, 1991). Stamens are isomerous and opposite sepals; they are rarely numerous (*Dialypetalanthus*, *Theligonum*). The insertion of anthers in the tube is variable, either at the corolla mouth, or at its base, more rarely at different levels of the tube (e.g. *Paederia*). The inferior ovary is topped by an epigynous nectary surrounding a single style. Two carpels occur in the majority of taxa, although five carpels are found in several genera (e.g. *Hamelia*; rarely up to sixteen in *Praravinia*: Robbrecht, 1988). Placentation is generally axile, apical, to basal. The gynoecium develops in two parts, a basal part producing the placenta and an apical part developing as an apical septum (e.g. Svoma, 1991; Ronse De Craene and Smets, 2000). Differential growth of the two parts leads to a variable

[b] Weakly monosymmetric.

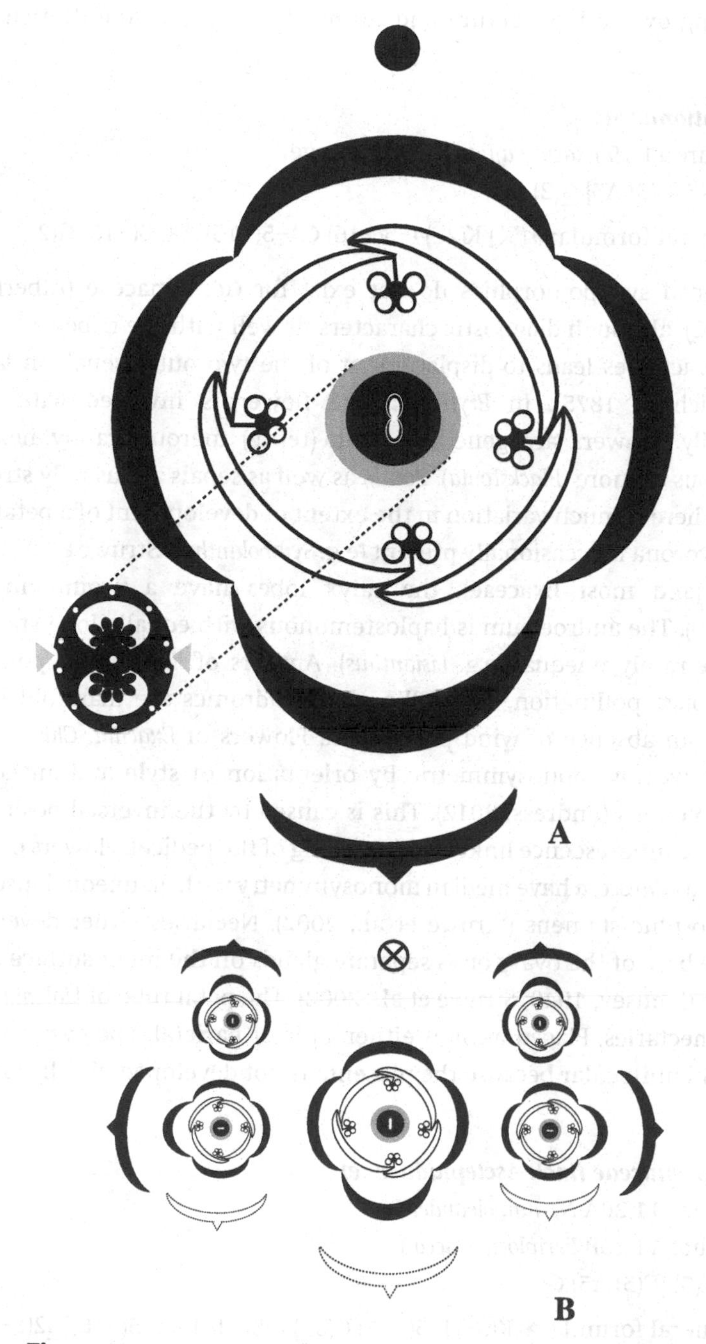

Figure 11.18 *Rondeletia laniflora* (Rubiaceae): A. flower; B. partial inflorescence

placentation. Ovules range from several on protruding placentae to a single basal ovule. The ovary of *Theligonum* is pseudomonomerous with a single basal ovule curving over a low septum and filling the empty locule (Rutishauser et al., 1998).

Gentianaceae

Figure 11.19 *Exacum affine* Balf.f. ex Regel

↖ K5 [C(5) A5] $\underline{G}$(2)

General formula: ✶(↖) K(2–)4–5(–16) C4–5(–16) A4–5(–16) $\underline{G}$(2

Clearly defined synapomorphies do not exist for Gentianaceae (Albert and Struwe, 2002), although diagnostic characters fit well with the tribes.

Loss of bracteoles leads to displacement of the two outer sepals in lateral position (Eichler, 1875). In *Erythraea*, the flower is inversed with sepal two abaxially. Flowers are generally penta-(tetra-) merous, rarely hexa- to twelve-merous or more (*Blackstonia*). Petals as well as sepals are usually strongly contorted. There is much variation in the extent of development of a petal tube and a basal corona is occasionally present (e.g. *Symbolanthus*: Struwe et al., 2002). In *Exacum* (and most Exaceae), the calyx lobes have a prominent keel (Figure 11.19). The androecium is haplostemonous with equally long stamens, or these are rarely unequal (e.g. *Lisianthus*). Anthers of *Exacum* are poricidal, suggesting buzz pollination, but pollination syndromes are manifold in the family with an absence of wind pollination. Flowers of *Exacum*, *Chironia* and *Orphium* are weakly monosymmetric by orientation of style and anthers in opposite directions (Endress, 2012). This is caused by the inversed position of flowers on the inflorescence linked with curving of the pedicel. Flowers of some genera such as *Canscora* have median monosymmetry with an unequal insertion of heteromorphic stamens (Struwe et al., 2002). Nectaries either develop as a disc at the base of the ovary or as separate glands on the inner surface of the corolla tube (Lindsey, 1940; Struwe et al., 2002). The petal tube of *Halenia* bears spurs with nectaries. Placentation is either axile or parietal. The ovary is occasionally semi-unilocular because the placenta is not developed distally (Struwe et al., 2002).

Apocynaceae (incl. Asclepiadaceae)

Figure 11.20A *Nerium oleander* L.

Figure 11.20B *Periploca graeca* L.

✶ K(5) [C(5) A5] $\underline{G}$(2)

General formula: ✶ K(3–)4–5(–16) C(3–)4–5(–16) A4–5(–16) $\underline{G}$2(3–8)

Although Asclepiadaceae were considered a separate family, recent research has shown that they represent a level of complexity derived within the Apocynaceae

Figure 11.19 *Exacum affine* (Gentianaceae): partial inflorescence. Note the dorsal appendages on sepal lobes, giving a contorted appearance to the calyx. Oblique orientation is caused by curvature of style and stamens.

(Endress et al., 1996). A synapomorphy are the distinct carpels connected by their styles and an expanded apical portion of the style (Judd and Olmstead, 2004).

The floral Bauplan is generally well conserved within the family, contrary to other Gentianales. Floral adaptations to pollinators are highly diverse and are superimposed on the highly synorganized flowers (Endress, 1994, 2016). Two floral morphs can usually be recognized: flowers with a long, tubular corolla and a constriction hiding the insertion of the stamens and carpels, and flowers with a short corolla tube and strongly developed corona (former Asclepiadaceae). While the flower is pentamerous, the ovary is bicarpellate and superior (very rarely up to eight carpels). The corolla has variable aestivation, but is often contorted as are Gentianaceae, hence the old name 'Contortae' used for Gentianales. A prominent corona may develop in alternipetalous position, often attached at the back of the anther. In *Asclepias* or *Periploca* this becomes a prominent petaloid organ that functions as a nectar recipient (Figure 11.20B; Endress, 1994). Stamens alternate with petals and have a short filament, which can be distinct or fused into a tube around the ovary. In

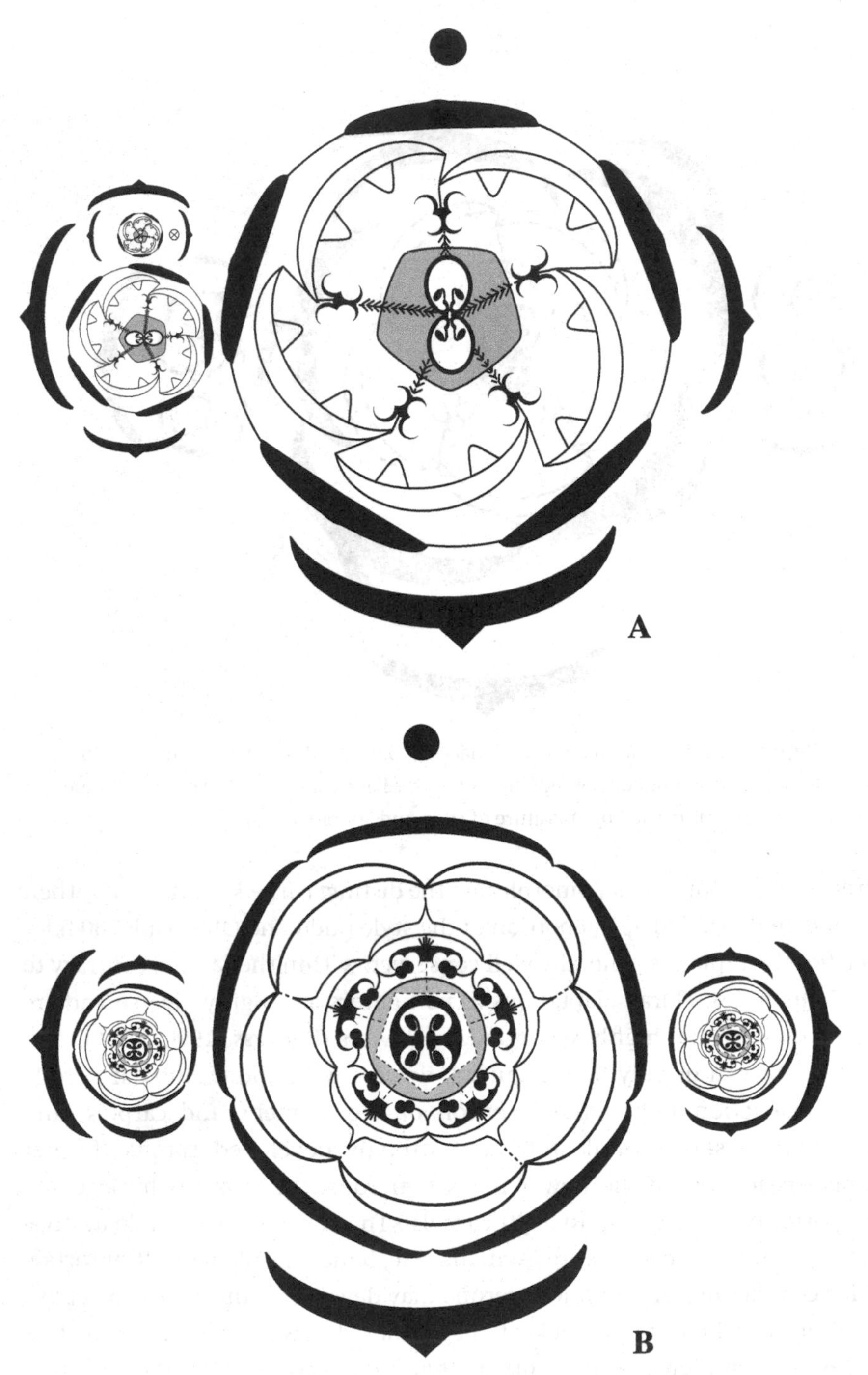

Figure 11.20 Apocynaceae: A. *Nerium oleander*: partial inflorescence. Note the connectives twining around the style. B. *Periploca graeca*: inflorescence

Asclepioideae the style develops as a pentagonal shield which is connected postgenitally with the anthers into a gynostegium. In *Periploca* and other Asclepioideae, each theca (with pollen sacs fused into a pollinium) is connected with that of a neighbouring anther through a clasp-like organ (translator). A visiting insect will pull out the pollinia. In *Nerium* the connectives are extended in long plumose appendages curling around the style (Figure 11.20A).

The ovary is surrounded by a disc nectary or two nectaries alternating with the carpels (Bernardello, 2007). Placentation is intruding parietal or axile, mostly with numerous ovules. In some genera the ovary becomes secondarily apocarpous (Endress, Jenny and Fallen, 1983), as is generally the case for the fruits. Floral diagrams of *Asclepias* and *Calotropis* are presented in Endress (1994, 2016).

Solanales

The order consists of five families (Solanaceae, Convolvulaceae, Hydroleaceae, Montiniaceae and Sphenocleaceae). Synapomorphies are generally few (e.g. radial symmetry, a plicate corolla tube, stamen number equaling the petals, persistent calyx: Judd and Olmstead, 2004). All Solanales have late sympetaly correlated with the occurrence of superior ovaries. Although there is a tendency in some families for a secondary loss of petal tubes (e.g. Montiniaceae: Ronse De Craene, Linder and Smets, 2000), petal tubes can be short or elongated with short petal lobes, or are trumpet shaped. Ovaries are generally superior and consist of two carpels (occasionally more through secondary increase). The oblique insertion of the ovary is a shared characteristic of Convolvulaceae and Solanaceae (Eichler, 1875; pers. obs.).

Solanaceae

Figure 11.21 A. *Cestrum parqui* L'Hérit

∗ K(5)[c] [C(5) A5] $\underline{G}$(2)

B. *Schizanthus hookeri* Gillies ex Graham

↙ K(5) [C(2:2:1) A1°:2:2°] $\underline{G}$(2)

General formula: ∗ (↙) K(4–)5(–10) C(4–)5(–10) A(2–)5(–7) G(1–)2(5)

Inflorescences are cymose and flowers generally have bracts and bracteoles (tendency for reduction in *Cestrum*). Flowers of Solanaceae are generally pentamerous, with occasional increases affecting the calyx (e.g. *Lycianthes*: D'Arcy, 1986) or the whole flower (e.g. *Solanum*). Flowers are polysymmetric except for the oblique gynoecium, more rarely obliquely monosymmetric with the symmetry line running through sepal one (e.g. *Schizanthus*, *Salpiglossis*:

[c] Presence of long abaxial slit.

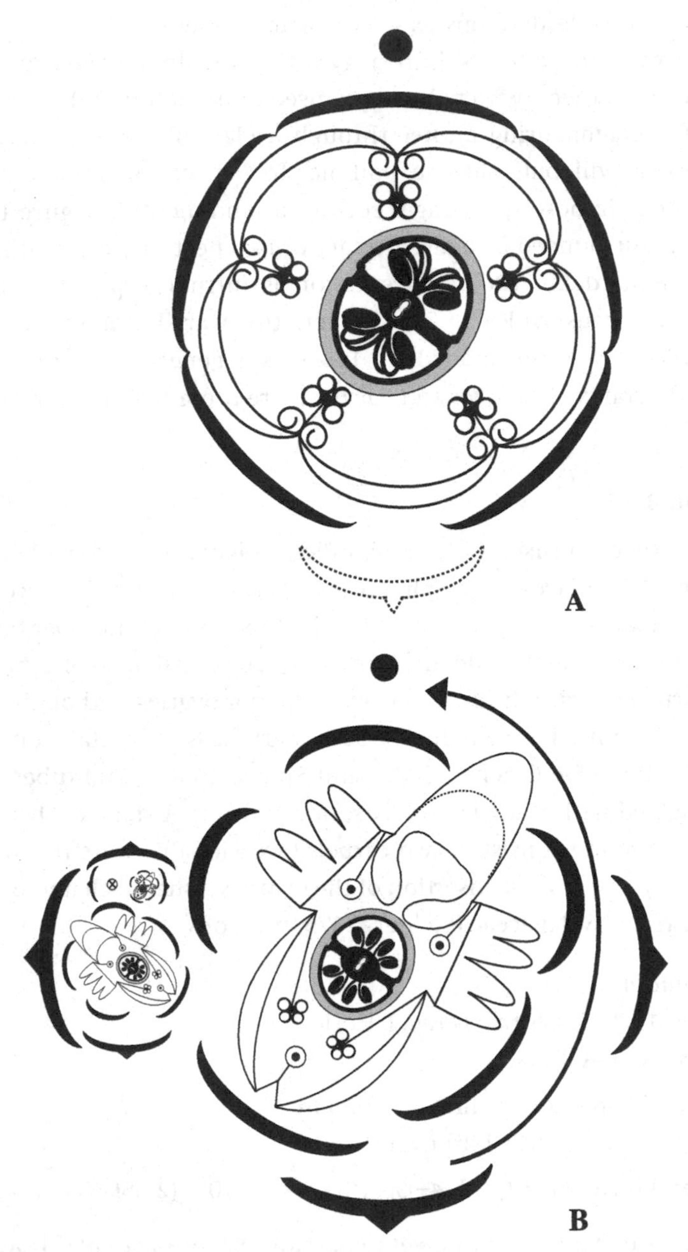

Figure 11.21 Solanaceae. A. *Cestrum parqui*. Note the valvate petals with involute margins. B. *Schizanthus hookeri*, partial inflorescence

Figure 11.21B; Eichler, 1875; Knapp, 2002; Stützel, 2006; Chinga and Pérez, 2016). Resupination of the flower of *Schizanthus* places the odd petal in a median adaxial position. This is a remarkable convergence with the papilionoid Leguminosae sharing a similar explosive pollination mechanism

(Westerkamp and Weber, 1999). Knapp (2002) argued that monosymmetry has arisen several times in the family, with a great variety of monosymmetric forms affecting several or a single organ. In *Salpiglossis* and related genera both corolla and androecium are monosymmetric, with the abaxial stamen sterile in *Browallia*. Petal tubes are either very long with shorter lobes, or short with reflexed lobes (*Solanum*, linked with buzz pollination). Aestivation is variable, ranging from quincuncial to contorted or valvate. Stamen number is generally five (haplostemony). Stamens can be unequal in length with one shorter stamen (e.g. *Nicotiana*), or stamens are of variable length (the one opposite sepal one shortest, the two neighbouring longest and the two posterior intermediate in size: e.g. *Petunia*, *Physalis*). In *Salpiglossis* there are two larger stamens, two smaller stamens and one staminode. In *Schizanthus* only the stamens alternating with the carpels are fertile (Figure 11.21B; Eichler, 1875; Knapp, 2002). The gynoecium is characteristically inserted obliquely, consisting of two carpels, rarely with five antepetalous carpels (*Nicandra*). In *Nolana* the number of carpels is increased by a multiplication from five carpels (Endress, 2014). Placentation is axile, often with many ovules, rarely with one or two ovules per carpel. The single style has an undifferentiated stigma, or if differentiated, stigmatic lobes are carinal. A nectary is generally developed at the base of the ovary, except in the buzz-pollinated species. Several floral diagrams are shown in Knapp (2002).

Boraginales

For consistency I follow APG IV (2016) in describing a single family for the order Boraginales, although the Boraginales Working Group (2016) recognizes eleven families that are morphologically distinct.

Boraginaceae (incl. Ehretiaceae, Heliotropiaceae, Namaceae, Wellstediaceae, Hydrophyllaceae, Lennoaceae, Cordiaceae, Hoplestigmataceae, Codonaceae, Coldeniaceae)
Figure 11.22 *Echium hierrense* Webb ex Bolle
↖ K(5) [C(5) A5] $\underline{G}$(2)

General formula: ∗(↖) K(4–)5(–20) C(4–)5(–20) A(4–)5(–20) G2(4–5)(–15)

The flowers are characteristically arranged in scorpioid cymes with a displaced bract relative to the flower. Flowers are mostly regular and pentamerous (up to octomerous in some *Cordia*). A dramatic increase in merism occurs in *Codon* (Jeiter, Danisch and Hilger, 2016), affecting all organs excluding the gynoecium (ranging from ten- to twenty- merous). The genus *Hoplestigma* with twenty to thirty stamens may have a similar merism increase. As in Solanales, monosymmetry is predominantly oblique and runs mainly through sepal four (Eichler, 1875; Stützel, 2006).

Echium is slightly monosymmetric through the size difference of sepals and stamens and oblique orientation of the style. Eichler (1875) and Leredde (1955) reported an unequal length of the stamens in the genus with a displacement in the upper part of the flower. Sepals are imbricate or valvate. Aestivation of the corolla is imbricate (2/5 arrangement: e.g. *Bourreria*), contorted (e.g. *Cordia*) or cochleate ascending (e.g. *Echium*). The corolla tube sometimes produces invaginations as well as scales that have occasionally been interpreted as staminodes, but these arise late in ontogeny (Jeiter, Langecker and Weigend, 2020). In *Codon* petal invaginations create specific nectar chambers that can only be accessed separately (Jeiter, Danisch and Hilger, 2016). The gynoecium is superior and surrounded by a disc nectary. The carpels range from one (*Rochelia*, pseudomonomerous: Hilger, 1984) to mostly two, or occasionally many more (*Lennoa*). Placentation is diverse, but basically axile with many ovules (Codoneae, Hydrophylleae) or two to four ovules. In Boraginoideae a false septum develops in the same way as in Lamiaceae by the late invagination of a partition, leading to four locules with two basally inserted ovules per carpel and a gynobasic style (e.g. Baillon, 1862; Gottschling, 2004; Jeiter et al., 2018; p. 393).

Lamiales

The order consists of twenty-four families and almost eighteen thousand species. The internal relationships of Lamiales are not well resolved, but families such as Oleaceae, Calceolariaceae and Gesneriaceae occupy a more basal position (Figure 11.17). A distinction of families is very difficult to make on a morphological basis, as major characters sweep through the order (Stevens, 2001 onwards). Olmstead et al. (2001) recognized that the large family Scrophulariaceae is polyphyletic and proposed its subdivision between at least eight lineages in a strongly altered Lamiales. However, the lack of any clear morphological synapomorphies to differentiate large clades, such as Plantaginaceae from Scrophulariaceae, remains an issue. The same problem that existed for Lamiaceae and Verbenaceae was partly solved by merging most genera of Verbenaceae in Lamiaceae (e.g. Cantino, 1992). A more global family approach with well-defined subfamilies, such as for Boraginaceae or Malvaceae, appears to be more appropriate.

Figure 11.22 shows the major steps in floral evolution of Lamiales. The calyx is generally well developed and often basally to strongly fused. A characteristic feature of the order is the presence of a monosymmetric bilabiate corolla mostly

Figure 11.22 *Echium hierrense* (Boraginaceae): A. flower; B. partial inflorescence. The broken line in the ovary shows the false septum.

constructed on a 3:2 pattern, and a superior bicarpellate ovary. The androecium fluctuates from one to five stamens, with various reductions and losses occurring in a median plane of the flower. The basic floral formula for the order is ↓ K(5) [C(5) A1–5] $\underline{G}$(2).

Although median monosymmetry is the most common feature, there are several reversals to polysymmetry (see p. 387; Figure 11.23H–I; e.g. Endress, 1998; Soltis et al., 2005, 2018). These flowers are either pentamerous or tetramerous and the shift is caused by different factors, which may be a genetic mutation (e.g. Gesneriaceae: Wang et al., 2010), or mechanical pressure causing fusion of the two posterior petals following the loss of the posterior stamen (e.g. Donoghue, Ree and Baum, 1998; Endress, 1999; Bello et al., 2004; Ronse De Craene, 2016). The polysymmetric Tetrachondraceae and Oleaceae are more basal to other Lamiales, although their floral structure is derived from more elaborate pentamerous precursors (Figure 11.23J–L; Mayr and Weber, 2006).

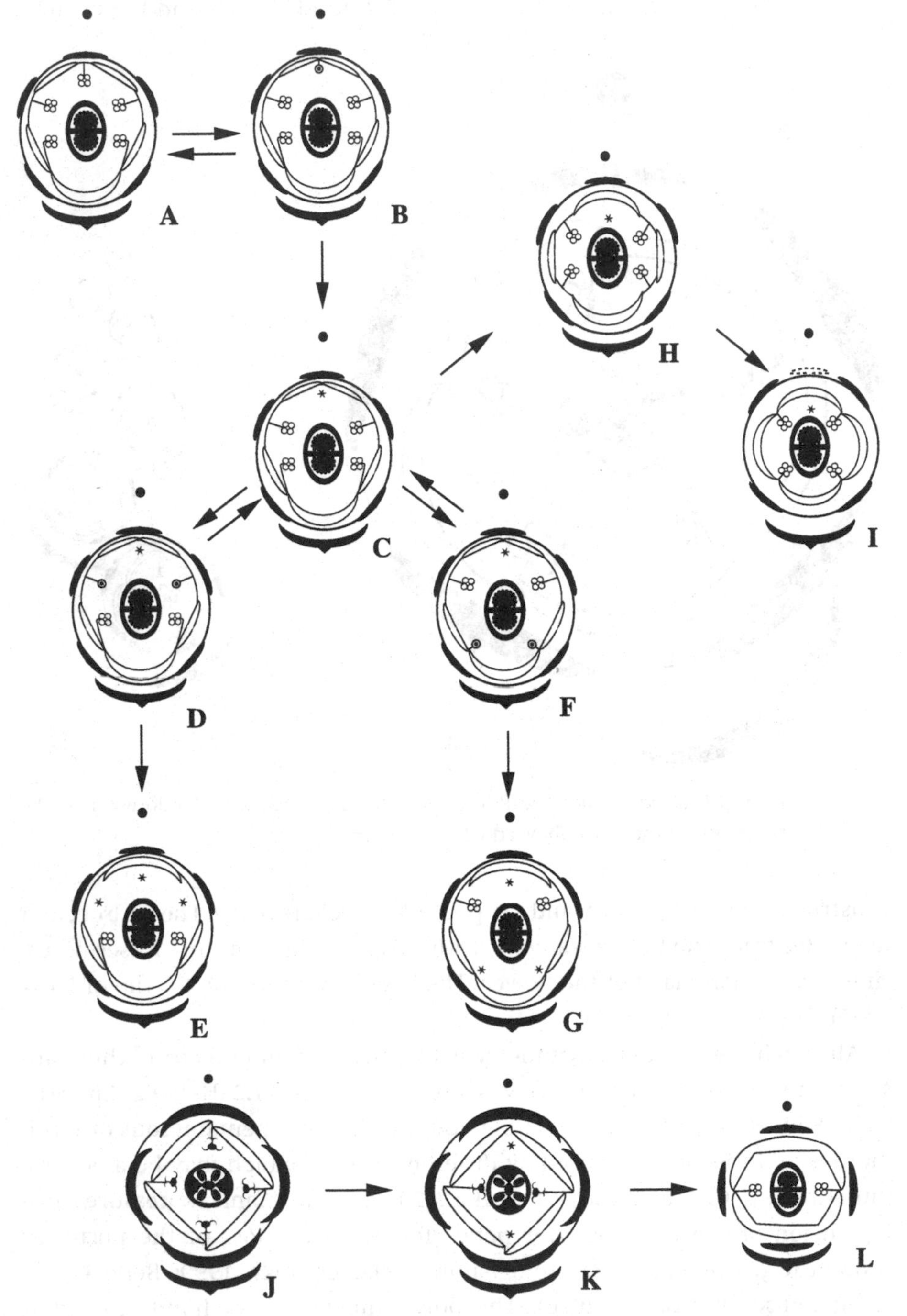

Figure 11.23 Floral evolution in Lamiales. A. weakly monosymmetric flower with five stamens; B. monosymmetric flower with adaxial staminode; C. adaxial staminode is lost; D. sterilization of lateral adaxial stamens; E. loss of adaxial stamens and fusion of posterior petals; F. sterilization of lateral abaxial stamens; G. loss of abaxial stamens and fusion of posterior petals; H. fusion of posterior petals in a single structure; I. tetramerous flower with loss of adaxial sepal; J. tetramerous flower with four stamens; K. Oleaceae with loss of median stamens and reorientation of the ovary; L. flower of Calceolariaceae with two stamens

The corolla is initiated at the same time or after the stamens (late sympetaly, cf Erbar, 1991); this development tends to be common for most Lamiales and is linked with a superior ovary. The reduction of the androecium always starts with the reduction or complete loss of the adaxial stamen (Figure 11.23B, C). The four remaining fertile stamens are arranged at two levels and are often connected by their anthers in pairs (didynamy). Further reductions lead to two stamens, either in latero-adaxial or in latero-abaxial position (Figure 11.23D–E, F–G). Walker-Larsen and Harder (2000, 2001) argued that the tendency for partial loss of stamens in monosymmetric Lamiales is a reversible process. They concluded that some Lamiales have become secondarily polysymmetric and that this is linked with the restoration of the lost stamen (e.g. *Verbena*, *Verbascum*, *Oroxylum*, as these taxa occur in derived clades). Endress (1998) argued that the limited extent of reduction of the staminode makes a reversal possible, as in some basal clades of Lamiales which have retained five stamens, or where the odd staminode is well developed. Gesneriaceae have the largest proportion of polysymmetric flowers with five stamens (Endress, 1998; Wang et al., 2010). It is the only family of Lamiales where the odd stamen is not missing from a particular genus (except for an occasional species). In Bignoniaceae, the odd staminode is mostly present and often larger than the stamens (only absent in *Tourettia*: Endress, 1992, 1998). The possibility of a reversed process of staminode fertility should be critically investigated in the order.

Flower development has been investigated in several representative species, indicating an early onset of monosymmetry (e.g. Endress, 1999). Depending on the direction of development, aestivation of the corolla is highly variable within a family (e.g. Plantaginaceae, Acanthaceae: Armstrong and Douglas, 1989; Scotland, Endress and Lawrence, 1994).

Contrary to the androecium the ovary is mostly bicarpellate and superior, often with a high number of ovules on a mainly axile placentation. Fruits are generally capsular, less often indehiscent.

Oleaceae

Figure 11.24 *Osmanthus delavayi* Franch

∗ K(4) [C(4) A2] $\underline{G}$(2)

General formula: ∗ K(0–)4(–15) C(0–)4(–12) A2(–4) $\underline{G}$(2)

Inflorescences are cymose. Contrary to most other Lamiales, flowers are polysymmetric and built on a tetramerous Bauplan. Merism of the perianth is rarely higher and variable in some *Jasminum* and *Nyctanthes* (Ronse De Craene, 2016). Most taxa have two stamens in transversal position. It is implied that these were derived from a tetramerous condition as is found in some *Chionanthus*. Sepals are

Figure 11.24 *Osmanthus delavayi* (Oleaceae)

usually four, although occasionally five (as in Figure 11.24, with two smaller lateral sepals). The superior ovary has a single style with carinal stigmatic lobes. Placentation is axile with two ovules per carpel. A nectary, if present, develops at the base of the ovary (gynoecial nectary). In the wind-pollinated species of the genus *Fraxinus*, the perianth is reduced or absent and flowers are unisexual (Eichler, 1875).

Calceolariaceae

Figure 11.25 *Jovellana violacea* (Cav.) G. Don

↓K(4)[C(2)A(2)] $\underline{G}$(2)

Flowers are arranged in compound dichasia. There are four sepals in median and transversal position and only two broad median petals alternating with two stamens. The ovary is bicarpellate with numerous ovules on axile placentation. Mayr and Weber (2006) have clearly demonstrated that the dimerous corolla is

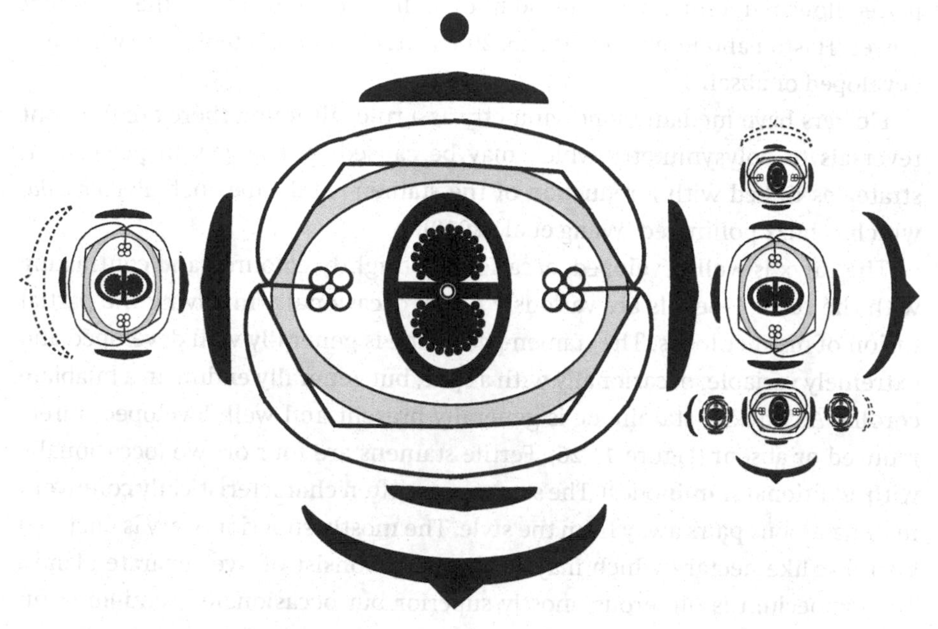

Figure 11.25 *Jovellana violacea* (Calceolariaceae): partial inflorescence

derived from a tetramerous condition as in Oleaceae, and not from pentamery. In Veroniceae a comparable reduction to tetramery is linked with loss of the median adaxial sepal and fusion of two petals in a single structure (see p. 387). There is no evidence for this in Calceolariaceae. An odd third stamen develops in an adaxial position in *Stemotria triandra* or as a teratology in *Calceolaria* and is linked with a bilobed lower lip. In *Jovellana*, the lower petal lip arises as a bilobed structure. Together with the presence of two lateral vascular bundles in the lower petal, this is evidence that the lower petal represents a composite structure linking Calceolariaceae to Oleaceae with four lateral petals (Mayr and Weber, 2006).

Calceolaria mostly has specialized oil flowers, with a much broader pouch-like lower lip covered with oil-secreting trichomes. In *Jovellana* both lips are equally developed and pollen is provided as reward.

Gesneriaceae

Figure 11.26 *Streptocarpus rexii* (Bowie ex Hook.) Lindl.

↓K(5)[C(3:2)A2:2°:1°] G(2)

General formula: ∗/↓ K(4–)5 C(4–)5 A4:1°(2:2°, 2, 5) G2(1:1°)

Inflorescences in the family are complex thyrses with mixtures of dichasial and monochasial branching (Weber, 2004). These are complex cymes ending in paired-flowered units, with an additional flower in front of the terminal flower (Haston and Ronse De Craene, 2007). Bracts and bracteoles are variously developed or absent.

Flowers have median monosymmetry as a rule, although there are frequent reversals to polysymmetry which may be caused by changes in pollination strategies linked with a reduction of the stamen-petal tube such as *Ramonda*, which is buzz pollinated (Wang et al., 2010).

The calyx is well developed, occasionally brightly coloured and contrasting with the corolla. Sepals are variously fused, occasionally involving the partial fusion of distinct lobes. The stamen-petal tube is generally well developed and extremely variable, occasionally with a spur, but generally ending in a bilabiate corolla. An adaxial staminode is generally present and well developed, rarely reduced or absent (Figure 11.26). Fertile stamens are four or two (occasionally with additional staminodes). The stamens are often characteristically connivent in didynamous pairs away from the style. The mostly superior ovary is enclosed by a disc-like nectary which may occasionally consist of five separate glands. The gynoecium is dimerous, mostly superior but occasionally (semi-)inferior. The ovary generally bears a well-developed style and both carpels are fertile, or the anterior carpel is occasionally sterile. There are usually two parietal T-shaped placentae, and this has traditionally been used as the criterium to distinguish Gesneriaceae from Scrophulariaceae s.l. (Weber, 2004). However, this distinction is not clear-cut as in both taxa there is a basal lower portion which is fused and an upper portion where the septum is incomplete; in Gesneriaceae, the upper portion is much more strongly developed, leading to the impression of a parietal placentation. However, exceptions exist on both sides. Ovules are generally numerous.

As this is a large family, several evolutionary trends have developed among the more than thirty-five hundred species (Weber, 2004). Flowers have evolved complex pollination syndromes, linked with the length of the stamen-petal tube and various constrictions within the tube (e.g. *Streptocarpus*: Harrison, Möller and Cronk, 1999).

Plantaginaceae (incl. Callitrichaceae, Globulariaceae p.p., Scrophulariaceae p.p., Hippuridaceae)

Figure 11.27A–B *Penstemon fruticosus* (Pursch.) Greene var. *scouleri* (Lindl.) Cronq.

↓K(5)[C(5)A4:1°] $\underline{G}$(2)

Figure 11.27C–D. *Hebe cheesemanii* (Buchanan) Cockayne and Allan

↓K(4)[C(4)A2] $\underline{G}$(2)

General formula: ↓(✱) K(0)4–5 C(0)4–5 A(1)2–4 (:1°) G(1)–2

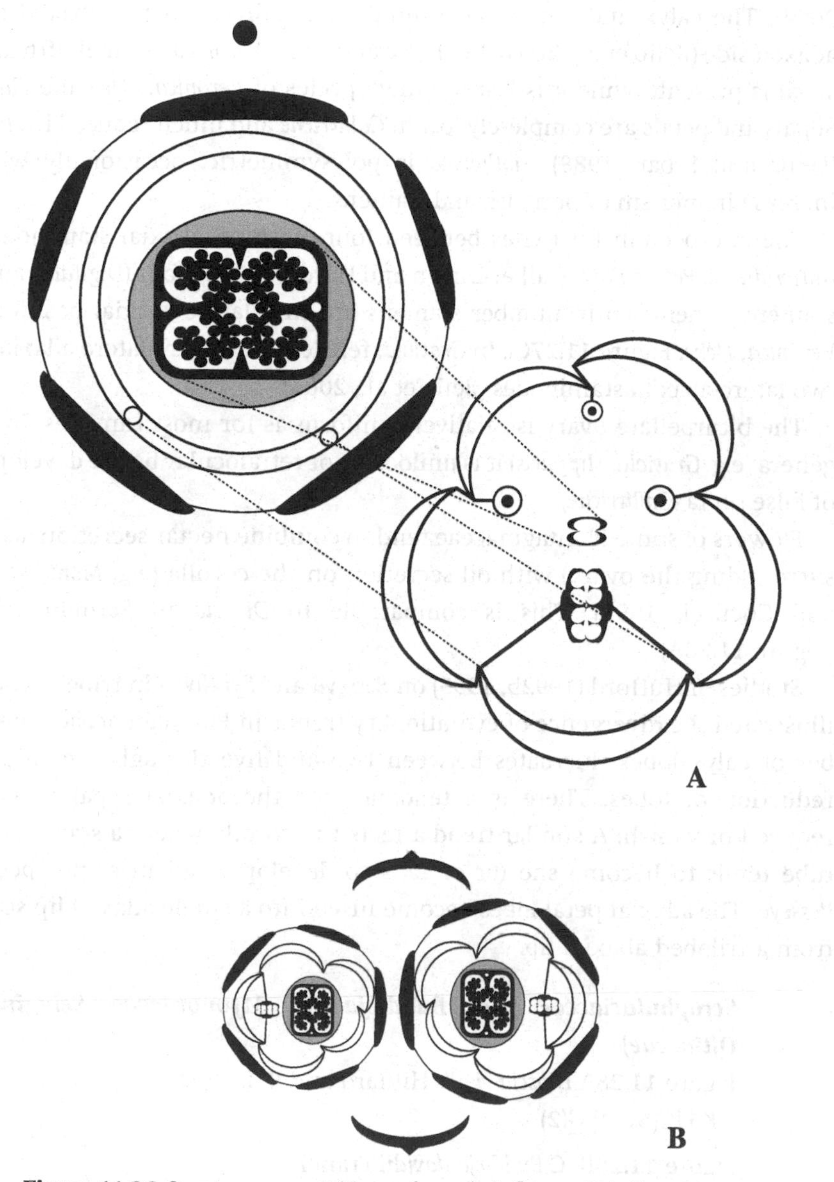

Figure 11.26 *Streptocarpus rexii* (Gesneriaceae): A. flower, B. inflorescence

As mentioned on page 380, it remains problematic to distinguish different families within the Scrophulariaceae-Plantaginaceae complex. Orobanchaceae appears more easily distinguished by the combination of a (hemi-)parasitic habit and general absence of an odd staminode (Endress, 1998).

Secondarily polysymmetric flowers are clearly derived from monosymmetric flowers in *Aragoa* and some other Plantaginaceae (e.g. Endress, 1999; Bello et al.

2004). The calyx and corolla arise unidirectionally from the abaxial to the adaxial side (Bello et al., 2004). In *Aragoa* and some *Veronica*, a small fifth adaxial sepal is present, while it is lost in other species of *Veronica*, *Hebe* and *Plantago*. Sepals and petals are completely lost in *Callitriche* and much reduced in *Hippuris* (Leins and Erbar, 1988). *Sibthorpia* is polysymmetric, occasionally with an increase in merism of perianth and stamens.

The androecium fluctuates between four (with an abaxial staminode: e.g. *Antirrhinum, Penstemon*: Walker-Larsen and Harder, 2000), four (*Digitalis*) and two stamens; when two in number stamens are in a latero-abaxial position (e.g. *Veronica, Hebe*: Figure 11.27C). In *Gratiola*, fertile stamens are latero-adaxial with two latero-abaxial staminodes (Bello et al., 2004).

The bicarpellate ovary is relatively uniform as for most Lamiales. In a few genera (e.g. *Gratiola, Hippuris*) it is unilocular or tetralocular by the development of false septa (*Callitriche*).

Flowers of some Plantaginaceae tend to combine nectar secretion (as a disc surrounding the ovary) with oil secretion on the corolla (e.g. *Monttea*: Sérsic and Cocucci, 1999). This is comparable to *Diascia* in Scrophulariaceae (Figure 11.28A).

Studies of Hufford (1992b, 1995) on *Besseya* and *Synthyris* in tribe Veroniceae illustrated the divergence of evolutionary trends in Plantaginaceae. The number of calyx lobes fluctuates between two and five through connation and reduction of lobes. There is a tendency for the adaxial sepal to become reduced or vanish. A similar trend affects the corolla where a stamen-corolla tube tends to become shorter or fails to develop at all in some species of *Besseya*. The adaxial petal lobes become fused into a single adaxial lip separate from a trilobed abaxial lip.

Scrophulariaceae (incl. Buddlejaceae, Myoporaceae, Selaginaceae, Oftiaceae)

Figure 11.28A *Diascia vigilis* Hilliard and B. L. Burtt

↓ K5 [C(5)A4] $\underline{G}$(2)

Figure 11.28B, C *Buddleja davidii* Franch

∗K(4) [C(4)A4] $\underline{G}$(2)

General formula: ↓(∗) K(2)4–5(–8) C(0)4–5(–8) A2–4–5(–8) G2

Scrophulariaceae were considerably reduced compared to earlier circumscriptions of the family. Weakly monosymmetric flowers are found in *Verbascum*, with the occasional presence of an adaxial fertile stamen. Most other members of the family are two-lipped with four or fewer fertile stamens and 3:2 is the most common petal arrangement (exc. *Mutisia*: 1:4).

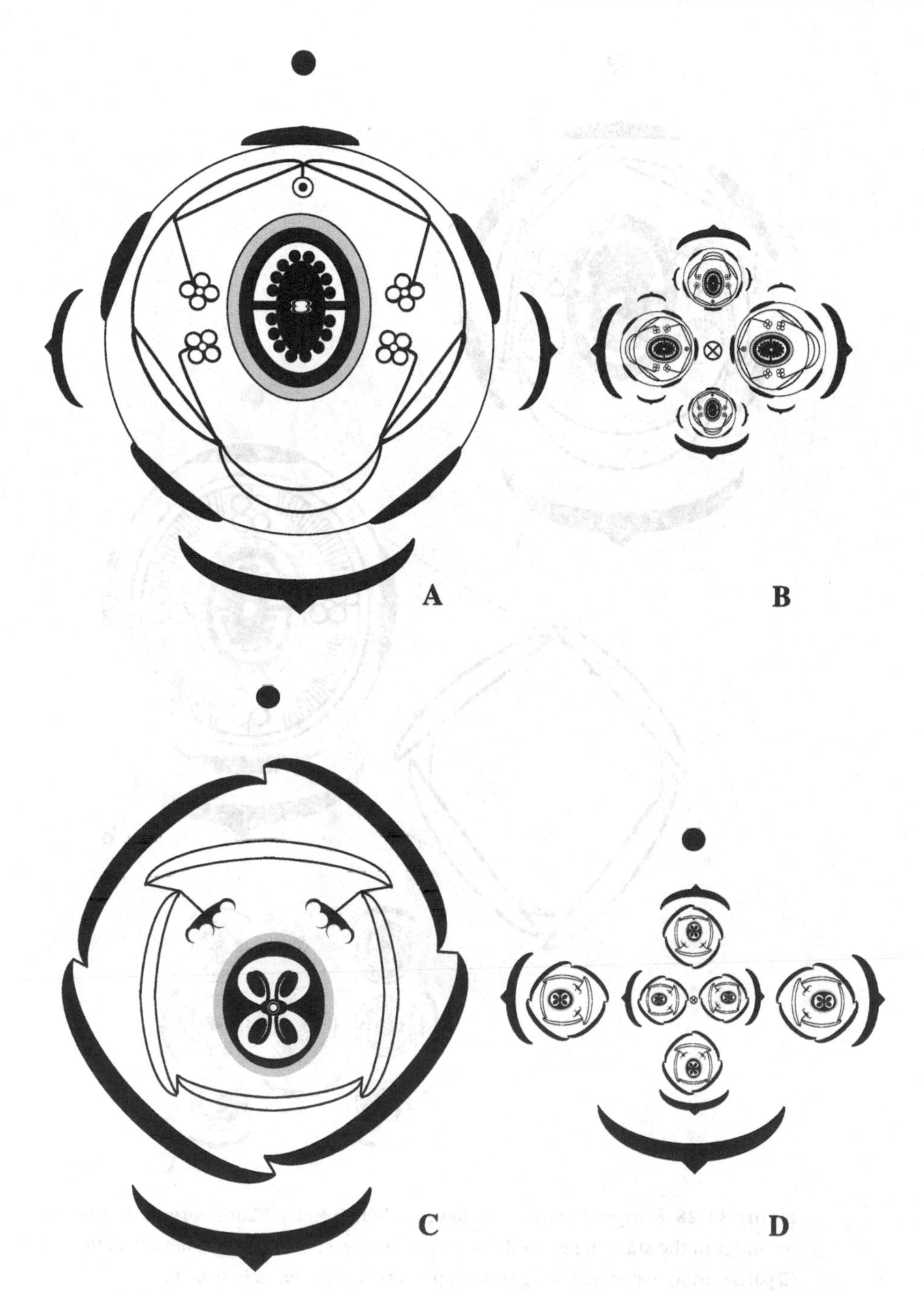

Figure 11.27 Plantaginaceae: *Penstemon fruticosus* var. *scouleri*, A. flower, B. partial inflorescence; *Hebe cheesemanii*, C. flower; D. partial inflorescence

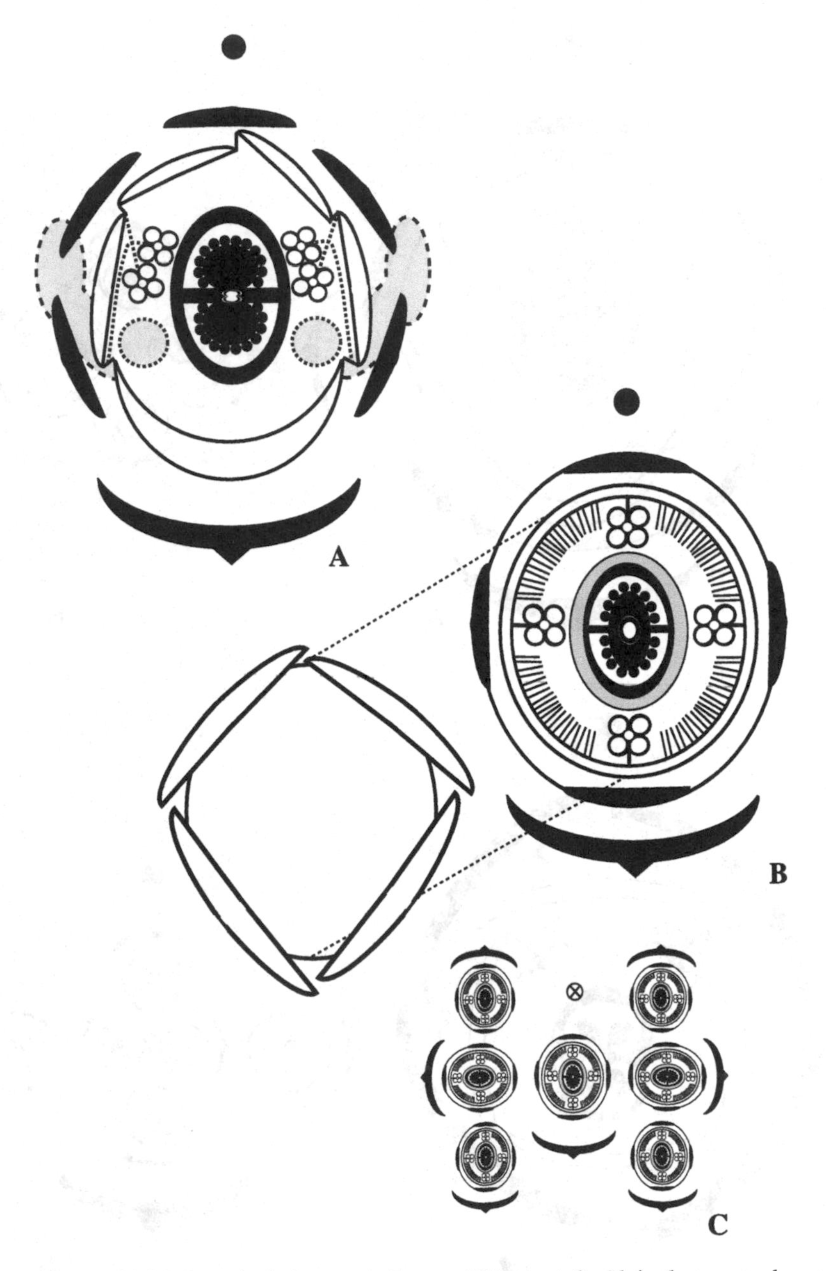

Figure 11.28 Scrophulariaceae: A. flower of *Diascia vigilis*. Elaiophores are shown as openings in the stamen-petal tube with underlying spurs. *Buddlejia davidii*. B. flower; C. partial inflorescence. A long stamen-petal tube with hairs is shown.

The adaxial staminode is elaborate in *Scrophularia* (reflected in *Penstemon* of Plantaginaceae: Endress, 1998; López et al., 2016). López et al. (2016) demonstrated how the extent of development of a large staminode in *Scrophularia* is linked to attraction of pollinators and successful pollination (e.g. *S. scorodonia*), although it is lost at least twice in the genus (e.g. *S. canina*). Flowers of *Diascia* have two petal spurs containing oils and no nectaries (Figure 11.28A). The bicarpellate ovary resembles Plantaginaceae to a great extent. The tetramerous and polysymmetric flowers of *Buddleja* (Figure 11.28B) are probably derived in a similar way as *Plantago* (see Bello et al., 2004). Petal lobes are slightly unequal with the abaxial longer.

Armstrong (1985) used floral anatomical characters to distinguish Scrophulariaceae *sensu lato* from Bignoniaceae. Scrophulariaceae have a simple axile placentation and only two large dorsal bundles in the ovary, while Bignoniaceae consistently have two to four distinct placental ridges and four large bundles, two opposite the septum and two dorsals. This characteristic should be studied further in the context of renewed phylogenetic relationships.

Acanthaceae (incl. Avicenniaceae)

Figure 11.29A–C *Odontonema strictum* Kuntze

↓ K(5) [C(5) A2:2°] $\underline{G}$(2)

General formula: ↓(✱) K(0)4–5 C4–5 A2–4(–5) $\underline{G}$2

The family is characterized by the presence of showy bracts and flowers arranged in decussate terminal inflorescences.

The calyx is generally well developed and basally fused. It is strongly reduced or absent in *Thunbergia*, where attraction and protection is transferred to the bracteoles (Rao, 1953; Schönenberger and Endress, 1998; Endress, 2008b). The calyx of *Thunbergia* can develop as a lobed rim resembling the calyx lobes in some Caprifoliaceae. The adaxial sepal is occasionally reduced or two sepal lobes are laterally fused. Petals and stamens of *Avicennia* become tetramerous by reduction, while the calyx remains pentamerous (Tomlinson, 1986). Flowers are rarely regular, more often monosymmetric. The monosymmetric flowers have a long petal tube with elaborate lower lip. Aestivation of the corolla tends to be highly variable (Scotland, Endress and Lawrence, 1994). Petals are generally arranged as 3:2, rarely 5:0 (e.g. *Sclerochiton*). In a few genera belonging to Strobilanthinae, flowers are resupinate through twisting of the bud and petals are arranged as 2:3 (Moylan, Rudall and Scotland, 2004). The posterior staminode is generally absent, although it can be initiated and pervade in the vasculature (Rao, 1953), and the androecium has four fertile didynamous stamens (e.g. *Acanthus, Thunbergia*), two stamens with two staminodes (Figure 11.29A, e.g. *Sanchezia*) or two latero-adaxial stamens (e.g. *Nelsonia*). The ovary is superior, surrounded by a conspicuous disc nectary, with few ovules compared to Plantaginaceae. *Mendoncia* has a pseudomonomerous ovary (Schönenberger and Endress, 1998). There is a single style with funnel-form stigma

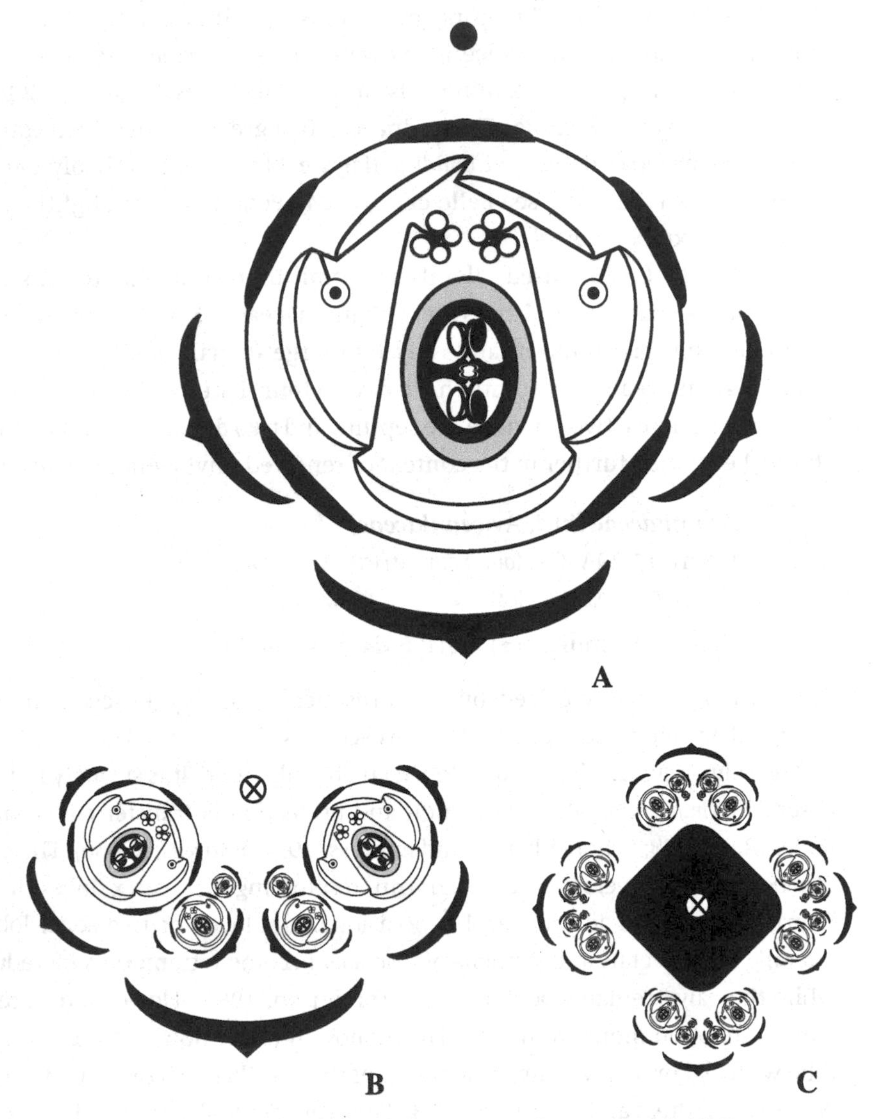

Figure 11.29 *Odontonema strictum* (Acanthaceae): A. flower; B. partial inflorescence; C. view of partial inflorescences at level of node

or with two stigmatic lobes (occasionally reduced to one). Fruits are characteristically explosively dehiscent and seeds are projected from the jaculator, an appendage derived from the funiculus.

In *Hemigraphis* and other Strobilantheae, an internal partition formed by an extension of the stamen-petal tube ('filament curtain') divides the internal flower in two compartments and leads pollinators to an abaxial chamber containing nectar (Moylan, Rudall and Scotland, 2004).

Lamiaceae (incl. Verbenaceae p.p.)
Figure 11.30A–B *Westringia fruticosa* (Willd.) Druce
Figure 11.30C *Clerodendrum petasites* (Lour.) A. Meeuse
↓K(5) [C(5) A4] $\underline{G}$(2)
General formula: ↓(✶) K(2–)4–5 C4–5 A(2–)4(–5) $\underline{G}$2(–5)

Lamiaceae (or Labiatae) used to be easily recognized by a combination of square stems with opposite leaves, verticillate inflorescences, bilabiate flowers with four stamens, a bicarpellate gynoecium with gynobasic style and a fruit with four nutlets. The circumscription of the family has become much extended following molecular systematics as several genera of Verbenaceae were displaced to Lamiaceae (Cantino, 1992). This makes an identification of Verbenaceae flowers easier, but adds complexity to the circumscription of Lamiaceae.

A calyx is well developed and variously fused. The corolla is monosymmetric and often bilabiate (Figure 11.30A), or less clearly so (*Clerodendrum*) with cochleate descending aestivation. The orientation of the lobes is mostly 3:2, more rarely 4:1 (e.g. *Perovskia*) or 5:0 (e.g. *Teucrium*) (see also Donoghue, Ree and Baum, 1998). The corolla tube is very long in *Clerodendrum* with imbricate lobes; the abaxial corolla lobe appears to be the upper one because of hanging flowers. *Symphorema* is unusual in being hexa- to sixteen-merous and polysymmetric (Byng, 2014). Stamens are mostly four in number and didynamous, and there is never a trace of an abaxial staminode. While the anterior stamens are sterile in *Westringia* (Figure 11.30A), in *Salvia*, there is further reduction to two latero-adaxial stamens with various elaborations, including a lever mechanism and the development of only one fertile theca (see Stützel, 2006, and review by Claßen-Bockhoff, Wester and Tweraser, 2003). The gynoecium is mostly bicarpellate (rarely five) and superior with either a terminal style, or a gynobasic style. Each locule is usually subdivided by a false septum separating the two erect ovules. The false partition is either incomplete (e.g. *Clerodendrum*) or complete, and the latter is correlated with a gynobasic style. Cantino (1992) argued that the gynobasic style must have evolved more than once in the family. A (semi-)annular disc nectary is present below the gynoecium, which is occasionally displaced in a unilateral position.

11.3 Campanulids: Aquifoliales, Dipsacales, Apiales and Asterales

Figure 11.31 shows the phylogenetic tree of the campanulids, based on Soltis et al. (2018). Besides the basal order Aquifoliales the clade contains seven well-defined orders. Bremer et al. (2001) listed epigynous flowers,

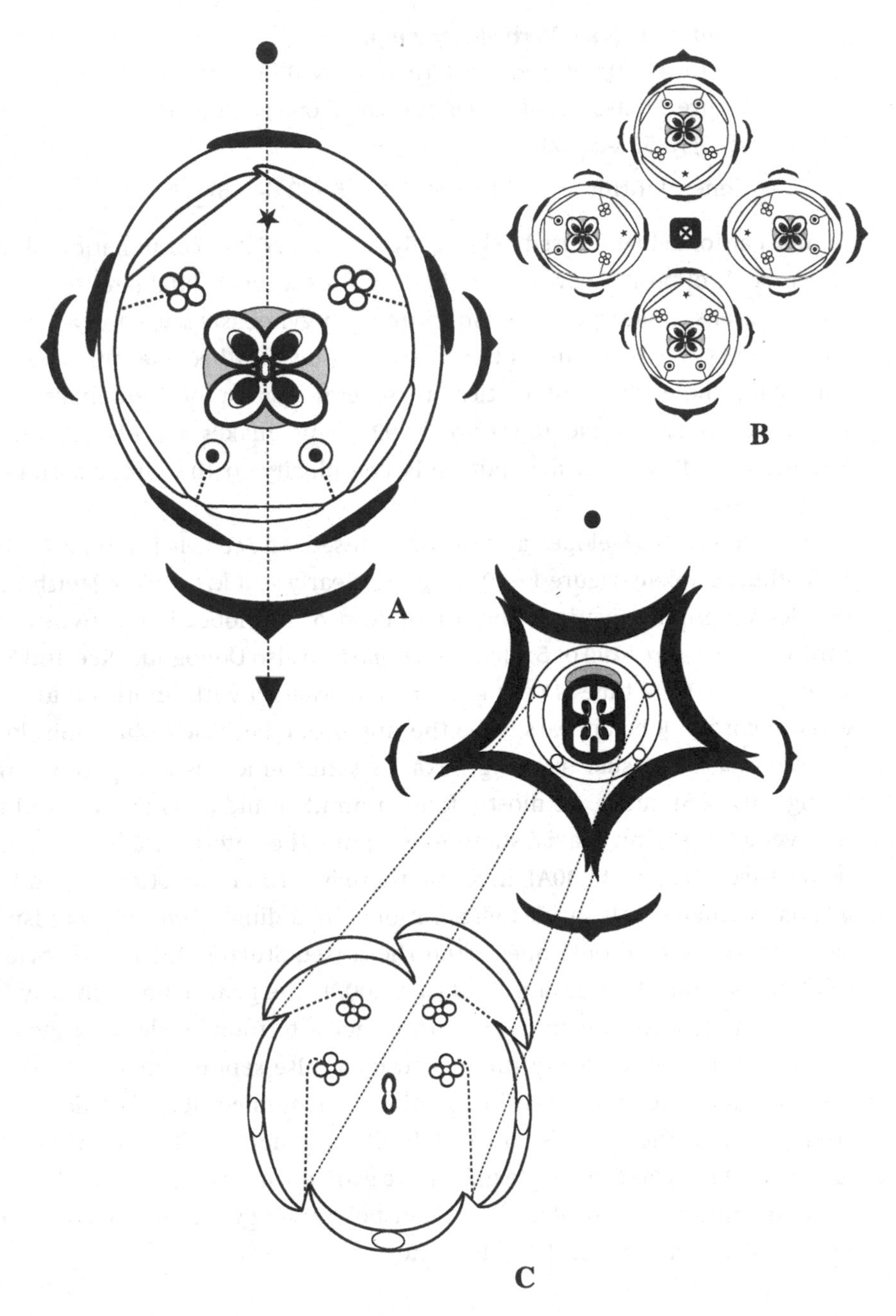

Figure 11.30 Lamiaceae. *Westringia fruticosa*: A. flower; B. partial inflorescence; *Clerodendron petasites*, C. flower

early sympetaly with distinct petal primordia, and alternate, dentate leaves as distinctive characters of campanulids. Several characters are recurrent and representative for campanulids, such as inferior ovaries with apical or

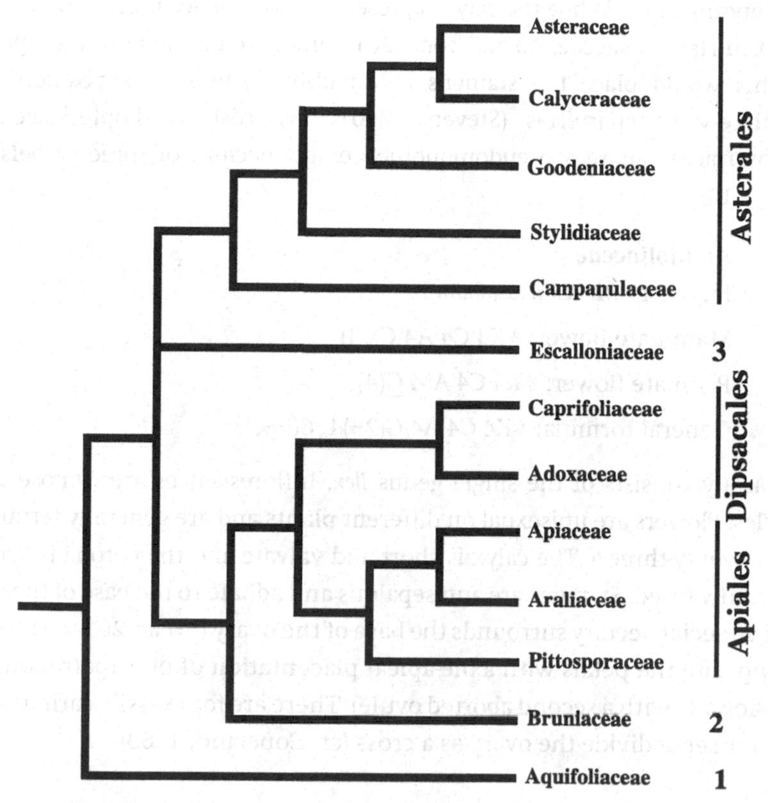

Figure 11.31 Phylogenetic tree of campanulids, based on Soltis et al. (2018). 1. Aquifoliales; 2. Bruniales; 3. Escalloniales

basal placentation and few ovules, frequent pseudomonomery, stamen numbers ranging from five to one, a tendency for reduction or loss of the sepals, reduction or loss of the stamen-petal tube and arrangement of flowers in pseudanthia. Reductions of stamens do not always follow the median axis. When flowers are monosymmetric, petals are often arranged as 5:0. nectaries are often an intrastaminal disc or part of the gynoecium.

I have not considered the smaller orders Paracryphiales, Escalloniales and Bruniales, consisting mainly of smaller families with uncertain taxonomic position.

Aquifoliales

The order consists of five families: Aquifoliaceae, Phyllonomaceae, Helwingiaceae, Cardiopteridaceae and Stemonuraceae. Shared characters are unisexual flowers, the absence of a style and a single (rarely two) apical-axile ovule per carpel. Inflorescences of *Phyllonoma* and *Helwingia* are shifted to the middle of a subtending bract. While the calyx is present in Aquifoliaceae and *Phyllonoma*, it is absent in Helwingiaceae. Ao and Tobe (2015) interpreted the flower as apetalous, but this would place the stamens in an obhaplostemonous position, which is unlikely in euasterids (Stevens, 2001 onwards). Cardiopteridaceae and Stemonuraceae share a pseudomonomerous gynoecium of three carpels (Kong et al., 2018).

Aquifoliaceae

Figure 11.32 *Ilex aquifolium* L.

Staminate flower: ∗K4 C4 A4 $\underline{G}^{\circ}(4)$

Pistillate flower: ∗K4 C4 A°4 $\underline{G}(4)$

General formula: ∗K4 C4 A4 G(2–)4–6(–∞)

The family consists of the single genus *Ilex*. Inflorescences are cymose axillary fascicles. Flowers are unisexual on different plants and are generally tetramerous (up to twenty-three?). The calyx is short and valvate and the corolla is imbricate and basally fused. Stamens are antesepalous and adnate to the base of the corolla.

A gynoecial nectary surrounds the base of the ovary (Erbar, 2014). The carpels are opposite the petals with axile-apical placentation of one apotropous ovule (occasionally with a second aborted ovule). There are four sessile carinal stigmas and four septa divide the ovary as a cross (cf. Copeland, 1963).

Dipsacales

Two families make up the Dipscacales. The circumscription of Caprifoliaceae and Adoxaceae is broad, containing several smaller families from previous classifications (Judd, Sanders and Donoghue, 1994). Dipsacales share several characters with other asterids, such as a stamen-corolla tube, monosymmetry with loss of adaxial stamens and an inferior ovary.

Common characteristics of Caprifoliaceae are the particular trichome nectaries, which are absent from Adoxaceae (Wagenitz and Laing, 1984). Roels and Smets (1994, 1996) implied that there is no early sympetaly *sensu* Erbar (1991) in Dipsacales. Floral organs arise on the margin of a depression on a ring-like zone and a petal tube is formed by upward growth of the marginal area. The septate ovary develops an apical placentation with pendent ovules curving into five-three locules (Erbar, 1994; Roels and Smets, 1994, 1996). The carpel wall curves as a roof above the placentation without connecting to it.

Caprifoliaceae (incl. Dipsacaceae, Morinaceae, Valerianaceae, Triplostegiaceae, Linnaeaceae, Diervillaceae)

Figure 11.33A–B *Lonicera giraldii* Rehder

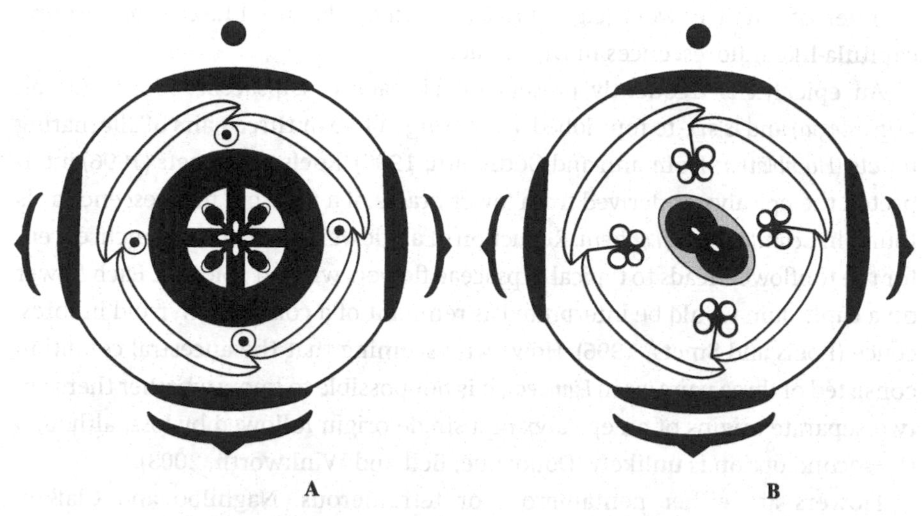

Figure 11.32 *Ilex aquifolium* (Aquifoliaceae). A. pistillate flower; B. staminate flower

↓ K5 [C(5) A5] Ğ(3)

Fig. 11.33 C-D. *Valeriana repens* Host

↙ K∞ [C(5) A3] Ğ(1:2°)

General formula: ↓ /↙/✱ K(0)4–5–∞ C4–5 A1–5 Ğ2–3(5–8)

Molecular phylogenies have shown that the traditional circumscription of Caprifoliaceae is paraphyletic. Important synapomorphies of Caprifoliaceae (against Adoxaceae) are: cymose inflorescences with dichasial subunits, monosymmetry (occasionally asymmetry), a long style, large capitate stigmas alternating with locules, presence of specific unicellular trichome nectaries, inferior ovary with a tendency for pseudomonomery, and tubular corolla (see Wagenitz and Laing, 1984; Judd, Sanders and Donoghue, 1994; Howarth and Donoghue, 2005, Landrein and Prenner, 2013), although the latter two characters are also present in *Viburnum*.

Different approaches to the intra-familial classification have been presented, which are not satisfactory (overview in Donoghue, Bell and Winkworth, 2003). Based on more recent phylogenies, it is possible to recognize four tribes: Diervilleae, Caprifolieae, Linnaeaeae and Valerianeae. Inflorescences are built on a monotelic pattern and consist of racemes with cymose lateral axes (thyrses), but there are several complex derivations of simple inflorescences that evolved separately (Landrein and Prenner, 2013). Partial inflorescences are dichasial with three flowers (e.g. *Leycesteria*), or are two flowered in *Lonicera* and *Linnaea* by abortion of the central flower; in *Symphoricarpos* only the central flower is formed. Other genera have a more complex pattern with a higher number of small flowers (e.g. *Patrinia, Valeriana*: Figure 11.32D), or forming capitula-like inflorescences in Dipsaceae.

An epicalyx is frequently present (in Linnaeeae, Dipsaceae, Morineae and *Triplostegia*) and is six- to four-lobed, consisting of two or three pairs of alternating bracts (*Hochblätter*: Hofmann and Göttmann, 1990). Roels and Smets (1996) interpreted the epicalyx as derived from lower bracts of a thyrsoid inflorescence as is found in *Centranthus* or *Lonicera*. Reduction of all flowers of a lateral branch, except for the topflower, leads to typical Dipsaceae flowers with an epicalyx. Each flower on a capitulum would be interpreted as remnant of a complex thyrsoid inflorescence (Roels and Smets, 1996). However, assuming that the ancestral condition consisted of three pairs (as in *Linnaea*), it is not possible to know whether there are two separate origins of an epicalyx or a single origin followed by loss, although the second option is unlikely (Donoghue, Bell and Winkworth, 2003).

Flowers are either pentamerous or tetramerous. Naghiloo and Claßen-Bockhoff (2017) demonstrated that a shift in merism is linked with proportional change of floral organs and floral meristem size in Dipsaceae, as is visible in the transition between five sepals and four petals. The calyx is rarely well developed; it is mostly cupulate or as a small rim, knobs (e.g. *Valeriana*: Figure 11.32C) or bristles, or is absent (in analogy with the Asteraceae). In *Lomelosia* and *Pterocephalus* the five or six sepal lobes subdivide in a series of parallel lobes that develop as elongated bristles (Naghiloo and Claßen-Bockhoff, 2017). Reduction of the calyx is accompanied by the development of the epicalyx, which takes over the protective and dispersal functions in combination with the calyx, or at the expense of calyx and ovary wall (see Donoghue, Bell and Winworth, 2003).

Monosymmetry is median (e.g. *Lonicera, Morina, Linnaeeae*: Figure 11.32A) or the flower is asymmetric by reduction of stamens and carpels (e.g. *Valeriana*: Figure 11.32C; *Centranthus*). Initiation of floral organs is unidirectional from the adaxial to the abaxial side in basal Caprifoliaceae (Roels and Smets, 1996; Landrein and Prenner, 2013), but becomes equal in some polysymmetric flowers of Dipsaceae or shifts in a transversal direction in asymmetric flowers (Naghiloo and Claßen-Bockhoff, 2017). Although symmetry patterns tend to be similar to those of Lamiales (Donoghue, Ree and Baum, 1998), changes in stamen number become decoupled from petal evolution (Donoghue, Bell and Winkworth, 2003). Stamen number is rarely five (e.g. *Patrinia pentandra*), mostly reduced to four (e.g. *Dipsacus, Patrinia*: loss of the adaxial stamen), three (e.g. *Valeriana*), two (e.g. *Fedia*) or one (e.g. *Centranthus*). In *Morina* the posterior pair is fertile while the anterior is staminodial. Reduction of stamens tends to follow the phylogeny with clades characterized by a given stamen number (Donoghue, Bell and Winkworth, 2003).

Petal arrangement is usually as 3:2 with three anterior petals (e.g. *Weigelia*, strongly resembling Lamiales flowers), rarely 1:4 (e.g. *Lonicera*, *Triostemum*). Most Dipsaceae are secondarily polysymmetric with four petals (by fusion of two posterior petals: e.g. *Dipscacus*, *Knautia*). In *Scabiosa*, five corolla lobes and four stamens are formed (Eichler, 1875; Roels and Smets, 1996). Howarth and Donoghue (2005) mentioned this for *Symphoricarpos* but tetramery was not detected in that genus, except for the ovary.

Trichome nectaries are inserted on the corolla tube, often concentrated in one area and occasionally in a spur (e.g. *Lonicera*: Figure 11.33A, *Centranthus*).

The gynoecium is always inferior. *Leycesteria* has five carpels in antepetalous position with ovules in two rows in each locule (Eichler, 1875). In *Symphoricarpos* the ovary is four carpellate, but the median carpels are positioned lower than the transveral carpels and bear single ovules instead of pairs in the latter (Eichler, 1875; Roels and Smets, 1996). In other genera, he ovary is mostly tricarpellate and inferior with three apical ovules, but two ovules are frequently sterile or absent (pseudomonomery). Eichler (1875) presented the orientation of the gynoecium of *Valeriana* with the odd carpel in adaxial position; I found the opposite pattern (Figure 11.33C). There is a complete transition series between well-developed carpels and sterile carpels that are barely visible (Hofmann and Göttmann, 1990). In Dipsaceae it is not clear how many carpels make up the ovary as the position of the single ovule is variable (see Hofmann and Göttmann, 1990; Roels and Smets, 1996). The single style has three stigmas in a commissural position.

Adoxaceae (incl. Viburnaceae)

Figure 11.34A–B *Viburnum grandiflorum* Wall. ex DC

∗ K(5) [C(5) A5] Ğ(1:2°)

The family is restricted to *Adoxa*, *Viburnum* and *Sambucus*, which have following characteristics in common: polysymmetry, a short style with lobed stigma (lobes carinal), a rotate corolla, three to five carpels and drupaceous fruits (e.g. Judd, Sanders and Donoghue, 1994; Howarth and Donoghue, 2005). Flowers contain five stamens in a haplostemonous arrangement. Sepals are short, valvate and fused at the base. The arrangement of lateral flowers of *Adoxa* is inversed with the odd petal adaxially (Erbar, 1994), and this was also reported for *Sambucus ebulus* by Eichler (1875). The corolla is imbricate. *Sambucus* and *Adoxa* have extrorse anthers and a short stamen-petal tube. The stamens of *Adoxa* arise as common primordia but are split to the base forming pairs of half-stamens. In the sister genera *Sinadoxa* and *Tetradoxa*, the filament is divided to the base, respectively to the middle (Donoghue, Bell and Winkworth, 2003). In *Viburnum*, stamens are occasionally of different length (e.g. *V. grandiflorum*). The ovary is inferior with apical insertion of ovules. In

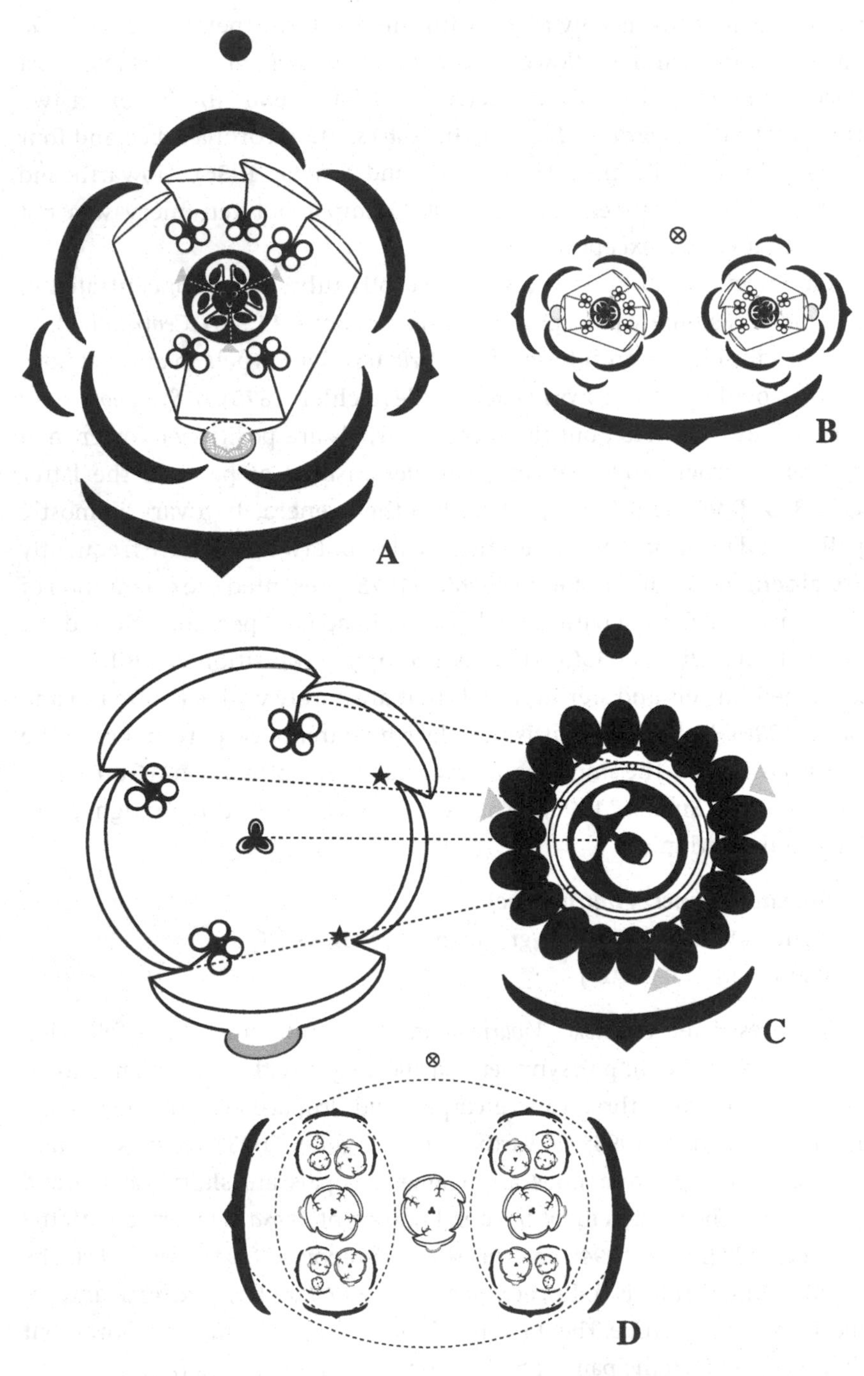

Figure 11.33 Caprifoliaceae: *Lonicera giraldii*, A. flower; B. partial inflorescence, *Valeriana repens*; C. flower; D. partial inflorescence

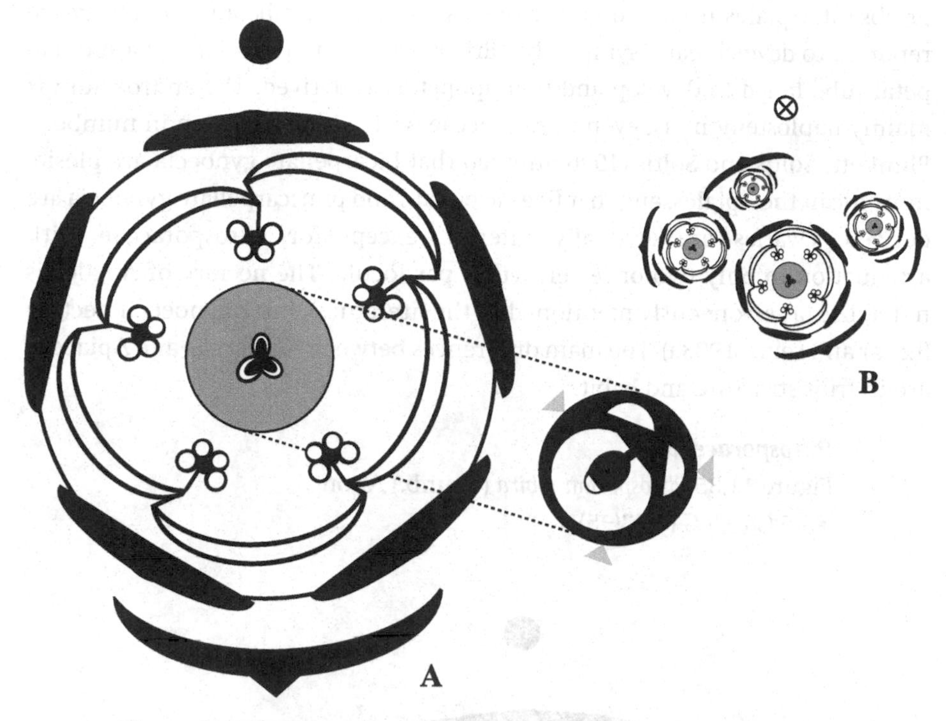

Figure 11.34 *Viburnum grandiflorum* (Adoxaceae): A. flower; B. partial inflorescence

Viburnum, the ovary is pseudomonomerous by reduction of two locules. Vestigial archesporial tissue is present at the base of the style of all Adoxaceae. This was interpreted as the remnant of a second ovule present in ancestral groups (Erbar, 1994).

In *Adoxa*, glandular multicellular trichomes are formed in groups at the base of petals. In *Viburnum*, a disc-like nectary sits on top of the ovary (cf. Erbar, 1994). *Sambucus* lacks nectaries and this is interpreted as a loss, but Vogel (1998b) mentioned the presence of glands associated with inflorescences of *S. javanica*. Some species of *Viburnum* (e.g. *V. opulus*) have a tendency to develop complex inflorescences with marginal sterile flowers that have accrescent calyces, analogous to Hydrangeaceae. A floral diagram of *Adoxa moschatellina* was provided by Erbar (1994).

Apiales

Apiales contain seven families, according to APG IV (2016), of which the largest are Apiaceae and Araliaceae. Araliaceae and Apiaceae are difficult to keep separated and some authors have implied that Apiaceae evolved from within a paraphyletic Araliaceae (e.g. Judd, Sanders and Donoghue, 1994).

Apiales share a combination of characters. The complex inflorescences are probably derived from cymes. There is a tendency for sepals to become reduced or absent. Apiales usually do not have a stamen-petal tube although they are reported to develop early sympetaly[d] (Erbar, 1991). It is assumed that a stamen-petal tube failed to develop and that apopetaly is derived. The androecium is mainly haplostemonous, even in Araliaceae with a high variation in numbers. Plunkett, Soltis and Soltis (1996) implied that bicarpellate gynoecia are plesiomorphic in the Apiales, and that five-carpellate and pluricarpellate gynoecia are derived. Ovaries are generally inferior, except for Pittosporaceae with a reduction to only two or fewer ovules per locule. The nectary of Apiales is not a disc (as erroneously mentioned in the literature), but a gynoecial nectary (Erbar and Leins, 1988a). The main differences between Araliaceae and Apiaceae are in fruit structure and habit.

Pittosporaceae

Figure 11.35 *Pittosporum tobira* (Thunb.) Aiton

∗ K5 C5 A5 $\underline{G}$ (2–)3(–5)

Figure 11.35 *Pittosporum tobira* (Pittosporaceae): partial inflorescence

[d] Sokoloff et al. (2018a) indicate that it is impossible to distinguish a petal ring primordium (early sympetaly) from the margins of the concave receptacle and conclude that petals arise as free lobes.

Flowers are grouped in compact panicles, with two bracteoles that may enclose additional flowers. Sepals and petals are erect, free and imbricate, but petals do not spread by the tight constriction of sepals. Placentation is broadly parietal with few ovules inserted in two irregular rows. A gynoecial nectary develops at the base of the abaxial carpel flanks and is visible as an area not covered by trichomes. Erbar and Leins (1995a) reported characters linking Pittosporaceae with Apiales, including early sympetaly and the gynoecial nectary in *Pittosporum tobira*. Plunkett (2001) mentioned the basal fusion of petals in some Pittosporaceae and Araliaceae – it is to be questioned whether this fusion is postgenital by compression between the erect sepals. He also implied that the superior ovary of Pittosporaceae is secondary, although this is not supported by floral ontogeny (Erbar and Leins, 1995a).

Araliaceae

Figure 11.36 *Schefflera* aff. *elliptica* (Blume) Harms

∗ K5 C5 A5 -G(5)-

General formula: ∗ K0–5 C5–10(–12) A5–10(–120) G(1)2–5–10(–200)

Figure 11.36 *Schefflera* aff. *elliptica* (Araliaceae)

Araliaceae flowers tend to be easily recognizable. Inflorescences are heads or umbels with small flowers. Erbar and Leins (1988) demonstrated the tendency for bracts to be lost in *Aralia elata* and *Hedera helix*. The position of outer sepals depends on presence/absence of the two bracteoles in *Hedera*, while it is more variable in *Aralia*. The calyx is rarely well developed, more often small or develops as a rim with loss of individuality of sepal lobes. Sepals are absent in *Meryta* and *Hydrocotyle* where they are not initiated (Erbar and Leins, 1985). Petals have generally a valvate (e.g. *Hedera*), occasionally imbricate (e.g. *Panax*) aestivation. Petals are usually five in number, rarely ten or more arising by lateral division (Philipson, 1970). There is considerable diversity in the perianth morphology of Araliaceae, with petals generally free and occasionally fused postgenitally, but rarely with a tubular corolla (*Tupidanthus*, *Sheffleria subintegra*: Sokoloff et al., 2007, 2018a; Nuraliev et al., 2014). A calyptra is frequently formed by postgenital fusion of the petal tips. Contrary to other haplostemonous Apiales, the androecium can be highly variable. Secondary polyandry is present in some genera (*Plerandra*, *Tetraplesandra*: Philipson, 1970; Nuraliev et al., 2010). In *Tetraplesandra* polyandry is derived from original haplostemony by the lateral increase from antesepalous primary primordia, while the stamens arise in continuous girdles on a concave receptacle in *Plerandra*. In *Tetraplesandra*, the androecium varies from haplostemony to eight times the petal number (Costello and Motley, 2004). Stamens are inflexed in bud, as in Apiaceae.

Merism is highly variable in Araliaceae, especially in the genus *Shefflera* (see Nuraliev et al., 2014). There is frequently a lateral increase of carpels, ranging from six to up to two hundred in *Tupidanthus* (Sokoloff et al., 2007). The multiplication of carpels is often linked with a meristic increase of stamens and petals, while sepals remain pentamerous or are reduced.

The ovary is usually inferior, rarely superior. While most *Tetraplesandra* species have inferior ovaries, Costello and Motley (2004) reported an upward expansion of the ovary resulting in a secondary superior position in *T. gymnocarpa*. When isomerous, carpels are antepetalous.

In *Seemannaralia*, the bicarpellate ovary becomes unilocular by formation of an incomplete septum (Burtt and Dickison, 1975). Some taxa are pseudomonomerous (e.g. *Diplopanax* and *Eremopanax*: Philipson, 1970). Only one apical hanging ovule develops in each locule. The other ovule on the placenta develops in a modified tapering organ, which is also found in Apiaceae (Philipson, 1970; Erbar and Leins, 1988; Karpunina et al., 2016). It was considered vestigial, although Philipson (1970) believed it has another function, maybe acting as an obturator as it is curved upwards into the symplicate zone of the ovary. A unicarpellate ovary is found in some *Polyscias* species with unstable position of the single carpel and without evidence of pseudomonomery, implying that the number of carpels can be abruptly reduced (Karpunina et al., 2016). As in Apiaceae, a stylopodium is generally formed on top of the ovary bearing a nectary.

Apiaceae

Figure 11.37 *Aegopodium podagraria* L.

∗ K5 C5 A5 Ğ(2)

General formula: ∗/↓ K(0–)5 C(0–)5 A5 G2

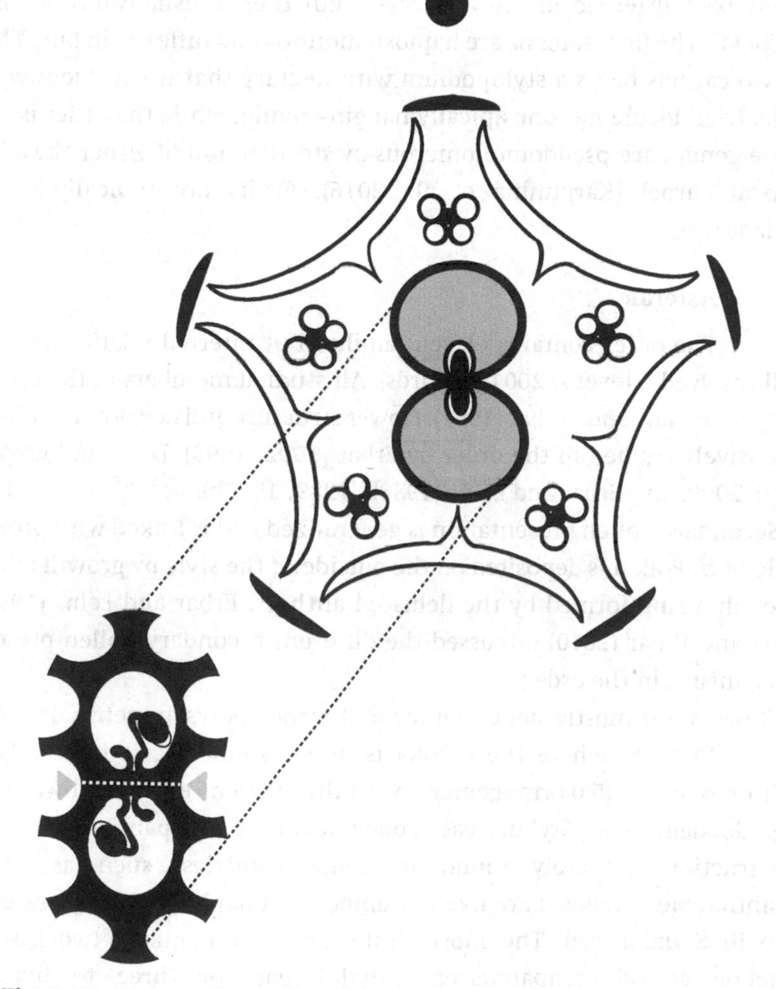

Figure 11.37 *Aegopodium podagraria* (Apiaceae)

The general floral structure is very similar to some Araliaceae. Inflorescences are umbels or heads, occasionally with showy basal bracts, forming pseudo-flowers (e.g. *Astrantia*). Flowers are bisexal or unisexual, often with complex systems of dichogamy. Flowers are either polysymmetric or bilaterally symmetric by unequal development of petal lobes, especially in the peripheral flowers of umbels. The calyx is usually reduced to teeth or is absent at maturity. Although the calyx is initiated in most cases (Leins and Erbar, 1985), it has a strong tendency for reduction with a delayed initiation (e.g. *Anthriscus*: Sattler, 1973). Petals are valvate and free, usually clawed, divided and with an inflexed tip (enclosing the anthers in bud); there is usually only a single vein per petal. The five stamens are haplostemonous and inflexed in bud. The ovary of two carpels bears a stylopodium with nectary that is confluent with each style. Each locule has one apically hanging ovule, while the other is reduced. Some genera are pseudomonomerous by sterilization of either the adaxial or abaxial carpel (Karpunina et al., 2016). Fruits are typically two-parted schizocarps.

Asterales

The order contains eleven families but internal relationships are not well resolved (Stevens, 2001 onwards). All studied members of the order share early sympetaly *sensu* Erbar (1991). Flower structure and development have been extensively studied in the order by Erbar (1992, 1993), Leins and Erbar (1987, 1989, 2000) and Erbar and Leins (1988b, 1989, 1995b).

Secondary pollen presentation is generalized and is linked with proterandry of flowers. Pollen is deposited on the outside of the style by growth of the style through a ring formed by the dehisced anthers. Erbar and Leins (1995b) and Leins and Erbar (2010) discussed the different secondary pollen presentation mechanisms in the order.

Flowers are mostly pentamerous and either polysymmetric or monosymmetric. In cases where the corolla is monosymmetric, it is usually deeply split on one side (5:0 arrangement) with the stamens exserted towards the slit (e.g. Goodeniaceae, Stylidiaceae, Lobelioideae of Campanulaceae). A similar construction is rarely found in some Lamiales, such as *Sclerochiton* (Acanthaceae). Stamens are five in number and haplostemonous, rarely fewer (two in Stylidiaceae). The mostly inferior ovary contains two (Asteraceae, Lobelioideae of Campanulaceae, Stylidiaceae) or three to five carpels (Campanuloideae of Campanulaceae, Calyceraceae) with axile, basal or apical placentation. With fewer fertile carpels there are clear indications of pseudomonomery in Calyceraceae, Goodeniaceae and Stylidiaceae (Erbar and Leins,

1988b; Erbar, 1992, 1993). An epigynous disc nectary is common. A single style is formed, generally with stigmatic lobes reflecting carpel numbers. Corolla aestivation is generally valvate (imbricate in Stylidiaceae). In Menyanthaceae and Goodeniaceae, petals are induplicatively valvate. The petal apex is reduced and has two lobes (Endress and Matthews, 2006b). This is reflected in the petal vasculature of Asteraceae where the median vein is also reduced. Gustafsson (1995) gave an overview of petal venation in Asterales and related families; he found that Asteraceae and the closely related families Calyceraceae, Menyanthaceae and Goodeniaceae share an unusual petal vasculature in which prominent marginal veins meet in the top of the petals. Stamens are strongly associated with the style, often forming a tube of connivent anthers surrounding the protruding style.

Campanulaceae (incl. Lobeliaceae, Cyphiaceae)

Figure 11.38 *Lobelia tupa* L.

↓ K(5) C(5) A(5) Ğ(2)

General formula: ∗/↓ K(3−)5(−10) C(3−)5(−10) A(3−)5(−10) G2−5(−10)

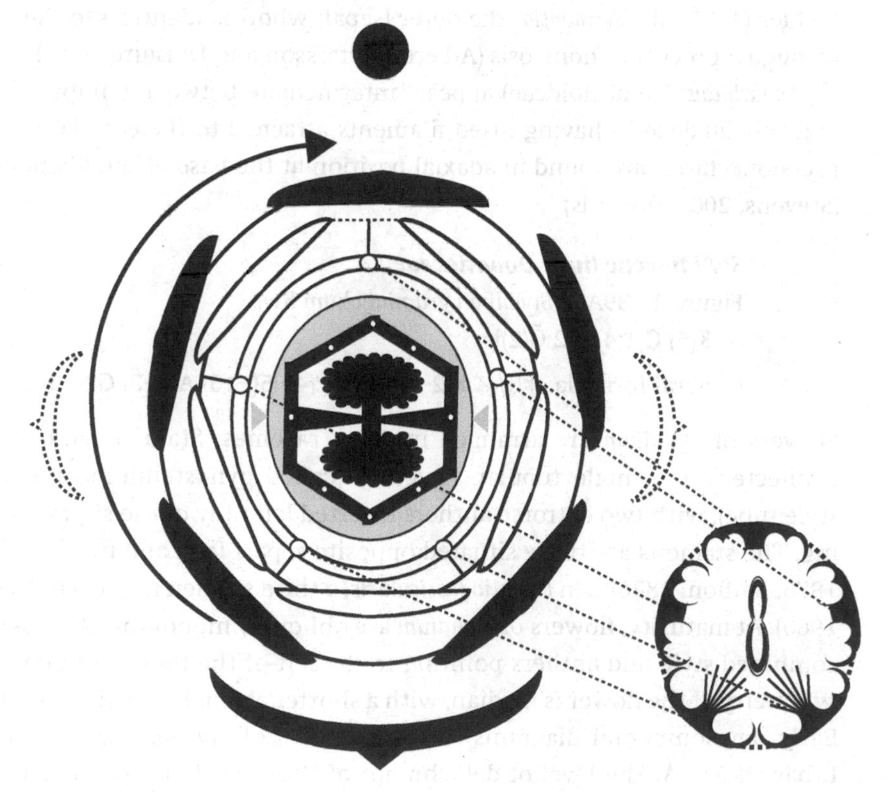

Figure 11.38 *Lobelia tupa* (Campanulaceae). The arrow points to resupination of 180 degrees.

Lobelioideae differ from Campanuloideae in the bicarpellate gynoecium versus three to five carpels, and the monosymmetric flowers with unusual orientation.

The odd petal of *Lobelia* is placed in an adaxial position (cf. Eichler (1875). However, the flower is resupinated at maturity, restoring the condition found in most other Pentapetalae. In other genera such as *Nemacladus* and *Cyphia* the petal arrangement remains as 2:3 (Byng, 2014). The corolla is split on the side where the fused stamens protrude. Anthers are broad and closely packed. The two adaxial anthers bear long hairs missing in the other anthers (cf. Payer, 1857). A broad septum divides the ovary in two locules and bears a placenta in its upper part covered with a high number of ovules. A nectary covers the ovary roof (Erbar and Leins, 1989).

Flowers are pentamerous, although merism can fluctuate in a number of genera (e.g. hexamery in *Canarina*; up to decamerous in *Michauxia*: Byng, 2014). In *Campanula* and related genera the base of the filament is strongly inflated and covers the top of the ovary as a nectar chamber (cf. Erbar and Leins, 1989). Access to the nectary is through spaces between the stamen bases. When isomerous, the ovary is either antesepalous or antepetalous according to Eichler (1875). In *Clermontia*, the outer (sepal) whorl is identical to the corolla through a process of homeosis (Albert, Gustafsson and Di Laurenzio, 1998).

Nemacladus (Nemacloideae) appears intermediate between Campanuloideae and Lobelioideae in having fused filaments attached to the corolla tube. Two pseudonectaries are found in adaxial position at the base of the filament tube (Stevens, 2001 onwards).

Stylidiaceae (incl. Donatiaceae)

Figure 11.39A–B *Stylidium graminifolium* Sw.

↙ K(5) C(1:4) [A2 Ğ(2)]

General formula: ✶/↓/↙ K(2–)5(–10) C(4–)5(–10) A2(–3) G2(–3)

Flowers of *Stylidium* are arranged in short racemes. Stamens and style are connected in a single tubular structure (called gynostemium or filament-style tube), with two extrorse anthers inserted laterally of the stigmatic opening. The stamens are those situated opposite sepals four and five (cf. Eichler, 1875; Baillon, 1876b). In *Donatia fascicularis*, a third stamen may occur (Carolin, 1960). At maturity, flowers of *Stylidium* are obliquely monosymmetric with the combined style and anthers pointing to the left of the flower. In bud, monosymmetry of the flower is median, with a shorter abaxial petal (Figure 11.39A). Early developmental diagrams for two species of *Stylidium* were shown by Erbar (1992). At the level of detachment of the petal lobes two erect appendages (auricles) are found. During development, the common stamen-style

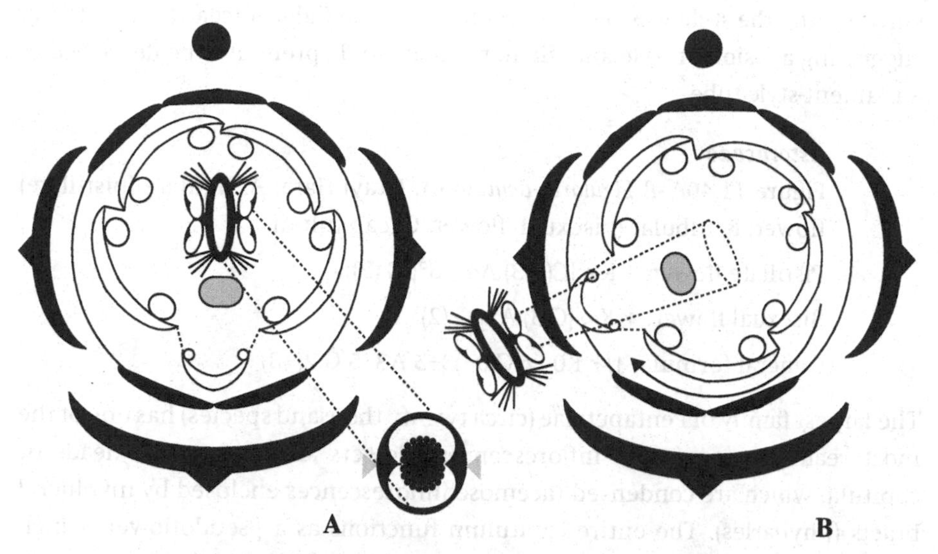

Figure 11.39 Stylidiaceae: *Stylidium graminifolium* A. flower in bud; B. flower at anthesis. Note the corona of globular protuberances opposite the petals. The arrow shows resupination of 45 degrees.

curves forward and fits between the two auricles of the shorter petal. This corresponds with a torsion of the pedicel of 45 degrees (not 90 degrees as Erbar, 1992 suggested) to the left (Figure 11.39B). As the small petal diverges much lower on the tube, a short slit is present. The ovary has septa only in the basal part of the ovary, connecting a globular central placentation. One of the carpels is sterile in *S. adnatum*, while the posterior locule can be shorter in *S. graminifolium* (Erbar, 1992). In *Donatia* septa are almost completely formed (Carolin, 1960). Two nectaries are reported to be in the median plane of the flower with the adaxial larger (Erbar, 1992). In my material only the adaxial nectary was well developed. Below the anther-stigma complex long hairs are present, functioning in the reception and distribution of pollen. A complete picture of the flower structure and pollination mechanism of *Stylidium adnatum* and *S. graminifolium* was given by Erbar (1992).

Except for Erbar, other authors did not mention oblique monosymmetry in the family. Baillon (1876b), Sattler (1973) and Erbar (1992) studied the floral development of *Stylidium*. The anthers differentiate earlier than the corolla and push the style rapidly upwards. In *Donatia* the anthers are not connected to the ovary. The origin of the stamen-stigma tube appears difficult to resolve because no clear carpel primordia are visible in *Stylidium* (only a narrow slit), although clear primordia are visible in *Levenhookia*. Erbar (1992) interpreted the tube as receptacular, lifting up carpels and stamens. A similar filament tube surrounding the style was seen in Lobelioideae and Calyceraceae (Erbar, 1993), suggesting a fusion of style and filaments that would preferably be described as a filament-style tube.

Asteraceae

Figure 11.40A–B *Ligularia dentata* (A. Gray) Hara. A. ligulate (pistillate) flower; B. tubular (bisexual) flower; C. capitulum

Pistillate flower: ↓ K∞ [C(0:3) A4°–5°] Ğ (2)

Bisexual flower: ✶ K∞ [C(5) A5] Ğ (2)

General formula: ↓/✶ K0–∞ C(0–)3–5 A3–5 Ğ 2(–3)

The largest family of Pentapetalae (circa twenty thousand species) has one of the most readily recognizable inflorescences. Flowers are grouped in heads or capitula, which are condensed racemose inflorescences enclosed by involucral bracts (phyllaries). The entire capitulum functions as a pseudoflower, which may consist of one (e.g. *Ligularia*: 11.40 C) or several capitula (e.g. *Echinops*, *Achillea*: Harris, 1995). Flowers (florets) are generally small, ranging from a few per capitulum to more than one thousand. Subtending bracts (palea or 'chaff') can be present or are absent. When present, they arise with the flower on a common primordium, which may be due to condensation of the inflorescence (Harris, 1995). Flowers of Asteraceae are homogenous throughout the family. Flowers are mostly bisexual - peripheral flowers may be pistillate (Figure 11.40A) or sterile. Unisexual capitula are rare (e.g. *Xanthium*). A distinction can often be made between peripheral ray florets (ligulate flowers) with monosymmetric petals and central polysymmetric tubular disc florets. Some capitula only have tubular florets (e.g. *Echinops*), or only ligulate florets (e.g. *Hieracium*). Sepals are generally reduced and replaced by scales, teeth or hairs (pappus). The ovary is bicarpellate and inferior with a single basal ovule (connected to the adaxial carpel). Petals are fused with valvate aestivation and are connected with a whorl of stamens. Anthers are connivent around the style and this arrangement is linked to the secondary pollination mechanism. The stigma spreads open as two lobes above the stamen ring and is covered abaxially with pollen. Ligulate flowers which function as petals in the pseudanthium can have three fully developed and two reduced lobes (3:2), three lobes only (Asteroideae; Figure 11.40A), a 5:0 (Cichorioideae) or a 1:4 arrangement (e.g. *Barnadesia*). Staminodes are present in *Ligularia* (Figure 11.40A), but they are rarely noticeable in mature flowers of Asteraceae (Harris, 1995).

Flowers develop acropetally along a helix. The ray florets lag behind in development compared to disc florets caused by the compression of involucral bracts. Flower parts arise on a concave apex (early sympetaly: Leins and Erbar,

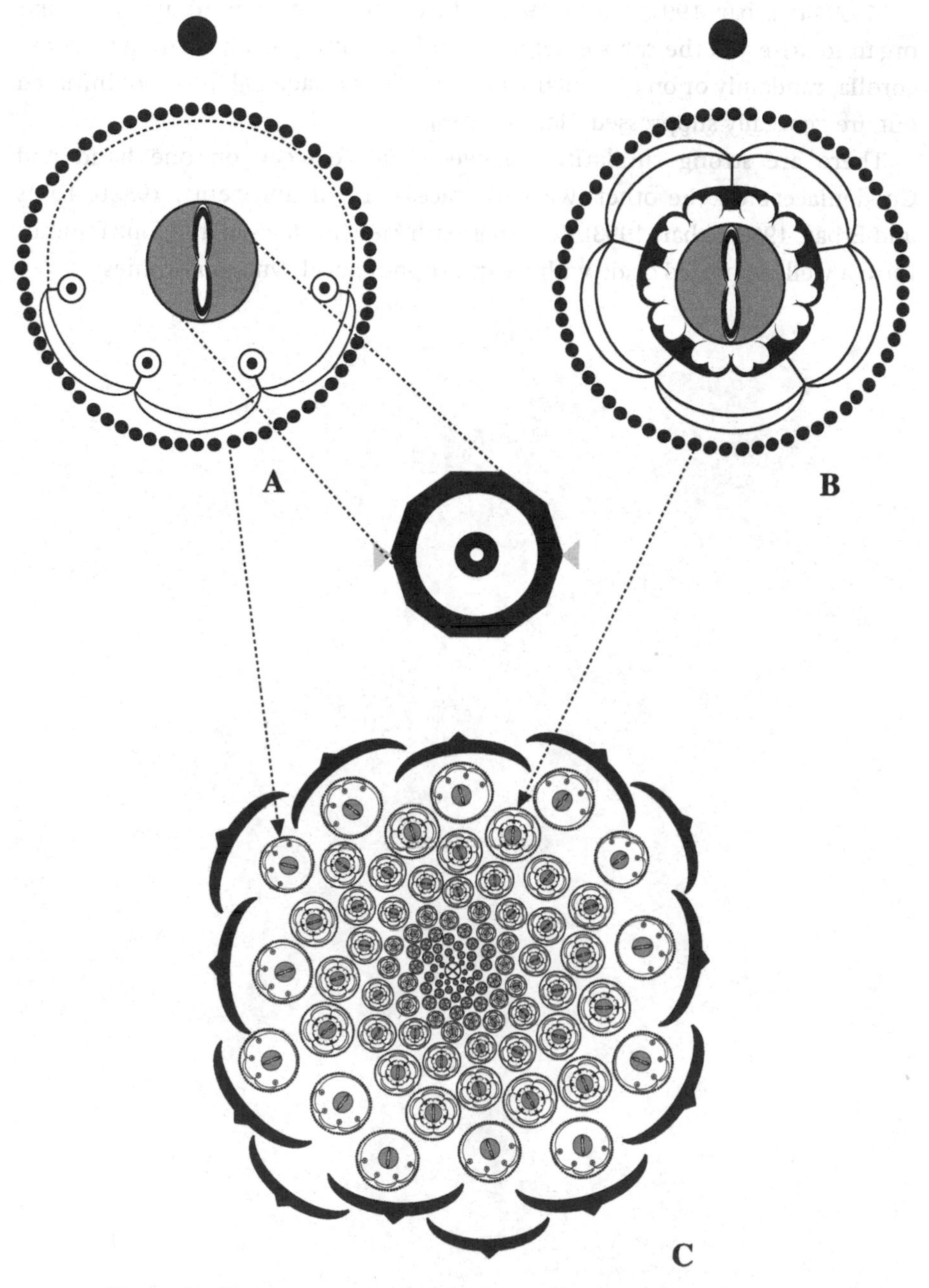

Figure 11.40 Asteraceae: *Ligularia dentata*. A. ligulate pistillate flower; B. tubular bisexual flower; C. capitulum. The ovary is the same for A and B.

1987, 2000; Erbar, 1991; Harris, 1995). The corolla ring primordium is the first organ to arise and the calyx emerges from five primordia alternating with the corolla, randomly or on a circular rim. In ray florets, adaxial lobes are initiated but are generally suppressed (Harris, 1995).

There are strong similarities between Calyceraceae on one hand and Goodeniaceae on the other, with Asteraceae (Erbar and Leins, 1988b; Leins and Erbar, 1989; Erbar, 1993). Together with Menyanthaceae, the four families form a well-supported clade with clear morphological synapomorphies.

PART III CONCLUSIONS

12

Distinctive Systematic Characters and Apomorphic Tendencies

Floral diagrams are the equivalent of descriptions of flowers with a high degree of detail. Although it is difficult to represent specific three-dimensional depth, the amount of information provided is considerable and may include developmental and anatomical evidence (see p. 45).

Floral diagrams allow the recognition of plants, at least to the family level, as major groups can be identified by their floral diagram, but are also a reflection of the morphological evolution of flowers in angiosperms. Major evolutionary changes in the floral Bauplan can be stressed by representing floral diagrams in the context of the phylogenetic tree of angiosperms. Floral diagrams make a comparison between divergent characters possible by stressing the positional relationships of floral structures.

Floral diagrams are not rigidly fixed in time, but are an expression of the developmental plasticity in flowers. Important morphological changes are often the result of subtle shifts in the primordial body during development, and this is also reflected at the genetic level. One aspect of morphological observations is that changes are often gradual, without clear-cut boundaries in characters between different clades. Several plant groups are characterized by 'apomorphic tendencies' (cf. Endress and Matthews, 2006a, 2012), defined as 'cryptic apomorphies' by Ronse De Craene (2010), viz. characters that may not be generalized in a clade, but occur on a much more frequent basis than in any related clades. These characters also include parallelisms and convergent evolution and are important for recognizing groups morphologically, although they are often rejected as homoplasies (Vasconcelos et al., 2017). Suites of morphological characters can be used to identify clades, even if they are not present in all taxa. There are several examples representing specific trends, visible in a given family or order. For example, in Ericales stamen increase is predominantly lateral in a girdle, and this may be linked to other features such as the lack of a well-developed hypanthium. The same order shows a tendency for the development of stamen-petal tubes, which are present in a number of families, but not generalized. A precondition for certain characters can be present in ancestors but is not expressed in all descendants. This has been identified as 'deep homology', where the genetic potential is present but not necessarily expressed within a clade, or expressed in selected lineages in parallel (see also Endress and Matthews, 2006b, Vasconcelos et al., 2017, 2020, and Ronse De Craene, 2018, for a discussion). Characters also appear to be correlated because one character change can have an effect on another character.

Throughout this book I recognized several apomorphic tendencies among different groups of plants that are also visible on floral diagrams. A good example is the predisposition for halving anthers that is characteristic for Malvaceae. Monothecate anthers have arisen convergently in *Durio* and Malvoideae (Bayer et al., 1999); it is not a synapomorphy for the clade and has arisen several times independently. Another apomorphic tendency is the repeated loss or reduction of sepals or the transformation of a calyx in bristles or hairs in several campanulids, correlated with elaboration of compact, small-flowered inflorescences (umbels or heads). Rasmussen, Kramer and Zimmer (2009) demonstrated for the Ranunculales that a genetic programme for petal identity is present in most families, even if petals are not developed or are highly divergent. There are several more examples throughout this book.

Syndromes of correlated characters are intimately linked with the Bauplan of the flower and can be detected through floral diagrams. These syndromes can be ultimately linked with pollination mechanisms (see Van der Niet et al., 2014; Rose and Sytsma, 2021). An example is the syndrome of synandria with extrorse anthers in fully unisexual flowers (e.g. Triuridaceae, Nepenthaceae, Myristicaceae, Menispermaceae). Another example is the correlation between early sympetaly and inferior ovaries, and late sympetaly and superior ovaries that reflect coherence between several interdependent characters linked through floral development.

Character mapping and its interpretation on phylogenetic trees remain controversial, especially in basal angiosperms, as we do not know how extensive morphological variability was at the lower nodes. Extant families represent tips of millions of years of evolution and extinction that can only be partially resolved by the fossil record. A good example is Proteales, grouping four families with extremely different morphologies. Because of the principle of parsimony and because of lack of intermediate taxa, character evolution is often referred to as 'reversals' in cases where more basal characteristics are found at the end of nodes (see e.g. Soltis et al., 2018). This can be reasonable in a cladistic context, but does not always help in understanding evolution of morphological characters. The repeated evolution of certain floral syndromes may also lead to wrong definitions of morphological characters that are not homologous and consequently lead to a wrong use of terminology. Vasconcelos et al. (2020) illustrated this with the calyptrate perianth of Myrtaceae. Understanding and defining morphological characters remains one of the major challenges for botanical research.

13

Floral Diagrams and Major Angiosperm Groups

This book is not exhaustive, as the total number of families has not been covered, although it is comprehensive in reflecting the floral diversity of the majority of angiosperms. Major groups of angiosperms are covered in order to reflect important floral evolutionary trends. The sections that follow summarize the most important floral attributes of major clades of angiosperms and are also a first step in the identification of a specific taxon.

13.1 Basal Angiosperms, Monocots and Early Diverging Eudicots

Basal Angiosperms (Figure 5.1)

Floral evolution of basal angiosperms has been studied in a phylogenetic context by several authors (e.g. Doyle and Endress, 2000, 2011; Ronse De Craene, Soltis and Soltis, 2003; Zanis et al., 2003; Endress and Doyle, 2009, 2015). Mapping of floral characters on phylogenetic trees has not clarified whether cyclic flowers have evolved from spiral flowers or the opposite (see Sauquet et al., 2017; Sokoloff et al., 2018b; Cavaleri De-Paula, Assis and Ronse De Craene, 2018), although a helical flower with relatively few stamens and carpels is found in *Amborella* and the ANA-grade (Austrobaileyales), with frequent shifts between spiral and whorled phyllotaxis in different orders (Nymphaeales, Laurales, Magnoliales). The existence of hemicyclic flowers illustrates the transitional nature between spirals and whorls. The most recent addition in Nymphaeales, Hydathellaceae, raises questions about the flowers of the earliest angiosperms (see Rudall et al., 2007; Endress and Doyle, 2009).

Flowers of basal angiosperms have a low synorganization with high plasticity in organ number and position (cf. Endress, 1990, 2001, 2008c). The transition between vegetative shoot and flower is often gradual with a progressive change from leaves to bracts and tepals within the inflorescence or at the base of the flower (Endress, 2003b; Remizowa and Sokoloff, 2003; Buzgo et al., 2007). This pattern of an unclear distinction between bracts and flowers is also present in early diverging eudicots and basal monocots. Taxa with spiral flowers look remarkably similar in the gradual transition of bracts to floral organs and transitions between different organ categories (e.g. Austrobaileyaceae, Schisandraceae, Eupomatiaceae, Calycanthaceae). Transitional organs include staminodes between fertile stamens and carpels, or between tepals and stamens, and have intermediate characteristics, explaining the difficulty in differentiating bracts from tepals, or tepals from staminodes (e.g. Endress, 1980b, 1990; Staedler and Endress, 2009). This is in contrast with core eudicots where an abrupt transition from the vegetative shoot to the flower is commonplace and is generally mediated by bracts and bracteoles.

Changes in phyllotaxis appear to be complex in basal angiosperms and early diverging eudicots, with mixtures of whorled, spiral or irregular patterns in several clades, such as Monimiaceae, Winteraceae, Ranunculaceae, and Nymphaeaceae (Endress, 1980a, 2001; Ronse De Craene, Soltis and Soltis, 2003; Staedler and Endress, 2009). Cyclic flowers are generally trimerous or dimerous, with far fewer parts, and cyclization starts from the perianth onwards, leading to hemicyclic flowers with spiral androecium and gynoecium (e.g. Magnoliaceae, Monimiaceae, Annonaceae, Ranunculaceae: Leins and Erbar, 2010). Cyclic flowers such as Lauraceae have an inner staminodial whorl which is involved in nectar production. Trimerous flowers are widespread (as in monocots and some early diverging eudicots) and linked with a whorled initiation. Trimerous and dimerous flowers often coexist in families such as Lauraceae, Hernandiaceae and Papaveraceae, although dimerous patterns may be a continuation of a decussate bract arrangement (e.g. Buxaceae: Von Balthazar and Endress, 2002a, 2002b). Number and arrangement of whorls is generally variable. Differentiation into sepals and petals is rarely clear-cut (e.g. *Cabomba*, Annonaceae), more often gradual or reflected in minor differences (colour, number of vascular traces) with little scope for specializations. A combination of fusion and monosymmetry is found only in Piperales (Aristolochiaceae) where the perianth is reduced to a single trimerous whorl. A much higher numbers of stamens and carpels can arise in different ways, linked with the reduction in size of stamens in some cyclic Annonaceae, by extension of a conical receptacle in Magnoliaceae or by growth of a hypanthium in Nymphaeaceae.

Austrobaileyales, Magnoliales and Laurales are proterogynous with separate pistillate and staminate phases developing in succession, or flowers are unisexual. This leads to specializations in floral structure where staminodes play a role in preventing self-pollination (Walker-Larsen and Harder, 2000). Staminodes can take up secondary reward functions, such as providing food-bodies, scent or visual attraction (Endress, 1984, 1994). Syncarpy is rare in basal angiosperms and Ranunculales. Syncarpy has evolved independently only on four occasions in basal angiosperms (Canellaceae, Piperales, *Takhtajania* of Winteraceae, *Isolona* and *Monodora* of Annonaceae: Endress and Igersheim, 2000b). In all cases placentation is parietal, and this probably represents an independent origin by postgenital fusion of initially free carpels, comparable to *Berberidopsis* (Ronse De Craene, 2017a). There is more generally a trend leading to unicarpellate gynoecia, sometimes in the same family with many carpels (e.g. *Idiospermum* in Calycanthaceae). However, single carpels are not derived through pseudomonomery, which involves syncarpy (see Sokoloff, 2016). Endress and Igersheim (2000b) discussed the evolution of early carpel closure, leading to the formation of a compitum and fused ovaries. A hypanthium is rare in basal angiosperms, except in families of Laurales characterized by elaborate cups enclosing floral organs.

A well-differentiated disc nectary is not found in basal angiosperms. Instead, nectarioles (viz. unusual nectar producing idioblasts or groups of cells on the surface of organs) are present in a few genera (e.g. *Cabomba* in Cabombaceae, *Chimonanthus* in Calycanthaceae: Vogel, 1998a; Endress, 2001, 2008c; Erbar, 2014). This may indicate that nectaries are in an early stage of differentiation.

Monocots (Figure 6.1)

Monocots are easily characterized by a number of characters reflected in their floral diagrams: flowers are trimerous with a more or less stable formula (P3+3 A3+3 G3). Only some families of Alismatales do not follow this strict arrangement, with variability in number of stamen and carpel whorls. Although the floral formula is superficially similar to some apetalous rosids (see p. 310), flowers are never spiral throughout and the origin of monocot flowers has to be sought in spiral or trimerous basal angiosperms. Iwamoto et al. (2018) suggested that polyandric flowers containing stamen pairs (Butomaceae, Alismataceae, Hydrocharitaceae, Tofieldiaceae) may be closer to the ancestral flowers of monocots. Monocots are structurally far less variable than dicots. Differences between groups are the result of subtle changes between and within whorls, rarely with secondary increases of organs or shifts in merism.

Trimerous flowers are prevalent among monocots. Pentamery is rare, although occasionally present in *Paris* (Figure 6.12) and only stable in the genus *Pentastemona* of Stemonaceae (Pandanales). Dimery is widespread and is often found together with trimery (e.g. Restionaceae, Asparagaceae, Stemonaceae: Ronse De Craene, 2016). Tetramery is more common although restricted to Cyclanthaceae, Arecaceae, Triuridaceae and Convallariaceae (Dahlgren, Clifford and Yeo, 1985). In monocots, syncarpy is generally the result of a postgenital fusion of carpels with marginal placentation (Remizowa et al., 2006, 2010), although apocarpous gynoecia are probably derived. Styles and stigmas are generally carinal, reflecting the lack of fusion or its late occurrence. Contrary to Pentapetalae where petals can have different homologies (see Ronse De Craene, 2007, 2008; Ronse De Craene and Brockington, 2013), the biseriate perianth of monocots is of a homologous nature. Differences between the inner tepals (petals) and outer tepals (sepals) were interpreted as the result of shifts of gene expression (sliding boundary hypothesis). Nonetheless, gene expression patterns in the perianth can be much variable (see overview in Ronse De Craene, 2007). A bipartite perianth remains constant throughout monocots, occasionally with further reductions in wind-pollinated Poales.

Unique floral or inflorescence characters for monocots include the presence of a single bracteole (adaxial or transversal), and septal nectaries. Septal nectaries are an early-diverging feature in monocots and are associated with postgenital carpel fusion (Dahlgren, Clifford and Yeo, 1985; Smets et al., 2000; Rudall, 2002; Remizowa, Sokoloff and Kondo, 2008). Flowers of monocots have undergone the same process of hypanthium growth and shifts to inferior ovaries as in Pentapetalae, reflecting comparable interactions with pollinators. A preliminary condition for an inferior ovary is the fusion of carpels. The shift to an inferior ovary position is strongly correlated with shifts of the septal nectaries and their increase in complexity (e.g. Simpson, 1998a; Rudall, 2002).

In earlier publications I emphasized the different structural nature for the androecium in monocots compared to eudicots and introduced a different terminology (dicycly and monocycly, compared to diplostemony and haplostemony: e.g. Ronse De Craene and Smets, 1993, 1998a; Figure 1.2). The prevalent trimery in monocots is related to a polycyclic ancestry found in the basal angiosperms and this is reflected in the evolution of the androecium (see Ronse De Craene and Smets, 1995a, 1998a; Iwamoto et al., 2018). Although *Acorus* represents the basalmost monocot, the genus has several unique apomorphies and probably a derived androecium. Multistaminate Alismataceae could have a floral arrangement closer to the ancestral monocots, and therefore be basal to the common two-whorled androecium. In some Alismataceae, the upper stamens and carpels tend to develop in irregular position or form parastichies (e.g. *Ranalisma*: Charlton, 1991, 2004). Two stamen whorls are most common in all major monocot clades, with the occasional loss of inner or outer whorl characterizing specific families (Ronse De Craene and Smets,

1995a). Secondary polyandry is rare in monocots and characterizes only a few families (Arecaceae, some Poaceae, Velloziaceae). In some Alismatales, it may be superimposed on existing polyandry (e.g. *Limnocharis*). The ovary is generally tricarpellate, except for Alismatales where the gynoecium may consist of several whorls of free carpels. Reductions within the tricarpellate ovary are linked to pseudomonomery and are concentrated in certain families, either as the result of a prevailing monosymmetry (e.g. Poaceae, Pontederiaceae), or as a general trend to a lower ovule number (e.g. Arecaceae). Parietal placentation (occasionally found in some families: Orchidaceae, Melanthiaceae, Hydrocharitaceae) probably has an independent origin from postgenitally fused apocarpous gynoecia with marginal placentae. Contraction of the margins led to parietal placentation, while intrusion and fusion led to axile placentation. In eudicots parietal placentation can have different origins, mostly as a derivation from an axile placentation but also by the marginal connection of free carpels (parasyncarpous; Figure 1.3).

Early Diverging Eudicots (Figure 7.1)

Early diverging or basal eudicots are a transitional grade between basal angiosperms and core eudicots, as they share morphological characters of both groups. There is a general tendency for flowers to evolve a weakly differentiated dimerous perianth with a single kind of organs (tepals) or no perianth at all, although there is a high diversification in certain clades (e.g. Proteales, Ranunculales). Basal eudicots were probably much more diverse in the past, as can be shown in florally diverse Platanaceae (e.g. Crepet, 2008; Von Balthazar and Schönenberger, 2009). Especially in the grade following the divergence of Ranunculales, there is a general shift to small dimerous flowers without petals and with questionable identity of the perianth (e.g. Proteaceae, Trochodendraceae, Buxaceae, Gunneraceae). As the grade immediately precedes the Pentapetalae, there has been much speculation about the origin of Pentapetalae flowers from such prototypes (e.g. Von Balthazar and Endress, 2002b; Soltis et al., 2003; Ronse De Craene, 2004, 2007, 2008, 2017a; Wanntorp and Ronse De Craene, 2005; Endress and Doyle, 2009; Doyle and Endress, 2011). Von Balthazar and Endress (2002b) interpreted the condition in Buxaceae, where flowers generally have bractlike phyllomes, as a precursory stage in the differentiation of a bipartite perianth, while Soltis et al. (2003) derived the pentamerous core eudicot flower from a dimerous Gunneraceae-like precursor. However, early diverging eudicots represent an evolutionary line with a gradual reduction of flowers and are not precursory to core eudicot floral diversity (cf. Wanntorp and Ronse De Craene, 2005). Early diverging eudicots, especially Ranunculales, represent a playground for floral diversification with a far greater variation in perianth forms than Pentapetalae, and this is reflected in higher diversity of *AP3* and *PI* paralogs (Kramer et al., 2006). In Ranunculaceae, several floral elaborations have evolved independently, such as spurred staminodial petals, cyclic flowers, zygomorphy, syncarpy and pentamery (Ronse De Craene, Soltis and Soltis, 2003).

The occurrence of pentamery and dimery appear to be linked with trimerous precursors, as stamens are inserted opposite tepals, not alternating with them (e.g. Von Balthazar and Endress, 2002b; Wanntorp and Ronse De Craene, 2006; Ronse De Craene, Quandt and Wanntorp, 2015b). Fossil pentamerous Platanaceae and Buxaceae have carpels or stamens opposite tepals, as pentamerous Sabiaceae. Pentamery is different in Ranunculaceae where it is derived from more complex spiral flowers.

Floral diagrams of early diverging eudicots including Gunneraceae illustrate a progressive pauperization of flower structures, with the exception of increased diversification in Ranunculales, Sabiaceae and Nelumbonaceae. Although our understanding of early diverging eudicots has dramatically improved in recent years, it is not possible to throw unequivocal light on the sudden transition to core eudicot diversity, based on the evidence of extant flower diversity.

13.2 Pentapetalae

Flowers of Pentapetalae generally have a pronounced distinction between bracts and flower, except in some basal orders, such as Berberidopsidales. Inclusion of bracts in the confines of the flower is a secondary process leading to an epicalyx and eventually to a secondary calyx. The basic floral formula in Pentapetalae appears to be K5 C5 A5+5 G5, found in all major clades of Superrosidae and Superasteridae, with repeated reductions of the carpel number to three or two, except for some smaller orders such as Berberidopsidales, Vitales and Dilleniales. The perianth is generally differentiated into sepals and petals (with a transition between undifferentiated perianth to sepals and petals in Berberidopsidales), and diplostemony is found in basal clades of all major orders, except euasterids where the antepetalous stamen whorl is irretrievably lost. The reader is referred to Ronse De Craene (2004, 2007, 2008, 2017a) and Ronse De Craene and Brockington (2013) for a discussion of the evolution of the bipartite perianth. The ovary shows a strong degree of fusion and is rarely apocarpous, in which case it has frequently evolved secondarily (e.g. Endress, Jenny and Fallen, 1983; Sokoloff, 2016; El Ottra et al., 2022).

Hypanthial growth strongly shapes the flowers of Pentapetalae (as in some monocots), with specific tendencies among the major clades. Rosids often have cuplike hypanthia lined with nectary tissue (e.g. Malpighiales, Rosales, Myrtales), generally with a superior ovary, rarely with an inferior ovary (e.g. Cucurbitales, Myrtaceae). In caryophyllids, hypanthial growth tends to be restricted to the area between androecium and gynoecium and is correlated with an intrastaminal nectary. Asterids have generalized stamen-petal tubes, with variable changes in the position of the ovary.

Nectaries have evolved among Pentapetalae in relationship with hypanthia. The greatest variation in nectary structures is found in rosids, ranging from extrastaminal nectaries associated with stamens and petals, shifting to intrastaminal glands or a receptacular disc (e.g. Geraniales, Malpighiales). Asterids generally have a broad disc surrounding the ovary base, shifting to an epigynous disc or a gynoecial nectary.

A diplostemonous androecium is found in all major clades of Pentapetalae except the more derived euasterids, even characterizing most taxa in major orders (e.g. Geraniales, Malvales, Sapindales, Oxalidales). One stamen whorl is often reduced to staminodes and consequently lost, leading to either obhaplostemony or haplostemony, with both conditions rarely together. Further reductions may lead to a single stamen in some groups (e.g. Ochnaceae, Cleomaceae, Euphorbiaceae). The occurrence of polyandry is widespread and always secondary (contrary to basal angiosperms) and is derived from a single or two whorls of stamens. Stamens tend to develop centrifugally on common primordia or on a ring primordium, more rarely centripetally and this is linked with hypanthial growth (Figures 1.1, 1.2; Ronse De Craene and Smets, 1991a, 1992a).

The gynoecium is basically syncarpous, arising congenitally united (Endress, 2006). Five carpels represent the basal structure, but trimerous and dimerous gynoecia are firmly settled in the clade. Contrary to basal angiosperms where unicarpellate gynoecia are derived from a reduction of free carpels, a single carpel is nearly always derived through pseudomonomery (Sokoloff, 2016), unless the ovary has undergone a shift to secondary apocarpy (e.g. Fabales). Multiplications of carpels tend to occur rarely and lead to irregular carpel closure and deformed symmetry of flowers (Endress, 2006, 2014).

The early evolution of Pentapetalae is shrouded in mystery, either because the radiation of the clade was extremely rapid, or because no clear transitional fossils are available (Soltis et al., 2018). A highly divergent floral evolution characterizes Santalales, Dilleniales, Berberidopsidales and Caryophyllales with distinct levels of specialization.

Superrosidae (Figure 9.1)

The larger a clade, the more difficult it becomes to define and circumscribe it on a morphological basis, especially because there is more scope for variation and because the group is morphologically largely understudied. Rosids are a highly diverse grouping of taxa in which many relationships are still poorly resolved, and many non-molecular characters have barely been investigated (Soltis et al., 2005).

A generalized floral diagram does not exist for rosids, although one kind of floral arrangement is more common and tends to be found almost exclusively in Superrosidae: K5 C5 A5+5 $\underline{G}$(5). Monosymmetry is uncommon, and when present, generally arises late in ontogeny and is oblique. Flowers are mostly pentamerous to tetramerous, often with a biseriate perianth. Petals are free, often small and clawed and an apomorphic tendency is that they are delayed in growth, becoming lost in several clades (Ronse De Craene, 2008). The androecium is (ob)diplostemonous, rarely obhaplostemonous, and when haplostemonous, often with one whorl of staminodes. Staminodes are rarely persistent or take up new functions except in Malvaceae (Walker-Larsen and Harder, 2000). Secondary polyandry is common and highly variable. A hypanthium is found on a frequent basis and lifts calyx, corolla and androecium, or only calyx and corolla (Figure 1.4). The development of the hypanthium is correlated with the development of secondary elaborations (coronas), proliferation of nectary tissue and the loss of petals. The gynoecium is mostly superior, of three to five fused carpels with axile placentation, although parietal placentation is found in several families of Brassicales and Malpighiales. The nectary is receptacular in most cases and usually develops as a defined disc or embedded in the hypanthial slope. Free extrastaminal nectar glands appear frequently in different clades (e.g. Geraniales, Malpighiales). Sympetaly and inferior ovaries are rare and restricted to a few families. There are several exceptions to this assemblage of characters as it is difficult to assign synapomorphies for such a large clade. Endress and Matthews (2006a) enumerated characters that are concentrated or exclusively found in certain orders of rosids. In malvids, there is a concentration of (andro)gynophores, petals with ventral elaborations, contorted petals, campylotropous ovules and monosymmetry. In the nitrogen-fixing clade there is a concentration of apetaly, wind pollination and gynoecia with single ovules. Cucurbitales share trimerous inferior ovaries with their sister group Fagales, as well as the presence of wind-pollinated flowers (Matthews and Endress, 2004). Characters shared by the COM-clade among others (see Matthews and Endress, 2005) are: rarity of monosymmetry, absence of petal tubes, fringed petals, two stamen whorls (diplostemony) with a tendency for secondary polyandry (only in

Plagiopteron of Celastraceae but abundant in Malphighiales), often two lateral ovules per carpel, and presence of a broad nectary extending outside the stamens. There are several convergences between Rosaceae and Myrtaceae, such as a deep hypanthium with intrastaminal nectary, centripetal stamen development and inferior or half-inferior ovary. Convergences between Rosaceae and Malvaceae include the valvate calyx and presence of an epicalyx.

Several rosids share the presence of paired stamens, mostly in antepetalous position (listed for twenty-four families by Ronse De Craene and Smets, 1996a). Paired stamens are often linked with obdiplostemony and isomerous gynoecia; they appear to be one way to avoid loss of antepetalous stamens by the pressure of the carpels, which are generally in alternisepalous position (see p. 16).

Caryophyllids (Figure 10.1)

Although the clade is relatively small in number of species, evolutionary trends are fascinating, especially in perianth and androecium, which hold more scope for diversity than any other major clade of angiosperms. Caryophyllids are basically pentamerous and often apetalous with a well-developed sepal whorl in a 2/5 arrangement (at least in Caryophyllales). Evidence points to the progressive reduction and loss of petals in the basal grade of Caryophyllales linked with a tendency towards wind pollination (Wei and Ronse De Craene, 2019). Major evolutionary trends are the reinvention of petaloid organs, stamen shifts, and reduction in ovary structure (Ronse De Craene, 2008, 2013, 2021; Brockington et al., 2009). The petaloid perianth often has pronounced sepal characteristics. In Aizoaceae the petaloid perianth shows no expression of *PI* or *AP3* genes, suggesting that petaloidy was independently derived in the 'globular inclusion clade' from ancestral wind-pollinated Caryophyllales (Brockington et al., 2012).

The basic floral formula of Caryophyllales is K5 C0–5 A5+5 G(5). The presence of a third stamen (or staminodial) whorl as found in Kewaceae (Figure 10.14) and Petiveriaceae (Figure 10.18) is not reflected in any other taxon of Pentapetalae, where petals appear to have the same origin as sepals. As discussed in Ronse De Craene (2018), one option is to accept that the outer alternisepalous stamen/staminode/petal was secondarily derived by radial division of a primary stamen. This assumption is supported by the initiation of common primordia and centrifugal development of outer stamens.

Two opposite trends can be recognized for the evolution of the androecium: reduction of the initial ten stamens to five or less, and centrifugal multiplication of stamens by development of complex primordia or a ring primordium

(Ronse De Craene, 2013). There is a strong interaction between carpel position and number, and the insertion and evolution of the androecium. The reduction of the upper stamen whorl tends to be correlated with space constraints due to pressures of carpels and sepals and follows a predictable pattern, especially in Caryophyllaceae (Ronse De Craene, 2013; Ronse De Craene and Wei, 2019). A secondary centrifugal stamen increase is widespread and appears to be superimposed on the basic androecial Bauplan.

The ovary is extremely variable in Caryophyllales, with a strong tendency to a free-central or basal placentation by reduction of septa, and strong fluctuations in carpel numbers (Ronse De Craene, 2021). A similar trend is reflected in Polygonales, leading to a single basal ovule. Near absence of monosymmetry tends to be generalized in caryophyllids (except for some Cactaceae). Occurrence of nectaries is linked with hypanthial growth between androecium and ovary. Nectaries are generally confined to the inside of the filaments or filament tube, occasionally extending centrifugally (see Zandonella, 1977; Smets, 1986, Bernardello, 2007).

While floral evolution within Caryophyllales is guided by this initial Bauplan, a similar evolution occurs in the 'Polygonaceae clade' of Polygonales, which shares a wealth of characters with Caryophyllales representing clear apomorphic tendencies. Similar sieve-tube plastid forms were found in some Polygonaceae and Amaranthaceae (Behnke, 1999), which may be another convergence. It is reasonable to accept that a basic diplostemonous flower is ancestral in other Polygonales, as present in Tamaricaceae or Dioncophyllaceae, but that loss of petals is one major apomorphic tendency expressed mainly in Caryophyllales.

Superasteridae (Figure 11.1)

Sympetaly seems to be one of the few characters that are predominantly found in asterids and that have led to the high success of the clade. Sympetaly is rare elsewhere and is found only in a number of genera of Crassulaceae, Rutaceae, Caricaceae, Plumbaginaceae and Cucurbitaceae.

As discussed on p. 42, two steps lead to sympetaly: confluent meristems of initiated petals and stamens, and intercalary elongation below the united petal bases. In Apiales only step one takes place, or it is shortened in some Lamiales and Solanales (e.g. Montiniaceae: Ronse De Craene, Linder and Smets, 2000; *Besseya* in Plantaginaceae: Hufford, 1995). The development of sympetaly appears early in the Superasteridae as an apomorphic tendency before being canalized in lamiids and campanulids (see also Erbar and Leins, 2011). Other micromorphological characters linking the families were enumerated by Soltis et al. (2005). It is often difficult to unequivocally determine the direction of character evolution (e.g. bitegmy to unitegmy) as the genetic pathways for reduction can be highly complex and reversals are possible. However, for certain structural characters evolution appears to be clearly unidirectional, as for stamen loss affecting major groups of angiosperms, especially in Lamiales (Figure 11.23).

Lower Asterids (Figure 11.2)

Cornales and Ericales appear somewhat anomalous among asterids, having more morphological similarities with rosids: (ob)diplostemony or obhaplostemony are frequent, as well as secondary stamen increases, a relative rarity of petal tubes or stamen-petal tubes, and presence of more than two carpels. Staminodial structures play an important role in the pollination of several Ericales and Cornales by acquiring novel functions (Walker-Larsen and Harder, 2000). There is a syndrome of stamen and carpel increase concentrated in the basal orders Ericales and Cornales (Endress, 2003a, 2006; Jabour, Damerval and Nadot, 2008). While monosymmetry appears to be restricted to few families of Ericales (e.g. Balsaminaceae), it is widespread in euasterids.

However, there are several shared tendencies announcing euasterids, such as (half-) inferior ovaries with few pendent ovules and elaborate pseudanthia in Cornales, and elaborations of stamen-petal tubes in Ericales. Leins and Erbar (2011) described early sympetaly in a number of Cornales and Ericales, even when there is no fused corolla at maturity. As several taxa have an inferior ovary, I believe this phenomenon is linked with a concave floral primordium creating a hypanthial rim in the periphery. Common basal growth leads to short stamen-petal tubes in a number of families of Ericales (e.g. Sapotaceae, Primulaceae, Theaceae), although it is likely that this phenomenon has arisen several times (as an apomorphic tendency; see p. 348).

Cornales appear to be closer to campanulids on floral evidence than any other clade of asterids: there is a similar tendency for small flowers to become associated in complex inflorescences surrounded by involucral bracts; both share a similar initiation of a concave apex early in the development (cf. early sympetaly by formation of a ring primordium); there is rapid petal growth overtaking the weaker sepals; the placentation is generally apical and axile on an inferior ovary with few ovules; the nectaries are epigynous.

Euasterids (Figure 11.1)

Euasterids share a number of specific floral morphological characters that make them easily identifiable as the 'Sympetalae'. The general floral formula is K5[C(5)A5] G(2). Merism is mostly five or four, rarely higher (e.g. *Codon*, *Hoplestigma* in Boraginaceae, *Sheffleria* in Araliaceae: see Ronse De Craene, 2016). Haplostemony is almost universal (with few exceptions of secondary increase in Araliaceae and Rubiaceae, but in all cases developing from antesepalous stamen initials or being chaotic). This indicates an important evolutionary step in the permanent loss of the antepetalous stamen whorl in ancestors of euasterids, without traces of staminodes. Hypanthia are variously developed but are rarely cup- or funnel-shaped and affect the gynoecium and other organs to different degrees. The position of the gynoecium ranges from superior to inferior by progressive invagination in the receptacle. Development of a stamen-petal tube, a petal tube or stamen tube occurs independently of the gynoecium and is rarely absent. The gynoecium consists of two median carpels in most cases (rarely with three to five carpels or a secondary increase) and is surrounded by a disc nectary or gynoecial nectary, or the nectary is epigynous in inferior ovaries. Placentation is generally axile, occasionally parietal or mixed axile-parietal. Other derived states are linked with reduction of ovules to a single basal (Asteraceae) or apical structure (Apiaceae). Pseudomonomery is frequently found, especially in basal orders of lamiids and campanulids, or evolving within certain orders to a great extent (e.g. Dipsacales, Apiales). The pseudomonomerous ovary of Asteraceae is comparable to the ovary of Poaceae.

Lamiids (Figure 11.17) can be identified by superior bicarpellate median ovaries (with notable exceptions), surrounded by a well-developed persistent calyx. Nectaries are generally intrastaminal discs or free glands. Late sympetaly appears to be widespread with few exceptions (Leins and Erbar, 1991) and is generally linked with superior ovaries (except Rubiaceae). It is generally assumed that ancestors of asterids were polysymmetric and that monosymmetry evolved several times independently within the clade (e.g. Donoghue, Ree and Baum, 1998). Lamiids contains orders with polysymmetric flowers and contorted petals ('Contortae' – Gentianales) or valvate to plicate petals (Solanales). Monosymmetric flowers are generalized in Lamiales and mostly constructed on a 3:2 (bilabiate) plan with two petals adaxially and sometimes fused into a single unit, and three abaxial petals often fused into a lower lip. It is not possible to differentiate between floral diagrams of several families of Lamiales (such as Scrophulariaceae, Orobanchaceae, Plantaginaceae, Gesneriaceae or Bignoniaceae), except on the basis of the extent of reduction of stamens or ovules, but there are several exceptions. Monosymmetry evolved early in Lamiales, with several reversals (see Endress, 1994; Soltis et al. 2005: 235). There appears to be a common tendency for a transition to tetramerous polysymmetric flowers by fusion of the posterior petals and loss of the adaxial stamen (Ronse De Craene and Smets, 1994; Donoghue, Ree and Baum, 1998;

Endress, 1999; Bello et al., 2004). However, polysymmetric flowers with five stamens are not that rare in Lamiales, occurring mainly in more basal families, such as Gesneriaceae and Bignoniaceae, where staminodes tend to be well developed. The case for a reversal of the posterior staminode in *Penstemon* was presented by Walker-Larsen and Harder (2000, 2001). The possibility for a reversal obviously depends on how far the staminode is reduced (Endress, 1999). While reversals to pentamerous polysymmetry are possible at this stage, they become impossible with further reduction of the staminode; the only option left is the fusion of two posterior petals. Other clades of lamiids (e.g. Boraginales, Solanales, Gentialales) are basically polysymmetric and monosymmetry is only incidental and mostly oblique when it occurs.

Campanulids (Figure 11.31) share several clearly observable characters. Small flowers are grouped in more or less dense heads or umbels, often surrounded by involucral bracts (e.g. Asteraceae, Brunoniaceae, Campanulaceae, Calyceraceae, Caprifoliaceae, Apiaceae, Araliaceae, Goodeniaceae). However, capitula in Asterales have a polytelic origin, while those of Apiales and Dipsacales are monotelic. All campanulids share early sympetaly (Erbar, 1991). As mentioned earlier (p. 42), the term is not well chosen as it reflects an early invagination of the apex forming a ring-like zone on which the petal lobes emerge (cf. Roels and Smets, 1996; Ronse De Craene, Linder and Smets, 2000; Ronse De Craene and Smets, 2000). There is a clear tendency to form one to several hypanthia and fusions in the flower (stamen-petal tube, inferior ovary). Petal tubes occasionally fail to develop, as in Apiales, Bruniales, Aquifoliales and Escalloniales. Sepals do not contribute to the protection of the bud, or do so marginally. As a consequence, there is a strong tendency for sepal loss (e.g. Apiales) or their replacement by hairs, scales or a pappus (Dipsacales, Asterales). Monosymmetry is less common than in lamiids, but appears to develop in a similar manner (preferably as a 5:0 arrangement in Asterales, often with a slit on one side of the corolla tube, or occasionally a 3:2 (4:1) arrangement in Dipsacales: e.g. Donoghue, Ree and Baum, 1998; Endress, 1999). In some Caprifoliaceae there is a strong convergence with Lamiales in the development of median monosymmetry and transitions to tetramery. The androecium is universally haplostemonous (except for secondary increases in Araliaceae and reductions in Caprifoliaceae from five to one stamen). Carpels are antepetalous when isomerous, with a frequent switch to three or two median carpels, rarely more by a secondary increase (e.g. Araliaceae). Pseudomonomery is common and is linked with rudimentary septa in Asterales or empty locules in Dipsacales. The number of ovules is often low, mostly as a single pendent ovule with a rudiment of the other ovule in the same locule (Apiales, Dipsacales), or as a basal ovule (Asterales). Placentation is initially axile with septa developing independently from the ovary wall. All taxa have a long locular space which is often not filled by the ovule. Nectaries are generally gynoecial, developing at the base or on top of the gynoecium (Asterales, Apiales), or are trichomatic on the tubular corolla (Dipsacales).

14

Outlook

Floral diagrams build the foundations for the understanding and identification of flowers. The process of constructing diagrams is comparable to an architect laying the foundations of a building. It allows for the understanding of the special relationships of organs in the flower and ultimately captures the information to predict relationships with pollinators and occurring evolutionary trends. It is not always an easy task to capture floral diversity by floral diagrams. To be fully comprehensive, several volumes would have to be written, comprising several hundred detailed drawings. Flowers represent dynamic entities prone to influences of the environment, interactions with pollinators, pressures during floral development and genetic shifts. How a flower looks at maturity is largely caused by the processes affecting the floral development, with subtle shifts in time and space causing major changes in the floral morphology (discussed in Ronse De Craene, 2018, 2021). These changes allow us to predict trends in the floral evolution and reflect the apomorphic tendencies found in different clades. However, certain characters on floral diagrams are conservative so as to reflect where a taxon belongs and can be used for identification at least to family level. All major clades can be identified on a series of clear, stable characteristics reflected in the Bauplan of the flower and including the phyllotaxis, fusions, number of whorls and merism.

Floral diagrams are increasingly important as a principal means to understand the complexity of flowers and leading to hitherto unexplored floral characters. While the phylogeny of angiosperms is being refined by worldwide collaborative research, the challenge for studying flowers and their hidden secrets becomes increasingly important. While writing this book, I made several new observations of flower structures by study of fresh flowers, while the information from the literature was either non-existent, too basic or restricted to obscure nineteenth-century work. While Eichler had broad access to a wide botanical knowledge, this knowledge has become progressively eroded during the twentieth century because of emphasis on other exciting areas of botany. Nowadays, we know much more about the genetic structure of plant groups than about their floral structure. The scope of morphological research is immense, especially in tropical families for which only a fraction of the diversity is currently known. A stable phylogeny brought about by genetic study enables us to revisit floral morphology from a totally new perspective. In the current biodiversity crisis, such studies are a race against time, with the certainty that pertinent scientific knowledge and aesthetic models are lost forever, before even being discovered.

It is high time that morphology receives the place it deserves as a central focus for systematics, ecology and evolutionary developmental genetics. Hopefully this book on floral diagrams will be inspirational as a synthesis of the floral diversity of this world.

Glossary

Abaxial: the side of an organ away from the axis to which it is attached. Corresponds to the dorsal side or lower surface (of a leaf) or the anterior side (of a flower).
Abortion: the process of arrested growth.
Accrescent: enlarging with age (e.g. calyx of Olacaceae).
Acropetalous: in the direction of the apex, opposite of basipetalous.
Actinomorphic: see polysymmetry.
Acuminate: gradually tapering to an acute apex.
Acyclic: spiral, not in whorls.
Adaxial: the side of an organ towards the axis to which it is attached. Corresponds to the ventral side or upper surface (of leaf).
Adhesion (adherent): attached but not fused, refers to different organs.
Adnate: attached to, refers to fusion of different organs, e.g. stamens and petals. Compare with connate.
Aestivation: the mutual positional relationship of perianth parts in bud.
Alternipetalous: alternating with the petals.
Alternisepalous: alternating with the sepals.
Anatropous: refers to a curved ovule with the micropyle close to the attachment region of the funiculus.
Androecium: the totality of the stamens in a flower.
Androgynophore: an extension of the receptacle bearing stamens and carpels.
Androphore: an extension of the receptacle bearing stamens in a staminate flower without gynoecium.
Ante(i)petalous: opposite a petal.
Ante(i)sepalous: opposite a sepal.
Anterior (side): see abaxial.
Anther: the pollen-bearing part of a stamen; consists generally of two thecae and four pollen sacs.
Anthesis: the flowering phase, when all organs are exposed, pollen is presented and the stigma is receptive.
Anthocarp: basal persistent part of the perianth in Nyctaginaceae homologous to a hypanthium and enclosing the fruit.

Anthophore: extension of the receptacle between the calyx and the rest of the organs in a flower (e.g. in some Caryophyllaceae).

Antitropous: refers to the curvature of the ovules in the opposite direction of the carpel curvature (cf. Endress, 1994).

Apert: aestivation of perianth parts not touching each other.

Apetalous (-y): without petals.

Apex: the tip or distal structure of a leaf or axis.

Apical placentation: type of placentation with ovules attached to the top of the locule.

Apocarpous: refers to a gynoecium with distinct carpels.

Apomorphy (-ic): refers to a character state that is derived from an ancestral state.

Apomorphic tendency: cf. cryptic apomorphy: a character shared by several but not all members of a clade, indicating a common trend that is genetically latent.

Ascending: kind of cochleate aestivation, with innermost organs adaxial.

Ascidiate: refers to young carpel that is congenitally urn shaped, opposed to plicate, a carpel that is similar to a folded leaf.

Asymmetrical: refers to a structure that cannot be divided into two equal halves (e.g. flower of *Centranthus*, Caprifoliaceae).

Axile placentation: type of placentation in a syncarpous ovary with septa in which the ovules are attached in the middle section of the locule.

Basal placentation: type of placentation with ovules attached at the base of the locule.

Basipetalous: in the direction away from the apex, opposite of acropetalous.

Blade: the expanded section of a leaflike organ. Synonymous to lamina.

Bract: a differentiated or reduced leaf associated with a shoot, such as an inflorescence or subtending a flower; synonym of pherophyll.

Bracteole: the first leaf of a lateral shoot in line with the flower and occurring singly (monocots) or in pairs (dicots); synonym of prophyll.

Bauplan: the form or structural construction of a flower that tends to be constant for larger taxa.

Botryoid: racemose inflorescence with a terminal flower preceding all others.

Buzz pollination: kind of pollination in which the vibration of bee wings triggers the release of pollen. Associated with a syndrome of characters in the flower such as poricidal anthers, reflexed petals and hanging flowers.

Caducous: falling off early.

Calyculus: rim-like structure homologous to a reduced calyx in Loranthaceae and some Olacaceae.

Calyptra(-te): refers to the partial fusion of perianth parts, falling off as a whole. Fusion can be proximally and congenital or distally and postgenital.

Calyx: the outer perianth whorl in flowers; collective term for sepals.

Campylotropous: refers to a curved ovule with the nucellus included in the curvature in a more or less right angle to the funiculus.

Capitulum: refers to an inflorescence with flowers densely packed in a globular structure or head (e.g. Asteraceae).

Capsule: a dry dehiscent fruit that develops from a syncarpous ovary. The capsule opens in the middle of the locule (loculicidal), at the level of the septa (septicidal), or both.

Carinal: refers to the position of style or stigma in line with the carpel or locule; opposite of commissural.

Carpel: the basic unit of the gynoecium, corresponding to a transformed megasporophyll enclosing the ovules.

Carpophore: an extension of the receptacle to which the carpels are attached (e.g. Apiaceae).

Centrifugal: refers to the sequence of development of floral organs from the centre towards the periphery.

Centripetal: refers to the sequence of development of floral organs from the periphery towards the centre.

Chasmogamous (-y): refers to a flower that opens and is generally cross-fertilized. As opposed to cleistogamous.

Cincinnus: see scorpioid cyme.

Clade: a group of taxa that share features derived from a common ancestor; a monophyletic evolutionary line.

Claw(-ed): a constricted base of a petal that can be short or very long (e.g. Caryophyllaceae).

Cleistogamous (-y): refers to flowers that do not open and are self-fertilizing. As opposed to chasmogamous.

Cochleate (cochlear): refers to the aestivation with one perianth member completely outside and one member completely inside, and the rest in between. The orientation of symmetry is either adaxial (ascending) or abaxial (descending).

Colleter: glandular structure on vegetative or floral buds that secretes a fluid protecting the developing organs from dehydration.

Commissural: refers to the position of style or stigma in alternation with the carpel or locule; opposite of carinal.

Compitum: a common zone shared by all carpels in a syncarpous ovary, functioning as pollen transmitting area.

Congenital fusion: fusion of structures from the onset of initiation (zonal growth).

Connate: refers to the same organs that are marginally coherent.

Contorted (convolute): refers to the aestivation with neighbouring organs overlapping each other on one side only, as tiles on a roof. The flower has a twisted appearance in some cases (e.g. Malvaceae, Apocynaceae).

Convergent (evolution): the independent acquisition of similar characteristics in two of more unrelated groups.

Common (complex) primordium: a single primordium dividing in two or more secondary primordia. Each secondary primordium divides further or develops into a single organ. Division of a common stamen primordium leads to an increase in stamens. A stamen-petal primordium results from the congenital fusion of petals and opposite stamen and is linked to a retardation of the petals.

Corolla: the second whorl of the perianth; collective term for petals.
Corona: a showy structure located between perianth and androecium, with variable origin (e.g. Passifloraceae, Amaryllidaceae).
Cucullate: hood-shaped.
Cupule: a cup-like structure that usually subtends a fruit and consists of sterile bracts (e.g. Fagaceae); see also involucre.
Cyathium: an inflorescence consisting of several naked staminate flowers and a single pistillate flower, enclosed in a cup-shaped involucre (Euphorbiaceae).
Cyclic: whorled (as opposed to spiral).
Cyme: a determinate inflorescence (monotelic), with each major branch ending in a flower.
Decamerous (-y): with parts in ten.
Decussate: refers to arrangement of opposite leaves on the stem, with successive pairs arising at right angles of the pair below (1/2 arrangement).
Dédoublement: the process of splitting of an organ in two equal parts that are identical to the original organ.
Descending: kind of cochleate aestivation, with innermost organs abaxial.
Dichasium (-al): cymose inflorescence with lateral branches developing equally.
Dichogamy: refers to a flower in which pollen is released and stigmas are receptive at different times. See proterandry and proterogyny.
Dicliny: refers to separate staminate and pistillate flowers. Similar to unisexual flowers. See also dioecious and monoecious.
Didynamous: with two pairs of stamens of different length (e.g. Gesneriaceae).
Dimerous (-y): with parts in two.
Dimorphic: see heteromorphic.
Dioecious (-y): with staminate and pistillate flowers on different plants. Compare with monoecious.
Diplostemony (-ous): arrangement of stamens in two whorls, with the outer opposite the sepals and the inner opposite the petals. Compare with obdiplostemony.
Disc nectary: The development of a fleshy nectariferous circular protrusion of the receptacle. The disc can be intrastaminal, interstaminal or extrastaminal.
Distal: at the top of an organ or structure or away from the place of attachment, opposite of proximal.
Distichous: two-ranked or in two rows, referring to the arrangement of leaves or bracts on the stem, more rarely in the flower (e.g. Gunneraceae).
Disymmetry (-ic): refers to a flower with two symmetry planes (median and transversal).
Divergence angle: the angle between two organs arising in succession on a spiral.
Dorsal bundle/trace: vascular bundle representing the midvein of the carpel, which is usually well developed.
Dorsal side: see abaxial.
Elaiophore: outgrowth in the flower that produces oils collected by specialized pollinators.

Enantiostyly: refers to flowers whose styles are oriented alternate to the right and left in mirror-image flowers.

Eusyncarpous: syncarpous ovary resulting from the fusion of free carpels in the centre, associated with axile placentation (opposed to parasyncarpous).

Epicalyx: an additional whorl of structures of different origin inserted below the flower that resembles an extra calyx.

Epigynous: refers to a flower with the floral parts inserted on top of the ovary.

Extant: still in existence; having living representatives, as opposed to fossil.

Extrastaminal: outside the stamen whorl.

Extrorse: directed outwards; refers to the dehiscence of an anther. Compare with introrse and latrorse.

False septum: partition in the ovary that does not include the margin of a carpel.

Fascicle: refers to a tight cluster arising from the same point; in flowers often linked to stamens (e.g. Clusiaceae).

Filament: the stalk of a stamen that bears the anther.

Fimbriate: fringes of multicellular trichomes on the margin of an organ.

Floral apex: the top part of the flower that gives rise to floral organs and is used up completely or persists as a residue.

Floret: in Asteraceae, a small flower of a capitulum; in Poaceae, a small flower of a spikelet.

Free-central placentation: type of placentation in a syncarpous ovary without septa in which the ovules are attached on a central column.

Funiculus: the stalk of the ovule.

Grade: a group of organisms without phylogenetic implication.

Gynobasic: refers to the position of the style at the base of the ovary (e.g. Chrysobalanaceae).

Gynoecium: the collective term for the female parts of the flower.

Gynophore: an extension of the receptacle lifting the gynoecium (e.g. Capparaceae).

Gynostegium: a common structure formed by postgenital fusion of stamens and stigma in Apocynaceae.

Gynostemium: a common structure formed by congenital fusion of stamens and stigma in Orchidaceae and Aristolochiaceae.

Halophytic: adapted to saline conditions.

Haplostemony: arrangement of stamens in a single whorl opposite the sepals. Compare with obhaplostemony.

Helicoid cyme: cymose monochasial inflorescence with lateral flowers developing on the same side in one plane, but alternating from flower to flower resulting in a zigzag pattern (compare with a scorpioid cyme).

Homeosis: the 1:1 replacement of an organ by another organ that is not necessarily homologous to it.

Hemicyclic (spirocyclic): the combination of a spiral and whorled phyllotaxis in the same flower.

Heptamerous (-y): with parts in seven.

Herkogamy: spatial separation of male and female organs in a flower, avoiding self fertilization.

Heteranth(er)y: refers to two distinct types of stamens in a flower; one type bears fertile pollen and the other sterile (fodder) pollen.

Heterochrony: a shift in the timing of organ initiation and growth; this includes a more precocious or a delayed appearance of primordia during (floral) development. The principle is generally applied to mature structures, recognizing a shortening period of development leading to juvenile structures (paedomorphosis), or a prolongation of development leading to more elaborate structures (peramorphosis).

Heteromorphic: refers to structures of the same organism or species that differ in form and size; includes dimorphic with two forms. Synonym of polymorphism.

Heterostyly: condition in which style and stamens differ in length among individuals of a species, favouring cross-pollination.

Hexamerous (-y): with parts in six.

Hydathode: water pore; modification in the epidermis allowing release of water through an opening.

Hypanthium: receptacular tube, often cup-shaped and bearing perianth and androecium. Also refers to any tubes developing in the flower.

Hyperstigma: stigmatic area produced outside the gynoecium, formed by the receptacle or reduced tepals (Monimiaceae, Siparunaceae)

Hypogynous: refers to a flower with the floral parts inserted below the ovary.

Imbricate: refers to the aestivation pattern with overlapping margins of neighbouring organs.

Integument: the outer wall of the ovule that later develops into the seed wall; it may be composed of two (bitegmic ovules) or a single layer (unitegmic ovule).

Interstaminal: positioned between the stamens, i.e. within the stamen whorl.

Intrastaminal: positioned inside the stamens; often refers to a nectary.

Introrse: directed inwards; refers to the dehiscence of an anther. Compare with extrorse and latrorse.

Involucre, involucral bract: a series of fused or overlapping bracts that subtend inflorescences, sometimes confused with cupules, which can have the same origin (e.g. Apiaceae, Asteraceae, Montiaceae).

Isomerous: refers to the number of parts of a whorl being the same as the other whorls in the flower.

Labellum: a lip-like petaloid structure, such as the inner petal of orchids or staminodial structure of Zingiberaceae.

Laminar-diffuse placentation: type of placentation without clearly localized placentae, with ovules often located on the lamina of the carpels (also called laminar placentation).

Latrorse: directed towards the sides; refers to the dehiscence of an anther. Compare with introrse and extrorse.

Lemma: the lower of two bracts enclosing the floret in Poaceae (with palea).

Ligule: a straplike organ, refers to ligulate flowers with a straplike corolla in Asteraceae or an outgrowth of the corolla or leaves.

Locule: the cavity (-ies) within the ovary, containing the ovules.

Loculicidal capsule: see capsule.

Lodicule: a small scale in the floret of Poaceae, corresponding to a reduced perianth.

Marginal placentation: type of placentation of apocarpous ovary, in which the ovules arise on the margins of the carpel. Compare with laminar or diffuse placentation.

Median monosymmetry: monosymmetry with the two equal halves running along the median line.

Megasporangium: structure bearing megaspores, the ovules in angiosperms.

Merism (merosity): refers to the number of parts of organs in flowers or within whorls, generally used as '-merous'.

Meristem: an undifferentiated area of tissue with the potential of cell division, differentiation and growth.

Micropyle: an opening between the integuments of the ovule through which the pollen tube reaches the nucellus.

Microsporangium: structure bearing microspores, the anthers in angiosperms.

Midrib: the primary vein of a leaf, bract, perianth part or carpel.

Mimicry: the close resemblance of an organism or a part of an organism to an other organism (e.g. pistillate flowers of Caricaceae and Begoniaceae resembling their staminate counterparts).

Monadelphous: refers to stamens united by their filaments in a tube.

Moniliform: refers to a hair or filament with constrictions at regular intervals, like beads on a string (e.g. Commelinaceae).

Monocarpellate gynoecium: with a single carpel.

Monochasium (-ial): cymose inflorescence with lateral branches developing unequally, one branch being generally reduced.

Monoecious (-y): with staminate and pistillate flowers in the same plant. Compare with dioecious.

Monophyletic group: a group comprising a common ancestor and all its descendants.

Monosymmetry (-ic): refers to a bilateral symmetry, dividing the flower in two equal halves; synonym of zygomorphic.

Monotypic: refers to a family/genus with a single genus/species.

Multilocular ovary: refers to a gynoecium with many locules (septate ovary).

Nectar glands: The development of separate nectariferous appendages. In flowers, these can be receptacular or of staminodial origin.

Nectar guide: markings on the petals that guide pollinators to hidden nectar.

Nectary: a tissue or structure that develops nectar. A distinction is made between extrafloral (extranuptial) and floral (nuptial) nectaries.

Nucellus: tissue of the ovule in which the embryo develops.

Obdiplostemony (-ous): arrangement of stamens in two whorls, with the outer opposite the petals and the inner opposite the sepals. Compare with diplostemony.

Obhaplostemony: arrangement of stamens in a single whorl opposite the petals or alternating with the sepals. Compare with haplostemony.

Oblique monosymmetry: monosymmetry with the two equal halves being divided obliquely in the flower.

Obturator: outgrowth of the placenta, style or ovule facilitating pollen transfer from the style to the ovule (e.g. Euphorbiaceae, Sapindaceae).

Octomerous (-ry): with parts in eight.

Ontogeny, ontogenetic: the course of initiation and development of an organism and organ until maturity.

Orthostichies: set of straight lines connecting organs in whorled flowers.

Orthotropous: refers to the straight orientation of an ovule with the funiculus and micropyle along the same straight line.

Osmophore: tissue in the flower or inflorescence that emanates scent (e.g. spadix of Araceae).

Ovary: the section of the gynoecium that contains the ovules

Ovule: the megasporangium of seed plants that becomes the seed after fertilization.

Palea: the upper of two bracts enclosing the floret in Poaceae (with lemma).

Panicle: a compound raceme, with lateral branches repeating the main axes but terminated by a flower.

Pappus: a modified calyx, consisting of hairs, bristles or scales typical of Asteraceae and some Caprifoliaceae.

Paraphyletic group: a group including a common ancestor, but not all its descendants.

Parastichies: set of contact spirals resulting from a helical initiation of organs in whorled and spiral flowers.

Parasyncarpous: syncarpy resulting from the marginal connection of free carpels, leading to a parietal placentation (opposed to eusyncarpous).

Parietal placentation: type of placentation found in syncarpous unilocular ovaries where the placentae are located on the fusion of carpel margins.

Pedicel: the stalk supporting a flower

Peduncle: the primary axis of an inflorescence.

Pentamerous (-y): with parts in five.

Perianth: the collective term for sepals, tepals and petals.

Pericarpel: the condition in Cactaceae where the ovary is embedded in an expanded vegetative axis.

Perigon(e): defines the perianth when no distinction can be made between sepals and petals; collective term for tepals.

Perigynous: refers to a flower with the floral parts inserted somewhere in the middle of the ovary.

Perisperm: nutrient-rich tissue derived from the nucellus in seeds of some angiosperms (e.g. Caryophyllales).

Petaliferous: with petals.

Petaloid: resembling a petal (with pigmentation).

Petiole: the stalk of a leaf.

Phyllotaxis: the arrangement pattern of leaves or floral organs on an axis, distinguished between spiral and whorled.

Phylogenetic tree (or evolutionary tree): showing the evolutionary relationships among groups of organisms.

Pistil: the female part of the flower, referring to ovary, style and stigma. See also gynoecium.

Pistillate: refers to female part, specifically to unisexual flowers with functional carpels and without functional stamens.

Pistillode: a reduced or aborted gynoecium.

Placenta (plur. -ae): the area in the ovary to which the ovules are attached.

Placentation: the insertion of ovules in the ovary.

Plastochron: the time interval between the initiation of two successive organs on a spiral or whorl.

Plesiomorphy (-ic character): ancestral condition.

Plicate: folded into longitudinal plaits, referring to the perianth (e.g. corolla of *Napoleonaea*), a single carpel folded along the midrib, or to the area in a syncarpous ovary above the symplicate zone where individual carpels become separate from each other.

Pluricarpellate: with numerous carpels in a flower, in syncarpous ovaries, as the result of a secondary increase (e.g. Araliaceae).

Pollen flower: flowers offering pollen as reward.

Pollinium: pollen mass from a pollen sac moved as a single unit during pollination (e.g. Orchidaceae, Apocynaceae).

Polyandry: the presence of numerous stamens (more than double the number of petals); primary polyandry refers to many stamens arising singly in a spiral or in whorls; secondary or complex polyandry refers to the division of primary (common) primordia into several stamens.

Polymerisation: increase in number of parts.

Polyphyletic group: a phylogenetically artificial group, not including the common ancestor of the group.

Polysymmetry (-ic): refers to the radially symmetrical flowers that can divided in two equal halves from any angle; synonym of actinomorphic.

Poricidal: opening by pores; refers to anther dehiscence.

Posterior (side): see adaxial.

Postgenital fusion: fusion of structures that were free from each other at the onset of initiation.

Prot(er)androus (-y): refers to a flower in which pollen is released before the stigmas are receptive (e.g. Campanulaceae).

Prot(er)ogynous (-y): refers to a flower in which stigmas are receptive before the stamens shed pollen (e.g. Calycanthaceae).

Proximal: at the base of an organ or structure, or at the place of attachment; opposite of distal.

Pseudanthium: an inflorescence of several small or reduced flowers resembling a flower.

Pseudomonocarpous gynoecium: gynoecium with single fertile carpel, derived by reduction of other carpels in a syncarpous gynoecium (pseudomonomery).

Pseudomonomery: see pseudomonocarpous gynoecium.

Pseudonectaries: a glistening structure resembling a nectary but lacking any sugary secretion (e.g. *Parnassia*).

Pseudostaminodes: receptacular emergences that resemble staminodial structures.

Pyxidium (pyxis): a capsule that breaks open like the lid of a box.

Quincuncial: refers to an imbricate aestivation in pentamerous flowers, with two organs completely outside, two completely inside, and one intermediate member (2/5 arrangement).

Raceme (-ose): an indeterminate inflorescence (polytelic), with the uppermost flowers the youngest.

Receptacle: the more or less expanded axis bearing the floral parts.

Replum: persistent placental ridge between the two compartments of the fruit in Brassicaceae, arising from a false septum.

Resupinate (-ion): turned 45, 90 to 180° during development by the twisting of the pedicel (e.g. Orchidaceae, Balsaminaceae).

Ring primordium (ring wall): the initiation of a circular primordium giving rise to many stamens or carpels; this is phylogenetically derived from the concrescence of separate common primordia.

Scorpioid cyme (cincinnus): cymose monochasial inflorescence with lateral flowers developing on the same side in one plane, repeated from flower to flower and leading to a coiled appearance (as opposed to a helicoid cyme but often confused with it).

Sepaloid: resembling a sepal (without pigmentation)

Septum: a partition derived from the carpel margins dividing the ovary in locules.

Septal (gynopleural) nectary: a nectary arising within the septa of the gynoecium, characteristic of monocots.

Spike: racemose inflorescence with sessile flowers.

Spikelet: partial inflorescence of grasses, cyperoids and restios; in grasses consisting of a variable number of florets subtended by bracts (palea and lemma).

Spiral: successively arising organs with a same divergence angle.

Sporophyll: spore-bearing leaf; it is understood that flowers have been derived from ancestral forms bearing spores on specialized leaves attached on an axis. In seed plants, there is a differentiation between microsporophylls (stamens), bearing microsporangia (anthers) with microspores (pollen grains) and megasporophylls (carpels), bearing megasporangia (ovules) with megaspores (embryosac).

Spur: tubular, generally nectariferous outgrowth in the flower, arising from the receptacle, calyx or corolla.

Stamen: the male organ in the flower, bearing microspores (pollen), usually comprised of anther and filament.

Stamen-petal tube (stapet): a kind of hypanthium lifting congenitally fused petals and stamens into a tube, often referred to as 'sympetaly with epipetalous stamens'.

Staminate: refers to male part, specifically to unisexual flowers with functional stamens and without functional gynoecium.

Staminode: sterile stamen, sometimes strongly modified.

Stigma: the part of the gynoecium which is receptive to pollen.

Stipule: a reduced leaflike appendage, single or in pairs at the base of the petiole of a leaf.

Stoma (plur. stomata): specialized epidermal cells which allow air passage or nectar through an opening.

Style: the extended part of the gynoecium between ovary and stigma.

Stylode: a generally free style branch, generally stigmatic on the adaxial side (e.g. Caryophyllales).

Stylopodium: enlargement of the base of the style in Apiaceae and Araliaceae (generally nectariferous).

Sympetalous (-y): having petals fused for at least part of their length.

Symplicate zone: describes the upper zone of a syncarpous gynoecium (as opposed to the synascidate zone) generally where septa become separate.

Synandrium: a columnar structure consisting of fused stamens in the absence of a gynoecium in staminate flowers (e.g. Myristicaceae).

Synascidiate zone: describes the lower zone of a syncarpous gynoecium (as opposed to the symplicate zone) where the ovules are attached on axile placentation.

Syncarpy: refers to a gynoecium with fused carpels.

Syndrome: a combination of characters or features.

Synorganization: spatial and functional connections between organs of the same or different kind leading to a homogenous functional structure.

Taxon (plur. taxa): taxonomic unit of any rank; name designating an organism or group of organisms.

Tepal: an undifferentiated perianth part, can be sepaloid or petaloid.

Tetradynamous (-y): the association of two outer shorter and four inner longer stamens (Brassicaceae).

Tetramerous (-y): with parts in four.

Theca: the chamber enclosing the pollen grains; two thecae generally make up an anther and one theca consists of two pollen sacs.

Thyrse: a compound indeterminate inflorescence consisting of cymose lateral branches.

Thyrsoid: thyrse-like.

Transversal monosymmetry: monosymmetry with the two equal halves being divided in the transversal plane of the flower.

Trichome (-atic): a hair-like outgrowth of the epidermis.

Trimerous (-y): with parts in three.

Torus: a receptacle swollen in a distinct cushion (e.g. Capparaceae, Ochnaceae).

Umbel: a flat inflorescence with all pedicels arising from the same point; can be derived from racemes or cymes.

Unidirectional development: developmental sequence of organs from one side of the flower (generally the abaxial side) to the other.

Unilocular ovary: refers to an ovary with a single locule (absence of septa).

Urceolate: urn-shaped, as a swollen tube contracted at the top and with a narrow rim.

Valvate: aestivation of perianth with margins touching but not overlapping. Also refers to opening of thecae by curved flaps of tissue.

Ventral side: see adaxial.

Whorl: group of organs arranged more or less at the same level and arising simultaneously or in a rapid spiral.

Zygomorphic: see monosymmetric

References

Abbe, E. C. (1935). Studies in the phylogeny of the Betulaceae I. Floral and inflorescence anatomy and morphology. *Bot. Gaz.* **95**, 1–67.

Abbe, E. C. (1938). Studies in the phylogeny of the Betulaceae II. Extremes in the range of variation of floral and inflorescence morphology. *Bot. Gaz.* **99**, 431–69.

Abbe, E. C. (1974). Flowers and inflorescences of the 'Amentiferae'. *Bot. Rev.* **40**, 159–261.

Albert, V. A., Gustafsson, M. H. G. and Di Laurenzio, L. (1998). Ontogenetic systematics, molecular developmental genetics, and the Angiosperm petal. In *Molecular systematics of plants II, DNA sequencing*, ed. D. E. Soltis, P. S. Soltis and J. J. Doyle. Boston: Kluwer Academic, pp. 349–74.

Albert, V. A., and Struwe, L. (2002). Gentianaceae in context. In *Gentianaceae. Systematics and natural history*, ed. L. Struwe and V. A. Albert. Cambridge: Cambridge University Press, pp. 1–20.

Álvarez-Buylla, E. R., Ambrose, B. A., Flores-Sandoval, E., Englund, M., Garay-Arroyo, A., Garcia-Ponce, B., de la Torre-Bárcena, E., Espinosa-Matías, S., Martínez E., Piñero-Nelson, A., Engström, P. and Meyerowitz, E. M. (2010). B-function expression in the flower center underlies the homeotic phenotype of *Lacandonia schismatica* (Triuridaceae). *Plant Cell* **22**, 3543–59.

Alverson, W. S., Karol, K. G., Baum, D. A., Chase, M. W., Swensen, S. M., McCourt, R. and Sytsma, K. J. (1998). Circumscription of the Malvales and relationships to other Rosidae, evidence from *rbcL* sequence data. *Am. J. Bot.* **85**, 876–87.

Alverson, W. S., Whitlock, B. A., Nyffeler, R., Bayer, C. and Baum, D. A. (1999). Phylogeny of the core Malvales, evidence from *ndhF* sequence data. *Am. J. Bot.* **86**, 1474–86.

Amara-Neto, L. P., Westerkamp, C. and Melo, G. A. R. (2015). From keel to inverted keel flowers: Functional morphology of 'upside down' papilionoid flowers and the behaviour of their bee visitors. *Plant Syst. Evol.* **301**, 2161–78.

Ambrose, B. A., Espinosa-Matìas, S., Vásquez-Santana, S., Vergara-Silva, F., Martìnez, E., Márquez- Guzmán, J. and Alvarez-Buylla, E. R. (2006). Comparative developmental series of the Mexican Triurids support a euanthial interpretation for the unusual reproductive axes of *Lacandonia schismatica* (Triuridaceae). *Am. J. Bot.* **93**, 15–35.

Anderberg, A. A., Rydin, C. and Källersjö, M. 2002. Phylogenetic relationships in the order Ericales s.l., analysis of molecular data from five genes from the plastid and mitochondrial genomes. *Am. J. Bot.* **89**, 677–87.

Anderberg, A. A., and Ståhl, B. (1995). Phylogenetic interrelationships in the order Primulales, with special emphasis on the family circumscriptions. *Can. J. Bot.* **73**, 1699–1730.

Angiosperm Phylogeny Group I (1998). An ordinal classification for the families of flowering plants. *Ann. Mo. Bot. Gard.* **85**, 531–53.

Angiosperm Phylogeny Group II. (2003). An update of the Angiosperm Phylogeny Group classification for the orders and families of flowering plants, APG II. *Bot. J. Linn. Soc.* **141**, 399–436.

Angiosperm Phylogeny Group. (2009). An update of the Angiosperm Phylogeny Group classification for the orders and families of flowering plants: APG III. *Bot. J. Linn. Soc.* **161**, 105–21.

Angiosperm Phylogeny Group. (2016). An update of the Angiosperm Phylogeny Group classification for the orders and families of flowering plants: APG IV. *Bot. J. Linn. Soc.* **181**: 1–20.

Ao, C., and Tobe, H. 2015. Floral morphology and embryology of *Helwingia* (Helwingiaceae, Aquifoliales): Systematic and evolutionary implications. *J. Plant Res.* **128**, 161–75.

Appel, O. (1996). Morphology and systematics of the Scytopetalaceae. *Bot. J. Linn. Soc.* **121**, 207–27.

Appleton, A. D., and Schenk, J. J. (2021). Evolution and development of staminodes in *Paronychia* (Caryophyllaceae). *Int. J. Plant Sci.* **182**, 377–88.

Arber, A. (1925). *Monocotyledons. A morphological study*. Cambridge: Cambridge University Press.

Arber, A. (1934). *The Gramineae*. Cambridge: Cambridge University Press.

Arber, E. A. N., and Parkin, J. (1907). The origin of Angiosperms. *Bot. J. Linn. Soc.* **38**, 29–80.

Armstrong, J. E. (1985). The delimitation of Bignoniaceae and Scrophulariaceae based on floral anatomy, and the placement of problem genera. *Am. J. Bot.* **72**, 755–66.

Armstrong, J. E., and Douglas, A. W. (1989). The ontogenetic basis for corolla aestivation in Scrophulariaceae. *Bull. Torrey Bot. Club* **116**, 378–89.

Armstrong, J. E., and Tucker, S. C. (1986). Floral development in *Myristica* (Myristicaceae). *Am. J. Bot.* **73**, 1131–43.

Armstrong, J. E., and Wilson, T. K. (1978). Floral morphology of *Horsfieldia* (Myristicaceae). *Am. J. Bot.* **65**, 441–9.

Bachelier, J. B., and Endress, P. K. (2007). Development of inflorescences, cupules, and flowers in *Amphipterygium* and comparison with *Pistacia* (Anacardiaceae). *Int. J. Plant Sci.* **168**, 1237–53.

Bachelier, J. B., and Endress, P. K. (2009). Comparative floral morphology and anatomy of Anacardiaceae and Burseraceae (Sapindales), with a special focus on gynoecium structure and evolution. *Bot. J. Linn. Soc.* **159**, 499–571.

Bachelier, J. B., Endress, P. K., and Ronse De Craene, L. P. (2011). Comparative floral structure and development of Nitrariaceae (Sapindales) and systematic implications. In *Flowers on the tree of life*, ed. L. Wanntorp and L. P. Ronse De Craene. Cambridge: Cambridge University Press, pp. 181–217.

Backlund, M., Oxelman, B., and Bremer, B. 2000. Phylogenetic relationships within the Gentianales based on *ndhF* and *rbcL* sequences, with particular reference to the Loganiaceae. *Am. J. Bot.* **87**, 1029–43.

Baillon, H. (1860). Observations organogéniques pour servir à l'histoire des Polygalées. *Adansonia* **1**, 174–80.

Baillon, H. (1862). Organogénie florale des Cordiacées. *Adansonia* **3**, 1–7.

Baillon, H. (1867–95). *Histoire des plantes* (13 vols.). Paris: Hachette.

Baillon, H. (1868a). Monographie des Dilléniacées. In *Histoire des Plantes I, 2*. Paris: Hachette, pp. 89–132.

Baillon, H. (1868b). Monographie des Magnoliacées. In *Histoire des Plantes I, 3*. Paris: Hachette, pp. 133–92.

Baillon, H. (1868c). Monographie des Annonacées. In *Histoire des plantes I, 4*. Paris: Hachette, pp. 193–288.

Baillon, H. (1870). Monographie des Elaeagnacées. In *Histoire des plantes II*. Paris: Hachette, pp. 487–95.

Baillon, H. (1871a). Du genre *Garcinia* et de l'origine de la gomme-gutte. *Adansonia* 10, 283–98.

Baillon, H. (1871b). Observations sur les Rutacées. *Adansonia* **10**, 299–333.

Baillon, H. (1874). Euphorbiacées. In *Histoire des plantes V, 41*. Paris: Hachette, pp. 105–76.

Baillon, H. (1876a). Traité du développement de la fleur et du fruit X. Castanéacées. *Adansonia* **12**, 1–17.

Baillon, H. (1876b). Traité du développement de la fleur et du fruit. XVI. Stylidiées. *Adansonia* **12**, 354–61.

Barabé, D., and Lacroix, C. (2000). Homeosis in Araceae flowers, the case of *Philodendron melinonii*. *Ann. Bot.* **86**, 479–91.

Barroca, C. (2014). Floral development of *Cuphea* (Lythraceae): Understanding the origin of monosymmetry and the epicalyx in the flower. University of Edinburgh: Unpubl. MSc. Dissertation.

Bateman, R. M., Hilton, J. and Rudall, P. J. (2006). Morphological and molecular phylogenetic context of the angiosperms: Contrasting the 'top-down' and 'bottom-up' approaches used to infer the likely characteristics of the first flowers. *J. Exper. Bot.* **57**, 3471–3503.

Batenburg, L. H., and Moeliono, B. M. (1982). Oligomery and vasculature in the androecium of *Mollugo nudicaulis* Lam. (Molluginaceae). *Acta Bot. Neerl.* **31**, 215–20.

Bauer, R. (1922). Entwicklungsgeschichtliche Untersuchungen an Polygonaceenblüten. *Flora* **115**, 273–92.

Bayer, C. (1999). The bicolor unit: Homology and transformation of an inflorescence structure unique to core Malvales. *Plant Syst. Evol.* **214**, 187–98.

Bayer, C., Fay, M. F., De Bruijn, A., Savolainen, V., Morton, C. M., Kubitzki, K., Alverson, W. S. and Chase, M. W. (1999). Support for an expanded family concept of Malvaceae within a recircumscribed order Malvales, a combined analysis of plastid *atpB* and *rbcL* DNA sequences. *Bot. J. Linn. Soc.* **129**, 267–303.

Bayer, C., and Hoppe, J. R. (1990). Die Blütenentwicklung von *Theobroma cacao* L. (Sterculiaceae). *Beitr. Biol. Pflanz.* **65**, 301–12.

Bayer, C., and Kubitzki, K. (2003). Malvaceae. In *The families and genera of vascular plants vol. V*, ed. K. Kubitzki and C. Bayer. Berlin: Springer, pp. 225–311.

Bechtel, A. R. (1921). The floral anatomy of the Urticales. *Am. J. Bot.* **8**, 386–410.

Behnke, H. D. (1999). P-type sieve-element plastids present in members of the tribes Triplareae and Coccolobeae (Polygonaceae) renew the links between the Polygonales and the Caryophyllales. *Plant Syst. Evol.* **214**, 15–27.

Bello, M. A., Hawkins, J. A. and Rudall, P. J. (2007). Floral morphology and development in Quillajaceae and Surianaceae (Fabales), the species-poor relatives of Leguminosae and Polygalaceae. *Ann. Bot.* **100**, 1491–1505.

Bello, M. A., Martínez-Asperilla, A. and Fuertes-Aguilar, J. (2016). Floral development of *Lavatera trimestris* and *Malva hispanica* reveals the nature of the epicalyx in the *Malva* generic alliance. *Bot. J. Linn. Soc.* **181**, 84–94.

Bello, M. A., Rudall, P. J., González, F. and Fernández-Alonso, J. L. (2004). Floral morphology and development in *Aragoa* (Plantaginaceae) and related members of the order Lamiales. *Int. J. Plant Sci.* **165**, 723–38.

Belsham, S. R., and Orlovich, D. A. (2003). Development of the hypanthium and androecium in *Acmena smithii* and *Syzygium australe* (*Acmena* alliance, Myrtaceae). *Aust. Syst. Bot.* **16**, 621–8.

Bennek, C. (1958). Die morphologische Beurteilung der Staub- und Blumenblätter der Rhamnaceen. *Bot. Jahrb. Syst.* **77**, 423–57.

Bensel, C. R., and Palser, B. F. (1975a). Floral anatomy in the Saxifragaceae *sensu lato* I. Introduction, Parnassioideae and Brexioideae. *Am. J. Bot.* **62**, 176–85.

Bensel, C. R., and Palser, B. F. (1975b). Floral anatomy in the Saxifragaceae *sensu lato* II. Saxifragoideae and Iteoideae. *Am. J. Bot.* **62**, 661–75.

Berger, A. (1930). Crassulaceae. In *Die natürlichen Pflanzenfamilien 18a*, ed. A. Engler and K. Prantl. Leipzig: W. Engelmann, pp. 352–483.

Bernardello, G. (2007). A systematic survey of floral nectaries. In *Nectaries and nectar*, ed. S. W. Nicolson, M. Nepi and E. Pacini. Dordrecht: Springer, pp. 19–128.

Bernhard, A. (1999). Flower structure, development, and systematics in Passifloraceae and in *Abatia* (Flacourtiaceae). *Int. J. Plant Sci.* **160**, 135–50.

Bernhard, A., and Endress, P. K. (1999). Androecial development and systematics in Flacourtiaceae s.l. *Plant Syst. Evol.* **215**, 141–55.

Bittrich, V. (1993). Caryophyllaceae. In *The families and genera of vascular plants vol. II*, ed. K. Kubitzki, J. G. Rohwer and V. Bittrich. Berlin: Springer, pp. 206–36.

Bittrich, V., and Amaral, M. C. E. (1996). Flower morphology and pollination biology of *Clusia* species from the Gran Sabana (Venezuela). *Kew Bull.* **51**, 681–94.

Bohte, A., and Drinnan, A. (2005). Floral development and systematic position of *Arillastrum, Allosyncarpia, Stockwellia* and *Eucalyptopsis* (Myrtaceae). *Plant Syst. Evol.* **251**, 53–70.

Boke, N. H. (1963). Anatomy and development of the flower and fruit of *Pereskia pititache*. *Am. J. Bot.* **50**, 843–58.

Boke, N. H. (1966). Ontogeny and structure of the flower and fruit of *Pereskia aculeata*. *Am. J. Bot.* **53**, 534–42.

Boraginales Working Group (2016). Familial classification of the Boraginales. *Taxon* **65**, 502–22.

Bowman, J. L., and Smyth, D. R. (1998). Patterns of petal and stamen reduction in Australian species of *Lepidium* L. (Brassicaceae). *Int. J. Plant Sci.* **159**, 65–74.

Box, M. S., and Rudall, P. J. (2006). Floral structure and ontogeny in *Globba* (Zingiberaceae). *Plant Syst. Evol.* **258**, 107–22.

Brandbyge, J. (1993). Polygonaceae. In *The families and genera of vascular plants vol. II*, ed. K. Kubitzki, J. G. Rohwer, and V. Bittrich. Berlin: Springer, pp. 531–44.

Bremer, K., Backlund, A., Sennblad, B., Swenson, U., Andreasen, K., Hjertson, M., Lundberg, J., Backlund, M. and Bremer, B. (2001). A phylogenetic analysis of 100+ genera and 50+ families of euasterids based on morphological and molecular data with notes on possible higher level morphological synapomorphies. *Plant Syst. Evol.* **229**, 137–69.

Brett, J. F., and Posluszny, U. (1982). Floral development of *Caulophyllum thalictroides* (Berberidaceae). *Can. J. Bot.* **60**, 2133–41.

Brockington, S. F., Dos Santos, P., Glover, B. and Ronse De Craene, L. P. (2013). Evolution of the androecium in Caryophyllales: Insights from a paraphyletic Molluginaceae. *Am. J. Bot.* **100**, 1757–78.

Brockington, S. F., Roolse A., Ramdial J., Moore, M. J., Crawley, S., Dhingra, A. Hilu, K., Soltis, D. E. and Soltis, P. S. (2009). Phylogeny of the Caryophyllales *sensu lato*: Revisiting hypotheses on pollination biology and perianth differentiation in the core Caryophyllales. *Int. J. Plant Sci.* **170**, 627–43.

Brockington, S. F., Rudall, P. J., Frohlich, M. W., Oppenheimer, D. G., Soltis, P. S. and Soltis, D. E. (2012). 'Living stones' reveal alternative petal identity programs within the core eudicots. *Plant J.* **69**, 193–203.

Brockington, S. F., Yang, Y., Gandia-Herrero, F., Covshoff, S., Hibberd, J. M., Sage, R. F., Wong, G. K. S., Moore, M. J. and Smith, S. A. (2015). Lineage-specific gene radiations underlie the evolution of novel betalain pigmentation in Caryophyllales. *New Phytol.* **207**, 1170–80.

Brown, D. K., and Kaul, R. B. (1981). Floral structure and mechanism in Loasaceae. *Am. J. Bot.* **68**, 361–72.

Brown, R. H., Nickrent, D. L. and Gasser, C. S. (2010). Expression of ovule and integument-associated genes in reduced ovules of Santalales. *Evol. Dev.* **12**, 231–40.

Bukhari, G., Zhang, J., Stevens, P. F. and Zhang, W. (2017). Evolution of the process underlying floral zygomorphy development in pentapetalous angiosperms. *Amer. J. Bot.* **104**, 1846–56.

Buendía-Monreal, M., and Gillmor, C. S. (2018). The times they are A-changin': Heterochrony in plant development and evolution. *Front. Plant Sci.* **9**,1349. doi: 10.3389/fpls.2018.01349

Bull-Hereñu K., and Ronse de Craene, L. P. (2020). Ontogenetic base for the variation of flowers in *Malesherbia* Ruiz & Pav. (Passifloraceae). *Front. Ecol. Evol.* **8**, 202. doi: 10.3389/fevo.2020.00202

Bull-Hereñu, K., Ronse de Craene L. P. and Pérez, F. (2018). Floral meristem size and organ number correlation in *Eucryphia* Cav. (Cunoniaceae). *J. Plant Res.* **131**, 429–41.

Burtt, B. L., and Dickison, W. C. (1975). The morphology and relationships of *Seemannaralia* (Araliaceae). *Notes Roy. Bot. Gard. Edinburgh* **33**, 449–66.

Busch, A., and Zachgo, S. (2007). Control of corolla monosymmetry in the Brassicaceae *Iberis amara*. *Proc. Natl. Acad. Sci. USA* **104**, 16714–19.

Buxbaum, F. (1961). Vorlaüfige Untersuchungen über Umfang, systematische Stellung und Gliederung der Caryophyllales (Centrospermae). *Beitr. Biol. Pflanz.* **36**, 1–56.

Buzgo, M. (2001). Flower structure and development of Araceae compared with Alismatids and Acoraceae. *Bot. J. Linn. Soc.* **136**, 393–425.

Buzgo, M., Chanderbali, A. S., Kim, S., Zheng, Z., Oppenheimer, D. G., Soltis, P. S. and Soltis, D. E. (2007). Floral developmental morphology of *Persea Americana* (avocado, Lauraceae), the oddities of staminate organ identity. *Int. J. Plant Sci.* **168**, 261–84.

Buzgo, M., and Endress, P.K. (2000). Floral structure and development of Acoraceae and its systematic relationships with basal Angiosperms. *Int. J. Plant Sci.* **161**, 23–41.

Buzgo, M., Soltis, P. S. and Soltis, D. E. (2004). Floral developmental morphology of *Amborella trichopoda* (Amborellaceae). *Int. J. Plant Sci.* **165**, 925–47.

Buzgo, M., Soltis, D. E., Soltis, P. S., Kim, S., Ma, H., Hauser, B. A., Leebens-Mack, J. and Johansen B. (2006). Perianth development in the basal monocot *Triglochin maritima* (Juncaginaceae). *Aliso* **22**, 107–25.

Byng, J. W. (2014). *The flowering plants handbook. A practical guide to families and genera of the world.* Hertford: Plant Gateway.

Caddick, L. R., Rudall, P. J. and Wilkin, P. (2000). Floral morphology and development in Dioscoreales. *Feddes Repert.* **111**, 189–230.

Cano Niklitschek, M. J. (2012). Evolution of the nectaries in the Primuloid clade (Ericales). Royal Botanic Garden Edinburgh. Unpublished master's thesis.

Cantino, P. D. (1992). Evidence for a polyphyletic origin of the Labiatae. *Ann. Mo. Bot. Gard.* **79**, 361–79.

Cao, L.-M., Liu J., Lin Q., Ronse De Craene, L. P. (2018). The floral organogenesis of *Koelreuteria bipinnata* and its variety *K. bipinnata* var. *integrifolia* (Sapindaceae): Evidence of floral constraints on the evolution of monosymmetry. *Plant Syst. Evol.* **304**, 923–35.

Cao, L.-M., Newman, M., Kirchoff, B., Ronse De Craene, L. P. (2019). Develomental evidence helps resolve the evolutionary origins of anther appendages in Globba. *Bot. J. Linn. Soc.* **189**, 63–82.

Cao, L.-M., Ronse De Craene, L. P., Wang, Z.-X. and Wang Y.-H. (2017). The floral organogenesis of *Eurycorymbus cavaleriei* (Sapindaceae) and its systematic implications. *Phytotaxa* **297**, 234–44.

Caris, P. (2013). Bloemontogenetische patronen in the Ericales *sensu lato*. Katholieke Universiteit Leuven, Belgium: Unpubl. Doctoral Thesis.

Caris, P. L., Geuten K. P., Janssens, S. B. and Smets, E. F. (2006). Floral development in three species of *Impatiens* (Balsaminaceae). *Am. J. Bot.* **93**, 1–14.

Caris, P., Ronse De Craene, L. P. Smets, E. F. and Clinckemaillie, D. (2000). Floral development of three *Maesa* species, with special emphasis on the position of the genus within Primulales. *Ann. Bot.* **86**, 87–97.

Caris, P. and Smets, E. F. (2004). A floral ontogenetic study on the sister group relationship between the genus *Samolus* (Primulaceae) and the Theophrastaceae. *Am. J. Bot.* **91**, 627–43.

Caris, P., Smets, E. F., De Coster, K. and Ronse De Craene, L. P. (2006). Floral ontogeny of *Cneorum tricoccon* L. (Rutaceae). *Plant Syst. Evol.* **257**, 223–32.

Carolin, R. C. (1960). Floral structure and anatomy in the family Stylidiaceae Swartz. *P. Linn. Soc. N. S. W.* **85**, 189–96.

Carolin, R. C. (1993). Portulacaceae. In *The families and genera of vascular plants vol. II*, ed. K. Kubitzki, J. G. Rohwer and V. Bittrich. Berlin: Springer, pp. 544–55.

Carrive, L., Domenech, B., Sauquet, H., Jabbour, F., Damerval, C. and Nadot, S. (2020). Insights into the ancestral flowers of Ranunculales. *Bot. J. Linn. Soc.* **194**, 23–46.

Carrucan, A. E., and Drinnan, A. N. (2000). The ontogenetic basis for floral diversity in the *Baeckea* sub-group. *Kew Bull.* **55**, 593–613.

Cavalari De-Paula, O., Assis, L. and Ronse De Craene, L. P. (2018). Unbuttoning the ancestral flower of angiosperms. *Trends Plant Sci.* **23**, 551–4.

Cavalari De-Paula, O., das Graças Sajo, M., Prenner, G., Cordeiro, I. and Rudall, P.J. (2011). Morphology, development and homologies of the perianth and floral nectaries in *Croton* and *Astraea* (Euphorbiaceae – Malpighiales). *Plant Syst. Evol.* **292**, 1–14.

Chakravarty, M. L. (1958). Morphology of the staminate flowers in the Cucurbitaceae with special reference to the evolution of the stamens. *Lloydia* **21**, 49–87.

Charlton, W. A. (1991). Studies in the Alismataceae. IX. Development of the flower in *Ranalisma humile*. *Can. J. Bot.* **69**, 2790–6.

Charlton, W. A. (1999a). Studies in the Alismataceae. X. Floral organogenesis in *Luronium natans* (L.) Raf. *Can. J. Bot.* **77**, 1560–8.

Charlton, W. A. (1999b). Studies in the Alismataceae. XI. Development of the inflorescence and flowers of *Wiesneria triandra* (Dalzell) Micheli. *Can. J. Bot.* **77**, 1569–79.

Charlton, W. A. (2004). Studies in the Alismataceae. XII. Floral organogenesis in *Damasonium alisma* and *Baldellia ranunculoides*, and comparisons with *Butomus umbellatus*. *Can. J. Bot.* **82**, 528–39.

Chartier, M., Jabbour, F., Gerber, S., Mitteroecker, P., Sauquet, H., Von Balthazar, M., Staedler, Y., Crane, P. R. and Schönenberger, J. (2014). The floral morphospace: A modern comparative approach to study angiosperm evolution. *New Phytol.* **204**, 841–53.

Chen, L., Ren, Y., Endress, P. K., Tian, X. H. and Zhang, X. H. (2007). Floral organogenesis in *Tetracentron sinense* (Trochodendraceae) and its systematic significance. *Plant Syst. Evol.* **264**, 183–93.

Chinga, J., and Pérez, F. (2016). Ontogenetic integration in two species of *Schizanthus* (Solanaceae): A comparison with static integration patterns. *Flora* **22**, 75–81.

Chinga, J., Pérez, F. and Claßen-Bockhoff, R. (2021). The role of heterochrony in *Schizanthus* flower evolution: A quantitative analysis. *Perspect. Plant Ecol.* **49**, 12559. doi: 10.1016/j.ppees.2021.125591

Christenhusz, M. J. M., Brockington, S. F., Christin, P.-A. and Sage, R. F. (2014). On the disintegration of Molluginaceae: A new genus and family (*Kewa*, Kewaceae) segregated from *Hypertelis*, and placement of *Macarthuria* in Macarthuriaceae. *Phytotaxa* **181**, 238–42.

Church, A. H. (1908). *Types of floral mechanism. A selection of diagrams and descriptions of common flowers. Part I.* Oxford: Clarendon.

Citerne, H., Jabbour, F., Nadot, S. and Damerval, C. (2010). The evolution of floral symmetry. In *Advances in botanical research*, ed. J. C. Kader and M. Delseny. London: Elsevier, pp. 85–137.

Citerne, H. L., Pennington, R. T. and Cronk, Q. C. B. (2006). An apparent reversal in floral symmetry in the legume *Cadia* is a homeotic transformation. *Proc. Natl. Acad. Sci. USA* **103**, 12017–20.

Claßen-Bockhoff, R. (1990). Pattern analysis in pseudanthia. *Plant Syst. Evol.* **171**, 57–88.

Claßen-Bockhoff, R. (2016). The shoot concept of the flower: Still up to date? *Flora* **221**, 46–53.

Claßen-Bockhoff, R., and Arndt, M. (2018). Flower-like heads from flower-like meristems: Pseudanthium development in *Davidia involucrata* (Nyssaceae). *J. Plant Res.* **131**, 443–58.

Claßen-Bockhoff, R., and Bull-Hereñu, K. (2013). Towards an ontogenetic understanding of inflorescence diversity. *Ann. Bot.* **112**, 1523–42.

Claßen-Bockhoff, R., and Frankenhäuser, H. (2020). The 'male flower' of *Ricinus communis* (Euphorbiaceae) interpreted as a multi-flowered unit. *Front. Cell Dev. Biol.* **8**: 313. doi: 10.3389/fcell.2020.00313

Claßen-Bockhoff, R., and Heller, A. (2006). Floral synorganization and secondary pollen presentation in four Marantaceae from Costa Rica. *Int. J. Plant Sci.* **169**, 745–60.

Claßen-Bockhoff, R. and Meyer, C. (2016). Space matters: Meristem expansion triggers corona formation in *Passiflora*. *Ann. Bot.* **117**, 277–90.

Claßen-Bockhoff, R., Wester, P. and Tweraser, E. (2003). The staminal lever mechanism in *Salvia* L. (Lamiaceae): A review. *Plant Biol.* **5**, 33–41.

Clinckemaillie, D., and Smets, E. F. (1992). Floral similarities between Plumbaginaceae and Primulaceae, systematic significance. *Belg. J. Bot.* **125**, 151–3.

Cocucci, A. E., and Anton, A. M. (1988). The grass flower, suggestions on its origin and evolution. *Flora* **181**, 353–62.

Coen, E. S., and Meyerowitz, E. M. (1991). The war of the whorls: Genetic interactions controlling flower development. *Nature* **353**, 31–7.

Cook, C. D. K. (1998). Hydrocharitaceae. In *The families and genera of vascular plants Vol. IV*, ed. K. Kubitzki. Berlin: Springer, pp. 234–48.

Copeland, H. F. (1963). Structural notes on hollies (*Ilex aquifolium* and *I. cornuta*, family Aquifoliaceae). *Phytomorphology* **13**, 455–64.

Costello, A., and Motley, T. J. (2004). The development of the superior ovary in *Tetraplasandra* (Araliaceae). *Am. J. Bot.* **91**, 644–55.

Couvreur, T. L. P., Richardson, J. E., Sosef, M. S. M., Erkens, R. H. J. and Chatrou, L. W. (2008). Evolution of syncarpy and other morphological characters in African Annonaceae: A posterior mapping approach. *Mol. Phyl. Evol.* **47**, 302–18.

Cox, C. D. K. (1998). Hydrocharitaceae. In *The families and genera of vascular plants, Vol. IV*, ed. K. Kubitzki. Berlin: Springer, pp. 234–48.

Crane, P. R., Friis, E. M. and Pedersen, K. R. (1994). Paleobotanical evidence on the early radiation of magnoliid Angiosperms. *Plant Syst. Evol., Suppl.* **8**, 51–72.

Crepet, W. L. (2008). The fossil record of Angiosperms: Requiem or renaissance? *Ann. Mo. Bot. Gard.* **95**, 3–33.

Crepet, W. L., Nixon, K. C. and Weeks, A. (2018). Mid-Cretaceous angiosperm radiation and an asteroid origin of bilaterality: diverse and extinct 'Ericales' from New Jersey. *Am. J. Bot.* **105**, 1412–23.

Cronk, Q. C. B., Needham, I. and Rudall, P. J. (2015). Evolution of catkins: Inflorescence morphology of selected Salicaceae in an evolutionary and developmental context. *Front. Plant Sci.* **5**, 1030. doi: 10.3389/fpls.2015.01030

Cronquist, A. (1981). *An integrated system of classification of flowering plants*. New York: Columbia University Press.

Cuénoud, P., Savolainen, V., Chatrou, L. W., Powell, M., Grayer, R. J. and Chase, M. W. (2002). Molecular phylogenetics of Caryophyllales based on nuclear 18S rDNA and plastid *rbc*L, *atp*B, and *mat*K DNA sequences. *Am. J. Bot.* **89**, 132–44.

Dahlgren, R. (1975). A system of classification of the Angiosperms to be used to demonstrate the distribution of characters. *Bot. Not.* **128**, 119–47.

Dahgren, R. (1983). General aspects of Angiosperm evolution and macrosystematics. *Nord. J. Bot.* **3**, 119–49.

Dahlgren, R., Clifford, T. H. and Yeo, P. (1985). *The families of the monocotyledons. Structure, evolution and taxonomy*. Berlin: Springer.

Dahlgren, R., and Thorne, R. F. (1984). The order Myrtales: Circumscription, variation and relationships. *Ann. Mo. Bot. Gard.* **71**, 633–99.

D'Arcy, W. G. (1986). The calyx in *Lycianthes* and some other genera. *Ann. Mo. Bot. Gard.* **73**, 117–27.

Dawson, M. L. 1936. The floral morphology of the Polemoniaceae. *Am. J. Bot.* **23**, 501–11.

de Barros, T. C., Pedersoli, G. D., Paulino, J. V. and Teixeira, S. P. (2017). In the interface of caesalpinioids and mimosoids: Comparative floral development elucidates shared characters in *Dimorphandra mollis* and *Pentaclethra macroloba* (Leguminosae). *Am. J. Bot.* **104**, 218–32.

de Laet, J., Clinckemaillie, D., Jansen, S. and Smets, E. F. (1995). Floral ontogeny in the Plumbaginaceae. *J. Plant Res.* **108**, 289–304.

de Maggio, A. E., and Wilson, C. L. (1986). Floral structure and organogenesis in *Podophyllum peltatum* (Berberidaceae). *Am. J. Bot.* **73**, 21–32.

de Menezes, N. L. (1980). Evolution in Velloziaceae, with special reference to androecial characters. In *Petaloid monocotyledons*, ed. C. D. Brickell, D. F. Cutler and M. Gregory. Hortic. and Botan. Research. Linnean Soc. Symposium Series 8. London: Academic Press, pp. 117–38.

de Olivera Franca R. and Cavalari De-Paula, O. (2017). Embryology of *Pera* (Peraceae, Malpighiales): Systematics and evolutionary implications. *J. Plant Res.* **130**, 709–21.

Deroin, T. (1985). Contribution à la morphologie comparée du gynécée des Annonaceae: Monodoroideae. *Bull. Mus. Hist. Nat. (Paris)* **Sér. IV**, **7**, 167–76.

Deroin, T. (1997). Confirmation and origin of the paracarpy in Annonaceae, with comments on some methodological aspects. *Candollea* **52**, 45–58.

Deroin, T. (2000). Floral anatomy of *Toussaintia hallei* Le Thomas, a case of convergence of Annonaceae with Magnoliaceae. In *Proceedings of the International Symposium on the Family Magnoliaceae*, ed. Y.-H. Liu, H.-M. Fan, Z.-Y. Chen, Q.-G. Wu and Q.-W. Zeng. Beijing: Science Press, pp. 168–76.

Deroin, T. (2007). Floral vascular pattern of the endemic Malagasy genus *Fenerivia* Diels (Annonaceae). *Adansonia* **Sér. 3**, **29**, 7–12.

Deroin, T. (2010). Floral anatomy of *Magnolia decidua* (Q. Y. Zheng) V. S. Kumar (Magnoliaceae): Recognition of a partial pentamery. *Adansonia* **Sér. 3**, **32**, 39–55.

Deroin, T., and Le Thomas, A. (1989). Sur la systématique et les potentialités évolutives des Annonacées: cas d'*Ambavia gerrardii* (Baill.) Le Thomas, espèce endémique de Madagascar. *C. R. Acad. Sci. Paris* **t. 309, Sér. III**, 647–52.

Derstine, K. S., and Tucker, S. C. (1991). Organ initiation and development of inflorescences and flowers of *Acacia baileyana*. *Am. J. Bot.* **78**, 816–32.

de Wilde, W. J. J. O. (1974). The genera of tribe Passifloreae (Passifloraceae) with special reference to flower morphology. *Blumea* **22**, 37–50.

Dickison, W. C. (1970). Comparative morphological studies in Dilleniaceae VI. Stamens and young stem. *J. Arnold Arbor.* **51**, 403–18.

Dickison, W. C. (1972). Observations on the floral morphology of some species of *Saurauia*, *Actinidia* and *Clematoclethra*. *J. Elisha Mitchell Sci. Soc.* **88**, 43–54.

Dickison, W. C. (1975). Studies on the floral anatomy of the Cunoniaceae. *Am. J. Bot.* **62**, 433–47.

Dickison, W. C. (1978). Comparative anatomy of Eucryphiaceae. *Am. J. Bot.* **65**, 722–35.

Dickison, W.C. (1986). Floral morphology and antomy of Staphyleaceae. *Bot. Gaz.* **147**, 312–26.

Dickison, W. C. (1993). Floral anatomy of the Styracaceae, including observations on intra-ovarian trichomes. *Bot. J. Linn. Soc.* **112**, 223–55.

Dilcher, D. L. (2000). Toward a new synthesis, major evolutionary trends in the Angiosperm fossil record. *Proc. Natl. Acad. Sci. USA* **97**, 7030–6.

Donoghue, M. J., Bell, C. D. and Winkworth, R. C. (2003). The evolution of reproductive characters in Dipsacales. *Int. J. Plant Sci.* **164 (5 Suppl.)**, S453–S464.

Donoghue, M. J., Ree, R. H. and Baum, D. A. (1998). Phylogeny and the evolution of flower symmetry in the Asteridae. *Trends Plant Sc.* **3**, 311–17.

Douglas, A. W., and Tucker, S. C. (1996a). Comparative floral ontogenies among Persoonioideae including *Bellendena* (Proteaceae). *Am. J. Bot.* **83**, 1528–55.

Douglas, A. W., and Tucker, S. C. (1996b). Inflorescence ontogeny and floral organogenesis in Grevilleoideae (Proteaceae), with emphasis on the nature of the flower pairs. *Int. J. Plant Sci.* **157**, 341–72.

Douglas, A. W., and Tucker, S. C. (1996c). The developmental basis of diverse carpel orientations in Grevilleoideae (Proteaceae). *Int. J. Plant Sci.* **157**, 373–97.

Douglas, A. W., and Tucker, S. C. (1997). The developmental basis of morphological diversification and synorganization in flowers of Conospermeae (*Stirlingia* and Conosperminae, Proteaceae). *Int. J. Plant Sci.* **158**, S13–S48.

Doust, A. N. (2002). Comparative floral ontogeny in Winteraceae. *Ann. Mo. Bot. Gard.* **87**, 366–79.

Doyle, J. A. (2008). Integrating molecular phylogenetic and paleobotanical evidence on origin of the flower. *Int. J. Plant Sci.* **169**, 816–43.

Doyle, J. A., and Endress, P. K. (2000). Morphological phylogenetic analysis of basal Angiosperms, comparison and combination with molecular data. *Int. J Plant Sci.* **161 (6 Suppl.)**, S121-S153.

Doyle, J. A., and Endress, P. K. (2011). Tracing the early evolutionary diversification of the angiosperm flower. In *Flowers on the tree of life*, ed. L. Wanntorp and L. P. Ronse De Craene. Cambridge: Cambridge University Press, pp. 88–119.

Drinnan, A. N., and Ladiges, P. Y. (1989). Corolla and androecium development in some *Eudesmia* eucalypts (Myrtaceae). *Plant Syst. Evol.* **165**, 239–54.

Eames, A. J. (1961). *Morphology of the Angiosperms*. New York: Mc Graw-Hill.

Eckert, G. (1966). Entwicklungsgeschichtliche und blütenanatomische Untersuchungen zum Problem der Obdiplostemonie. *Bot. Jahrb. Syst.* **85**, 523–604.

Eckardt, T. (1937). Untersuchungen über Morphologie, Entwicklungsgeschichte und systematische Bedeutung des pseudomonomeren Gynoeceums. *Nova Acta Leopold. N. F.* **5**, 1–112.

Eckardt, T. (1974). Vom Blütenbau der Centrospermen-Gattung *Lophiocarpus* Turcz. *Phyton (Austria)* 16: 1–4.

Ecklund, H. (2000). Lauraceous flowers from the Late Cretaceous of North Carolina, U.S.A. *Bot. J. Linn. Soc.* **132**, 397–428.

Edgell, T. (2004). Floral studies of *Brexia madagascariensis* Thouars (Celastraceae). UK: Royal Botanic Garden Edinburgh: Unpubl. MSc thesis.

Eichler, A. W. (1875). *Blüthendiagramme vol. I.* Leipzig: Wilhelm Engelmann.

Eichler, A. W. (1878). *Blütendiagramme vol. II.* Leipzig: Wilhelm Engelmann.

El, E. S., Remizowa, M. V. and Sokoloff, D. D. (2020). Developmental flower and rhizome morphology in *Nuphar* (Nymphaeales): An interplay of chaos and stability. *Front. Cell Dev. Biol.* **8**, 303. doi: 10.3389/fcell.2020.00303

Eliasson, U. H. 1988. Floral morphology and taxonomic relations among the genera of Amaranthaceae in the New World and the Hawaiian Islands. *Bot. J. Linn. Soc.* **96**, 235–83.

El Ottra, J. H. L., Demarco, D. and Pirani, J. R. (2019). Comparative floral structure and evolution in Galipeinae (Galipeeae: Rutaceae) and its implications at different systematic levels. *Bot. J. Linn. Soc.* **191**, 30–101.

El Ottra, J. H. L., Mello-de-Pina, G. F. de A., Demarco, D., Pirani, J. R. and Ronse De Craene, L. P. (2022). Gynoecium structure in Sapindales: A review of selected features and a case study of *Trichilia pallens* C. DC. (Meliaceae). *J. Plant. Res.*

El Ottra, J. H. L., Pirani, J. R. and Endress, P. K. (2013). Fusion within and between whorls of floral organs in Galipeinae (Rutaceae): Structural features and evolutionary implications. *Ann. Bot.* **111**, 821–37.

Endress, M. E., and Bittrich, V. (1993). Molluginaceae. In *The families and genera of vascular plants, vol. II*, ed. K. Kubitzki, J. G. Rohwer and V. Bittrich, Berlin: Springer, pp. 419–26.

Endress, M. E., Sennblad, B., Nilsson, S., Civeyrel, L., Chase, M. W., Huysmans, S., Grafström, E. and Bremer, B. (1996). A phylogenetic analysis of Apocynaceae s. str. and some related taxa in Gentianales, a multidisciplinary approach. *Opera Bot. Belg.* **7**, 59–102.

Endress, P. K. (1967). Systematische Studien über die verwandschaftlichen Beziehungen zwischen den Hamamelidaceen und Betulaceen. *Bot. Jahrb. Syst.* **87**, 431–525.

Endress, P. K. (1976). Die Androeciumanlage bei polyandrischen Hamamelidaceen und ihre systematische Bedeutung. *Bot. Jahrb. Syst.* **97**, 436–57.

Endress, P. K. (1977). Evolutionary trends in the Hamamelidales-Fagales group. *Plant Syst. Evol.* **Suppl. 1**, 321–47.

Endress, P. K. (1978). Blütenontogenese, Blütenabgrenzung und Systematische Stellung der perianthlosen Hamamelidoideae. *Bot. Jahrb. Syst.* **100**, 249–317.

Endress, P. K. (1980a). Ontogeny, function and evolution of extreme floral construction in Monimiaceae. *Plant Syst. Evol.* **134**, 79–120.

Endress, P. K. (1980b). The reproductive structures and systematic position of the Austrobaileyaceae. *Bot. Jahrb. Syst.* **101**, 393–433.

Endress, P. K. (1984). The role of inner staminodes in the floral display of some relic Magnoliales. *Plant Syst. Evol.* **146**, 269–82.

Endress, P. K. (1986). Floral structure, systematics, and phylogeny in Trochodendrales. *Ann. Mo Bot. Gard.* **73**, 297–324.

Endress, P. K. (1987). Floral phyllotaxis and floral evolution. *Bot. Jahrb. Syst.* **108**, 417–38.

Endress, P. K. (1989). Chaotic floral phyllotaxis and reduced perianth in *Achlys* (Berberidaceae). *Bot. Acta* **102**, 159–63.

Endress, P. K. (1990). Patterns of floral construction in ontogeny and phylogeny. *Biol. J. Linn. Soc.* **39**, 153–75.

Endress, P. K. (1992). Evolution and floral diversity, the phylogenetic surroundings of *Arabidopsis* and *Antirrhinum*. *Int. J. Plant Sci.* **153**, S106–S122.

Endress, P. K. (1994). *Diversity and evolutionary biology of tropical flowers*. Cambridge: Cambridge University Press.

Endress, P. K. (1995a). Major evolutionary traits of monocot flowers. In *Monocotyledons, systematics and evolution*, ed. P. J. Rudall, D. F. Cribb and C. J. Humphries. Kew: Royal Botanic Gardens, pp. 43–79.

Endress, P. K. (1995b). Floral structure and evolution in Ranunculanae. *Plant Syst. Evol.* **Suppl. 9**, 47–61.

Endress, P. K. (1997). Relationships between floral organization, architecture, and pollination mode in *Dillenia* (Dilleniaceae). *Plant Syst. Evol.* **206**, 99–118.

Endress, P. K. (1998). *Antirrhinum* and Asteridae: Evolutionary changes of floral symmetry. *Soc. Exp. Biol. Sem. Ser.* **51**, 133–40.

Endress, P. K. (1999). Symmetry in flowers: Diversity and evolution. *Int. J. Plant Sci.* **160 (6 Suppl.)**, S3–S23.

Endress, P. K. (2001). The flower in extant basal Angiosperms and inferences on ancestral flowers. *Int. J. Plant Sci.* **162**, 1111–40.

Endress, P. K. (2003a). Morphology and Angiosperm systematics in the molecular era. *Bot. Rev.* **68**, 545–70.

Endress, P. K. (2003b). Early floral development and nature of the calyptra in Eupomatiaceae (Magnoliales). *Int. J. Plant Sci.* **164**, 489–503.

Endress, P. K. (2006). Angiosperm floral evolution, morphological developmental framework. *Adv. Bot. Res.* **44**, 1–61.

Endress, P. K. (2008a). My favourite flowering image. *J. Exper. Bot. FNL 2008*: 1–3.

Endress, P. K. (2008b). The whole and the parts, relationships between floral architecture and floral organ shape, and their repercussions on the interpretation of fragmentary floral fossils. *Ann. Mo. Bot. Gard.* **95**, 101–20.

Endress, P. K. (2008c). Perianth biology in the basal grade of extant angiosperms. *Int. J. Plant Sci.* **169**, 844–62.

Endress, P. K. (2010a). Synorganisation without organ fusion in the flowers of *Geranium robertianum* (Geraniaceae) and its not so trivial obdiplostemony. *Ann. Bot.* **106**, 687–95.

Endress, P. K. (2010b). Disentangling confusions in inflorescence morphology: Patterns and diversity of reproductive shoot ramification in angiosperms. *J. Syst. Evol.* **48**, 225–39.

Endress, P. K. (2011). Evolutionary diversification of the flowers in angiosperms. *Amer. J. Bot.* **98**, 370–96.

Endress, P.K. (2012). The immense diversity of floral monosymmetry and asymmetry across angiosperms. *Bot. Rev.* **78**, 345–97.

Endress, P. K. (2014). Multicarpellate gynoecia in angiosperms: Occurrence, development, organization and architectural constraints. *Bot. J. Linn. Soc.* **174**, 1–43.

Endress, P. K. (2016). Development and evolution of extreme synorganization in angiosperm flowers and diversity: A comparison of Apocynaceae and Orchidaceae. *Ann. Bot.* **117**, 749–67.

Endress, P. K. (2019). The morphological relationship between carpels and ovules in angiosperms: Pitfalls of morphological interpretation. *Bot. J. Linn. Soc.* **189**, 201–27.

Endress, P. K., Davis, C. C. and Matthews, M. L. (2013) Advances in the floral structural characterization of the major subclades of Malpighiales, one of the largest orders of flowering plants. *Ann. Bot.* **111**, 969–85.

Endress, P. K., and Doyle, J. A. (2007). Floral phyllotaxis in basal Angiosperms, development and evolution. *Curr. Opin. Plant Biol.* **10**, 52–7.

Endress, P. K., and Doyle, J. A. (2009). Reconstructing the ancestral angiosperm flower and its initial specializations. *Am. J. Bot.* **96**, 22–66.

Endress, P. K., and Doyle, J. A. (2015). Ancestral traits and specializations in the flowers of the basal grade of living angiosperms. *Taxon* **64**, 1093–1116.

Endress, P. K., and Hufford, L. D. (1989). The diversity of stamen structures and dehiscence patterns among Magnoliidae. *Bot. J. Linn. Soc.* **100**, 45–85.

Endress, P. K., and Igersheim, A. (1997). Gynoecium diversity and systematics of the Laurales. *Bot. J. Linn. Soc.* **125**, 93–168.

Endress, P. K., and Igersheim, A. (2000a). Gynoecium structure and evolution in basal Angiosperms. *Int. J. Plant Sci.* **161 (6 Suppl.)**, S211–S223.

Endress, P. K., and Igersheim, A. (2000b). The reproductive structures of the basal Angiosperm *Amborella trichopoda* (Amborellaceae). *Int. J. Plant Sci.* **161 (6 Suppl.)**, S237–S248.

Endress, P. K., Jenny, M. and Fallen, M. (1983). Convergent elaboration of apocarpous gynoecia in higher advanced dicotyledons. *Nord. J. Bot.* **3**, 293–300.

Endress, P. K., and Lorence, D. H. (2004). Heterodichogamy of a novel type in *Hernandia* (Hernandiaceae) and its structural basis. *Int. J. Plant Sci.* **165**, 753–63.

Endress, P. K., and Matthews, M. L. (2006a). First steps towards a floral structural characterization of the major Rosid subclades. *Plant Syst. Evol.* **260**, 223–51.

Endress, P. K., and Matthews, M. L. (2006b). Elaborate petals and staminodes in eudicots, diversity, function, and evolution. *Org. Div. Evol.* **6**, 257–93.

Endress, P. K., and Matthews, M. L. (2012). Progress and problems in the assessment of flower morphology in higher-level systematics. *Plant Syst. Evol.* **298**, 257–76.

Endress, P. K., and Rapini, A. (2014). Floral structure of *Emmotum* (Icacinaceae *sensu stricto* or Emmotaceae), a phylogenetically isolated genus of lamiids with a unique pseudotrimerous gynoecium, bitegmic ovules and monosporangiate thecae. *Ann. Bot.* **114**, 945–59.

Engler, A., and Krause, K. (1935). Loranthaceae. In *Die natürlichen Pflanzenfamilien 16b*, ed. A. Engler and K. Prantl. 2nd edn. Leipzig: W. Engelmann, pp. 98–203.

Engler, A., and Prantl, K., eds. (1887–1909). *Die natürlichen Pflanzenfamilien I–IV*, Leipzig: W. Engelmann.

Erbar, C. (1986). Untersuchungen zur Entwicklung der spiraligen Blüte von *Stewartia pseudocamellia* (Theaceae). *Bot. Jahrb. Syst.* **106**, 391–407.

Erbar, C. (1991). Sympetaly: A systematic character? *Bot. Jahrb. Syst.* **112**, 417–51.

Erbar, C. (1992). Floral development of two species of *Stylidium* (Stylidiaceae) and some remarks on the systematic position of the family Stylidiaceae. *Can. J. Bot.* **70**, 258–71.

Erbar, C. (1993). Studies on the floral development and pollen presentation in *Acicarpha tribuloides* with a discussion of the systematic position of the family Calyceraceae. *Bot. Jahrb. Syst.* **115**, 325–50.

Erbar, C. (1994). Contributions to the affinities of *Adoxa* from the viewpoint of floral development. *Bot. Jahrb. Syst.* **116**, 259–82.

Erbar, C. (1998). Coenokarpie ohne und mit Compitum, ein Vergleich der Gynoeceen von *Nigella* (Ranunculaceae) and *Geranium* (Geraniaceae). *Beitr. Biol. Pflanz.* **71**, 13–39.

Erbar, C. (2014) Nectar secretion and nectaries in basal angiosperms, magnoliids and non-core eudicots and a comparison with core eudicots. *Plant Div. Evol.* **131/2**, 63–143.

Erbar, C., Kusma, S. and Leins, P. (1998). Development and interpretation of nectary organs in Ranunculaceae. *Flora* **194**, 317–32.

Erbar, C., and Leins, P. (1981). Zur spirale in Magnolienblüten. *Beitr. Biol. Pflanz.* **56**, 225–41.

Erbar, C., and Leins, P. (1983). Zur sequenz von Blütenorganen bei einigen Magnoliiden. *Bot. Jahrb. Syst.* **103**, 433–49.

Erbar, C., and Leins, P. (1985). Studien zur Organsequenz in Apiaceen-Blüten. *Bot. Jahrb. Syst.* **105**, 379–400.

Erbar, C., and Leins, P. (1988a). Blütenentwicklungsgeschichtliche Studien an *Aralia* und *Hedera* (Araliaceae). *Flora* **180**, 391–406.

Erbar, C., and Leins, P. (1988b). Studien zur Blütenentwicklung und Pollenpräsentation bei *Brunonia australis* Smith (Brunoniaceae). *Bot. Jahrb. Syst.* **110**, 263–82.

Erbar, C., and Leins, P. (1989). On the early floral development and the mechanisms of secondary pollen presentation in *Campanula*, *Jasione* and *Lobelia*. *Bot. Jarhb. Syst.* **111**, 29–55.

Erbar, C., and Leins, P. (1994). Flowers in Magnoliidae and the origin of flowers in other subclasses of the Angiosperms. I. The relationships between flowers of Magnoliidae and Alismatidae. *Plant Syst. Evol., suppl.* **8**, 193–208.

Erbar, C., and Leins, P. (1995a). An analysis of the early floral development of *Pittosporum tobira* (Thunb.) Aiton and some remarks on the systematic position of the family Pittosporaceae. *Feddes Repert.* **106**, 463–73.

Erbar, C., and Leins, P. (1995b). Portioned pollen release and the syndromes of secondary pollen presentation in the Campanulales-Asterales-complex. *Flora* **190**, 323–38.

Erbar, C., and Leins, P. (1997). Different patterns of floral development in whorled flowers, exemplified by Apiaceae and Brassicaceae. *Int. J. Plant Sci.* **158 (Suppl. 6)**, S49–S64.

Erbar, C., and Leins, P. (2011). Synopsis of some important, non-DNA character states in the Asterids with special reference to sympetaly. *Plant Div. Evol.* **129**, 93–123.

Ernst, W. R. (1967). Floral morphology and systematics of *Platystemon* and its allies *Hesperomecon* and *Meconella* (Papaveraceae, Platystemonoideae). *Univ. Kansas Sci. Bull.* **47**, 25–70.

Etchevery, A. V., Alemán, M. M. and Fleming, T. F. (2008). Flower morphology, pollination biology and mating system of the complex flower of *Vigna Caracalla* (Fabaceae: Papilionoideae). *Ann. Bot.* **102**, 305–16.

Evans, R. C., and Dickinson, T. A. (1996). North American black-fruited hawthorns. II. Floral development of 10- and 20- stamen morphotypes in *Crateaegus* section *douglasii* (Rosaceae, Maloideae). *Am. J. Bot.* **83**, 961–78.

Evans, R. C., and Dickinson, T. A. (2005). Floral ontogeny and morphology in *Gillenia* ('Spiraeoideae') and subfamily Maloideae C. Weber (Rosaceae). *Int. J. Plant Sci.* **166**, 427–47.

Eyde, R. H. (1977). Reproductive structures and evolution in *Ludwigia* (Onagraceae). I. androecium, placentation, merism. *Ann. Mo. Bot. Gard.* **64**, 644–55.

Eyde, R. H., and Morgan, J. T. (1973). Floral structure and evolution in Lopezieae (Onagraceae). *Am. J. Bot.* **60**, 771–87.

Faden, R. B. (2000). Floral biology of Commelinaceae. In *Monocots, systematics and Evolution*, ed. K. L. Wilson and D. A. Morrison. Melbourne: CSIRO, pp. 309–17.

Falcão, M. J. A., Paulino, J. V., Kochanowski, F. J., Figgueiredo, R. C., Basso-Alves, J. P. and Mansano, V. F. (2020). Development of inflorescences and flowers in Fabaceae subfamily Dialioideae: An evolutionary overview and complete ontogenetic series for *Apuleia* and *Martiodendron*. *Bot. J. Linn. Soc.* **193**, 19–46.

Farrar J., and Ronse De Craene, L. P. (2013). To be or not to be a staminode: The floral development of *Sauvagesia* (Ochnaceae) reveals different origins of presumed staminodes. In *Flowers, morphology, evolutionary diversification and implications for the environment*, ed. T. Berntsen and K. Alsvik. New York: Nova Science, pp. 89–103.

Fey, B. S., and Endress, P. K. (1983). Development and morphological interpretation of the cupule in Fagaceae. *Flora* **173**, 451–68.

Fiedler, H. (1910). Beiträge zur Kenntnis der Nyctaginaceen. *Engl. Bot. Jahrb.* **44**, 572–605.

Flores-Olvera, H., Vrijdaghs, A., Ochoterena, H. and Smets, E. (2011). The need to re-investigate the nature of homoplastic characters: An ontogenetic case study of the 'bracteoles' in Atripliceae (Chenopodiaceae). *Ann. Bot.* **108**, 847–65.

Frame, D., and Durou, S. (2001). Morphology and biology of *Napolaeona vogellii* (Lecythidaceae) flowers in relation to the natural history of insect visitors. *Biotropica* **33**, 458–71.

Franz, E. (1908). Beiträge zur Kenntnis der Portulacaceen und Basellaceen. *Bot. Jahrb. Syst.* **42, Beibl. 97**, 1–28.

Freitas L., Bernardello, G., Galetto, L. and Paoli, A. A. S. (2001). Nectaries and reproductive biology of *Croton sarcopetalus* (Euphorbiaceae). *Bot. J. Linn. Soc.* **136**, 267–77.

Friedman, J. (2011). Gone with the wind: Understanding evolutionary transitions between wind and animal pollination in the angiosperms. *New Phytol.* **191**, 911–13.

Friedman, J., and Barrett, S. C. (2008). A phylogenetic analysis of the evolution of wind pollination in the Angiosperms. *Int. J. Plant Sci.* **169**, 49–58.

Friedrich, H.-C. (1956). Studien über die natürliche Verwandtschaft der Plumbaginales und Centrospermae. *Phyton* (Austria) **6**, 220–63.

Friis, E. M. (1984). Preliminary report of Upper Cretaceous Angiosperm reproductive organs from Sweden and their level of organisation. *Ann. Mo Bot. Gard.* **71**, 403–18.

Friis, E. M., Pedersen, K. R. and Crane, P. R. (2006). Cretaceous Angiosperm flowers, innovation and evolution in plant reproduction. *Paleogeogr. Palaeoclimat. Palaeoecol.* **232**, 251–93.

Friis, E. M., Pedersen, K. R. and Crane, P. R. (2016). The emergence of core eudicots: New floral evidence from the earliest late Cretaceous. *Proc. Roy. Soc.* **B283**, 20161325.

Friis, E. M., Pedersen, K. L., and Schönenberger, J. (2006). *Normapolles* plants, a prominent component of the Cretaceous Rosid diversification. *Plant Syst. Evol.* **260**, 107–40.

Fu, L., Zen, Q.-W., Liao, J.-P. and Xu, F.-X. (2009). Anatomy and ontogeny of unisexual flowers in dioecious *Woonyoungia septentrionalis* (Dandy) Law (Magnoliaceae). *J. Syst. Evol.* **47**, 263–72.

Fukuoka, N., Ito, M. and Iwatsuki, K. (1986). Floral anatomy of the mangrove genus *Lumnizera* (Combretaceae). *Acta Phytotax. Geobot.* **37**, 69–81.

Gagliardi, K. B., Cordeiro, I. and Demarco, D. (2018). Structure and development of flowers and inflorescences in Peraceae and the evolution of pseudanthia in Malpighiales. PLOS ONE. doi: 10.1371/journal.pone.0203954

Gagliardi, K. B., de Souza, L. A. and Albiero, A. L. M. (2014). Comparative fruit development in some Euphorbiaceae and Phyllanthaceae. *Plant Syst. Evol.* **300**, 775–82.

Gallant, J. B., Kemp, J. R. and Lacroix, C. R. (1998). Floral development of dioecious staghorn sumac, *Rhus hirta* (Anacardiaceae). *Int. J. Plant Sci.* **159**, 539–49.

Galle, P. (1977). Untersuchungen zur Blütenentwicklung der Polygonaceen. *Bot. Jahrb. Syst.* **98**, 449–89.

Gandhi, K. N., and Dale Thomas, R. (1983). A note on the androecium of the genus *Croton* and flowers in general of the family Euphorbiaceae. *Phytologia* **54**, 6–8.

Gandolfo, M. A., Nixon, K. C. and Crepet, W. L. (1998). *Tylerianthus crossmanensis* gen. et sp. nov. (Aff. Hydrageaceae) from the Upper Cretaceous of New Jersey. *Am. J. Bot.* **85**, 376–86.

Gauthier, R., and Arros, J. (1963). L'anatomie de la fleur staminée de l'*Hillebrandia sandwicensis* Oliver et la vascularisation de l'étamine. *Phytomorphology* **13**, 115–27.

Ge, L.-P., Lu, A.-M. and Gong, C.-R. (2007). Ontogeny of the fertile flower in *Platycrater arguta* (Hydrangeaceae). *Int. J. Plant Sci.* **168**, 835–44.

Geitler, L. (1929). Zur Morphologie der Blüten von *Polygonum*. *Österr. Bot. Zeit.* **78**, 229–41.

Gelius, L. (1967). Studien zur Entwicklungsgeschichte an Blüten der Saxifragales sensu lato mit besonderer Berücksichtigung des Androeceum. *Bot. Jahrb. Syst.* **87**, 253–303.

Gemmeke, V. (1982). Entwicklungsgeschichtliche Untersuchungen an Mimosaceen-Blüten. *Bot. Jahrb. Syst.* **103**, 185–210.

Gerrath, J. M., Lacroix, C. R. and Posluszny, U. (1990). The developmental morphology of *Leea guineensis* II. Floral development. *Bot. Gaz.* **151**, 210–20.

Geuten, K., Becker, A., Kaufmann, K., Caris, P., Janssens, S., Viaene, T., Theißen, G. and Smets, E. (2006). Petaloidy and petal identity MADS-box genes in the balsaminoid genera *Impatiens* and *Marcgravia*. *Plant J.* **47**, 501–18.

Gilg, E. (1894). Studien über die Verwandtschaftsverhältnisse der Thymelaeales und über die 'Anatomische Methode'. *Bot. Jahrb. Syst.* **18**: 488–574.

Glinos, E., and Cocucci, A. A. (2011). Pollination biology of *Canna indica* (Cannaceae) with particular reference to the functional morphology of the style. *Plant Syst. Evol.* **291**, 49–58.

Glover, B. J. (2007). *Understanding flowers and flowering. An integrated approach.* Oxford: Oxford University Press.

Glover, B. J., Airoldi, C. A., Brockington, S. F., Fernandez-Mazuecos, M., Martinez-Perez, C., Mellers, G., Moyroud, E. and Taylor, L. (2015). How have advances in comparative floral development influenced our understanding of floral evolution? *Int. J. Plant Sci.* **176**, 307–23.

Gonzalez, A. M. (2016). Floral structure, development of the gynoecium, and embryology in *Schinopsis balansae* Engler (Anacardiaceae) with particular reference to aporogamy. *Int. J. Plant Sci.* **177**, 326–38.

González, F., and Bello, M. A. (2009). Intraindividual variation variation of flowers of *Gunnera* (Gunneraceae) and proposed apomorphies for Gunnerales. *Bot. J. Linn. Soc.* **160**, 262–83.

González, F., and Rudall, P. (2010). Flower and fruit characters in the early-divergent Lamiid family Metteniusaceae, with particular reference to the evolution of pseudomonomery. *Am. J. Bot.* **97**, 191–206.

González, F., and Stevenson, D. W. (2000a). Perianth development and systematics of *Aristolochia*. *Flora* **195**, 370–91.

González, F., and Stevenson, D. W. (2000b). Gynostemium development in *Aristolochia* (Aristolochiaceae). *Bot. Jahrb. Syst.* **122**, 249–91.

Gottschling, M. (2004). Floral ontogeny *in Bourreria* (Ehretiaceae, Boraginales). *Flora* **199**, 409–23.

Graf, J. (1975). *Tafelwerk zur Pflanzensystematik mit euartiger Bildmethode*. München: J. F. Lehmanns.

Greenberg, A. K., and Donoghue, M. J. (2011). Molecular systematics and character evolution in Caryophyllaceae. *Taxon* **60**, 1637–52.

Grey-Wilson, C. (1980). Some observations on the floral vascular antomy of *Impatiens* (Studies in Balsaminaceae VI). *Kew Bull.* **35**, 221–7.

Groeninckx, I., Vrijdaghs, A., Huysmans, S., Smets, E. and Dessein, S. (2007). Floral ontogeny of the Afro-Madagascan genus *Mitrasacmopsis* with comments on the development of superior ovaries in Rubiaceae. *Ann. Bot.* **100**, 41–9.

Guédès, M. (1979). *Morphology of seed-plants*. Vaduz, Liechtenstein: J. Cramer.

Guo, X., Thomas, D. C. and Saunders, R. M. K. (2018). Organ homologies and perianth evolution in the *Dasymaschalon* alliance (Annonaceae): Inner petal loss and its functional consequences. *Front. Plant Sci.* **9**, 174. doi: 10.3389/fpls.2018.00174

Gustafsson, M. H. G. (1995). Petal venation in Asterales and related orders. *Bot. J. Linn. Soc.* **118**, 1–18.

Gustafsson, M. H. G. (2000). Floral morphology and relationships of *Clusia gundlachii* with a discussion of floral organ identity and diversity in the genus Clusia. *Int. J. Plant Sci.* **161**, 43–53.

Gustafsson, M. H. G., and Albert, V. A. (1999). Inferior ovaries and Angiosperm diversification. In *Molecular systematics and plant evolution*, ed. P. M. Hollingsworth, R. M. Bateman and R. J. Gornall. London: Taylor and Francis, pp. 403–31.

Gustafsson, M. H. G., and Bittrich, V. (2002). Evolution of morphological diversity and resin secretion in flowers of *Clusia* (Clusiaceae), insights from ITS sequence variation. *Nord. J. Bot.* **22**, 183–203.

Gustafsson, M. H. G., Bittrich, V., and Stevens, P. F. (2002). Phylogeny of Clusiaceae based on *rbc*L sequences. *Int. J. Plant Sci.* **163**, 1045–54.

Haas, P. (1976). Morphologische, anatomische und entwicklungsgeschichtliche Untersuchungen an Blüten und Früchten hochsukkulenter Mesembryanthemaceen-Gattungen – ein Beitrag zu ihrer Systematik. *Diss. Bot.* **33**, 1–256.

Haber, J. M. (1966). The comparative anatomy and morphology of the flowers and inflorescences of the Proteaceae III. Some African taxa. *Phytomorphology* **16**, 490–527.

Hall, J. C., Sytsma, K. J. and Iltis, H. H. (2002). Phylogeny of Capparaceae and Brassicaceae based on chloroplast sequence data. *Am. J. Bot.* **89**, 1826–42.

Hardy, C. R., and Stevenson, D. W. (2000a). Development of the gametophytes, flower, and floral vasculature in *Cochliostema odoratissimum* (Commelinaceae). *Bot. J. Linn. Soc.* **134**, 131–57.

Hardy, C. R., and Stevenson, D. W. (2000b). Floral organogenesis in some species of *Tradescantia* and *Callisia* (Commelinaceae). *Int. J. Plant Sci.* **161**, 551–62.

Hardy, C. R., Davis, J. R. and Stevenson, D. W. (2004). Floral organogenesis in *Plowmanianthus* (Commelinaceae). *Int. J. Plant Sci.* **165**, 511–19.

Harris, E. M. (1995). Inflorescence and floral ontogeny in Asteraceae, a synthesis of historical and current concepts. *Bot. Rev.* **61**, 94–278.

Harrison, C. J., Möller, M. and Cronk, Q. C. B. (1999). Evolution and development of floral diversity in *Streptocarpus* and *Saintpaulia*. *Ann. Bot.* **84**, 49–60.

Haston, E. and Ronse De Craene, L. P. (2007). Inflorescence and floral development in *Streptocarpus* and *Saintpaulia* (Gesneriaceae) with particular reference to the impact of bracteole suppression. *Plant Syst. Evol.* **265**, 13–25.

Hayes, V., Schneider, E. L. and Carlquist, S. (2000). Floral development of *Nelumbo nucifera* Nelumbonaceae). *Int. J. Plant Sci.* **161 (6 Suppl.)**, S183–S191.

Heinig, K. H. (1951). Studies in the floral morphology of the Thymelaeaceae. *Am. J. Bot.* **38**, 113–32.

Hiepko, P. (1964). Das zentrifugale Androeceum der Paeoniaceae. *Ber. Dtsch. Bot. Ges.* **77**, 427–35.

Hiepko, P. (1965). Vergleichend-morphologische und entwicklungsgeschichtliche Untersuchungen über das Perianth bei den Polycarpicae. *Bot. Jahrb. Syst.* **84**, 359–508.

Hiepko, P. (1966). Zur Morphologie, Anatomie und Funktion des Diskus der Paeoniaceae. *Ber. Dtsch. Bot. Ges.* **79**, 233–45.

Hileman, L. C., Kramer, E. M. and Baum, D. A. (2003). Differential regulation of symmetry genes and the evolution of floral morphologies. *Proc. Natl. Acad. Sci. USA* **100**, 12814–19.

Hilger, H. H. (1984). Wachstum und Ausbildungsformen des Gynoeceums von *Rochelia* (Boraginaceae). *Plant Syst. Evol.* **146**, 123–39.

Hofmann, U. (1973). Centrospermen-Studien 6, Morphologische Untersuchungen zur Umgrenzung und Gliederung der Aizoaceen. *Bot. Jahrb. Syst.* **93**, 247–324.

Hofmann, U. (1993). Flower morphology and ontogeny. In *Caryophyllales. Evolution and Systematics*, ed. H.-D. Behnke and T. J. Mabry. Berlin: Springer, pp. 123–66.

Hofmann, U., and Göttmann, J. (1990). *Morina* L. und *Triplostegia* Wall. ex DC. im Vergleich mit Valerianaceae und Dipsacaceae. *Bot. Jahrb. Syst.* **111**, 499–553.

Horn, J. W. (2007). Dilleniaceae. In *The families and genera of vascular plants, vol. IX*, ed. K. Kubitzki. Berlin: Springer, pp. 132–54.

Howarth, D. G., and Donoghue, M. J. (2005). Duplications in *cyc*-like genes from Dipsacales correlate with floral form. *Int. J. Plant Sci.* **166**, 357–70.

Hufford, L. D. (1989a). Structure of the inflorescence and flower of *Petalonix linearis* (Loasaceae). *Plant Syst. Evol.* **163**, 211–26.

Hufford, L. D. (1989b). The structure and potential loasaceous affinities of *Schismocarpus*. *Nord. J. Bot.* **9**, 217–27.

Hufford, L. D. (1990). Androecial development and the problem of monophyly of Loasaceae. *Can. J. Bot.* **68**, 402–19.

Hufford, L. D. (1992a). Rosidae and their relationships to other non-magnoliid Dicotyledons, a phylogenetic analysis using morphological and chemical data. *Ann. Mo Bot. Gard.* **79**, 218–48.

Hufford, L. D. (1992b). Floral structure of *Besseya* and *Synthyris* (Scrophulariaceae). *Int. J. Plant Sci.* **153**, 217–29.

Hufford, L. D. (1995). Patterns of ontogenetic evolution in perianth diversification of *Besseya* (Scrophulariaceae). *Am. J. Bot.* **82**, 655–80.

Hufford, L. D. (1998). Early development of androecia in polystemonous Hydrangeaceae. *Am. J. Bot.* **85**, 1057–67.

Hufford, L. D. (2001). Ontogeny and morphology of the fertile flowers of *Hydrangea* and allied genera of tribe Hydrangeeae (Hydrangeaceae). *Bot. J. Linn. Soc.* **137**, 139–87.

Hufford, L. D. (2003). Homology and developmental transformation, models for the origins of the staminodes of Loasaceae subfamily Loasoideae. *Int. J. Plant Sci.* **164 (5 Suppl.)**, S409–S439.

Ickert-Bond, S. M., Gerrath, J. and Wen, J. (2014). Gynoecial structure of Vitales and implications for the evolution of placentation in the Rosids. *Int. J. Plant Sci.* **175**, 998–1032.

Igersheim, A., Buzgo, M. and Endress, P. K. (2008). Gynoecium diversity and systematics in basal monocots. *Bot. J. Linn. Soc.* **136**, 1–65.

Igersheim, A., Puff, C., Leins, P. and Erbar, C. (1994). Gynoecial development of *Gaertnera* Lam. and of presumably allied taxa of the Psychotrieae (Rubiaceae), secondarily 'superior' vs. inferior ovaries. *Bot. Jahrb. Syst.* **116**, 401–14.

Ihlenfeldt, H. D. (1960). Entwicklungsgeschichtliche, morphologische und systematische Untersuchungen an Mesembryanthemen. *Feddes Repert.* **63**, 1–104.

Innes, R. L., Remphrey, W. R. and Lenz, L. M. (1989). An analysis of the development of single and double flowers in *Potentilla fruticosa*. *Can. J. Bot.* **67**, 1071–9.

Irish, V. F. (2009). Evolution of petal identity. *J. Exp. Bot.* **60**, 2517–27.

Ito, M. (1986a). Studies in the floral morphology and anatomy of Nymphaeales III. Floral anatomy of *Brasenia schreberi* Gmel. and *Cabomba caroliniana* A. Gray. *Bot. Mag. Tokyo* **99**, 169–84.

Ito, M. (1986b). Studies in the floral morphology and antomy of Nymphaeales IV. Floral anatomy of *Nelumbo nucifera*. *Acta Phytotax. Geobot.* **37**, 82–96.

Iwamoto, A., Ichigooka, S., Cao, L. and Ronse De Craene, L.P. (2020). Floral development reveals the existence of a fifth staminode on the labellum of basal Globbeae. *Front. Ecol. Evol.* **8**, 133. doi: 10.3389/fevo.2020.00133

Iwamoto, A., Izumidate, R. and Ronse De Craene, L. P. (2015). Floral anatomy and vegetative development in *Ceratophyllum demersum* (Ceratophyllaceae): Morphological picture of an 'unsolved plant'. *Am. J. Bot.* **102**, 1578–89.

Iwamoto, A., Nakamura, A., Kurihara, S., Otani, A. and Ronse De Craene, L. P. (2018). Floral development of petaloid Alismatales as an insight into the origin of the trimerous Bauplan in monocot flowers. *J. Plant Res.* **131**, 395–407.

Iwamoto, A., Shimizu, A. and Ohba, H. (2003). Floral development and phyllotactic variation in *Ceratophyllum demersum* (Ceratophyllaceae). *Am. J. Bot.* **90**, 1124–30.

Jabour, F., Damerval, C. and Nadot, S. (2008). Evolutionary trends in the flowers of Asteridae, is polyandry an alternative to zygomorphy? *Ann. Bot.* **102**, 153–65.

Jabour, F., Ronse De Craene, L. P., Nadot, S. and Damerval, C. (2009). Establishment of zygomorphy on an ontogenic spiral and evolution of perianth in the tribe Delphinieae (Ranunculaceae). *Ann. Bot.* **104**, 809–22.

Jäger-Zürn, I. (1966). Infloreszenz- und blütenmorphologische, sowie embryologische Untersuchungen an *Myrothamnus Welw. Beitr. Biol Pflanz.* **42**, 241–71.

Janka, H., Von Balthazar, M., Alverson, W. S., Baum, D. A., Semir, J. and Bayer, C. (2008). Structure, development and evolution of the androecium in Adansonieae (core Bombacoideae, Malvaceae s.l.). *Plant Syst. Evol.* **275**, 69–91.

Jansen, R. K., Cai, Z., Raubeson, L. A., Daniell, H., dePamphilis, C. W., Leebens-Mack, J., Müller, K. Guisinger-Bellian, M. Haberle, R. C., Hansen, A. K., Chumley, T. W., Lee, S.-B., Peery, R., McNeal, J. R., Kuehl, J. V. and Boore, J. L. (2007). Analysis of 81 genes from 64 plastid genomes resolves relationships in angiosperms and identifies genome-scale evolutionary patterns. *Proc. Natl. Acad. Sci. USA* **104**, 19369–74.

Janssens, S. B., Smets, E. F. and Vrijdaghs, A. (2012). Floral development of *Hydrocera* and *Impatiens* reveals evolutionary trends in the most early diverged lineages of the Balsaminaceae. *Ann. Bot.* **109**, 1285–96.

Jaramillo, M. A., and Kramer, E. M. (2004). *APETALA3* and *PISTILLATA* homologs exhibit novel expression patterns in the unique perianth of *Aristolochia* (Aristolochiaceae). *Evolution and Development* **6**, 449–58.

Jaramillo, M. A., and Manos, P. S. (2001). Phylogeny and patterns of floral diversity in the genus *Piper* (Piperaceae). *Am. J. Bot.* **88**, 706–16.

Jaramillo, M. A., Manos, P. and Zimmer, E. A. (2004). Phylogenetic relationships of the perianthless Piperales, reconstructing the evolution of floral development. *Int. J. Plant Sci.* **165**, 403–16.

Jeiter, J., Danisch, F. and Hilger, H. H. (2016). Polymery and nectary chambers in *Codon* (Codonaceae): Flower and fruit development in a small, capsule-bearing family of Boraginales. *Flora* **220**, 94–102.

Jeiter, J., Langecker, S. and Weigend, M. (2020). Towards an integrative understanding of stamen-corolla tube modifications and floral architecture in Boraginaceae s.s. (Boraginales). *Bot. J. Linn. Soc.* **193**, 100–24.

Jeiter, J., Staedler, Y. M., Schönenberger, J., Weigend, M. and Luebert, F. (2018). Gynoecium and fruit development in *Heliotropium* sect. *Heliothamnus* (Heliotropiaceae). *Int. J. Plant Sci.* **179**, 275–86.

Jeiter, J., Weigend, M. and Hilger, H. H. (2017). Geraniales flowers revisited: Evolutionary trends in floral nectaries. *Ann. Bot.* **119**, 395–408.

Jenny, M. (1988). Different gynoecium types in Sterculiaceae, ontogeny and functional aspects. In *Aspects of floral development*, ed. P. Leins, S. C. Tucker and P. K. Endress. Berlin: J. Cramer, pp. 225–36.

Judd, W. S., Campbell, C. S., Kellogg, E. A., Stevens, P. F. and Donoghue, M. J. (2002). *Plant systematics, a phylogenetic approach.* 2nd ed. Sunderland, Mass.: Sinauer.

Judd, W. S., and Olmstead, R. G. (2004). A survey of tricolpate (eudicot) phylogenetic relationships. *Am. J. Bot.* **91**, 1627–44.

Judd, W. S., Sanders, R. W. and Donoghue, M. J. (1994). Angiosperm family pairs, preliminary phylogenetic analyses. *Harvard Pap. Bot.* **5**, 1–51.

Juncosa, A. M. (1988). Floral development and character evolution in Rhizophoraceae. In *Aspects of floral development*, ed. P. Leins, S. C. Tucker and P. K. Endress. Vaduz (Liechtenstein): Cramer, pp. 83–101.

Juncosa, A. M., and Tomlinson, P. B. (1987). Floral development in mangrove Rhizophoraceae. *Amer. J. Bot.* **74**, 1263–79.

Källersjö, M., Bergqvist, G. and Anderberg, A. A. (2000). Generic realignment in primuloid families of the Ericales s.l., a phylogenetic analysis based on DNA sequences from three chloroplast genes and morphology. *Am. J. Bot.* **87**, 1325–41.

Kania, W. (1973). Entwicklungsgeschichtliche Untersuchungen an Rosaceenblüten. *Bot. Jahrb. Syst.* **93**, 175–246.

Kanno, A., Nakada, M., Akita, Y. and Hirae, M. (2007). Class B gene expression and the modified ABC model in nongrass monocots. *The Scientific World Journal* **7**, 268–79.

Karpunina, P. V., Oskolski, A. A., Nuraliev, M. S., Lowry II, P. P., Degtjavera, G. V., Samigullin, T. H., Valiejo-Roman, C. M. and Sokoloff, D. D. (2016). Gradual vs abrupt reduction of carpels in syncarpous gynoecia: A case study from *Polyscias* subg. *Arthrophyllum* (Araliaceae: Apiales). *Am. J. Bot.* **103**, 2028–57.

Karrer, A. B. (1991). Blütenentwicklung und systematische Stellung der Papaveraceae und Capparaceae. University of Zürich (Switzerland): Unpublished thesis.

Kaul, R. B. (1968). Floral morphology and phylogeny in the Hydrocharitaceae. *Phytomorphology* **18**, 13–35.

Keller, J. A., Herendeen, P. S. and Crane, P. R. (1996). Fossil flowers and fruits of the Actinidiaceae from the Campanian (late Cretaceous) of Georgia. *Am. J. Bot.* **83**, 528–41.

Kelly, L. M. (2001). Taxonomy of *Asarum* section *Asarum* (Aristolochiaceae). *Syst. Bot.* **26**, 17–53.

Kelly, L. M., and González, F. (2003). Phylogenetic relationships in Aristolochiaceae. *Syst. Bot.* **28**, 236–49.

Kessler, P. J. A. (1993). Menispermaceae. In *The families and genera of vascular plants vol. II*, ed. K. Kubitzki, J. G. Rohwer and V. Bittrich. Berlin: Springer, pp. 402–18.

Kim, S., Yoo, M.-J., Kong, H., Hu, Y., Ma, H., Soltis, P. S., and Soltis, D. E. (2005). Expression of floral MADS-box genes in basal Angiosperms, implications for the evolution of floral regulators. *Plant J.* **43**, 724–44.

Kirchoff, B. K. (1983). Floral organogenesis in five genera of the Marataceae and in *Canna* (Cannaceae). *Am. J. Bot.* **70**, 508–23.

Kirchoff, B. K. (1988). Inflorescence and flower development in *Costus scaber* (Costaceae). *Can. J. Bot.* **62**, 339–45.

Kirchoff, B. K. (1992). Ovary structure and anatomy in the Heliconiaceae and Musaceae (Zingiberales). *Can. J. Bot.* **70**, 2490–2508.

Kirchoff, B. K. (1997). Inflorescence and flower development in the Hedychieae (Zingiberaceae), *Hedychium. Can. J. Bot.* **75**, 581–94.

Kirchoff, B. K. (2000). Hofmeister's rule and primordium shape, influences on organ position in *Hedychium coronarium* (Zingiberaceae). In *Monocots, systematics and evolution*, ed. K. L. Wilson and D. A. Morrison. Melbourne: CSIRO, pp. 75–83.

Kirchoff, B. K., Liu, H. and Liao, J.-P. (2020). Inflorescence and flower development in *Orchidantha chinensis* T. L. Wu (Lowiaceae; Zingiberales): Similarities to inflorescence structure in the Strelitziaceae. *Int. J. Plant Sci.* **181**, 716–31.

Klopfer, K. (1973). Florale Morphogenese und Taxonomie der Saxifragaceae sensu lato. *Feddes Repert.* **84**, 475–516.

Knapp, S. (2002). Floral diversity and evolution in the Solanaceae. In *Developmental genetics and plant evolution*, ed. Q. C. B. Cronk, R. M. Bateman and J. A. Hawkins. London: Taylor and Francis, pp. 267–97.

Kocyan, A., and Endress, P. K. (2001a). Floral structure and development and systematic aspects of some 'lower' Asparagales. *Plant Syst. Evol.* **229**, 187–216.

Kocyan, A., and Endress, P. K. (2001b). Floral structure and development of *Apostasia* and *Neuwiedia* (Apostasioideae) and their relationships to other Orchidaceae. *Int. J. Plant Sci.* **162**, 847–67.

Koethe, S., Bloemer, J. and Lunau, K. (2017). Testing the influence of gravity on flower symmetry in five *Saxifraga* species. *Sci. Nat.* **104**, 37. doi: 10.1007/s00114-017-1458-4

Köhler, E. (2003). Simmondsiaceae. In *The families and genera of vascular plants vol. II*, ed. K. Kubitzki and C. Bayer. Berlin: Springer, pp. 355–8.

Kong, D.-R., Schori, M., Li, L. and Peng, H. (2018). Floral development of *Gonocaryum* with emphasis on the gynoecium. *Plant Syst. Evol.* **304**, 327–41.

Kosuge, K. (1994). Petal evolution in Ranunculaceae. *Plant Syst. Evol.* **Suppl. 8**, 185–91.

Kramer, E. M, Di Stilio, V. S. and Schlüter, P. M. (2003). Complex patterns of gene duplication in the *apetala3* and *pistillata* lineages of the Ranunculaceae. *Int. J. Plant Sci.* **164**, 1–11.

Kramer, E. M., Su, H.-J. Wu, C.-C., Hu, J.-M. (2006). A simplified explanation for the frameshift mutation that created a novel C-terminal motif in the *APETALA3* gene lineage. *BMC Evolutionary Biology* **6**, 30.

Kress, W. J. (1990). Phylogeny and classification of Zingiberales. *Ann. Mo. Bot. Gard.* **77**, 698–721.

Kron, K. A. Judd, W. S., Stevens, P. F., Crayn, D. M., Anderberg, A. A., Gadek, P. A., Quinn, C. J. and Luteyn, J. L. (2002). A phylogenetic classification of Ericaceae, molecular and morphological evidence. *Bot. Rev.* **68**, 335–423.

Krosnick, S. E., Harris, E. M. and Freudenstein, J. V. (2006). Patterns of anomalous floral development in the Asian *Passiflora* (subgenus *Decaloba*, supersection *Disemma*). *Am. J. Bot.* **93**, 620–36.

Krüger, H., and Robbertse, P. J. (1988). Floral ontogeny of *Securidaca longepedunculata* Fresen (Polygalaceae) including inflorescence morphology. In *Aspects of floral development*, ed. P. Leins, S. C. Tucker and P.K. Endress. Berlin: J. Cramer, pp. 159–67.

Kshetrapal, S. (1970). A contribution to the vascular anatomy of the flower of certain species of the Salvadoraceae. *J. Indian Bot. Soc.* **49**, 92–9.

Kubitzki, K. (1969). Monographie der Hernandiaceen. *Bot. Jahrb. Syst.* **89**, 78–148.

Kubitzki, K. (1987). Origin and significance of trimerous flowers. *Taxon* **36**, 21–8.

Kubitzki, K. (1993). Hernandiaceae. In *The families and genera of vascular plants vol. II*, ed. K. Kubitzki, J. G. Rohwer and V. Bittrich. Berlin: Springer, pp. 334–8.

Kubitzki, K. (2003) Salvadoraceae. In *The families and genera of vascular plants vol. V*, ed. K. Kubitzki. Berlin: Springer, pp. 342–4.

Kubitzki, K. (2007). Berberidopsidaceae. In *The families and genera of vascular plants vol. IX*, ed. K. Kubitzki. Berlin: Springer, pp. 33–5.

Kuijt, J. (2013). Prophyll, calyculus, and perianth in Santalales. *Blumea* **57**, 248–52.

Kümpers, B. M. C., Richardson, J. E, Anderberg, A. A., Wilkie, P. and Ronse De Craene, L. P. (2016). The significance of meristic changes in the flowers of Sapotaceae. *Bot. J. Linn. Soc.* **180**, 161–92.

Kuzoff, R. K., Hufford, L. and Soltis, D. E. (2001). Structural homology and developmental transformations associated with ovary diversification in *Lithophragma* (Saxifragaceae). *Am. J. Bot.* **88**, 196–205.

Lamb-Frye, A. S. and Kron, K. A. (2003). Phylogeny and character evolution in Polygonaceae. *Syst. Bot.* **21**, 17–29.

Landis, J. B., Barnett, L. L. and Hileman, L. C. (2012). Evolution of petaloid sepals independent of shifts in B-class MADS box gene expression. *Dev. Genes Evol.* **222**, 19–28.

Landrein, S., and Prenner, G. (2013). Unequal twins? Inflorescence evolution in the twinflower tribe Linnaeeae (Caprifoliaceae s.l.). *Int. J. plant Sci.* **174**, 200–33.

Laubengayer, R. A. (1937). Studies in the anatomy and morphology of the Polygonaceous flower. *Am. J. Bot.* **24**, 329–43.

Legume Phylogeny Working Group (2017). A new subfamily classification of the leguminosae based on a taxonomically comprehensive phylogeny. *Taxon* **66**: 44–77.

Lehmann, N. L., and Sattler, R. (1992). Irregular floral development in *Calla palustris* (Araceae) and the concept of homeosis. *Am. J. Bot.* **79**, 1145–57.

Lehmann, N. L., and Sattler, R. (1993). Homeosis in floral development of *Sanguinaria canadensis* and *S. canadensis* 'Multiplex' (Papaveraceae). *Am. J. Bot.* **80**, 1323–35.

Lehmann, N. L., and Sattler, R. (1994). Floral development and homeosis in *Actaea rubra* (Ranunculaceae). *Int. J. Plant Sci.* **155**, 658–71.

Lei, L.-G., and Liang, H.-X. (1998). Floral development of dioecious species and trends of floral evolution in *Piper sensu lato*. *Bot. J. Linn. Soc.* **127**, 225–37.

Leinfellner, W. (1950). Der Bauplan des synkarpen Gynoeceums. *Österr. Bot. Zeit.* **97**, 403–36.

Leins, P. (1964a). Entwicklungsgeschichtliche Studien an Ericales-Blüten. *Bot. Jahrb. Syst.* **83**, 57–88.

Leins, P. (1964b). Die frühe Blütenentwicklung von *Hypericum hookerianum* Wight et Arn. und *H. aegypticum* L. *Ber. Dtsch. Bot. Ges.* **77**, 112–23.

Leins, P. (1967). Die frühe Blütenentwicklung von *Aegle marmelos* (Rutaceae). *Ber. Dtsch. Bot. Ges.* **80**, 320–5.

Leins, P. (1988). Das zentripetale Androeceum von *Punica*. *Bot. Jahrb. Syst.* **109**, 555–61.

Leins, P., and Erbar, C. (1985). Ein Beitrag zur Blütenentwicklung der Aristolochiaceen, einer Vermittlergruppe zu den Monokotylen. *Bot. Jahrb. Syst.* **107**, 343–68.

Leins, P., and Erbar, C. (1987). Studien zur Blütenentwicklung an Compositen. *Bot. Jahrb. Syst.* **108**, 381–401.

Leins, P., and Erbar, C. (1988). Einige Bemerkungen zur Blütenentwicklung und systematische Stellung der Wasserpflanzen *Callitriche, Hippuris* und *Hydrostachys*. *Beitr. Biol. Pfl.* **63**, 157–78.

Leins, P., and Erbar, C. (1989). Zur Blütenentwicklung und sekundären Pollenpräsentation bei *Selliera radicans*. Cav. (Goodeniaceae). *Flora* **182**, 43–56.

Leins, P., and Erbar, C. (1991). Fascicled androecia in Dilleniidae and some remarks on the *Garcinia* androecium. *Bot. Acta* **104**, 336–44.

Leins, P., and Erbar, C. (1995). Das frühe Differenzierungsmuster in den Blüten von *Saruma henryi* Oliv. (Aristolochiaceae). *Bot. Jahrb. Syst.* **117**, 365–76.

Leins, P., and Erbar, C. (1996). Early floral developmental studies in Annonaceae. In *Reproductive morphology in Annonaceae*, ed. W. Morawetz and H. Winkler. Akademie der Wissenschaften, Biosystematics and Ecology Series 10. Wien: Österr, pp. 1–27.

Leins, P., and Erbar, C. (2000). Die frühesten Entwicklungsstadien der Blüten bei den Asteraceae. *Bot. Jahrb. Syst.* **122**, 503–15.

Leins, P., and Erbar, C. (2010). *Flower and fruit. Morphology, ontogeny, phylogeny, function and ecology*. Stuttgart: Schweizerbart Science.

Leins, P., Erbar, C. and Van Heel, W. A. (1988). Note on the floral development of *Thottea* (Aristolochiaceae). *Blumea* **33**, 357–70.

Leins, P., and Galle, P. (1971). Entwicklungsgeschichtliche Untersuchungen an Cucurbitaceen-Blüten. *Österr. Bot. Zeit.* **119**, 531–48.

Leins, P., and Schwitalla, S. (1985). Studien an Cactaceen-Blüten I. Einige Bermerkungen zur Blütenentwicklung von *Pereskia*. *Beitr. Biol. Pflanz.* **60**, 313–23.

Leins, P., and Stadler, P. (1973). Entwicklungsgeschichtliche Untersuchungen am Androecium der Alismatales. *Österr. Bot. Zeit.* **122**, 145–65.

Leins, P., and Winhard, W. (1973). Entwicklungsgeschichtliche Studien an Loasaceenblüten. *Österr. Bot. Zeit.* **122**, 145–65.

Leite, V. G., Mansano, V. F. and Teixeira, S. P. (2018). Floral development of Moraceae species with emphasis on the perianth and androecium. *Flora* **240**, 116–32.

Leme, F. M., Staedler, Y. M., Schönenberger, J. and Teixeira, S. P. (2018). Ontogeny and vascularization elucidate the atypical floral structure of *Ampelocera glabra*, a tropical species of Ulmaceae. *Int. J. Plant Sci.* **179**, 461–76.

Leredde, C. (1955). Sur la position des étamines chez quelques *Echium. Bull. Soc. Hist. Nat. Toulouse* **90**, 369–72.

Le Roux, L. G., and Kellogg, E. A. (1999). Floral development and the formation of unisexual spikelets in the Andropogoneae (Poaceae). *Am. J. Bot.* **86**, 354–66.

Levyns, M. R. (1949). The floral morphology of some South African members of Polygalaceae. *J. S. Afr. Bot.* **15**, 79–92.

Leyser, O., and Day, S. (2003). *Mechanisms in plant development*. Oxford: Blackwell.

Li, P., and Johnston, M. O. (2000). Heterochrony in plant evolutionary studies through the twentieth century. *Bot. Rev.* **66**, 57–88.

Liang, H.-X., and Tucker, S. C. (1989). Floral development in *Gymnotheca chinensis*. *Am. J. Bot.* **76**, 806–19.

Lin, R.-Z., Li, R.-Q., Lu, A.-M., Zhu, Y.-Y. and Chen Z.-D. (2016). Comparative flower development of *Juglans regia*, *Cyclocarya paliurus* and *Engelhardia spicata*: Homology of floral envelopes in Juglandaceae. *Bot. J. Linn. Soc.* **181**, 279–93.

Lin, R.-Z., Zeng, J. and Chen, Z.-D. (2010). Organogenesis of reproductive structures in *Betula alnoides* (Betulaceae). *Int. J. Plant Sci.* **171**, 586–94.

Lindenhofer, A., and Weber, A. (1999a). Polyandry in Rosaceae, evidence for a spiral origin of the androecium in Spiraeoideae. *Bot. Jahrb. Syst.* **121**, 553–82.

Lindenhofer, A., and Weber, A. (1999b). The spiraeoid androecium of Pyroideae and Amygdaloideae (Rosaceae). *Bot. Jahrb. Syst.* **121**, 583–605.

Lindenhofer, A., and Weber, A. (2000). Structural and developmental diversity of the androecium of Rosoideae (Rosaceae). *Bot. Jahrb. Syst.* **122**, 63–91.

Linder, H. P. (1991). A review of the southern African Restionaceae. *Contr. Bolus Herb.* **13**, 209–64.

Linder, H. P. (1992a). The gynoecia of Australian Restionaceae, morphology, anatomy and systematic implications. *Aust. Syst. Bot.* **5**, 227–45.

Linder, H. P. (1992b). The structure and evolution of the pistillate flower of the African Restionaceae. *Bot. J. Linn. Soc.* **109**, 401–25.

Linder, H. P. (1998). Morphology and the evolution of wind pollination. In *Reproductive biology in systematics, conservation and economic botany*, ed. S. J. Owens and P. J. Rudall. Kew: Royal Botanic Gardens, pp. 123–35.

Linder, H. P. and Rudall, P. J. (2005). Evolutionary history of Poales. *Ann. Rev. Ecol. Syst.* **36**, 107–24.

Lindsey, A. A. (1940). Floral anatomy in the Gentianaceae. *Am. J. Bot.* **27**, 640–52.

Litt, A., and Kramer, E. M. (2010). The ABC model and the diversification of floral organ identity. *Sem. Cell Dev. Biol.* **21**, 129–37.

Litt, A., and Stevenson, D. W. (2003a). Floral development and morphology of Vochysiaceae. I. The structure of the gynoecium. *Am. J. Bot.* **90**, 1533–47.

Litt, A., and Stevenson, D. W. (2003b). Floral development and morphology of Vochysiaceae. II. The position of the single fertile stamen. *Am. J. Bot.* **90**, 1548–59.

Löfstrand, S. D., and Schönenberger, J. (2015). Comparative floral structure and systematics in the sarracenioid clade (Actinidiaceae, Roridulaceae and Sarraceniaceae) of Ericales. *Bot. J. Linn. Soc.* **178**, 1–46.

López, J., Rodríguez-Riaño, T., Valtueña, F. J., Pérez, J. L., González, M. and Ortega-Olivencia, A. (2016). Does the *Scrophularia* staminode influence female and male functions during pollination? *Int. J. Plant Sci.* **177**, 671–81.

Lorence, D. H. (1985). A monograph of the Monimiaceae (Laurales) in the Malagasy region (Southwest Indian Ocean). *Ann. Mo. Bot. Gard.* **72**, 1–165.

Lüders, H. (1907). Systematische Untersuchungen über die Caryophyllaceen mit einfachem Diagramm. *Bot. Jahrb. Syst.* **40, Beibl. 91**, 1–37.

Luo, D., Carpenter, R., Vincent, C., Copsey, L. and Coen, E. (1996). Origin of floral asymmetry in *Antirrhinum. Nature* **383**, 794–9.

Ma, O. S. W., and Saunders, R. M. K. (2003). Comparative floral ontogeny of *Maesa* (Maesaceae), *Aegiceras* (Myrsinaceae) and *Embelia* (Myrsinaceae), taxonomic and phylogenetic implications. *Plant Syst. Evol.* **243**, 39–58.

Maas, P. J. M., and Rübsamen, T. (1986). *Triuridaceae, Flora neotropica no. 40*. New York: Hafner.

Mabberley, D. (2000). *Arthur Harry Church. The anatomy of flowers*. London: The Natural History Museum.

Mabry, T. J. (1977). The order Centrospermae. *Ann. Mo Bot. Gard.* **64**, 210–20.

Macfarlane, J. M. (1908). Nepenthaceae. In *Das Pflanzenreich IV, 3*, ed. A. Engler. Leipzig: W. Engelmann, pp. 1–92.

MacMahon, M., and Hufford, L. (2005). Evolution and development in the Amorphoid clade (Amorpheae: Papilionoideae: Leguminosae): Petal loss and dedifferentiation. *Int. J. Plant Sci.* **166**, 383–96.

Magallón, S. (2007). From fossils to molecules, phylogeny and the core eudicot floral groundplan in Hamamelidoideae (Hamamelidaceae, Saxifragales). *Syst. Bot.* **32**, 317–47.

Malécot, V., and Nickrent, D. L. (2008). Molecular phylogenetic relationships of Olacaceae and related Santalales. *Syst. Bot.* **33**, 97–106.

Manchester, S. R., Dilcher, D. L., Judd, W. S., Corder, B. and Basinger, J. F. (2018). Early eudicot flower and fruits: *Dakotanthus* gen. nov. from the Cretaceous Dakota Formation of Kansas and Nebraska, USA. *Acta Palaeobot.* **58**, 27–40.

Manning, W. E. (1948). The morphology of the flowers of the Juglandaceae. III. The staminate flowers. *Am. J. Bot.* **35**, 606–21.

Marazzi, B., and Endress, P. K. (2008). Patterns and development of floral asymmetry in *Senna* (Leguminosae, Cassiinae). *Am. J. Bot.* **95**, 22–40.

Martínez-Millán, M., Crepet, W. L. and Nixon, K. C. (2009). *Pentapetalum trifasciculandricus* gen. et sp. nov., a thealean fossil flower from the Raritan Formation, New Jersey, USA (Turonian, Late Cretaceous). *Am. J. Bot.* **96**, 933–49.

Matthews, M. L., Amaral, M. C. E. and Endress, P. K. (2012). Comparative floral structure and systematics in Ochnaceae *s.l.* (Ochnaceae, Quiinaceae and Medusagynaceae; Malpighiales). *Bot. J. Linn. Soc.* **170**, 299–392.

Matthews, M. L., and Endress, P. K. (2002). Comparative floral structure and systematics in Oxalidales (Oxalidaceae, Connaraceae, Brunelliaceae, Cephalotaceae, Cunoniaceae, Elaeocarpaceae, Tremandraceae). *Bot. J. Linn. Soc.* **140**, 321–81.

Matthews, M. L., and Endress, P. K. (2004). Comparative floral structure and systematics in Cucurbitales (Corynocarpaceae, Coriariaceae, Tetramelaceae, Datiscaceae, Begoniaceae, Cucurbitaceae, Anisophylleaceae). *Bot. J. Linn. Soc.* **145**, 129–85.

Matthews, M. L., and Endress, P. K. (2005). Comparative floral structure and systematics in Celastrales (Celastraceae, Parnassiaceae, Lepidobotryaceae). *Bot. J. Linn. Soc.* **149**, 129–94.

Matthews, M. L., and Endress, P. K. (2008). Comparative floral struture and systematics in Chrysobalanaceae s.l. (Chrysobalanaceae, Dichapetalaceae, Euphroniaceae, Trigoniaceae; Malpighiales). *Bot. J. Linn. Soc.* **157**, 249–309.

Matthews, M. L., and Endress, P. K. (2011). Comparative floral structure and systematics in Rhizophoraceae, Erythroxylaceae and the potentially related Ctenolophonaceae, Linaceae, Irvingiaceae and Caryocaraceae. *Bot. J. Linn. Soc.* **166**, 331–416.

Mayr, B. (1969). Ontogenetische Studien an Myrtales-Blüten. *Bot. Jahrb. Syst.* **89**, 210–71.

Mayr, E. M., and Weber, A. (2006). Calceolariaceae: Floral development and systematic implications. *Am. J. Bot.* **93**, 327–43.

Medan, D., and Hilger, H. H. 1992. Comparative flower and fruit morphogenesis in *Colubrina* (Rhamnaceae) with special reference to *C. asiatica*. *Am. J. Bot.* **79**, 809–19.

Melchior, H. (1925). Violaceae. In *Die natürlichen Pflanzenfamilien XXI*, 2nd edn. ed. A. Engler and K. Prantl. Leipzig: Wilhelm Engelmann, pp. 329–77.

Melchior, H. (1964). *Engler's Syllabus der Pflanzenfamilien*. Berlin: Gebr. Borntraeger.

Melville, R. (1984). The affinity of *Paeonia* and a second genus of Paeoniaceae. *Kew Bull.* **38**, 87–105.

Melzer, R., Wang, Y.-Q. and Theißen, G. (2010). The naked and the dead: The ABCs of gymnosperm reproduction and the origin of the angiosperm flower. *Sem. Cell Dev. Biol. 21*, 118–28.

Meng, A., Zhang, Z., Li, J., Ronse De Craene, L. and Wang, H. (2012). Floral development of *Stephania* (Menispermaceae): Impact of organ reduction on symmetry. *Int. J. Plant Sci.* **173**, 861–74.

Merckx, V., Schols, V., Maas-van de Kamer, H., Maas, P., Huysmans, S., and Smets, E. (2006). Phylogeny and evolution of Burmanniaceae (Dioscoreales) based on nuclear and mitochondrial data. *Am. J. Bot.* **93**, 1684–98.

Merino Sutter, D., Foster, P. I. and Endress, P. K. (2006). Female flowers and systematic position of Picodendraceae (Euphorbiaceae s.l., Malpighiales). *Plant Syst. Evol.* **261**, 187–215.

Michaelis, P. (1924). Blütenmorphologische Untersuchungen an den Euphorbiaceen, unter besonderer Berücksichtigung der Phylogenie der Angiospermenblüte. *Goebel Bot. Abhandl.* **3**, 1–150.

Milby, T. H. (1980). Studies in the floral anatomy of *Claytonia*. *Am. J. Bot.* **67**, 1046–50.

Mione, T., and Bogle, A. L. (1990). Comparative ontogeny of the inflorescence and flower of *Hamamelis virginiana* and *Loropetalum chinense* (Hamamelidaceae). *Am. J. Bot.* **77**, 77–91.

Mitchell, C. H., and Diggle, P. K. (2005). The evolution of unisexual flowers: Morphological and functional convergence results from diverse developmental transitions. *Am. J. Bot.* **92**, 1068–76.

Monniaux, M., and Vandenbussche, M. (2018). How to evolve a perianth: A review of cadastral mechanisms for perianth identity. *Front. Plant Sci.* **9**, 1573. doi: 10.3389/fpls.2018.01573

Moody, M., and Hufford, L. (2000a). Floral development and structure of *Davidsonia* (Cunoniaceae). *Can. J. Bot.* **78**, 1034–43.

Moody, M., and Hufford, L. (2000b). Floral ontogeny and morphology of *Cevallia, Fuertesia,* and *Gronovia* (Loasaceae subfamily Gronovioideae). *Int. J. Plant Sci.* **161**, 869–83.

Moore, H. E. (1973). The major groups of palms and their distribution. *Gentes Herb.* **11**: 27–141.

Moore, H. E., and Uhl, N. W. (1982). Major trends of evolution in palms. *Bot. Rev.* **48**, 1–69.

Moore, M. J., Bell, C. D., Soltis, P. S., and Soltis, D. E. (1997). Using plastid genome-scale data to resolve enigmatic relationships among basal angiosperms. *Proc. Natl. Acad. Sci. USA* **104**, 19363–8.

Moore, M. J., Soltis, P. S., Bell, C. D., Burleigh, G. and Soltis, D. E. (2010). Phylogenetic analysis of 83 plastid genes further resolves the early diversification of eudicots. *Proc. Nat. Acad. Sci. USA* **107**, 4623–8.

Morgan, D. R., and Soltis, D. E. (1993). Phylogenetic relationships among members of Saxifragaceae *sensu lato* based on *rbc*L sequence data. *Ann. Mo Bot. Gard.* **80**, 631–60.

Mort, M. E., Soltis, D. E., Soltis, P. S., Francisco-Ortega, J. and Santos-Guerra, A. (2001). Phylogenetic relationships and evolution of Crassulaceae inferred from *matK* sequence data. *Am. J. Bot.* **88**, 76–91.

Moylan, E. C., Rudall, P. J. and Scotland, R. W. (2004). Comparative floral anatomy of Strobilanthinae (Acanthaceae), with particular reference to internal partitioning of the flower. *Plant Syst. Evol.* **249**, 77–98.

Murbeck, S. (1912). Üntersuchungen über den Blütenbau der Papaveraceen. *Kungl. Sv. Vet. Akad. Handl.* **50**, 1–168.

Naghiloo, S. (2020). Patterns of symmetry expression in angiosperms: Developmental and evolutionary lability. *Front. Ecol. Evol.* **8**: 104. doi: 10.3389/fevo.2020.00104

Naghiloo, S., and Claßen-Bockhoff, R. (2017). Developmental changes in time and space promote evolutionary diversification of flowers: A case study in Dipsacoideae. *Front. Pl. Sci.* **8**, 1665. doi: 10.3389/fpls.2017.01665

Nair, N. C., and Abraham, V. (1962). Floral morphology of a few species of Euphorbiaceae. *Proc. Indian Acad. Sci.* **B,56**, 1–12.

Nandi, O. I. (1998). Floral development and systematics of Cistaceae. *Plant Syst. Evol.* **212**, 107–34.

Narayana, R. (1958a). Morphological and embryological studies in the family Loranthaceae: II. *Lysiana exocarpi* (Behr.) van Tieghem. *Phytomorphology* **8**, 147–68.

Narayana, R. (1958b). Morphological and embryological studies in the family Loranthaceae: III. *Nuytsia floribunda* (Labill.) R. Br. *Phytomorphology* **8**, 306–23.

Narayana, L. L., and Rao, D. (1969). Contributions to the floral anatomy of Linaceae 1. *J. Jap. Bot.* **44**, 289–94.

Narayana, L. L., and Rao, D. (1973). Contributions to the floral anatomy of Linaceae 5. *J. Jap. Bot.* **48**, 205–8.

Narayana, L. L., and Rao, D. (1976). Contributions to the floral anatomy of Linaceae 6. *J. Jap. Bot.* **51**, 92–6.

Narita, M., and Takahashi, H. (2008). A comparative study of shoot and floral development in *Paris tetraphylla* and *P. verticillata* (Trilliaceae). *Plant Syst. Evol.* **272**, 67–78.

Nepi, M., and Pacini, E. (1993). First observations on nectaries and nectar of *Cucurbita pepo*. *Giorn. Bot. Ital.* **127**, 1208–10.

Newman, S. W. H., and Kirchoff, B. K. (1992). Ovary structure in the Costaceae (Zingiberales). *Int. J. Plant Sci.* **153**, 471–87.

Ng, F. (1991). The relationship of the Sapotaceae within the Ebenales. In *The genera of Sapotaceae*, ed. T. D. Pennington. Kew: Royal Botanic Garden; Bronx: New York Botanic Garden, pp. 1–14.

Nicholas, A., and Baijnath, H. (1994). A consensus classification for the order Gentianales with additional details on the suborder Apocynineae. *Bot. Rev.* **60**, 440–82.

Nickrent, D. L., Malécot, V., Vidal-Russell, R. and Der, J. P. (2010). A revised classification of Santalales. *Taxon* **59**, 538–58.

Niedenzu, F. (1897). Malpighiaceae. In *Die natürlichen Pflanzenfamilien III, 4*, 1st edn., ed. A. Engler and K. Prantl. Leipzig: W. Engelmann, pp. 41–74.

Niedenzu, F. (1925). Frankeniaceae. In *Die natürlichen Pflanzenfamilien 21*, 2nd edn., ed. A. Engler and K. Prantl. Leipzig: W. Engelmann, pp. 276–81.

Nishino, E. (1988). Early floral organogenesis in *Tripetaleia* (Ericaceae). In *Aspects of floral development*, ed. P. Leins, S. C. Tucker and P. K. Endress. Berlin: J. Cramer, pp. 181–90.

Nooteboom, H. P. (1993). Magnoliaceae. In *The families and genera of vascular plants II*, ed. K. Kubitzki, J. G. Rohwer and V. Bittrich. Berlin: Springer, pp. 391–401.

Nuraliev, M. S., Degtajareva, G. V., Sokoloff, D. D., Oskolski, A. A., Samigullin, T. H. and Valiejo-Roman, C. M. (2014). Flower morphology and relationships of *Schefflera subintegra* (Araliaceae, Apiales): An evolutionary step towards extreme floral polymery. *Bot. J. Linn. Soc.* **175**, 553–97.

Nuraliev, M. S., Oskolski, A. A., Sokoloff, D. D. and Remizowa, M. V. (2010). Flowers of Araliaceae: Structural diversity, developmental and evolutionary aspects. *Plant Div. Evol.* **128**, 247–68.

Nyffeler, R., and Eggli, U. (2010). Disintegrating Portulacaceae: A new familial classification of the suborder Portulacineae (Caryophyllales) based on molecular and morphological data. *Taxon* **59**, 227–40.

Ochoterena, H., Vrijdaghs, A., Smets, E. and Claßen-Bockhoff, R. (2019). The search for common origin: Homology revisited. *Syst. Biol.* **68**, 767–80.

Oh, S.-H., and Manos, P. S. (2008). Molecular phylogenetics and cupule evolution in Fagaceae as inferred from nuclear *CRABS CLAW* sequences. *Taxon* **57**, 434–51.

Okamoto, M. (1983). Floral development of *Castanopsis cuspidata* var. *sieboldii*. *Acta Phytotax. Geobot.* **34**, 10–17.

Okamoto, M., Kosuge, K. and Fukuoka, N. (1992). Pistil development and parietal placentation in the pseudomomerous ovary of *Zelkova serrata* (Ulmaceae). *Am. J. Bot.* **79**, 921–7.

Olmstead, R. G., DePamphilis, C. W., Wolfe, A. D., Young, N. D., Elisons, W. J. and Reeves, P. A. (2001). Disintegration of the Scrophulariaceae. *Am. J. Bot.* **88**, 348–61.

Olson, M. E. (2003). Ontogenetic origins of floral bilateral symmetry in Moringaceae (Brassicales). *Am. J. Bot.* **90**, 49–71.

Ono, A., Dohzono, I. and Sugawara, T. (2008). Bumblebee pollination and reproductive biology of *Rhododendron semibarbatum* (Ericaceae). *J. Plant Res.* **121**, 319–27.

Orlovich, D. A., Drinnan, A. N. and Ladiges, P. Y. (1996). Floral development in the *Metrosideros* group (Myrtaceae) with special emphasis on the androecium. *Telopea* **6**, 689–719.

Orlovich, D. A., Drinnan, A. N. and Ladiges, P. Y. (1999). Floral development in *Melaleuca* and *Callistemon* (Myrtaceae). *Aust. Syst. Bot.* **11**, 689–710.

Pabón-Mora, N., and González, F. (2008). Floral ontogeny of *Telipogon spp.* (Orchidaceae) and insights on the perianth symmetry in the family. *Int. J. Plant Sci.* **169**, 1159–73.

Pacini, E., Nepi, M. and Vesprini, J. L. (2003). Nectar biodiversity: A short review. *Plant Syst. Evol.* **238** 7–21.

Pai, R. M. (1965). Morphology of the flower in the Cannaceae. *J. Biol. Sci.* **8**, 4–8.

Pai, R. M., and Tilak, V. D. (1965). Septal nectaries in the Scitamineae. *J. Biol. Sci.* **8**, 1–3.

Palazzesi, L., Gottschling, M., Barreda, V. and Weigend, M. (2012). First Miocene fossils of Vivianiaceae shed new light on phylogeny, divergence times, and historical biogeography of Geraniales. *Biol. J. Linn. Soc.* **107**, 67–85.

Patchell, M. J., Bolton, M. C., Mankowski, P. and Hall, J. C. (2011). Comparative floral development in Cleomaceae reveals two distinct pathways leading to monosymmetry. *Int. J. Plant Sci.* **172**, 352–65.

Paulino, J. V., Prenner, G., Mansano, V. F. and Teixeira, S. P. (2014). Comparative development of rare cases of a polycarpellate gynoecium in an otherwise monocarpellate family, Leguminosae. *Am. J. Bot.* **101**, 572–86.

Pauwels, L. (1993). *Nzayilu N'ti. Guide des arbres et arbustes de la région de Kinshasa-Brazzaville*. Meise: Jardin Botanique National de Belgique.

Pauzé, F., and Sattler, R. (1978). L'Androcée centripète *d'Ochna atropurpurea.Can. J. Bot.* **56**, 2500–11.

Payer, J. B. (1857). *Traité d'organogénie comparée de la fleur*. Paris: Victor Masson.

Pennington, T. D. (2004). Sapotaceae. In *The families and genera of vascular plants VI*, ed. K. Kubitzki. Berlin: Springer, pp. 390–421.

Petersen, G., Seberg, O., Cuenca, A., Sevenson, D. W., Thadeo, M., Davis, J. I., Graham, S. and Ross, T. G. (2016). Phylogeny of the Alismatales (Monocotyledons) and the relationship of *Acorus* (Acorales?). *Cladistics* **32**, 141–59.

Philipson, W. R. (1970). Constant and variable features of the Araliaceae. *Bot. J. Linn. Soc.* **63 (Suppl. 1)**, 87–100.

Philipson, W. R. (1985). Is the grass gynoecium monocarpellary? *Am. J. Bot.* **72**, 1954–61.

Philipson, W. R. (1993). Monimiaceae. In *The families and genera of vascular plants vol. II*, ed. K. Kubitzki, J. G. Rohwer and V. Bittrich. Berlin: Springer, pp. 426–37.

Pilger, R. (1935). Santalaceae. In *Die natürlichen Pflanzenfamilien 16b*, ed. A. Engler and K. Prantl, Leipzig: W. Engelmann, pp. 52–91.

Plunkett, G. M. (2001). Relationship of the order Apiales to subclass Asteridae: A re-evaluation of morphological characters based on insights from molecular data. *Edinburgh J. Bot.* **58**, 183–200.

Plunkett, G. M., Soltis, D. E. and Soltis, P. S. (1996). Higher level relationships of Apiales (Apiaceae and Araliaceae) based on phylogenetic analysis of *rbc*L sequences. *Am. J. Bot.* **83**, 499–515.

Pluys, T. (2002). Bloemontogenetische studie van de Rosaceae, Dipsacaceae en Malvaceae met bijzondere aandacht voor de bijkelk. Katholieke Universiteit Leuven (Belgium): Unpublished Dissertation.

Prance, G. T., and Mori, S. A. (2004). Lecythidaceae. In *The families and genera of vascular plants, vol. VI*, ed. K. Kubitzki, Berlin: Springer, pp. 221–32.

Prenner, G. (2004a). New aspects in floral development of Papilionoideae, initiated but suppressed bracteoles and variable initiation of sepals. *Ann. Bot.* **93**, 537–45.

Prenner, G. (2004b). Floral development in *Polygala myrtifolia* (Polygalaceae) and its similarities with Leguminosae. *Plant Syst. Evol.* **249**, 67–76.

Prenner, G. (2004c). Floral ontogeny in *Calliandra angustifolia* (Leguminosae, Mimosoideae, Ingeae) and its systematic implications. *Int. J. Plant Sci.* **165**, 417–26.

Prenner, G. (2014). Floral ontogeny in *Passiflora lobata* (Malpighiales, Passifloraceae) reveals a rare pattern in petal formation and provides new evidence for interpretation of the tendril and corona. *Plant Syst. Evol.* **300**: 1285–97.

Prenner, G., Bateman, R. M. and Rudall, P. J. (2010). Floral formulae updated for routine inclusion in formal taxonomic descriptions. *Taxon* **59**, 241–50.

Prenner G., Cardoso, D., Zartman, C. E. and de Quieroz, L. P. (2015). Flowers of the early-branching papilionoid legume *Petaladenium urceoliferum* display unique morphological and ontogenetic features. *Am. J. Bot.* **102**, 1780–93.

Prenner, G., and Klitgaard, B. B. (2008). Towards unlocking the deep nodes of Leguminosae: Floral development and morphology of the enigmatic *Duparquetia orchidacea* (Leguminosae, Caesalpinioideae). *Am. J. Bot.* **95**: 1349–65.

Prenner, G., and Rudall, P. (2007). Comparative ontogeny of the cyathium in *Euphorbia* (Euphorbiaceae) and its allies, exploring the organ-flower-inflorescence boundary. *Am. J. Bot.* **94**, 1612–29.

Prenner, G., and Rudall, P. (2008). The branching stamens of *Ricinus* and the homologies of the Angiosperm stamen fascicle. *Int. J. Plant Sci.* **169**, 735–44.

Proctor, M., Yeo, P. and Lack, A. (1996). *The natural history of pollination*. Portland, Oreg.: Timber Press.

Puff, C., and Igersheim, A. (1991). The flowers of *Paederia* L. (Rubiaceae-Paederieae). *Opera Bot. Belg.* **3**, 55–75.

Qiu, Y.-L., Lee, J., Bernasconi-Quadroni, F., Soltis, D. E., Soltis, P. S., Zanis, M., Zimmer, E. A. et al. (1999). The earliest angiosperms: Evidence from mitochondrial, plastid and nuclear genomes. *Nature* **402**, 404–7.

Rama Devi, D. (1991a). Floral anatomy of *Hypseocharis* (Oxalidaceae) with a discussion on its systematic position. *Plant Syst. Evol.* **177**, 161–4.

Rama Devi, D. (1991b). Floral anatomy of six species of *Impatiens*. *Feddes Repert.* **102**, 395–8.

Ramirez-Domenech, J. I., and Tucker, S. C. (1990). Comparative ontogeny of the perianth in Mimosoid legumes. *Am. J. Bot.* **77**, 624–35.

Rao, V. S. (1953). The floral anatomy of some bicarpellatae 1. Acanthaceae. *J. Univ. Bombay* **21**, 1–34.

Rao, V. S. (1974). The nature of the perianth in *Elaeagnus* on the basis of floral anatomy with some comments on the systematic position of Elaeagnaceae. *J. Indian Bot. Soc.* **53**, 156–61.

Rao, V.S., Karnik, H. and Gupte, K. (1954). The floral anatomy of some Scitamineae: Part I. *J. Indian Bot. Soc.* **33**, 118–47.

Rasmussen, D. A., Kramer, E. M. and Zimmer, E. A. (2009). One size fits all? Molecular evidence for a commonly inherited petal identity program in Ranunculales. *Am. J. Bot.* **96**, 96–109.

Reardon, R., Gallagher, P., Nolan, K. M., Wright, H., Cruz Cardeñosa-Rubio, M., Bragalini C., Lee, C.-S., Fitzpatrick, D. A., Corcoran, K., Wolff, K. and Nugent, J. M. (2014). Different outcomes for the MYB floral symmetry genes *DIVARICATA* and *RADIALIS* during the evolution of derived actinomorphy in *Plantago*. *New Phytol.* doi: 10.1111/nph.12682

Reinheimer, R., Pozner, R. and Vegetti, A. C. (2005). Inflorescence, spikelet, and floral development in *Panicum maximum* and *Urochloa plantaginea* (Poaceae). *Am. J. Bot.* **92**, 565–75.

Remizowa, M. V., Rudall, P., Choob, V. and Sokoloff, D. D. (2012). Racemose inflorescences of monocots: Structural and morphogenetic interaction at the flower/inflorescence level. *Ann. Bot.* **112**, 1553–66.

Remizowa, M. and Sokoloff, D. (2003). Inflorescence and floral morphology in *Tofieldia* (Tofieldiaceae) compared with Araceae, Acoraceae and Alismatales s.str. *Bot. Jahrb. Syst.* **124**, 255–71.

Remizowa, M., Sokoloff, D. and Kondo, K. (2008). Floral evolution in the monocot family Nartheciaceae (Dioscoreales), evidence from anatomy and development in *Metanarthecium luteo-viride* Maxim. *Bot. J. Linn. Soc.* **158**, 1–18.

Remizowa, M., Sokoloff, D. and Rudall, P. J. (2006). Evolution of the monocot gynoecium, evidence from comparative morphology and development in *Tofieldia*, *Japanolirion*, *Petrosavia* and *Narthecium*. *Plant Syst. Evol.* **258**, 183–209.

Remizowa, M. V., Sokoloff, D. D. and Rudall, P. J. (2010). Evolutionary history of the monocot flower. *Ann. Mo Bot. Gard.* **97**, 617–45.

Ren, Y., Li, H.-F., Zhao, L. and Endress, P. K. (2007). Floral morphogenesis in *Euptelea* (Eupteleaceae, Ranunculales). *Ann. Bot.* **100**, 185–93.

Renner, S. S. (1999). Circumscription and phylogeny of the Laurales, evidence from molecular and morphological data. *Am. J. Bot.* **86**, 1301–15.

Reynders, M., Vrijdaghs, A., Larridon, I., Huygh, W., Leroux, O., Muasya, A. M. and Goetghebeur, P. (2012). Gynoecial anatomy and development in Cyperoideae (Cyperaceae, Poales): Congenital fusion of carpels facilitates evolutionary modifications in pistil structure. *Plant Ecol. Evol.* **145**, 96–125.

Richardson, F.C. (1969). Morphological studies of the Nymphaeaceae IV. Structure and development of the flower of *Brasenia schreberi* Gmel. *Univ. Calif. Publ. Bot.* **47**, 1–101.

Richardson, J. E., Fay, M. F., Cronk, Q. C. B., Bowman, D. and Chase, M. W. (2000). A phylogenetic analysis of Rhamnaceae using *RbcL* and *trnL-F* plastid DNA sequences. *Am. J. Bot.* **87**, 1309–24.

Robbrecht, E. (1988). Tropical woody Rubiaceae. *Opera Bot. Belg.* **1**, 1–271.

Rodman, J. E., Soltis, P. S., Soltis, D. E., Sytsma, K. J. and Karol, K. G. (1998). Parallel evolution of glucosinolate biosynthesis inferred from congruent nuclear and plastid gene phylogenies. *Am. J. Bot.* **85**, 997–1006.

Roels, P., Ronse De Craene, L. P. and Smets, E. F. (1997). A floral ontogenetic investigation of the Hydrangeaceae. *Nord. J. Bot.* **17**, 235–54.

Roels, P., and Smets, E. F. (1994). A comparative floral ontogenetical study between *Adoxa moschatellina* and *Sambucus ebulus*. *Belg. J. Bot.* **127**, 157–70.

Roels, P., and Smets, E. F. (1996). A floral ontogenetic study in the Dipsacales. *Int. J. Plant Sci.* **157**, 203–18.

Rohrer, J. R., Robertson, K. R. and Phipps, J. B. (1994). Floral morphology of Maloideae (Rosaceae) and its systematic relevance. *Am. J. Bot.* **81**, 574–81.

Rohweder, O. (1965). Centrospermen-Studien 2, Entwicklung und morphologische Deutung des Gynöciums bei *Phytolacca*. *Bot. Jahrb. Syst.* **84**, 509–26.

Rohweder, O., and Huber, K. (1974). Centrospermen-Studien 7. Beobachtungen und Anmerkungen zur Morphologie und Entwicklungsgeschichte einiger Nyctaginaceen. *Bot. Jahrb. Syst.* **94**, 327–59.

Rohwer, J. (1993a). Lauraceae. In *The families and genera of vascular plants vol. II*, ed. K. Kubitzki, J. G. Rohwer and V. Bittrich. Berlin: Springer, pp. 366–91.

Rohwer, J. (1993b). Phytolaccaceae. In *The families and genera of vascular plants vol. II*, ed. K. Kubitzki, J. G. Rohwer and V. Bittrich. Berlin: Springer, pp. 506–15.

Ronse De Craene, L. P. (1988). Two types of ringwall formation in the development of complex polyandry. *Bull. Soc. Roy. Bot. Belg.* **121**, 122–4.

Ronse De Craene, L. P. (1989a). The flower of *Koenigia islandica* L. (Polygonaceae), an interpretation. *Watsonia* **17**, 419–23.

Ronse De Craene, L. P. (1989b). The floral development of *Cochlospermum tinctorium* and *Bixa orellana* with special emphasis on the androecium. *Am. J. Bot.* **76**, 1344–59.

Ronse De Craene, L. P. (1990). Morphological studies in Tamaricales I: Floral ontogeny and anatomy of *Reaumuria vermiculata* L. *Beitr. Biol. Pflanz.* **65**, 181–203.

Ronse De Craene, L. P. (2003). The evolutionary significance of homeosis in flowers, a morphological perspective. *Int. J. Plant Sci.* **164 (5 Suppl.)**, S225–S235.

Ronse De Craene, L. P. (2004). Floral development of *Berberidopsis corallina*: A crucial link in the evolution of flowers in the core eudicots. *Ann. Bot.* **94**, 1–11.

Ronse De Craene, L. P. (2005). Floral developmental evidence for the systematic position of *Batis* (Bataceae). *Am. J. Bot.* **92**, 752–60.

Ronse De Craene, L. P. (2007). Are petals sterile stamens or bracts? The origin and evolution of petals in the core eudicots. *Ann. Bot.* **10**, 621–30.

Ronse De Craene, L. P. (2008). Homology and evolution of petals in the core eudicots. *Syst. Bot.* **33**, 301–25.

Ronse De Craene, L. P. (2010). *Floral diagrams. An aid to understanding flower morphology and evolution*, 1st edition. Cambridge: Cambridge University Press.

Ronse De Craene, L. P. (2011). Floral development of *Napoleonaea* (Lecythidaceae), a deceptively complex flower. In *Flowers on the tree of life*, ed. L. Wanntorp and L. P. Ronse De Craene. Cambridge: Cambridge University Press, pp. 279–95.

Ronse De Craene, L. P. (2013). Reevaluation of the perianth and androecium in Caryophyllales: Implications for flower evolution. *Plant Syst. Evol.* **299**, 1599–1636.

Ronse De Craene, L. P. (2016). Meristic changes in flowering plants: How flowers play with numbers. *Flora* **221**, 22–37.

Ronse De Craene, L. P. (2017a). Floral development of *Berberidopsis beckleri* (Berberidopsidaceae): Unusual species or key to understanding the origin of the floral Bauplan in the core eudicots? *Ann. Bot.* **119**, 599–610.

Ronse De Craene, L. P. (2017b). Floral development of the endangered genus *Medusagyne* (Medusagynaceae-Malpighiales): Spatial constraints of stamen and carpel increase. *Int. J. Plant Sci.* **178**, 639–49.

Ronse De Craene, L. P. (2018) Understanding the role of floral development in the evolution of angiosperm flowers: Clarifications from a historical and physico-dynamic perspective. *J. Plant Res.* **131**: 367–93.

Ronse De Craene, L. P. (2021). Gynoecium structure and development in Caryophyllales. A matter of proportions. *Bot. J. Linn. Soc.* **195**, 437–66.

Ronse De Craene, L. P., and Akeroyd, J. R. (1988). Generic limits in *Polygonum* and related genera (Polygonaceae) on the basis of floral characters.*Bot. J. Linn. Soc.* **98**, 321–71.

Ronse De Craene, L. P., and Brockington, S. (2013). Origin and evolution of petals in the angiosperms. *Plant Ecol. Evol.* **146**, 5–25.

Ronse De Craene, L. P., and Bull-Hereñu, K. (2016). Obdiplostemony: The occurrence of a transitional stage linking robust flower configurations. *Ann. Bot.* **117**, 709–24.

Ronse De Craene, L. P., Clinckemaillie, D. and Smets, E. F. (1993). Stamen-petal complexes in Magnoliatae. *Bull. Jard. Bot. Nat. Belg.* **62**, 97–112.

Ronse De Craene, L. P., De Laet, J. and Smets, E. F. (1996). Morphological studies in Zygophyllaceae. II. The floral development and vascular anatomy of *Peganum harmala*. *Am. J. Bot.* **83**, 201–15.

Ronse De Craene, L. P., De Laet, J. and Smets, E. F. (1998). Floral development and anatomy of *Moringa oleifera* (Moringaceae): What is the evidence for a capparalean or sapindalean affinity? *Ann. Bot.* **82**: 273–84.

Ronse De Craene, L. P., and Haston, E. (2006). The systematic relationships of glucosinolate-producing plants and related families, a cladistic investigation based on morphological and molecular characters. *Bot. J. Linn. Soc.* **151**, 453–94.

Ronse De Craene, L. P., Hong, S.-P. and Smets, E. F. (2004). What is the taxonomic status of *Polygonella*? Evidence from floral morphology. *Ann. Mo. Bot. Gard.* **91**, 320–45.

Ronse De Craene, L. P., Iwamoto, A., Bull-Hereñu, K., Dos Santos, P., Luna-Castro, J. and Farrar, J. (2014). Understanding the structure of flowers: The wonderful tool of floral formulae. A response to Prenner & al. *Taxon* **63**, 1103–11.

Ronse De Craene, L. P., Linder, H. P., Dlamini, T. and Smets, E. F. (2001). Evolution and development of floral diversity of Melianthaceae, an enigmatic Southern African family. *Int. J. Plant Sci.* **162**: 59–82.

Ronse De Craene, L. P., Linder, H. P. and Smets, E. F. (2000). The questionable relationship of *Montinia* (Montiniaceae), evidence from a floral ontogenetic and anatomical study. *Am. J. Bot.* **87**, 1408–24.

Ronse De Craene, L. P., Linder, H. P. and Smets, E. F. (2001). Floral ontogenetic evidence in support of the *Willdenowia* clade of South African Restionaceae. *J. Plant Res.* **114**, 329–42.

Ronse De Craene, L. P., Linder, H. P. and Smets, E. F. (2002). Ontogeny and evolution of the flower of South African Restionaceae with special emphasis on the gynoecium. *Plant Syst. Evol.* **231**, 225–58.

Ronse De Craene, L. P. and Miller, A. G. (2004). Floral development and anatomy of *Dirachma socotrana* (Dirachmaceae), a controversial member of the Rosales. *Plant Syst. Evol.* **249**, 111–27.

Ronse De Craene, L. P., Quandt, D. and Wanntorp, L. (2015a). Flower morphology and anatomy of *Sabia* (Sabiaceae): Structural basis of an advanced pollination system among basal eudicots. *Plant Syst. Evol.* **301**, 1543–53.

Ronse De Craene, L. P., Quandt, D. and Wanntorp, L. (2015b). Floral development of *Sabia* (Sabiaceae): Evidence for the derivation of pentamery from a trimerous ancestry. *Am. J. Bot.* **102**, 336–49.

Ronse De Craene, L. P., and Smets, E. F. (1987). The distribution and the systematic relevance of the androecial characters Oligomery and Polymery in the Magnoliophytina. *Nord. J. Bot.* **7**, 239–53.

Ronse De Craene, L. P., and Smets, E. F. (1990a). The floral development of *Popowia whitei* (Annonaceae). *Nord. J. Bot.* **10**, 411–420. [Correction in *Nord. J. Bot.* **11** (1991), 420].

Ronse De Craene, L. P., and Smets, E. F. (1990b). The systematic relationship between Begoniaceae and Papaveraceae, a comparative study of their floral development. *Bull. Jard. Bot. Nat. Belg.* **60**, 229–73.

Ronse De Craene, L. P., and Smets, E. F. (1991a). The impact of receptacular growth on polyandry in the Myrtales. *Bot. J. Linn. Soc.* **105**, 257–69.

Ronse De Craene, L. P., and Smets, E. F. (1991b). The floral nectaries of *Polygonum* s.l. and related genera (Persicarieae and Polygoneae), position, morphological nature and semophylesis. *Flora* **185**, 165–85.

Ronse De Craene, L. P., and Smets, E. F. (1991c). The floral ontogeny of some members of the Phytolaccaceae (subfamily Rivinoideae) with a discussion of the evolution of the androecium in the Rivinoideae. *Biol. Jb. Dodonaea* **59**, 77–99.

Ronse Decraene, L. P., and Smets, E. F. (1991d). Androecium and floral nectaries of *Harungana madagascariensis* (Clusiaceae). *Plant Syst. Evol.* **178**, 179–94.

Ronse De Craene, L. P., and Smets, E. F. (1991e). Morphological studies in Zygophyllaceae I. The floral development and vascular anatomy of *Nitraria retusa*. *Am. J. Bot.* **78**, 1438–48.

Ronse De Craene, L. P., and Smets, E. F. (1992a). Complex polyandry in the Magnoliatae, definition, distribution and systematic value. *Nord. J. Bot.* **12**, 621–49.

Ronse De Craene, L. P., and Smets, E. F. (1992b). An updated interpretation of the androecium of the Fumariaceae. *Can. J. Bot.* **70**, 1765–76.

Ronse De Craene, L. P., and Smets, E. F. (1993). The distribution and systematic relevance of the androecial character polymery. *Bot. J. Linn. Soc.* **113**, 285–350.

Ronse De Craene, L. P., and Smets, E. F. (1994). Merosity, definition, origin and taxonomic significance. *Plant Syst. Evol.* **191**, 83–104.

Ronse De Craene, L. P., and Smets, E. F (1995a). The androecium of monocotyledons. In *Monocotyledons. Systematics and evolution*, ed. P. J. Rudall, P. Cribb, D. F. Cutler and C. J. Hymphries. Kew: Royal Botanic Gardens, pp. 243–54.

Ronse De Craene, L. P., and Smets, E. F. (1995b). The distribution and systematic relevance of the androecial character oligomery. *Bot. J. Linn. Soc.* **118**, 193–247.

Ronse De Craene, L. P., and Smets, E. F. (1995c). Evolution of the androecium in the Ranunculiflorae. *Plant Syst. Evol.* **suppl. 9**, 63–70.

Ronse Decraene, L. P., and Smets, E. F. (1996a). The morphological variation and systematic value of stamen pairs in the Magnoliatae. *Feddes Repert.* **107**, 1–17.

Ronse Decraene, L. P., and Smets, E. F. (1996b). The floral development of *Neurada procumbens* L. (Neuradaceae). *Acta Bot. Neerl.* **45**, 229–41.

Ronse De Craene, L. P., and Smets, E. F. (1997a). A floral ontogenetic study of some species of *Capparis* and *Boscia*, with special emphasis on the androecium. *Bot. Jahrb. Syst.* **119**, 231–55.

Ronse De Craene, L. P., and Smets, E. F. (1997b). Evidence for carpel multiplications in the Capparaceae. *Belg. J. Bot.* **130**, 59–67.

Ronse De Craene, L. P., and Smets, E. F. (1998a). Notes on the evolution of androecial organisation in the Magnoliophytina (Angiosperms). *Bot. Acta* **111**, 77–86.

Ronse De Craene, L. P., and Smets, E. F. (1998b). Meristic changes in gynoecium morphology, exemplified by floral ontogeny and anatomy. In *Reproductive biology in systematics, conservation and economic botany*, ed. S. J. Owens and P. J. Rudall. Kew: Royal Botanic Gardens, pp. 85–112.

Ronse De Craene, L. P., and Smets, E. F. (1999a). The floral development and anatomy of *Carica papaya* (Caricaceae). *Can. J. Bot.* **77**, 582–98.

Ronse De Craene, L. P., and Smets E. F. (1999b). Similarities in floral ontogeny and anatomy between the genera *Francoa* (Francoaceae) and *Greyia* (Greyiaceae). *Int. J. Plant Sci.* **160**, 377–93.

Ronse De Craene, L. P., and Smets E. F. (2000). Floral development of *Galopina tomentosa* with a discussion of sympetaly and placentation in the Rubiaceae. *Syst. Geogr. Plants* **70**, 155–70.

Ronse De Craene, L. P., and Smets, E. F. (2001a). Staminodes. Their morphological and evolutionary significance. *Bot. Rev.* **67**, 351–402.

Ronse De Craene, L. P., and Smets, E. F. (2001b). Floral developmental evidence for the systematic relationships of *Tropaeolum* (Tropaeolaceae). *Ann. Bot.* **88**, 879–92.

Ronse De Craene, L. P., Smets, E. F. and Clinckemaillie, D. (1995). The floral development and floral anatomy of *Coris monspeliensis*. *Can. J. Bot.* **73**, 1687–98.

Ronse De Craene, L. P., Smets, E. F. and Clinckemaillie, D. (2000). Floral ontogeny and anatomy in *Koelreuteria* with special emphasis on monsymmetry and septal cavities. *Plant Syst. Evol.* **223**, 91–107.

Ronse De Craene, L. P., Smets, E. F. and Vanvinckenroye, P. (1998). Pseudodiplostemony, and its implications for the evolution of the androecium in the Caryophyllaceae. *J. Plant Res.* **111**, 25–43.

Ronse De Craene, L. P., Soltis, P. S. and Soltis, D. E. (2003). Evolution of floral structures in basal Angiosperms. *Int. J. Plant Sci.* **164 (5 Suppl.)**, S329–S363.

Ronse De Craene, L. P., Tréhin, C., Morel, P. and Negrutiu, I. (2011). Carpeloidy in flower evolution and diversification: A comparative study in *Carica papaya* and *Arabidopsis thaliana*. *Ann. Bot.* **107**: 1453–63.

Ronse De Craene, L. P., and Stuppy, W. (2010). Floral development and anatomy of *Aextoxicon punctatum* (Aextoxicaceae – Berberidopsidales): An enigmatic tree at the base of core eudicots. *Int. J. Plant Sci.* **171**, 244–57.

Ronse De Craene, L. P., Vanvinckenroye, P. and Smets, E. F. (1997). A study of the floral morphological diversity in *Phytolacca* (Phytolaccaceae) based on early floral ontogeny. *Int. J. Plant Sci.* **158**, 56–72.

Ronse De Craene, L. P., Volgin, S. A. and Smets, E. F. (1999). The floral development of *Pleuropetalum darwinii*, an anomalous member of the Amaranthaceae. *Flora* **194**, 189–99.

Ronse De Craene, L. P., and Wanntorp, L. (2006). Evolution of floral characters in *Gunnera* (Gunneraceae). *Syst. Bot.* **31**, 671–88.

Ronse De Craene, L. P., and Wanntorp, L. (2008). Morphology and anatomy of the flower of *Meliosma* (Sabiaceae), implications for pollination biology. *Plant Syst. Evol.* **271**, 79–91.

Ronse De Craene, L. P., and Wanntorp, L. (2009). Floral development and anatomy of Salvadoraceae. *Ann. Bot.* **104**, 913–23.

Ronse De Craene, L. P., and Wei, L. (2019). Floral development and anatomy of *Macarthuria australis* (Macarthuriaceae): Key to understanding the unusual initiation sequence of Caryophyllales. *Aust. Syst. Bot.* **32**, 49–60.

Ronse De Craene, L. P., Yang, T. Y., Schols, P. and Smets, E. F. (2002). Floral anatomy and systematics of *Bretschneidera* (Bretschneideraceae). *Bot. J. Linn. Soc.* **139**, 29–45.

Rosas-Reinhold, I., Piñeyro-Nelson, A., Rosas, U. and Arias, S. (2021). Blurring the boundaries between a branch and a flower: Potential developmental venues in Cactaceae. *Plants* **10**, 1134. doi: org/10.3390/plants10061134

Rose, J. P., and Sytsma, K. J. (2021). Complex interactions underlie the correlated evolution of floral traits and their association in a clade with diverse pollination systems. *Evolution* **75–76**, 1431–49.

Ross, R. (1982). Initiation of stamens, carpels and receptacle in the Cactaceae. *Am. J. Bot.* **69**, 369–79.

Rothwell, G. Z., Escapa, I. H. and Tomescu, A. M. (2018). Tree of death: The role of fossils in resolving the overall pattern of plant phylogeny. *Am. J. Bot.* **105**, 1239–42.

Rudall, P. J. (2002). Homologies of inferior ovaries and septal nectaries in monocotyledons. *Int. J. Plant Sci.* **163**, 261–76.

Rudall, P. J. (2003). Monocot pseudanthia revisited, floral structure of the mycoheterotrophic family Triuridaceae. *Int. J. Plant Sci.* **164 (5 Suppl.)**, S307–S320.

Rudall, P. J., Alves, M. and Sajo, M. das Graças (2016). Inside-out flowers of *Lacandonia brasiliana* (Triuridaceae) provide new insights into fundamental aspects of floral patterning. *Peer J.* doi: 10.7717/peerj.1653

Rudall, P. J., and Bateman, R. M. (2003). Evolutionary change in flowers and inflorescences, evidence from naturally occurring terata. *Trends Pl. Sci.* **8**, 76–82.

Rudall, P. J., and Bateman, R. M. (2004). Evolution of zygomorphy in monocot flowers, iterative patterns and developmental constraints. *New Phytol.* **162**, 25–44.

Rudall, P. J., Bateman, R. M., Fay, M. F. and Eastman, A. (2002). Floral anatomy and systematics of Alliaceae with particular reference to *Gilliesia*, a presumed insect mimic with strongly zygomorphic flowers. *Am. J. bot.* **89**, 1867–83.

Rudall, P. J., Sokoloff, D. D., Remizowa, M. V., Conran, J. G., Davis, J. I., Macfarlane, T. D. and Stevenson, D. W. (2007). Morphology of Hydatellaceae, an anomalous aquatic family recently recognized as an early-divergent Angiosperm lineage. *Am. J. Bot.* **94**, 1073–92.

Rudall, P. J., Stuppy, W., Cunniff, J., Kellogg, E. A. and Briggs, B. G. (2005). Evolution of reproductive structures in grasses (Poaceae) inferred by sister-group comparison with their putative closest living relatives, Ecdeiocoleaceae. *Am. J. Bot.* **92**, 1432–43.

Ruhfel, B. R., Bittrich, V., Bove, C. P., Gustafsson, M. H. G., Philbrick, C. T., Rutishauser, R., Xi, Z. and Davis, C. C. (2011). Phylogeny of the clusioid clade (Malpighiales): Evidence from the plastid and mitochondrial genomes. *Am. J. Bot.* **98**, 306–25.

Rümpler, F., and Theissen, G. (2019). Reconstructing the ancestral flower of extant angiosperms: The 'war of the whorls' is heating up. *J. Exper. Bot.* **70**, 2615–22.

Rutishauser, R., Ronse De Craene, L. P., Smets, E. F. and Mendoza-Heuer, I. (1998). *Theligonum cynocrambe*, developmental morphology of a peculiar rubiaceous herb. *Plant Syst. Evol.* **210**, 1–24.

Sajo, M. G., de Mello-Silva, R. and Rudall, P. J. (2010). Homologies of floral structures in Velloziaceae with particular reference to the corona. *Int. J. Plant Sci.* **171**, 595–606.

Sajo, M. G., Longhi-Wagner, H. and Rudall, P. J. (2007). Floral development and embryology in the early-divergent grass *Pharus*. *Int. J. Plant Sci.* **168**, 181–91.

Sajo, M. G., Longhi-Wagner, H. M. and Rudall, P. J. (2008). Reproductive morphology of the early-divergent grass *Streptochaeta* and its bearing on the homologies of the grass spikelet. *Plant Syst. Evol.* **275**, 245–55.

Sajo, M. G., Moraes, P. L. R., Assis, L. C. S. and Rudall, P. J. (2016). Comparative floral anatomy and development in neotropical Lauraceae. *Int. J.Plant Sci.* **177**, 579–89.

Sampson, F. B. (1969). Studies on the Monimiaceae II. Floral morphology of *Laurelia novae-zelandiae* A. Cunn. (Subfamily Atherospermoideae). *New Zeal. J. Bot.* **7**, 214–40.

Sanchez, A., and Kron, K. A. (2008). Phylogenetics of Polygonaceae with an emphasis on the evolution of Eriogonoideae. *Syst. Bot.* **33**, 87–96.

Sánchez-Del Pino, I., Vrijdaghs, A., De Block, P., Flores-Olvera, H., Smets, E. and Eliasson, U. (2019). Floral development in Gomphrenoideae (Amaranthaceae) with a focus on androecial tube and appendages. *Bot. J. Linn. Soc.* **190**, 315–32.

Sastri, R. L. N. (1952). Studies in Lauraceae: I. Floral anatomy of *Cinnamomum iners* Reiw. and *Cassytha filiformis* Linn. *J. Indian Bot. Soc.*, **31**: 240–6.

Sattler, R. (1962). Zur frühen Infloreszenz und Blütenentwicklung der Primulales sensu lato mit besonderer Berücksichtigung der Stamen-Petalum-Entwicklung. *Bot. Jahrb. Syst.* **81**, 385–96.

Sattler, R. (1973). *Organogenesis of flowers. A photographic text-atlas*. Toronto and Buffalo: University of Toronto Press.

Sattler, R. (1974). A new conception of the shoot of higher plants. *J. Theor. Biol.* **47**, 367–82.

Sattler, R. (1977). Kronröhrenentstehung bei *Solanum dulcamara* L. und 'kongenitale Verwachsung'. *Ber. Dtsch. Bot. Ges.* **90**: 29–38.

Sattler, R. (1978). 'Fusion' and 'continuity' in floral morphology. *Notes Roy. Bot. Gard. Edinburgh* **36**: 397–405.

Sattler, R., and Perlin, L. (1982). Floral development of *Bougainvillea spectabilis* Willd., *Boerhaavia diffusa* L. and *Mirabilis jalapa* L. (Nyctaginaceae). *Bot. J. Linn. Soc.* **84**, 161–82.

Sattler, R., and Singh, V. (1973). Floral development of *Hydrocleis nymphoides*. *Can. J. Bot.* **51**, 2455–8.

Sattler, R., and Singh, V. (1977). Floral organogenesis of *Limnocharis flava*. *Can. J. Bot.* **55**, 1076–86.

Sattler, R., and Singh, V. (1978). Floral organogenesis of *Echinodorus amazonicus* Rataj and floral construction of the Alismatales. *Bot. J. Linn. Soc.* **77**, 141–56.

Saunders, E. R. (1937, 1939). *Floral morphology. A new outlook with special reference to the interpretation of the gynoecium*. Vols I and II. Cambridge: W. Heffer and Sons.

Saunders, R. M. K. (2010). Floral evolution in the Annonaceae: Hypotheses of homeotic mutations and functional convergence. *Biol. Rev.* **85**, 571–91.

Sauquet, H. (2003). Androecium diversity and evolution in Myristicaceae (Magnoliales), with a description of a new Malagasy genus, *Doyleanthus* gen. nov. *Am. J. Bot.* **90**, 1293–1305.

Sauquet, H., Von Balthazar, M., Magallón, S., Doyle, J. A., Endress P. K. et al. (2017). The ancestral flower of angiosperms and its early diversification. *Nature Comm.* **8**, 16047. doi: 10.1038/ncomms16047

Schaeppi, H. (1976). Über die männlichen Blüten einiger Menispermaceen. *Beitr. Biol. Pflanz.* **52**, 207–15.

Schindler, A. K. (1905). Halorrhagaceae. In *Das Pflanzenreich IV, 225*, ed. A. Engler. Leipzig: W. Engelmann, pp. 1–133.

Schmid, R. (1980). Comparative anatomy and morphology of *Psiloxylon* and *Heteropyxis*, and the subfamilial and tribal classification of Myrtaceae. *Taxon* **29**, 559–95.

Schmidt, E. (1928). Untersuchungen über Berberidaceen. *Beih. Bot. Centralbl.* **45**, 329–96.

Schneider, E. L. (1976). The floral anatomy of *Victoria* Schomb. (Nymphaeaceae). *Bot. J. Linn. Soc.* **72**, 115–48.

Schneider, E. L., Tucker, S. C. and Williamson, P. S. (2003). Floral development in the Nymphaeales. *Int. J. Plant Sci.* **164 (5 Suppl.)**, S279–S292.

Schöffel, K. (1932). Untersuchungen über den Blütenbau der Ranunculaceen. *Planta* **17**, 315–71.

Schönenberger, J., (2009). Comparative floral structure and systematics of Fouquieriaceae and Polemoniaceae (Ericales). *Int. J. Plant Sci.* **170**, 1132–67.

Schönenberger, J., Anderberg, A. A. and Systsma, K. J. (2005). Molecular phylogenetics and patterns of floral evolution in the Ericales. *Int. J. Plant Sci.* **166**, 265–88.

Schönenberger, J., and Conti, E. (2003). Molecular phylogeny and floral evolution of Penaeaceae, Oliniaceae, Rhychocalycaceae, and Alzateaceae (Myrtales). *Am. J. Bot.* **90**, 293–309.

Schönenberger, J., and Endress, P. K. (1998). Structure and development of the flowers in *Mendoncia, Pseudocalyx*, and *Thunbergia* (Acanthaceae) and their systematic implications. *Int. J. Plant Sci.* **159**, 446–65.

Schönenberger, J., and Friis, E. M. (2001). Fossil flowers of ericalean s.l. affinity from the Late Cretaceous of southern Sweden. *Am. J. Bot.* **88**, 467–80.

Schönenberger, J., Friis, E. M., Matthews, M. L. and Endress, P. K. (2001). Cunoniaceae in the Cretaceous of Europe, evidence from fossil flowers. *Ann. Bot.* **88**, 423–37.

Schönenberger, J., and Grenhagen, A. (2005). Early floral development and androecium organization in Fouquieriaceae (Ericales). *Plant Syst. Evol.* **254**, 233–49.

Schönenberger, J., and Von Balthazar, M. (2006). Reproductive structures and phylogenetic framework of the Rosids: Progress and prospects. *Plant Syst. Evol.* **260**, 87–106.

Schönenberger, J., Von Balthazar, M., Lopez Martinez, A., Albert, B., Prieu, C., Magallón, S. and Sauquet, H. (2020). Phylogenetic analysis of fossil flowers using an angiosperm-wide data set: Proof-of-concept and challenges ahead. *Am. J. Bot.* **107**, 1–16.

Schönenberger J., Von Balthazar, M., Takahashi M., Xiao X., Crane P. R. and Herendeen, P. S. (2012). *Glandulocalyx upatoiensis*, a fossil flower of Ericales (Actinidiaceae/Clethraceae) from the Late Cretaceous (Santonian) of Georgia, USA. *Ann. Bot.* **109**, 921–36.

Scotland, R. W., Endress, P. K. and Lawrence, T. J. (1994). Corolla ontogeny and aestivation in the Acanthaceae. *Bot. J. Linn. Soc.* **114**, 49–65.

Scribailo, R. W., and Posluszny, U. (1985). Floral development of *Hydrocharis morsus-ranae* L. (Hydrocharitaceae). *Am. J. Bot.* **72**, 1578–89.

Sérsic, A. N., and Cocucci, A. A. (1999). An unusual kind of nectary in the oil flowers of *Monttea*, its structure and function. *Flora* **194**, 393–404.

Setoguchi, H., Ohba, H., Tobe, H. (1996). Floral morphology and phylogenetic analysis in *Crossostylis* (Rhizophoraceae). *J. Plant Res.* **109**, 7–19.

Sharma, B., Guo, C., Kong, H., and Kramer, E. M. (2011). Petal-specific subfunctionalization of an *APETALA3* paralog in the Ranunculales and its implications for petal evolution. *New Phytol.* **191**, 870–83.

Simmons, M. P. (2004). Celastraceae. In *The families and genera of vascular plants vol. VI*, ed. K. Kubitzki. Berlin: Springer, pp. 29–64.

Simpson, M. G. (1990). Phylogeny and classification of Haemodoraceae. *Ann. Mo. Bot. Gard.* **77**, 722–84.

Simpson, M. G. (1998a). Reversal in ovary position from inferior to superior in the Haemodoraceae, evidence from floral ontogeny. *Int. J. Plant. Sci.* **159**, 466–79.

Simpson, M. G. (1998b). Haemodoraceae. In *The families and genera of vascular plants vol. IV*, ed. K. Kubitzki. Berlin: Springer, pp. 212–22.

Simpson, M. G. (2006). *Plant systematics*. Amsterdam: Elseviers.

Simpson, N., in collaboration with Barnes, P. G. (2016). *Nuphar lutea – botanical images for the digital documentation of a taxon*. Visual Botany, UK.

Singh, V., and Sattler, R. (1973). Nonspiral androecium and gynoecium of *Sagittaria latifolia*. *Can. J. Bot.* **51**, 1093–5.

Singh, V., and Sattler, R. (1977a). Development of the inflorescence of *Sagittaria cuneata*. *Can. J. Bot.* **55**, 1087–1105.

Singh, V., and Sattler, R. (1977b). Floral development of *Aponogeton natans* and *A. undulatus*. *Can. J. Bot.* **55**, 1106–20.

Sinjushin, A. A., and Ploshinskaya, M. E. (2020). Flower development in *Lythrum salicaria* L., *Cuphea ignea* A. DC and *C. hyssopifolia* Kunth (Lythraceae): The making of monosymmetry in hexamerous flowers. *Wulfenia* **27**: 303–20.

Sleumer, H. (1935). Olacaceae. In *Die natürlichen Pflanzenfamilien 16b*, ed. A. Engler and K. Prantl. Leipzig: W. Engelmann, pp. 5–32.

Smets, E. (1986). Localization and systematic importance of the floral nectaries in the Magnoliatae (Dicotyledons). *Bull. Jard. Bot. Nat. Belg.* **56**, 51–76.

Smets, E. (1988). La présence des 'nectaria persistentia' chez les Magnoliophytina (Angiospermes). *Candollea* **43**, 709–16.

Smets, E. F., Ronse De Craene, L. P., Caris, P. and Rudall, P. J. (2000). Floral nectaries in monocotyledons, distribution and evolution. In *Monocots, systematics and evolution*, ed. K. L. Wilson and D. A. Morrison. Melbourne: CSIRO, pp. 230–40.

Smyth, D. R. (2018). Evolution and genetic control of the floral ground plan. *New Phytol.* doi: 10.1111/nph.15282

Sobick, U. (1983). Blütenentwicklungsgeschichtliche Untersuchungen an Resedaceen unter besonderer Berücksichtigung von Androeceum und Gynoeceum. *Bot. Jahrb. Syst.* **104**, 203–48.

Soetiarto, S. R., and Ball, E. (1969). Ontogenetical and experimental studies of the floral apex of *Portulaca grandiflora* I. Histology of transformation of the shoot apex into the floral apex. *Can. J. Bot.* **47**, 133–40.

Sokoloff, D. D. (2016). Correlations between gynoecium morphology and ovary position in angiosperm flowers: Roles of developmental and terminological constraints. *Biol. Bull. Rev.* **6**, 84–95.

Sokoloff, D., Oskolski, A. A., Remizowa, M. V. and Nuraliev, M. S. (2007). Flower structure and development in *Tupidanthus calyptratus* (Araliaceae), an extreme case of polymery among asterids. *Plant Syst. Evol.* **268**, 209–34.

Sokoloff, D. D., Remizowa, M. V., Bateman, R. M. and Rudall, P. J. (2018b). Was the ancestral angiosperm flower whorled throughout? *Am. J. Bot.* **105**, 5–15.

Sokoloff, D. D., Remizowa, M. V., Timonin, A. C., Oskolski, A. A. and Nuraliev, M. S. (2018a). Types of organ fusion in angiosperm flowers (with examples from Chloranthaceae, Araliaceae and monocots). *Biol. Serb.* **40**, 16–46.

Sokoloff, D., Rudall, P. J. and Remizowa, M. (2006). Flower-like terminal structures in racemose inflorescences, a tool in morphogenetic and evolutionary research. *J. Exper. Bot.* **57**, 3517–30.

Sokoloff, D. D., Von Mering, S., Jacobs, S. W. L. and Remizowa, M. W. (2013). Morphology of *Maundia* supports its isolated phylogenetic position in the early divergent monocot order Alismatales. *Bot. J. Linn. Soc.* **173**, 12–45.

Soltis, D. E., Senters, A. E., Zanis, M. J., Kim, S., Thompson, J. D., Soltis, P. S, Ronse De Craene, L. P., Endress, P. K. and Farris, J. S. (2003). Gunnerales are sister to other core eudicots, implications for the evolution of pentamery. *Am. J. Bot.* **90**, 461–70.

Soltis, D. E., Smith, S. A., Cellinese, N., Wurdack, K. J., Tank, D. C., Brockington, S. F., Refulio-Rodriguez, N. F., Walker, J. B., Moore, M. J., Carlsward, B. S., Bell, C. D., Latvis, M., Crawley, S., Black, C., Diouf, D., Xi, Z., Rushworth, C. A., Gitzendanner, M. A., Sytsma, K. J., Y.-L. Qiu, Y.-L., Hilu, K. W., Davis, C. C., Sanderson, M. J., Beaman, R. S., Olmstead, R. G., Judd, W. S., Donoghue, M. J. and Soltis, P. S. (2011). Angiosperm phylogeny: 17 genes, 640 taxa. *Am. J. Bot.* **98**, 704–30.

Soltis, P. S., and Soltis, D. E. (2004). The origin and diversification of Angiosperms. *Am. J. Bot.* **91**, 1614–26.

Soltis, D. E., Soltis, P. S., Endress, P. K. and Chase, M. W. (2005). *Phylogeny and evolution of angiosperms*. Sunderland, Mass.: Sinauer.

Soltis, D. E., Soltis, P. S., Endress, P. K., Chase, M. W., Manchester, S., Judd, W., Majure, L. and Mavrodiev, E. (2018). *Phylogeny and evolution of angiosperms*. Revised and updated edition. Chicago: University of Chicago Press.

Specht, C. D., Yockteng, R., Almeida, A. M., Kirchoff, B. K. and Kress, W. J. (2012). Homoplasy, pollination, and emerging complexity during the evolution of floral development in the tropical gingers (Zingiberales). *Bot. Rev.* **78**, 440–62.

Spichiger, R.-E., Savolainen, V. V., Figeat, M. and Jeanmonod, D. (2002). *Botanique systématique des plantes à fleurs*. 2nd edn. Lausanne (Switzerland): Presses polytechniques et universitaires Romandes.

Stace, C. A. (2007). Combretaceae. In *The families and genera of vascular plants. Vol. IX*, ed. K. Kubitzki. Berlin: Springer, pp. 67–82.

Staedler, Y. M., and Endress, P. K. (2009). Diversity and lability of floral phyllotaxis in the pluricarpellate families of core Laurales (Gomortegaceae, Atherospermataceae, Siparunaceae, Monimiaceae). *Int. J. Plant Sci.* **170**, 522–50.

Staedler, Y. M., Weston, P. H. and Endress, P. K. (2007). Floral phyllotaxis and floral architecture in Calycanthaceae (Laurales). *Int. J. Plant Sci.* **168**, 285–306.

Stauffer, F. W., and Endress, P. K. (2003). Comparative morphology of female flowers and systematics in Geonomeae (Arecaceae). *Plant Syst. Evol.* **242**, 171–203.

Stauffer, F. W., Rutishauser, R. and Endress, P. K. (2002). Morphology and development of the female flowers in *Geonoma interrupta* (Arecaceae). *Am. J. Bot.* **89**, 220–9.

Steeves, T. A., Steeves, M. W. and Randall Olson, A. (1991). Flower development in *Amelanchier alnifolia* (Maloideae). *Can. J. Bot.* **69**, 844–57.

Steinecke, H. (1993). Embryologische, morphologische und systematische Untersuchungen ausgewählter Annonaceae. *Diss. Bot.* **205**, 1–237.

Stern, K. (1917). Beiträge zur Kenntnis der Nepenthaceae. *Flora* **109**, 213–82.

Stevens, P. F. (2001 onwards). Angiosperm Phylogeny Website. Version 14, July 2017 [and more or less continuously updated since]. www.mobot.org/MOBOT/research/APweb

Stone, D. E. (1989). Biology and evolution of temperate and tropical Juglandaceae. In *Evolution, systematics, and fossil history of the Hamamelidae, Vol. 2: 'Higher' Hamamelidae*, ed. P. R. Crane and S. Blackmore. Oxford: Clarendon, pp. 117–45.

Strange, A., Rudall, P. J. and Prychid, C. J. (2004). Comparative floral anatomy of Pontederiaceae. *Bot. J. Linn. Soc.* **144**, 395–408.

Struwe, L., Kadereit, J. W., Klackenberg, J., Nilsson, S., Thiv, M., Von Hagen, K. B. and Albert, V. A. (2002). Systematics, character evolution, and biogeography of Gentianaceae, including a new tribal and subtribal classification. In *Gentianaceae. Systematics and natural history*, ed. L. Struwe and V. A. Albert. Cambridge: Cambridge University Press, pp. 21–309.

Stützel, T. (2006). *Botanische Bestimmungsübungen*. 2nd edn. Stuttgart: Ulmer.

Suaza-Gaviria, V., González, F. and Pabón-Mora, N. (2017). Comparative inflorescence development in selected Andean Santalales. *Am. J. Bot.* **104**, 24–38.

Suaza-Gaviria, V., Pabón-Mora, N. and González, F. (2016). Development and morphology of flowers in Loranthaceae. *Int. J. Plant Sci.* **177**, 559–78.

Suessenguth, K. (1938). *Neue Ziele der Botanik. Über das Vorkommen getrennter Kronblätter bei den Sympetalen*. München-Berlin, pp. 32–6.

Sugiyama, M. (1991). Scanning electron microscopy observation on early ontogeny of the flower of *Camellia japonica* L. *J. Japan. Bot.* 66, 295–9.

Sugiyama, M. (1995). Floral anatomy of *Camelia japonica* (Theaceae). *J. Plant Res.* **110**, 45–54.

Sutter, D., and Endress, P. K. (1995). Aspects of gynoecium structure and macrosystematics in Euphorbiaceae. *Bot. Jahrb. Syst.* **116**, 517–36.

Svoma, E. (1991). The development of the bicarpellate gynoecium of *Paederia* L. species (Rubiaceae-Paederieae). *Opera Bot. Belg.* **3**, 77–86.

Sweeney, P. W. (2008). Phylogeny and floral diversity in the genus *Garcinia* (Clusiaceae) and relatives. *Int. J. Plant Sci.* **169**, 1288–1303.

Sweeney, P. W. (2010). Floral anatomy in *Garcinia nervosa* and *G. xanthochymus* (Clusiaceae): A first step toward understanding the nature of nectaries in garcinia. *Bull. Peabody Mus. Nat. Hist.* **51**, 157–68.

Takahashi, H. (1994). A comparative study of floral development in *Trillium apetalon* and *T. camtschaticum* (Liliaceae). *J. Plant Res.* **107**, 237–45.

Takhtajan, A. (1997). *Diversity and classification of flowering plants*. New York: Columbia University Press.

Terabayashi, S. (1983). Studies in the morphology and systematics of Berberidaceae VI. Floral anatomy of *Diphylleia* Michx., *Podophyllum* L. and *Dyosma* Woodson. *Acta Phytotax Geobot.* **34**, 27–47.

Thaowetsuwan, P. (2020). Evolution of floral diversity in the genus *Croton* and related genera (Crotonoideae, Euphorbiaceae). University of Edinburgh and Royal Botanic Garden Edinburgh (UK): Unpublished Doctoral Thesis.

Thaowetsuwan, P., Honorio Coronado, E. N. and Ronse De Craene, L. P. (2017). Floral morphology and anatomy of *Ophiocaryon*, a paedomorphic genus of Sabiaceae. *Ann. Bot.* **120**, 819–32.

Thaowetsuwan, P., Ritchie, S., Riina, R. and Ronse De Craene, L. P. (2020). Divergent developmental pathways in dimorphic flowers of *Croton* L. (Euphorbiaceae) with special emphasis on petals. *Front. Ecol. Evol.* **8**, 253. doi: 10.3389/fevo.2020.00253

Theißen, G., Becker, A., Winter, K.-U., Münster, T., Kirchner, C. and Saedler, H. (2002). How the land plants learned their floral ABCs, the role of MADS box genes in the evolutionary origin of flowers. In *Developmental genetics and plant evolution*, ed. A. C. B. Cronk, R. M. Bateman and J. A. Hawkins. London: Taylor and Francis, pp. 173–205.

Thorne, R. F. (1992). An updated phylogenetic classification of the flowering plants. *Aliso* **13**, 365–89.

Tiagi, Y. D. (1969). Vascular anatomy of the flower of certain species of the Combretaceae. *Bot. Gaz.* **130**, 150–7.

Tillson, A. H. (1940). The floral anatomy of the Kalanchoideae. *Am. J. Bot.* **27**, 595–600.

Tobe, H. (2012). Floral structure of *Cardiopteris* (Cardiopteridaceae) with special emphasis on the gynoecium: Systematic and evolutionary implications. *J. Plant Res.* **125**, 361–9.

Tobe, H., Graham, S. A. and Raven, P. H. (1998). Floral morphology and evolution in Lythraceae *sensu lato*. In *Reproductive biology in systematics, conservation and economic botany*, ed. S. J. Owens and P. J. Rudall. Kew: Royal Botanic Gardens, pp. 329–44.

Tobe, H., Huang, Y.-L., Kadokawa, T. and Tamura, M. N. (2018). Floral structure and development in Nartheciaceae (Dioscoreales), with special reference to ovary position and septal nectaries. *J. Plant Res.* **131**, 411–28.

Todzia, C. A. (1993). Ulmaceae. In *The families and genera of vascular plants vol. II*, ed. K. Kubitzki, J. G. Rohwer and V. Bittrich. Berlin: Springer, pp. 603–11.

Tokuoka, T. (2008). Molecular phylogenetic analysis of Violaceae (Malpighiales) based on plastid and nuclear DNA sequences. *J. Plant Res.* **121**, 253–60.

Tokuoka, T., and Tobe, H. (2006). Phylogenetic analyses of Malpighiales using plastid and nuclear DNA sequences, with particular reference to the embryology of Euphorbiaceae *sensu stricto*. *J. Plant Res.* **119**, 599–616.

Tomlinson, P. B. (1986). *The botany of mangroves*. Cambridge: Cambridge University Press.

Trimbacher, C. (1989). Der Aussenkelch der Rosaceen. In *9. Symposium Morphologie, Anatomie und Systematik, Zusammenfassungen der Vorträge*, ed. A. Weber, E. Vitek and M. Kiehn. Vienna: Institute of Botany, University of Vienna, p. 66.

Troll, W. (1956). Die Urbildlichkeit der organischen Gestaltung und Goethes prinzip der 'Variablen Proportionen'. *Neue Hefte zur Morphologie* **2**, 64–76.

Tsou, C.-H. (1998). Early floral development of Camellioideae (Theaceae). *Am. J. Bot.* **85**, 1531–47.

Tsou, C.-H., and Mori, S. A. (2007). Floral organogenesis and floral evolution of the Lecythidoideae. *Am. J. Bot.* **94**, 716–36.

Tucker, S. C. (1984). Unidirectional organ initiation in leguminous flowers. *Am. J. Bot.* **71**, 1139–48.

Tucker, S. C. (1985). Initiation and development of inflorescence and flower in *Anemopsis californica* (Saururaceae). *Am. J. Bot.* **72**, 20–31.

Tucker, S. C. (1988a). Dioecy in *Bauhinia* resulting from organ suppression. *Am. J. Bot.* **75**, 1584–97.

Tucker, S. C. (1988b). Loss versus suppression of floral organs. In *Aspects of floral development*, ed. P. Leins, S. C. Tucker and P. K. Endress. Berlin: J. Cramer, pp. 69–82.

Tucker, S. C. (1992). The developmental basis for sexual expression in *Ceratonia siliqua* (Leguminosae, Caesalpinioideae, Cassieae). *Am. J. Bot.* **79**, 318–27.

Tucker, S. C. (1996). Trends in evolution of floral ontogeny in *Cassia sensu stricto*, *Senna*, and *Chamaecrista* (Leguminosae, Caesalpinioideae, Cassieae, Cassiinae): A study in convergence. *Am. J. Bot.* **83**, 687–711.

Tucker, S. C. (1997). Floral evolution, development, and convergence, the hierarchical-significance hypothesis. *Int. J. Plant Sci.* **158 (6 Suppl.)**, S143–S161.

Tucker, S. C. (1998). Floral ontogeny in Legume genera *Petalostylis*, *Labichea*, and *Dialium* (Caesalpinioideae, Cassieae), a series in floral reduction. *Am. J. Bot.* **85**, 184–208.

Tucker, S. C. (1999a). The inflorescence, introduction. *Bot. Rev.* **65**, 303–16.

Tucker, S. C. (1999b). Evolutionary lability of symmetry in early floral development. *Int. J. Plant Sci.* **160 (6 Suppl.)**, S25–S39.

Tucker, S. C. (2000a). Evolutionary loss of sepals and/or petals in Detarioid legume taxa (*Aphanocalyx*, *Brachystegia*, and *Monopetalanthus* (Leguminosae, Caesalpinioideae). *Am. J. Bot.* **87**, 608–24.

Tucker, S. C. (2000b). Floral development in tribe Detarieae (Leguminosae, Caesalpinioideae), *Amherstia, Brownea*, and *Tamarindus*. *Am. J. Bot.* **87**, 1385–1407.
Tucker, S. C. (2000c). Floral development and homeosis in *Saraca* (Leguminosae, Caesalpinioideae, Detarieae). *Int. J. Plant Sci.* **161**, 537–49.
Tucker, S. C. (2001a). The ontogenetic basis for missing petals in *Crudia* (Leguminosae, Caesalpinioideae, Detarieae). *Int. J. Plant Sci.* **162**, 83–9.
Tucker, S. C. (2001b). Floral development in *Schotia* and *Cynometra* (Leguminosae, Caesalpinioideae, Detarieae). *Am. J. Bot.* **88**, 1164–80.
Tucker, S. C. (2002a). Floral ontogeny in Sophoreae (Leguminosae, Papilionoideae). III. Radial symmetry and random petal aestivation in *Cadia purpurea*. *Am. J. Bot.* **89**, 748–57.
Tucker, S. C. (2002b). Comparative floral ontogeny in Detarieae (Leguminosae, Caesalpinioideae). 2. zygomorphic taxa with petal and stamen suppression. *Am. J. Bot.* **89**, 888–907.
Tucker, S. C. (2003a). Floral development in Legumes. *Plant Physiol.* **131**: 911–26.
Tucker, S. C. (2003b). Floral ontogeny in *Swartzia* (Leguminosae, Papilionoideae, Swartzieae): Distribution and role of the ring meristem. *Am. J. Bot.* **90**, 1274–92.
Tucker, S. C. (2003c). Comparative floral ontogeny in Detarieae (Leguminosae: Caesalpinoideae). III. Adaxially initiated whorls in *Julbernardia* and *Sindora*. *Int. J. Plant Sci.* **164**, 275–86.
Tucker, S. C., and Bernhardt, P. (2000). Floral ontogeny, pattern formation, and evolution in *Hibbertia* and *Adrastea* (Dilleniaceae). *Am. J. Bot.* **87**, 1915–36.
Tucker, S. C., and Douglas, A. W. (1996). Floral structure, development, and relationships of paleoherbs, *Saruma, Cabomba, Lactoris* and selected Piperales. In *Flowering plant origin, evolution and phylogeny*, ed. D. W. Taylor and L. J. Hickey. New York: Chapman and Hall, pp. 141–75.
Tucker, S. C., Douglas, A. W. and Liang, H.-X. (1993). Utility of ontogenetic and conventional characters in determining phylogenetic relationships of Saururaceae and Piperaceae (Piperales). *Syst. Bot.* **18**, 614–41.
Uhl, N. W. (1976a). Developmental studies in *Ptychosperma* (Palmae). I. The inflorescence and flower cluster. *Am. J. Bot.* **63**, 82–96.
Uhl, N. W. (1976b). Developmental studies in *Ptychosperma* (Palmae). II. The staminate and pistillate flowers. *Am. J. Bot.* **63**, 97–109.
Uhl, N. W., and Moore, H.E. (1977). Centrifugal stamen initiation in phytelephantoid palms. *Am. J. Bot.* **64**, 1152–61.
Uhl, N. W., and Moore, H. E. (1980). Androecial development in six polyandrous genera representing five major groups of palms. *Ann. Bot.* **45**, 57–75.
Urban, I. (1892). Blüthen- und Fruchtbau der Loasaceen. *Ber. Dtsch. Bot. Ges.* **10**, 259–65.
Vaes, E., Vrijdaghs, A., Smets, E. F. and Dessein, S. (2006). Elaborate petals in Australian *Spermacoce* (Rubiaceae) species, morphology, ontogeny and function. *Ann. Bot.* **98**, 1167–78.
Van der Niet, T., Peakall, R. and Johnson, S. D. (2014). Pollinator-driven ecological speciation in plants: New evidence and future perspectives. *Ann. Bot.* **113**, 199–211.

Van Heel, W. A. (1966). Morphology of the androecium in Malvales. *Blumea* **13**, 177–394.

Van Heel, W. A. (1978). Morphology of the pistil in Malvaceae - Ureneae. *Blumea* **24**, 123–37.

Van Heel, W. A. (1987). Androecium development in *Actinidia chinensis* and *A. melanandra* (Actinidiaceae). *Bot. Jahrb. Syst.* **109**, 17–23.

Van Heel, W. A. (1995). Morphology of the gynoecium of *Kitaibelia vitifolia* Willd. and *Malope trifida* L. (Malvaceae-Malopeae). *Bot. Jahrb. Syst.* **117**, 485–93.

Vanvinckenroye, P., Cresens, E., Ronse De Craene, L. P. and Smets, E. F. (1993). A comparative floral developmental study in *Pisonia, Bougainvillea* and *Mirabilis* (Nyctaginaceae) with special emphasis on the gynoecium and floral nectaries. *Bull. Jard. Bot. Nat. Belg.* **62**, 69–96.

Vanvinckenroye, P., Ronse De Craene, L. P. and Smets, E. F. (1997). The floral development of *Monococcus echinophorus* (Phytolaccaceae). *Can. J. Bot.* **75**, 1941–50.

Vanvinckenroye, P., and Smets, E. F. (1996). Floral ontogeny of five species of *Talinum* and of related taxa (Portulacaceae). *J. Plant Res.* **109**, 387–402.

Vanvinckenroye, P., and Smets, E. F. (1999). Floral ontogeny of *Anacampseros* subg. *Anacampseros* sect. *Anacampseros* (Portulacaceae). *Syst. Geogr. Pl.* **68**, 173–94.

Vasconcelos, T. N. C, Lucas, E. J., Conejero, M., Giaretta, A. and Prenner, G. (2020). Convergent evoution in calyptrate flowers of Syzygieae (Myrtaceae). *Bot. J. Linn. Soc.* **192**, 498–509.

Vasconcelos, T. N. C., Prenner, G., Buenger, M. O., De-Carvalho, P. S., Wingler, A. and Lucas, E. J. (2015). Systematic and evolutionary implications of stamen position in Myrteae (Myrtaceae). *Bot. J. Linn. Soc.* **179**, 388–402.

Vasconcelos, T. N. C., Prenner, G., Santos, M. F., Wingler, A. and Lucas, E. J. (2017). Links between parallel evolution and systematic complexity in angiosperms: A case study of floral development in *Myrcia* s.l. (Myrtaceae). *Perspect. Plant Ecol.* **24**, 11–24.

Venkata Rao, C. (1952). Floral anatomy of some Malvales and its bearing on the affinities of families included in the order. *J. Indian Bot. Soc.* **31**, 171–203.

Venkata Rao, C. (1963). On the morphology of the calyculus. *J. Indian Bot. Soc.* **42**, 618–28.

Vergara-Silva, F. Espinosa-Matías, S., Ambrose, B. A., Vázquez-Santana, S., Martínez-Mena, A., Márquez-Guzmán, J., Martínez, E., Meyerowitz, E. M. and Alvarez-Buylla, E. R. (2003). Inside-out flowers characteristic of *Lacandonia schismatica* evolved at least before its divergence from a closely related taxon, *Triuris brevistylis*. *Int. J. Plant Sci.* **164**, 345–57.

Vogel, S. (1977). Nektarien und ihre ökologische Bedeutung. *Apidology* **8**, 321–35.

Vogel, S. (1997). Remarkable nectaries, structure, ecology, organophyletic perspectives I. Substitutive nectaries. *Flora* **192**, 305–33.

Vogel, S. (1998a). Remarkable nectaries, structure, ecology, organophyletic perspectives II. Nectarioles. *Flora* **193**, 1–29.

Vogel, S. (1998b). Remarkable nectaries, structure, ecology, organophyletic perspectives III. Nectar ducts. *Flora* **193**, 113–31.

Vogel, S. (1998c). Remarkable nectaries, structure, ecology, organophyletic perspectives IV. Miscellaneous cases. *Flora* **193**, 225–48.

Vogel, S. (2000). The floral nectaries of Malvaceae *sensu lato*: A conspectus. *Kurtziana* **28**, 155–71.

Von Balthazar, M., Alverson, W. S., Schönenberger, J. and Baum, D. A. (2004). Comparative floral development and androecium structure in Malvoideae (Malvaceae s.l.). *Int. J. Plant Sci.* **165**, 445–73.

Von Balthazar, M., and Endress, P. K. (2002a). Reproductive structures and systematics of Buxaceae. *Bot. J. Linn. Soc.* **140**, 193–228.

Von Balthazar, M., and Endress, P. K. (2002b). Development of inflorescences and flowers in Buxaceae and the problem of perianth interpretation. *Int. J. Plant Sci.* **163**, 847–76.

Von Balthazar, M., and Schönenberger, J. (2009). Floral structure and organization in Platanaceae. *Int. J. Plant Sci.* **170**, 210–25.

Von Balthazar, M., and Schönenberger, J. (2013). Comparative floral structure and systematics in the balsaminoid clade including Balsaminaceae, Marcgraviaceae and Tetrameristaceae (Ericales). *Bot. J. Linn. Soc.* **173**, 325–86.

Von Balthazar, M., Schönenberger, J., Alverson, W. S., Janka, H., Bayer, C. and Baum, D. A. (2006). Structure and evolution of the androecium in the *Malvatheca* clade (Malvaceae s.l.) and implications for Malvaceae and Malvales. *Plant Syst. Evol.* **260**, 171–97.

Vrijdaghs, A., Caris, P., Goetghebeur, P. and Smets, E. (2005). Floral ontogeny in *Scirpus*, *Eriophorum* and *Dulichium* (Cyperaceae), with special reference to the perianth. *Ann. Bot.* **95**, 1199–1209.

Vrijdaghs, A., Flores-Olvera, H. and Smets, E. (2014). Enigmatic floral structures in *Alternanthera*, *Iresine*, and *Tidestromia* (Gomphrenoideae, Amaranthaceae). A developmental homology assessment. *Plant Syst. Evol.* **147**, 49–66.

Wagenitz, G., and Laing, B. (1984). The nectaries of the Dipsacales and their systematic significance. *Bot. Jahrb. Syst.* **104**, 483–507.

Wagner, K. A., Rudall, P. J. and Frohlich, M. W. (2009). Environmental control of sepalness and petalness in perianth organs of waterlilies: A new mosaic theory for the evolutionary origin of a differentiated perianth. *J. Exp. Bot.* **60**, 3559–74.

Walker-Larsen, J., and Harder, L. D. (2000). The evolution of staminodes in Angiosperms, patterns of stamen reduction, loss, and functional re-invention. *Am. J. Bot.* **87**, 1367–84.

Walker-Larsen, J., and Harder, L. D. (2001). Vestigial organs as opportunities for functional innovation: The example of the *Penstemon* staminode. *Evolution* **55**, 477–87.

Wallnöfer, B. (2004). Ebenaceae. In *The families and genera of vascular plants vol. VI*, ed. K. Kubitzki and C. Bayer. Berlin: Springer, pp. 125–30.

Walter, H. (1906). Die Diagramme der Phytolaccaceen. *Bot. Jahrb. Syst.* **37, Beibl. 85**, 1–57.

Wang, H., Meng, A., Li, J., Feng, M., Chen, Z. and Wang, W. (2006). Floral organogenesis of *Cocculus orbiculatus* and *Stephania dielsiana* (Menispermaceae). *Int. J. Plant Sci.* **167**, 951–60.

Wang, H., Moore, M. J., Soltis, P. S., Bell, C. D., Brockington, S. F., Alexandre, R., Davis, C. C., Latvis, M., Manchester, S. R. and Soltis, D. E. (2009). Rosid radiation and the rapid rise of angiosperm-dominated forests. *Proc. Natl. Acad. Sci. USA* **106**, 3853–8.

Wang, J.-R., Wang, X., Li, Q.-J., Zhang, X.-H., Ma, Y.-P., Zhao, L., Ginefra Toni, J. F. and Ronse De Craene, L. P. (2020). Floral morphology and morphogenesis of *Sanguisorba* (Rosaceae) and its systematic significance. *Bot. J. Linn. Soc.* **193**, 47–63.

Wang, Y.-Z., Liang, R.-H., Wang, B.-H., Li, J.-M., Qiu, Z.-J., Li, Z.-Y. and Weber, A. (2010). Origin and phylogenetic relationships of the Old World Gesneriaceae with actinomorphic flowers inferred from ITS and *trnL-trnF* sequences. *Taxon* **59**, 1044–52.

Wang, X., Wang J. R., Xie, S., Zhang X.-H., Chang Z.-Y., Zhao, L., Ronse De Craene, L. P. and Wen J. (2021). Floral morphogenesis of the *Maddenia* and *Pygeum* groups of Prunus (Rosaceae). *J. Syst. Evol.* doi: 10.1111/jse.12748

Wannan, B. C., and Quinn, C. J. (1991). Floral structure and evolution in the Anacardiaceae. *Bot. J. Linn. Soc.* **107**, 349–85.

Wanntorp, L., Puglisi, C., Penneys, D. and Ronse De Craene, L. P. (2011). Multiplications of floral organs in flowers: A case study in *Conostegia* (Melastomataceae, Myrtales). In *Flowers on the tree of life*, ed. L. Wanntorp and L. P. Ronse De Craene. Systematics Association Special Volume Series 80. Cambridge: Cambridge University Press, pp. 218–35.

Wanntorp, L., and Ronse De Craene, L. P. (2005). The *Gunnera* flower, key to eudicot diversification or response to pollination mode? *Int. J. Plant Sci.* **166**, 945–53.

Wanntorp, L., and Ronse De Craene, L. P. (2007). Floral development of *Meliosma* (Sabiaceae). Evidence for multiple origins of pentamery in the eudicots. *Am. J. Bot.* **94**, 1828–36.

Wanntorp, L., and Ronse De Craene, L. P. (2009). Perianth evolution in the Sandalwood order Santalales: How does morphology relate to molecular phylogenetics? *Am. J. Bot.* **96**, 1361–71.

Wanntorp, L., Ronse De Craene, L. P., Peng, C.-I. and Anderberg, A. A. (2012). Floral ontogeny and morphology of *Stimpsonia* and *Ardisiandra*, two aberrant genera of the primuloid clade of Ericales. *Int. J. Plant Sci.* **173**, 1023–35.

Waters, M. T., Tiley, A. M. M., Kramer, E. M., Meerow, A. W., Langdale, J. A. and Scotland R. W. (2013). The corona of the daffodil *Narcissus bulbocodium* shares stamen-like identity and is distinct from the orthodox floral whorls. *Plant J.* **74**, 615–25.

Weber, A. (2004). Gesneriaceae. In *The families and genera of vascular plants*, ed. K. Kubitzki and J. W. Kadereit. Berlin: Springer, pp. 63–158.

Weberling, F. (1989). *Morphology of flowers and inflorescences*. Cambridge: Cambridge University Press.

Wei, L., and Ronse De Craene, L. P. (2019). What is the nature of petals in Caryophyllaceae? Developmental evidence clarifies their evolutionary origin. *Ann. Bot.* **124**, 281–95.

Wei, L., and Ronse De Craene, L. P. (2020). Hofmeister's rule's paradox: The explanation of the changeable carpel position in Caryophyllaceae. *Int. J. Plant Sc.* **181**, 911–25.

Weigend, M. (2007). Grossulariaceae. In *The families and genera of vascular plants vol. IX*, ed. K. Kubitzki and C. Bayer. Berlin: Springer, pp. 168–76.

Westerkamp, C., and Weber, A. (1999). Keel flowers of the Polygalaceae and Fabaceae: A functional comparison. *Bot. J. Linn. Soc.* **129**, 207–21.

Whittall, J. B., and Hodges, S. A. (2007). Pollinator shifts drive increasingly long nectar spurs in columbine flowers. *Nature* **447**, 706–10.

Williams, S. E., Albert, V. A. and Chase, M. W. (1994). Relationships of Droseraceae, a cladistic analysis of *rbcL* sequence and morphological data. *Am. J. Bot.* **81**, 1027–37.

Williamson, P. S., and Moseley, M. F. (1989). Morphological studies of the Nymphaeaceae *sensu lato* XVII. Floral anatomy of *Ondinea purpurea* ssp. *purpurea* (Nymphaeaceae). *Am. J. Bot.* **76**, 1779–94.

Wróblewska, M., Dolzblasz, A. and Zagórska-Marek, B. (2016). The role of ABC genes in shaping perianth phenotype in the basal angiosperm *Magnolia*. *Plant Biol.* **18**, 230–8.

Wu, H.-C., Su, H.-J. and Hu, J.-M. (2007). The identification of A-, B-, C-, and E-class Mads-Box genes and implications for perianth evolution in the basal eudicot *Trochodendron aralioides* (Trochodendraceae). *Int. J. Plant Sci.* **168**, 775–99.

Wurdack, K. J., and Davis, C. C. (2009). Malpighiales phylogenetics: Gaining ground on one of the most recalcitrant clades in the angiosperm tree of life. *Am. J. Bot.* **96**, 1551–70.

Xi, Z., Ruhfel, B. R., Schaefer, H., Amorim, A. M., Sugumaran, M., Wurdack, K. J., Endress, P. K., Matthews, M. l., Stevens, P. F., Mathews, S. and Davis, C. C. (2012). Phylogenomics and a posteriori data partitioning resolve the Cretaceous angiosperm radiation Malpighiales. *Proc. Nat. Acad. Sci. USA* **109**, 17519–24.

Xu, F.-X. (2006). Floral ontogeny of two species of *Magnolia* L. *J. Integr. Plant Biol.* **48**, 1197–1203.

Xu, F.-X., and Ronse De Craene, L. P. (2010a). Floral ontogeny of Annonaceae: Evidence for major floral diversity in Magnoliales. *Ann. Bot.* **106**, 591–605.

Xu F.-X., and Ronse De Craene, L. P. (2010b). Floral ontogeny of *Knema* and *Horsfieldia* (Myristicaceae): Evidence for a complex androecial evolution. *Bot. J. Linn. Soc.* **164**, 42–52.

Xu F.-X., and Rudall, P. J. (2006). Comparative floral anatomy and ontogeny in Magnoliaceae. *Plant Syst. Evol.* **258**, 1–15.

Xue, L.-L., Jian, H.-L., Yun, F.-Y. and Jun Y.-Z. (2017). Floral development of *Gymnospermium microrhynchum* (Berberidaceae) and its systematic significance in the Nandinoideae. *Flora* **228**, 10–16.

Yao, G., Jin, J.-J., Li, H.-T., Yang, J.-B., Mandala, V. S., Croley, M., Mostow, R., Douglas, N. A., Chase, M. W., Christenhusz, M. J. M., Soltis, D. E., Soltis, P. S., Smith, S. A., Brockington, S. F., Moore, M. J., Yi, T.-S., Li, D.-Z. (2019). Plastid phylogenomic insights in the evolution of Caryophyllales. *Mol. Phyl. Evol.* **134**, 74–86.

Zalko, J., Frachon, S., Morel, A., Deroin, T., Espinosa, F., Xiang, K.-L., Wang, W., Zhang W.-G., Lang, S., Dixon, L. Pinedo-Castro, M. and Jabbour, F. (2021). Floral organogenesis and morphogenesis of *Staphisagria* (Ranunculaceae): Implications for the evolution of synorganized floral structures in Delphinieae. *Int. J. Plant Sci.* **182**, 59–70.

Zandonella, P. (1977). Apports de l'étude comparée des nectaires floraux à la conception phylogénétique de l'ordre des Centrospermes. *Ber. Dtsch. Bot. Ges.* **90**, 105–25.

Zanis, M. J., Soltis, P. S., Qiu, Y. L., Zimmer, E. and Soltis, D. E. (2003). Phylogenetic analyses and perianth evolution in basal Angiosperms. *Ann. Mo Bot. Gard.* **90**, 129–50.

Zeng, L., Zhang, N., Zhang, Q., Endress, P. K., Huang, J. and Ma, H. (2017). Resolution of deep eudicot phylogeny and their temporal diversification using nuclear genes from transcriptomic and genomic datasets. *New Phytol.* **214**, 1338–54.

Zhang, R.J., and Schönenberger J. (2014). Early floral development of Pentaphylaceae (Ericales) and its systematic implications. *Plant Syst. Evol.* **300**, 1547–60.

Zhang, Z.-G., Meng, A.-P., Li, J.-Q., Ye, Q.-G., Wang, H.-C. and Endress, P. K. (2012). Floral development of *Phyllanthus chekiangensis* (Phyllanthaceae), with special reference to androecium and gynoecium. *Plant Syst. Evol.* **298**, 1229–38.

Zhang, R., Guo, C., Zhang, W., Wang, P., Li, L., Duan, X., Du, Q., Zhao, L., Shan, H., Hodges, S. A., Kramer, E. M., Ren, Y. and Kong, H. (2013). Disruption of the petal identity gene *APETALA3-3* is highly correlated with loss of petals within the buttercup family (Ranunculaceae). *Proc. Nat. Acad. Sci. USA* **110**, 5074–9.

Zhang, X.-H., and Ren, Y. (2008). Floral morphology and development in *Sargentodoxa* (Lardizabalaceae). *Int. J. Plant Sci.* **169**, 1148–58.

Zhang X.-H., and Ren Y. (2011). Comparative floral development in Lardizabalaceae (Ranunculales). *Bot. J. Linn. Soc.* **166**, 171–84.

Zini, L. M., Galati, B. G. and Ferrucci, M. S. (2017). Perianth organs in Nymphaeaceae: Comparative study on epidermal and structural characters. *J. Plant Res.* **130**, 1047–60.

Zohary, M., and Baum, B. (1965). On the androecium of *Tamarix* flower and its evolutionary trends. *Israel J. Bot.* **14**, 101–11.

Zúñiga, J. D. (2015). Phylogenetics of Sabiaceae with emphasis on *Meliosma* based on nuclear and chloroplast data. *Syst. Bot.* **30**, 761–75.

Index